BEHAVIOR GENETICS
Principles and Applications

BEHAVIOR GENETICS
Principles and Applications

Edited by
JOHN L. FULLER
State University of New York at Binghamton

EDWARD C. SIMMEL
Miami University
Oxford, Ohio

with the editorial assistance
of Marjorie Bagwell

1983

LAWRENCE ERLBAUM ASSOCIATES, PUBLISHERS
Hillsdale, New Jersey   London

Lawrence Erlbaum Associates, Inc., Publishers
365 Broadway
Hillsdale, New Jersey 07642

Library of Congress Cataloging in Publication Data

Main entry under title:

Behavior genetics.

  Bibliography: p.
  Includes indexes.
  1. Behavior genetics. I. Fuller, John L.
II. Simmel, Edward C. [DNLM: 1. Genetics, Behavioral.
QH 457 B4185]
QH457.B423 1983   591.51   82-18247
ISBN 0-89859-211-9

Printed in the United States of America
10 9 8 7 6 5 4 3 2 1

Contents

Preface

Behavior Genetics: Principles and Applications is a collection of chapters by active research workers. It includes reviews of special areas within the field and discussions of interactions with other behavioral sciences such as psychology, ethology, and sociobiology. Applications to medicine, psychiatry, and education are also considered. One might say that the purpose of this book is to help in dealing with the problem implied by DeFries and Plomin (1978): "If a successful area is one in which the rate of publication is so great that no one person could keep up with the current literature and wouldn't understand all of it if he tried, then behavioral genetics is a rousing success [p. 473]." Our aim is to present summaries of the "state of the art" in a form that will be useful to researchers, teachers, and students in a variety of disciplines. Contributors were encouraged to integrate history, present knowledge, and projections for the future.

Although the book is not divided into sections there is some grouping of related chapters. Ehrman and Probber briefly summarize the fundamentals of genetics and evolutionary theory. Their contribution is particularly useful for readers desiring to brush up on biological basics. Claridge and Mangan, reviewing Western and Pavlovian approaches to human nervous system functioning, summarize a somewhat neglected literature on genes, neurophysiology, and personality. Simmel and Bagwell take up the present status of heritable variation in activity and exploration in animals, while introducing the more general problem of the definition and measurement of behavioral phenotypes. The chapter by Horowitz and Dudek deals primarily with experimental psychopharmacogenetics in animals, an area that has importance for the understanding of the biological base of individual differences among humans in response to psychoactive drugs.

Omenn's review of medical genetics ranges over a variety of inherited meta-bolic and structural deviations with effects on behavior. He emphasizes chro-mosomal aberrations and single locus effects. Diederen summarizes the present status of the genetics of schizophrenia, a disorder that runs in families but does not fit simple Mendelian models. Continuing on aspects of human behavior, Scarr and Carter-Saltzman take up the genetics of intelligence, one of the tradi-tional areas of contention in the nature–nurture debate. The emphasis in Fuller's chapter on ethology is the possible role of genes in the behavioral variability of free-living species. Data from the laboratory are introduced where pertinent. A general survey of the genetics of animal social behavior by Scott reflects his long association with this field of research. Two of his former students review specific areas: caretaking and care soliciting (Gurski), and agonistic and sexual behavior (Hyde). Finally, Fuller looks at the relationship between behavior genetics and sociobiology and concludes that each would benefit from closer interaction with the other.

A bit of history concludes this preface. In 1977 Sonja Haber and Robert Dworkin conceived and planned a *Handbook of Behavior Genetics* to serve as a basic reference for specialists. Difficulties were encountered: target dates for manuscripts were not met; a number of persons withdrew after accepting an invitation. In the meantime a number of excellent contributions remained in limbo. Dworkin withdrew from the project in 1980 for personal reasons, and Haber found it necessary to do so in 1981. She asked two of her former mentors to consider taking over the project. Following conversations with the publisher we acceded to the stipulation that the format be changed from a "handbook" to a "current advances" familiar in other disciplines. Should the present volume serve a need, we envision sequels covering other important areas omitted here.

In addition to our debt to Haber and Dworkin we acknowledge the editorial aid of Marjorie Bagwell and Ruth Fuller.

John L. Fuller
Edward C. Simmel

List of Contributors

Marjorie Bagwell, *Behavior Genetics Laboratory, Miami University, Oxford, Ohio 45056*

Louise Carter-Saltzman, *Department of Psychology, University of Washington, Seattle, Washington 98195*

Gordon Claridge, *Department of Experimental Psychology, Oxford University, Oxford, Englad OX1 3UD*

Ingrid Diederen, *Department of Psychology, Miami University, Oxford, Ohio 45056*

Bruce C. Dudek, *Department of Psychology, State University of New York at Albany, Albany, New York 12203*

Lee Ehrman, *Division of Natural Sciences, State University of New York College at Purchase, Purchase, New York 10577*

John L. Fuller, *P.O. Box 543, York, Maine 03909*

John C. Gurski, *Department of Psychology, Fort Hays State University, Hays, Kansas 67601*

Gary P. Horowitz, *Department of Psychology, State University of New York at Binghamton, Binghamton, New York 13901*

Janet S. Hyde, *Department of Psychology, Denison University, Granville, Ohio 43023*

Gordon Mangan, *Department of Experimental Psychology, Oxford University, Oxford, England OX1 3UD*

Gilbert S. Omenn, *School of Medicine, University of Washington, Seattle, Washington 98195*

Joan Probber, *Division of Natural Sciences, State University of New York College at Purchase, Purchase, New York 10577*

Sandra Scarr, *Department of Psychology, Yale University, New Haven, Connecticut 06520*

J. Paul Scott, *Center for Research on Social Behavior, Bowling Green State University, Bowling Green, Ohio 43403*

Edward C. Simmel, *Behavior Genetics Laboratory, Miami University, Oxford, Ohio 45056*

BEHAVIOR GENETICS
Principles and Applications

1 Fundamentals of Genetic and Evolutionary Theories

Lee Ehrman
Joan Probber
State University of New York at Purchase

Behavior genetics—from where do diverse scientific disciplines emerge, develop, and finally coalesce? Psychology, which attempts the description, prediction, and ultimately the understanding of behavior, began in the universities of Europe. One tradition emerged from the 18th- and 19th-century German philosophers as exemplified by Leibniz (Boring, 1950), and was given an experimental basis by Wundt (later in the United States by his student Titchener). The other, emanating from England, was based on the principles of John Locke and the *philosophes* of the French enlightenment. This Anglo–French influence came to America in 1882 with Herbert Spencer, who was enamored of Charles Darwin's theory of natural selection. Although these traditions are diffuse they help provide an orientation for understanding the development of psychology and its attempts to provide a firm biological basis for the study of behavior.

Situations in which stimuli could be controlled and responses objectively measured were the foundation of the Wundtian school. The Lockean prescription offered a more comfortable cultural fit to our society, and its view of mental process as the product of nothing but sensation furnished a comfortable ambience for the growth of the pragmatically oriented behaviorist school.

PRINCIPLES OF GENETICS

Mendelian Genetics

Genetics formal beginnings are found in the work of Gregor Mendel (1822–1884) whose elegant experiments with garden peas and mice provided the foundation for the science of genetics. In the garden of a monastery in Moravia (now

1

Czechoslovakia), of which he eventually became abbot, Mendel crossbred varieties of the garden pea that had maintained relatively constant differences in physical characteristics over generations. This hereditary constancy occurred despite the occasional, now fully scientifically explained, disappearance of a given trait during several generations and its subsequent reemergence.

Some of the factors Mendel analyzed were tall versus dwarf plants, red or white colored blossoms, seed color and shape, pod forms, and the position of flowers on stems. He theorized that each plant contained two factors for a particular characteristic such as color (one of each of these factors contributed by each parental plant) and that these units were transmitted to descendants in adherence to statistical predictions. These two factors (now called genes) might be identical, in which case the organism was said to be homozygous for that gene, or they might differ and the organism would be heterozygous.

From his plant-breeding experiments and observations, Mendel derived two laws that serve as the foundation, both historically and functionally, of the science of genetics. The first law concerns segregation. Genes are particulate units of inheritance that may exist in two or more variant forms (alleles) at the same position (locus) on a chromosome. Members of a chromosome pair separate when sex cells are produced. If an individual is heterozygous at the A-locus, the gametes will be of two types—A or a. The second law deals with independent assortment of alleles on different chromosomes. When the chromosomes separate to form eggs and sperm, these variant alleles (Aa and Bb, for example) are transmitted independently of one another to the next generation. When one allele is always expressed phenotypically in the organism bearing it, it is called the dominant allele (A); a recessive allele, expressed only when the dominant allele is absent, is designated a. A dominant trait is expressed when one or two dominant alleles are present (AA or Aa) but one can see only the recessive phenotype when both recessives (aa) are present. Some alleles are not expressed as either dominant or recessive but are intermediate, and they produce a phenotype somewhere between the phenotypes of the parents (red, pink, white). Suppose that two individuals differing in genes at only three different loci (genetically active sites on chromosomes) are crossed: $AABBCC \times aabbcc$. The sex cells (gametes) are carrying ABC or abc, whereas the first filial generation (F_1) or offspring, is entirely $AaBbCc$. Each F_1 offspring can produce eight different types of gametes: ABC, ABc, AbC, aBC, Abc, aBc, abC, and abc. Figure 1.1 indicates the number of possible genotypes when three genes are involved. For an idea of what occurs when more than three gemes are involved see Table 1.1. Inasmuch as each chromosome has many thousands of genes the number of possible combinations is astronomical.

The word chromosome comes from the Greek *chroma,* which means color, and *soma*—body, because these gene-carrying units can be routinely observed microscopically when stained with basic dyes. Genes are arranged in linear order on these chromosomes, which are found in the nucleus of eucaryotic cells.

	EGGS							
	ABC	ABc	AbC	aBC	abC	aBc	Abc	abc
ABC	AABBCC	AABBCc	AABbCC	AaBBCC	AaBbCC	AaBBCc	AABbCc	AaBbCc
ABc	AABBCc	AABBcc	AABbCc	AaBBCc	AaBbCc	AaBBcc	AABbcc	AaBbcc
AbC	AABbCC	AABbCc	AAbbCC	AaBbCC	AabbCC	AaBbCc	AAbbCc	AabbCc
aBC	AaBBCC	AaBBCc	AaBbCC	aaBBCC	aaBbCC	aaBBCc	AaBbCc	aaBbCc
abC	AaBbCC	AaBbCc	AabbCC	aaBbCC	aabbCC	aaBbCc	AabbCc	aabbCc
aBc	AaBBCc	AaBBcc	AaBbCc	aaBBCc	aaBbCc	aaBBcc	AaBbcc	aaBbcc
Abc	AABbCc	AABbcc	AAbbCc	AaBbCc	AabbCc	AaBbcc	AAbbcc	Aabbcc
abc	AaBbCc	AaBbcc	AabbCc	aaBbCc	aabbCc	aaBbcc	Aabbcc	aabbcc

(Left margin label: SPERM)

FIG. 1.1. A tri-hybrid cross: each cell represents a different genotype.

Organisms are classified as eucaryotes if their genetic material is contained in the nucleus of their cells. This includes all life above the level of bacteria and blue–green algae. The latter—called procaryotes—are considered to be earlier stages in the evolution of life on this earth. Procaryotes have no nuclei and their genetic material is distributed throughout their cells. In eucaryotes, each individual of a species has a characteristic chromosome number—46 in humans, 40 in the house mouse, 20 in corn, 48 in the potato,and 8 in the fruit fly, (*Drosophila melanogaster*).

Sexually reproducing individuals often have two different alleles (are heterozygous) for many of their genes. Thus, the possible combinations of parental genotypes greatly outnumber the actually ever-realized genotypes. The probability of siblings acquiring the same genetic endowment from both or even one of their parents is infinitely small. This probability decreases as more and more pairs of genes are involved. So, each and every human individual (except for identical twins) is and has been genetically unique.

Genotype and Phenotype

The observable physical and behavioral aspects of an organism constitute its *phenotype*. The phenotype is the joint product of the genes that an organism possesses (its genotype) plus the effects of the environments in which it developed and now lives. It has become increasingly obvious with the predominance of the behaviorist school of psychology throughout the 1950s that the environment is of particular significance in the development of behavioral phenotypes (Beach, 1950).

TABLE 1.1

Gregor Mendel's Laws of Segregation and Independent Assortment
As Applied to Numbers of Differing Pairs of Genes.
F_1 Heterozygosity Is Assumed As Is Complete Dominance

n GENES	F_1 GAMETES 2^n	$F_1 \times F_2$ ZYGOTES 4^n	F_2 GENOTYPES 3^n	HOMOZYGOUS F_2 GENOTYPES 2^n	HETEROZYGOUS F_2 GENOTYPES $3^n - 2^n$	F_2 PHENOTYPES 2^n
1	2	4	3	2	1	2
2	4	16	9	4	5	4
3	8	64	27	8	19	8
4	16	256	81	16	65	16
5	32	1,024	243	32	211	32
10	1,024	1,084,576	59,049	1,024	58,025	1,024

A prime example of how behavior may be altered via the environment is phenylketonuria—a genetic disease more fully discussed elsewhere in this chapter, arising from an inborn error of metabolism. The enzyme that normally breaks down phenylalanine—an essential amino acid—is lowered or missing due to gene malfunction. Unmetabolized phenylalanine accumulates in the brain, and one of the effects is a lowered intelligence quotient (IQ). If the metabolic error is circumvented by a special phenylalanine-free diet, IQ is ostensibly improved.

Chromosomal Genetics

It might be best to consider the nature of the genotype at this point so that genetic principles are understood before we consider the complexities of the effects due to environmental fluctuations. As noted before, the physical bases of heredity are the genes that are borne on the chromosomes carried in cell nuclei. These chromosomes can be observed during mitosis (when the cell divides and reproduces an exact replica of its genetic content, and in meiosis (including reduction division), where the chromosome number is halved to produce the gametes (spermatozoan or ovum) involved in sexual reproduction.

The human species' 46 chromosomes are arranged in 23 pairs; in females all 23 pairs are homologous (correspond structurally and derivatively); in males, there are 22 homologous pairs plus a nonhomologous X and Y chromosome (Fig. 1.2). The chromosome pairs, when properly stained, differ in length and appearance so that they can be individually recognized, much as one might recog-

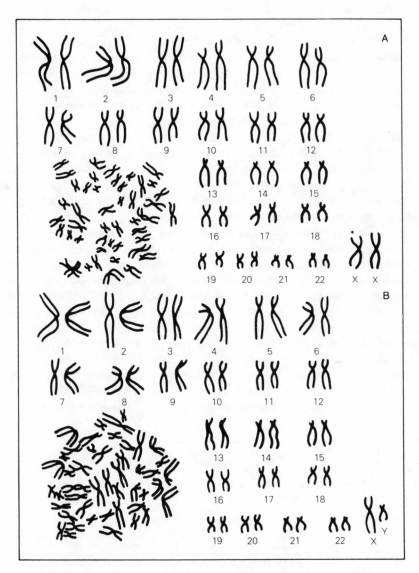

FIG. 1.2. Human chromosomes. A. Normal female cell with 46 chromosomes and the normal female karyotype (XX). B. Normal male cell with 46 chromosomes and the normal male karyotype (XY). (Courtesy of Professor Raymond Turpin.)

nize the familiar face of a friend. During meiosis, when the gametes are formed and the chromosome number is halved, all the chromosomes in a single gamete differ because each gamete has only one member of each chromosome pair. At fertilization, two gametes unite, each containing a set of 23 chromosomes to form the fertilized cell (zygote) with 23 pairs or 46 chromosomes once again. This process is diagrammed in Fig. 1.3. The number of chromosomes contained in gametes (23) is referred to as the haploid number and that of zygotes ($2 \times 23 = 46$) is the diploid number—written as n and $2n$, respectively.

The process of meiosis affords a significant opportunity for increasing genetic variability. The most important consistent mechanism of change is crossing over, which takes place during the early phases of meiosis. Homologous chromosomes, lying next to each other, break, exchange equivalent portions of their genes, and reanneal to produce combinations that differ in arrangement from the genetic endowment provided by parents. This phenomenon is omnipresent in the normal meiotic process; when it fails to occur in at least one parent, the resulting organism is often sterile. Crossing over is the most consistent of a number of processes that provide the raw material upon which natural selection may act.

A coarser and more sporadic mechanism is individualized chromosome rearrangement, of which there are four possible types: deletion, duplication, inversion, and translocation. Deletion is the sometimes lethal removal of a gene or sequence of genes. Death may occur because a biochemical sequence has been excised, resulting in an intolerable structural or functional gap in the organism.

Duplication of a chromosome or part of a chromosome may cause an imbalance of gene activity, reducing the viability of an organism. However, because some organisms can tolerate duplications of genetic material, these duplications might play an evolutionary role if part or all of one of the duplications

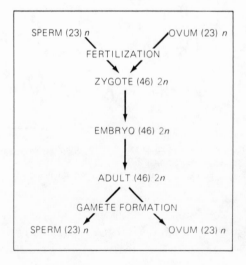

FIG. 1.3. Human chromosome number changes during gamete and zygote formation. This can be generalized for sexually reproducing organisms, which will be designated in this book by n, where n is the haploid number and $2n$ the diploid number.

mutated and functioned differently from the original. Indeed, this is one way in which evolutionary change is postulated to occur. Greater genetic variability gives rise (in minute quanta, to be sure) to physiologic and morphologic differences that endow the organism with an enlarged behavioral repertoire so that it can more successfully cope with an ever-changing environment.

An inversion occurs when a chromosome breaks in one or two places and the segment between the break(s) rotates 180° leading to a reversal of gene order and a resulting change in biochemical product. A translocation occurs when two chromosomes from different pairs break simultaneously and exchange segments.

According to current knowledge, not all these chromosomal and gene changes are equally important in their effects on behavior. To date, inversions and chromosome number changes appear to form the main categories of importance. But, crucial to the meiotic process (which reproductive systems evolved to support) is the shuffling and mingling of genetic material into infinite numbers of combinations.

Mutation and Mutants

Until now this discussion has centered on the shuffling of genetic material in sexually reproducing species. Genetic variability within a species has been discussed with emphasis on the variety of gene combinations attributable to meiosis, crossing over, duplication, and translocation. All these simply involve the reassortment of a set of variable genes indigenous to a population. In the following section we consider the appearance of novelty in the genes themselves.

The essence of creating novelty is the process of mutation. These are alterations in the chemical structure of a gene that lead to changes in gene products that may ramify throughout the organism. Mutations at some loci occur spontaneously on the average of one in a million newly arisen gametes. The frequency is increased by radiation and by some chemicals known as mutagens.

A veritable storehouse of genetic mutations in mice is found at the Jackson Memorial Laboratory in Bar Harbor, Maine. Through selective inbreeding, when a mutant trait of interest is found, a strain of mice each of whose members possess the trait is established and bred for research uses. Many of the known mutant genes affect the nervous system during its development and produce a range of deficits in behavior from minor disturbances to severe defects. One of these found in the mouse is a mutation on chromosome five, an altered gene called *fidget*. When this gene is present on both members of the fifth chromosome, it produces a behavioral phenotype in which the mouse's head moves from side to side continuously. The gene is written *fi* and the mouse is said to have an *fifi* genotype. In most mice the fidget gene is not present at this locus but the normal alternative gene, written as +, is. If the mouse is genotypically either *fi* + or + +, the behavioral alteration is not seen. Only if there are two *fi* genes present does the fidget phenotype become visible, as the wild type (+) gene is dominant to the recessive *fi*). However, dominance and recessivity are not necessarily complete, because heterozygotes (individuals with nonidentical alleles) are often distinguishable from both homozygotes (individuals with identical alleles). Even if at first sight dominance appears to be complete, detailed biochemical tests may reveal differences between the heterozygote and the homozygote.

Untreated phenylketonuria in man provides one such example. Superficially viewed, the disease is controlled by a recessive gene *p* with afflicted individuals being *pp* and normals *p+* or *++*. However, at the beochemical level, *p+* and *++* may be distinguishable because the *p+* individual tends to have more phenylalanine in blood serum. Dominance, in this case is incomplete. So, depending on which phenotype component is being examined, different conclusions as to the dominance may be drawn.

If a simple dominant–recessive relationship is assumed with two pairs of alleles at two loci (i.e., *A* and *B* are dominant to *a* and *b*), what might ensue? If a double heterozygote (*AaBb*) is crossed with a double homozygote (*aabb*) what are the statistical expectations? The double recessive homozygote produces only *ab* gametes. From the double heterozygote each separate locus yields equal numbers of *AB*, *Ab*, *aB* and *ab* gametes according to Mendel's principles of segregation and of independent assortment of alleles during gamete formation. If these gametes are fertilized with the recessive *ab* gametes, the results, assuming two independent pairs of alleles and simple dominant–recessive relationships, will yield a 1:1:1:1 phenotype ratio. The backcross to a recessive parent (one with an *aabb* genotype) is often used as a testcross in practical breeding programs because it can determine the actual genotype of an organism whose recessive alleles may be obscured by dominant ones. The use of this wholly recessive dihybrid parent unmasks recessive alleles in the alternate parental genotype. If we return to Mendel's garden peas and specify physical traits in a dihybrid cross (*A* is the gene for yellow and *a* the gene for green; *B* is the gene for a smooth surface and *b* the gene for a wrinkled one), we emerge with four types of gametes (*AB, Ab, aB, ab*) produced in approximately equal numbers, which yield the basic Mendelian ratio of 9:3:3:1.

Recombination and Genetic Maps

There is an important exception to Mendel's second law. Assortment of genes is not generally independent if gene loci are on the same chromosome. The closer genes are to each other along a given chromosome, the more closely linked they are during gamete formation. Although genes on the same chromosome may recombine with each other during meiosis, the frequency of the recombination depends on the spatial distance between the relevant loci. Generally, the further apart the loci are, the greater the possibility of recombination.

From the frequency of these recombinational events, chromosome maps have been constructed for individual chromosomes. Genes located on the same chromosome are said to belong to the same linkage group. Thus, in humans we have 23 linkage groups (one for each pair of chromosomes, i.e., the haploid number of chromosomes [n] corresponds to the number of linkage groups). The house mouse, possessing 40 chromosomes, has 20 linkage groups; in *Drosophila melanogaster,* an organism of historical importance in behavior genetics, the numbers are 8 and 4 respectively. During the formation of human eggs and sperm (gametes), the chromosome number is halved (meiosis), leaving the egg with 22 autosomes and an X chromosome, and the sperm with 22 autosomes and an X or a Y sex chromosome. Males produce four sperm from each specialized cell in the testes and X- and Y-bearing sperm occur in about equal numbers.

Gametes and Fertilization

In ovaries, meiosis also produces four products, but three of these become polar bodies, whereas the fourth becomes the chromosome-containing nucleus of the developing egg. Human females are born with their entire reproductive quotient already present and the limits are set for the number of eggs they can produce (about 400 during a reproductive life span). Males, however, continue to produce sperm from sexual maturation until senescence.

Differences in gametic size and function are extreme. The head of the sperm is about 3 to 5 microns long and 2 to 3 microns wide. The tail, which propels it, is 10 times the length of the head. Essentially, it is a pared-down packet of miraculously organized DNA (genetic material) with a launching and traveling apparatus to respond to and fertilize ova.

The human egg, in contrast, is one of the largest cells in the body—a sphere about 130 microns in diameter. It moves only passively when it is essentially shoved along, taking a few days to progress from ovary to uterus. As noted by Hartl (1977, cited in Probber and Ehrman, 1978): "All the sperm that gave rise to all the people that ever lived could be carried in a teaspoon . . . the eggs, would require a small bucket. [p. 18]." The egg is comparatively enormous because in addition to its complement of chromosomes, its cytoplasm provides the nourishment for, and biochemical assistance to the developing embryo for translation of the information carried in its genes. Unlike the female gamete the function of the sperm as a cell is complete when the sperm contacts and penetrates the egg (having shed its tail), and sperm and egg fuse.

With fertilization of the egg, formation of the embryo begins, but it is not until 6 weeks afterward that sex chromosomes become operative. At this point the Y chromosome, if present, will cause the inner layer of the gonads to develop into testes. If, however, there are two X chromosomes, at about 12 weeks the outer layer of the gonads will elaborate into ovaries.

X-Chromosome Inactivation

A further complication relates to sex. Although every human embryo is predisposed to develop as a female, its eventual sex phenotype will be determined by the types of chromosomes borne by the sperm. Our species has 23 pairs of chromosomes, each of which carries thousands of genes that compose the genotype that develops and unfolds to become the visible organism, the phenotype. In humans, 22 of the total number of chromosomes are autosomes; that is, two copies of these chromosomes are present in both sexes. The 23rd pair is made up of the unique sex chromosomes; the female has two of the X chromosomes; the male has one X and one Y chromosome. These were originally called X from X the unknown factor, and Y because it follows X alphabetically.

The X chromosome was first described in 1891 from insect material as "a peculiar chromatin body" but its function was not known then or for a long time afterward. Although the female had two doses of all the genes on the X chromosome and the male only one, there seemed to be no significant differences in X-chromosome activity between the sexes. It was not until 1962 that a basis for dosage compensation emerged. This has been named the Lyon hypothesis after Mary F. Lyon, who first presented detailed studies on the subject. She hypothesized that in the normal female, one X chromosome is inactivated in each somatic body cell early in embryonic development. Whether the maternal or paternal X is rendered inert is a matter of chance and the result is that patches of cells with X-chromosomal lineages emerge. The deactivated X becomes a tightly packed mass usually found close to the inner surface of the nuclear membrane of somatic cells in normal female mammals. It is called a Barr body after its discoverer. Accordingly, each cell in a male or a female has only one operative X chromosome. Because a female's cells are of two types (maternal X and paternal X), she is a genetic mosaic. A male, however, has only his maternal X chromosome to express.

What are the implications of this double X-chromosome dose for the female of our species? At the molecular level the X chromosome is a much more information packed and therefore biochemically active chromosome than the Y, which is much smaller and, except for male sex determination and fertility, inert. Recent work by McKusick and Ruddle (1977) cited more than 100 genes in the human X-chromosome map. The Y chromosome bears only two known genes, one controlling the TDF (Testis Determining Factor) and the other coding for the H–Y antigen. The H–Y antigen plays an important role in the expression of male physiological and morphological characteristics, specifically for the differentiation of the testis. According to Wachtel (1977): "Further male differentiation is imposed on the embryo by the action of testicular hormones against the inherent tendency toward the female phenotype [p. 798]." It is also known that males with their X and Y chromosomal complement are relatively unprotected from the expression of unpaired deleterious genes on their single X chromosome. Everything on their maternal X chromosome is expressed if not modified by genes on the autosomes. This is demonstrated by the much greater percentage of males afflicted by X-linked defects such as color blindness and hemophilia. Genes on the X chromosome are said to be sex linked. In females the principles of heterozygosity and homozygosity apply to genes on the sex chromosomes just as they do for the genes on autosomes. In males, loci on the X are not matched by loci on the nearly inert Y chromosome; this condition is known as hemizygosity.

EPIGENESIS

The process of development involves constant interaction and coaction of an organism's genes and its environment. Given identical genotypes as in monozygotic human twins or highly inbred mice and *Drosophila,* phenotypes are usually very similar. Even under optimal circumstances however, phenotypic variation occurs because of perturbations in the environment. In a given situation one can speak of the average phenotype of individuals with the same genotype as a *norm of reaction*. The *range of reaction* covers the spread of phenotypes that could potentially develop from a given fertilized ovum. Clearly, the range of reaction increases if the environment is not carefully controlled. The *epigenetic* view of development postulates that genotype and environment are involved at every stage from fertilization to birth and beyond. At fertilization, these actions begin at a molecular level.

Enzyme activities and changes in the intracellular environment are initiated. The egg provides a source of energy for the newly formed zygote and a well-primed machinery for protein synthesis, which initiates the primary series of cleavages and invaginations that produce the embryo, fetus, and finally, the neonate. The control mechanisms fostering maturation are encoded within the genes, in the internal and external environment of the DNA, and in the interrelations and interactions between them. At every level—molecular, cellular, and organismic—there are functional gene–environment interactions and products.

Role of the Biosphere: Origins of Life

But what is the raw material that comprises the stuff of life—that initiates and directs the master plan for the architecture of all the organisms with which we share or have shared this planet? It is a flexible thread that binds the most minute of submicroscopic flora and fauna to the most magnificent (and threatened) product par excellence of evolution—*Homo sapiens*. How did genetic material, those miracle molecules, originate, organize, and evolve?

The condition of primitive Earth (about 4 billion years ago) provided an atmosphere rich in hydrogen, methane, ammonia, and water. In 1952 Harold Urey and and his associates posited the first steps in the evolution of life from inorganic compounds by nonbiologic means. Possible energy sources for the synthesis of these molecules might have been ultraviolet radiation from the sun, lightning, radioactive decay in the earth's crust, and heat from volcanoes and hot springs. Laboratory experiments undertaken by Stanley L. Miller and others (Miller & Orgel, 1974) since 1953 have demonstrated the possibility (by simulating primitive Earth atmosphere) of production of many amino acids, purines, pyrimidines, and sugars—such as ribose and deoxyribose—by nonbiologic means. So, the primeval organic "soup" postulated by J. B. S. Haldane (1929) was probably present on primitive Earth. As the earth evolved chemically, the atmosphere changed. Most of its hydrogen was lost to outer space and the released oxygen thus gained was converted to ozone and formed above the earth a layer that absorbed ultraviolet light. After the formation of this layer, the main sources of energy were visible light and heat from the sun.

As more synthesis of these primary molecules occurred—especially in the oceans—more and more intermediate products accumulated in the thickening organic soup. An ingredient of the soup was adenine—one of DNA's bases. It was plentiful in that broth because it was easily synthesized; a simple compound of hydrogen cyanide. The other bases—cytosine, guanine, and thymine—are more complex structures. Irradiation of solutions of adenine, ribose, and phosphate compounds with ultraviolet light (2400–2900 Å) has led to the synthesis of AMP (adenosine monophosphate), ADP (adenosine diphosphate) and ATP (adenosine triphosphate)—the basic energy currency of cells. This means that these high-energy phosphate compounds would have been part of the "soup" when life initially evolved and that these adenosine compounds would have been the most common. It is interesting to note three steps that are a precondition for the formation of DNA:

1. The union of base and sugar to form the nucleoside.
2. The union of nucleoside and phosphoric acid to form the nucleotide.
3. The union of nucleotides to form DNA (nucleic acid).

Finally, the union of amino acids to form polypeptides is accomplished simply by the removal of water.

The next step in evolution toward life from monomers to polymers (and collections of polymers) came with the development of interdependence of these chemical systems. Proteins and nucleic acids became dependent on each other in their evolution and for their existence. Nucleic acid can function nonenzymatically to replicate itself; this is known as autocatalysis. Such autocatalytic function means that a pool of raw nucleotide materials would more easily synthesize copies of a given polynucleotide than it would synthesize, by complementation, different polynucleotides. Natural selection would then occur and the most abundant polynucleotide would be the most stable and quickest to replicate.

With the nucleic acids, proteinoids that were present on early Earth underwent a somewhat altered evolution. The formation of protein polymers is accomplished simply, but the polymerization of amino acids to protein chains is not spontaneous. It can become so if the concentrations of reacting substances are sufficiently high. Haldane (1929) proposed that this was accomplished by the concentration of amino acids in drying pools of water near seashores. Nearly all the 20 amino acids common to proteins were included in these proteinoids, which were very similar to currently known proteins and polypeptides of corresponding size and all showed at least some catalytic activity. Although natural proteins are catalysts, unlike nucleic acids they are poor autocatalysts, so the products and diversity of protein catalysis are determined more by the amount and type of reactants involved than by the qualities of the catalyst. In 1965, Fox discovered a distinct propensity on the part of proteinoids to form microspheres about 2 microns in diameter when hot concentrated solutions of them are slowly cooled. These microspheres show a double layer boundary resembing a membrane (but without lipids) and swell or shrink as ambient salt concentrations vary. They are not alive or are not even direct precursors of living cells, but they demonstrate the effects of forces that would have operated on primitive living systems as they evolved. If allowed to stand for several weeks these microspheres absorb more proteinoid material from solution, produce buds, and sometimes divide to produce second-generation microspheres. This is not the ancestral paradigm of reproduction. It simply illustrates how one of the functions usually associated with living organisms can be mimicked by the remarkable self-organizing properties of simpler chemical systems. Thus, it appears at least possible to speculate that the immediate precursors of life were capsules of chemical reactions called *coacervate drops* by Oparin (1968). He saw them as a model for the evolutionary development of cell membranes, as coacervate preparations form droplets with an acqueous solution surrounding them. These drops will form in many kinds of solutions of polymers including proteins or nucleic acids. Although this droplet-forming property is also a function of purely chemical systems, the behavior of these sorts of molecules sheds light on the origin of life. More stable coacervates could survive longer and perhaps grow by adsorbing the remains of less stable droplets. As a result of selection, accelerated rates of reaction would become advantageous and more and more efficient and elaborate catalysts would be developed and retained until evolution of such a system produced the ultimate catalyst—the organic enzyme. This led to what Oparin (1968) described as protobionts—organized metabolizing systems. These protobionts delineated by Oparin lacked one crucial quality—they were not capable of reproduction; they did not have the capacity to replicate themselves and to persevere as a species. The mechanism for doing this, the molecules of instruction for self-duplication, were still evolving in the polynucleotides of DNA.

BIOCHEMICAL GENETICS

DNA (known fully as deoxyribonucleic acid) is present in almost all cells of a given organism in the same amount and type. The genes (which are mostly organized in chromosomes) are segments of a DNA molecule that specify the structure of a single type of protein. Although genes ultimately determine all the characteristics of a cell, they do not do so directly. DNA exerts its control over a cell by synthesizing the closely related ribonucleic acid—RNA—which in turn directs the synthesis of proteins. Experiments with microorganisms clearly show that DNA contains all information necessary to make daughter cells that are essentially identical to parent cells. The amount of DNA per cell is not large; in the human cell there is about 6×10^{-12} g. In spite of this minute quantity, the amount of information contained is enormous—sufficient to direct the synthesis of a human individual.

A DNA molecule is made up of chemical units called nucleotides, each consisting of:

1. A base that is a purine [adenine (A) or guanine (G)] or a pyrimidine [cytosine (C) or thymine (T)].
2. A pentose (5-carbon sugar), deoxyribose.
3. A phosphate group.

In a given DNA molecule, the phosphate and sugar groups are identical and only the bases vary. The information determining heredity resides in the arrangement of the bases, A, G, C, and T. The quantity of these bases per cell is constant within a given species but varies among species. However in every species, the quantity of A$=$T and that of G$=$C, proving that A is always paired with T and G with C.

DNA was chemically known in the 1800s but the first comprehensive elaboration of the structure and coding of genes was demonstrated in 1953 by Watson and Crick. The bases are attached to a sugar–phosphate backbone forming a chain of nucleotides:

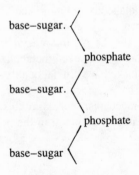

Watson and Crick found DNA to be a double chain of nucleotides:

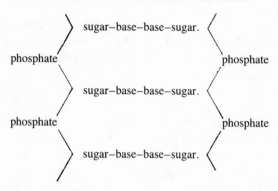

The two chains are joined by the bases, pairing by the rule stated previously and are twisted around each other in a double helix (Fig. 1.4). The nucleotide pairs are separated by 3.4 Angstrom units (Å; an Angstrom unit equals 10^{-7}mm), and the whole structure repeats itself in 10 pairs at intervals of 34 Å. Thus, by applying the pairing rules (A≡T and G≡C), if we know the sequence of bases on one helix, we also know the sequence on the other.

According to the Watson–Crick theory, the linear sequence of nucleotides is characteristic for a given species although there are minor heritable variations that lead to variations in proteins that have as their primary structure a chain of amino acids. Therefore, the linear sequence of nucleotides can be regarded as a

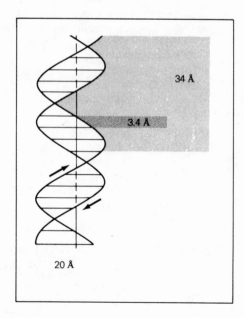

FIG. 1.4. The double helix of DNA.

genetic code specifying the proteins that are essential components of all living organisms.

There are 20 essential amino acids specified by the genetic code. Because there are four possible bases (A, T, G, and C), one- and two-letter codes are inadequate, specifying 4 and $4^2 = 16$ combinations only. A triplet code specifying $4^3 = 64$ combinations is a necessary and apparently sufficient minimum. The nucleotide triplet is referred to as a *codon,* and because a triplet code gives 64 different combinations, or words, of which only 20 are needed for amino acid synthesis, the code is referred to as a *degenerate code.* In fact some amino acids are coded by more than one triplet (Fig. 1.5). The names and abbreviations of the 20 essential amino acids are:

Alanine	Ala	Leucine	Leu
Arginine	Arg	Lysine	Lys
Asparagine	AspN	Methionine	Met
Aspartic acid	Asp	Phenylalanine	Phe
Cysteine	Cys	Proline	Pro
Glutamic acid	Glu	Serine	Ser
Glutamine	GluN	Threonine	Thr
Glycine	Gly	Tryptophan	Tryp
Histidine	His	Tyrosine	Tyr
Isoleucine	Ileu	Valine	Val

From the DNA code a linear message of triplets or codons is *transcribed* to a form of RNA (or ribonucleic acid) called *messenger* RNA (*m*RNA). RNA is chemically very similar to DNA except that: (1) its sugar is ribose rather than deoxyribose; (2) it has the base uracil (U) substituting for thymine; and (3) it is single- rather than double-stranded. In *transcription* from DNA to *m*RNA, the following pairing rules therefore occur:

Base in DNA	*Base in mRNA*
A	U
T	A
C	G
G	C

There are three kinds of RNA on which protein synthesis depends: messenger RNA, transfer RNA, and ribosomal RNA, all carrying codons that correspond to DNA.

SECOND LETTER

		U	C	A	G	
FIRST LETTER	U	UUU } Phe UUU UUA } UUG } Leu	UCU } UCC UCA } Ser UCG }	UAU } Tyr UAC } UAA Chain End UAG Chain End	UGU } Cys UGC } UGA Chain End UGG Try	U C A G
	C	CUU } CUC CUA } Leu CUG }	CCU } CCC CCA } Pro CCG }	CAU } His CAC } CAA } Gln CAG }	CGU } CGC CGA } Arg CGG }	U C A G
	A	AUU } AUC } Ilu AUA } AUG } Met	ACU } ACC ACA } Thr ACG }	AAU } Asn AAC } AAA } Lys AAG }	AGU } Ser AGC } AGA } Arg AGG }	U C A G
	G	GUU } GUC GUA } Val GUG }	GCU } GCC GCA } Ala GCG }	GAU } Asp GAC } GAA } Glu GAG }	GGU } GGC GGA } Gly GGG }	U C A G

THIRD LETTER

FIG. 1.5. The genetic code. See text for explanation.

Messenger RNA (*m*RNA) has been described previously.

Before amino acids are assembled into protein chains they must be "activated" by the attachment of a special phosphoric acid group. They are then attached to another type of RNA called *transfer* RNA (*t*RNA). In fact, there are as many varieties of *t*RNA molecules as there are triplets determining amino acids.

Alignment of the *t*RNA and *m*RNA so as to assemble protein chains in an orderly way is mediated by special particles in the cytoplasm of the cell called *ribosomes,* which are made from a third form of RNA called *ribosomal* RNA (*r*RNA).

The process of the formation of protein from the code carried by the *m*RNA is referred to as *translation,* so we can summarize what happens as:

transcription   translation

DNA ⎯⎯⎯⎯⎯⎯⎯⎯⎯→ *m*RNA ⎯⎯⎯⎯⎯⎯⎯⎯⎯→ protein

The important point then is that the amino acid sequence in proteins is directly determined by the genetic code carried by the DNA molecules even though the mechanism by which this occurs is complex. A great deal of additional information about this process can be found in many texts, but the details of the process, largely established in microorganisms, are only of preliminary interest to behavior geneticists at the present stage in the development of our field. On the other hand, with time there will certainly be a trend toward metabolic explanations of behavioral processes, so that an understanding of the biochemical basis of gene actions will assume progressively more importance.

Regulatory Processes

So far as protein synthesis itself is concerned, most proteins are made only when they are useful. Therefore, there must be a means of *regulation*. In fact, *regulator genes* have been discovered in microorganisms. These regulator genes determine whether the genes that specify types of proteins (*structural genes*) are active or otherwise. The regulator genes themselves are controlled by events taking place in the cytoplasm of the cell and thus are open to environmental influences; for example, if a certain amino acid is necessary for growth and is present in the environment, then the cells can cease making the amino acid as well as the enzymes necessary for its synthesis (enzyme repression). Undoubtedly this process of regulation of protein synthesis must be the basis of *differentiation,* which is the development of specialized cells and tissues. At all stages of development, it is clear that different portions of DNA are active in each type of cell and tissue. Such temporal regulation of gene action must be carefully studied in order to understand behavior. It is clear that genes act in sequence during development so that a given gene may start one event which in turn activates other genes, leading to a series of developmental steps. Gene–hormone interactions, for example, are probably involved in sex differentiation, the onset of puberty, and the development of learning in man.

The other important consequence of a knowledge of the physiological processes of a gene substitution is that their pattern can be modified by appropriate treatment. A good example of environmental influence on genes in man is phenylketonuria, which we have already considered in other respects. Untreated individuals homozygous for this recessive gene usually have IQs below 30. Their skin and hair pigmentation is generally lighter than that of the population from which they arise. Phenylketonuria is due to a deficiency or absence of the enzyme phenylalanine hydroxylase, which is necessary in the metabolism of phenylalanine, an amino acid that is an essential dietary constituent. Normally, phenylalanine $\rightarrow$ tyrosine $\rightarrow \cdots \rightarrow$ various metabolic breakdown products. In phenylketonuria, this first step is blocked; phenylalanine accumulates to a level 40 to 50 times that found in unaffected individuals and this excess leads to mental deficiency. Hence, a likely treatment would be to feed a phenylalanine-free diet. This poses problems, as all known proteins contain phenylalanine. However, such a diet can be synthesized by breaking down protein and reconstituting it without phenylalanine, but still containing other essential amino acids. The diet must begin early in life to have a major effect on IQ and is likely to be proportionately less effective if delayed beyond infancy. The treatment must strike a delicate balance between malnutrition (i.e., inadequacy of the essential amino acid, phenylalanine) and intoxication. More recently (Ambrus, Ambrus, Horvath, Pederson, Sharma, Kant, Mirand, Guthrie, & Paul, 1978), management of phenylalanine levels by regulating enzyme levels in the blood has been successfully achieved and may lead to even more effective control of this once irreversible and untreatable genetic disease. (See Chapter 5 for further discussion.)

QUANTITATIVE GENETICS

The focus of classical Mendelian genetics concerned individual traits as reflected in the phenotypic characteristics (color, structural, and chemical differences) that resulted from the genotype—a product of segregating particulate units that assorted independently. Inferences concerning the gene were derived from data produced from breeding experiments. The distribution of phenotypes was seen as discrete (i.e., falling into distinct classes such as red, white, or pink, according to the dominance or recessiveness of specific alleles). Genetic variation was conceived of as being under the control of specific genes assignable to specific loci on chromosomes. This approach is often not applicable to traits that are quantitative and do not segregate into discrete classes. An example of this in humans is IQ within a population. This does not mean that specific genes affecting these traits are not present, only that they cannot be precisely identified with techniques currently available. There is evidence, however, that IQ is influenced by genes at known loci and by specific chromosomal anomalies. (See Chapter 5 for further discussion).

Quantitive genetics is the study of variation in those traits where genes responsible for variability in phenotype cannot routinely be individually recognized. This type of inheritance, leading to continuous variation in complex characteristics (such as skin color, disease resistance, or IQ) is called polygenic because it involves the summation of the action of many genes at different loci that are often cumulative in their effect on a particular trait.

Quantification of genetics began even before the rediscovery of Mendel's work in 1900 by Hugo de Vries, Carl Correns, and Erich von Tschermak. It probably had its origins in the work of Adolphe Quetelet, a Belgian mathematician and sociologist (1796–1894) who was best known for his application of statistics and probability theory to social phenomena. He conceived of the idea of the "average man" as a central value about which measurements of a human trait may be grouped according to the normal probability curve. A further development of mathematical method apropos genetics was provided by Sir Frances Galton—a cousin of Charles Darwin and one of the first to recognize the implications of Darwin's theory of evolution for human traits (including cognitive ones). Although Galton is perhaps best known by behavior geneticists for his book *Hereditary Genius* (1869) in which he argued that mental and physical features are both inherited, he concerned himself for more than three decades with the improvement of standards of measurement in a variety of disciplines. This concern was carried further in the work of his disciple, Karl Pearson (1857–1936), one of the founders of modern statistical science. Influenced by Galton's evolutionary writings, Pearson became enamored with the notion of applying statistics to the biologic problems of heredity and evolution. These efforts in turn led to an assortment of methods that are part of the core of statistics: the chi-square (χ^2) test of statistical significance and the correlation coefficient or Pearson r.

For a time this resulted in a rather acrimonious confrontation between the biometricians led by Pearson, who touted the quantitative aspects of heredity, and the Mendelians, inspired by William Bateson—founder of the science of genetics—who argued for the unit character of inheritance and discrete classes of traits. Finally, a resolution of these two views of inheritance—discrete versus continuous—came about through the experimental work of G. H. Shull and others via inbreeding experiments in maize. Such applied plant breeding was influenced by the work of W. L. Johannsen (1857–1927), a Danish biologist who coined the word "gene." He stressed the use of pure lines to clarify the concepts of inheritance and, in doing so, supported Galton's regression theory, that parents who are above or below the population average tend to produce offspring who approach the mean. Using pure lines of beans (Princess variety of *Phaseolus vulgaris*) that produced large or small plants from beans of corresponding size, he crossed tall with dwarf plants and demonstrated that the mean value of a measureable trait—in this case, weight—of the beans produced by the progeny of these crosses tended to approach the mean of the total population and not to reflect either of the parental extremes.

Although these experiments provided a basis for the notion of a continuously varying heredity, the accommodation of this theory with the Mendelian concept of transmission of discrete units of inheritance was furnished by the work of a Swedish geneticist, Nilsson-Ehle, who discovered a hereditary trait with distinguishable classes—the depth of red pigment in the hull of a wheat, *Triticum vulgare*. This plant exhibited a graded series of redness from deep red to white (with five intermediate shades), which formed a distribution encompassing easily distinguished discontinuous groups. The variation was controlled by three independent loci with pairs of red or nonred alleles at each locus and with each allele for pigment contributing additively to the total phenotype. The documentation of discrete qualitative differences within an otherwise continuous distribution, that of seed color in wheat, was explained by Nilsson-Ehle in terms of a continuous graduated character determined by three Mendelian pairs of alleles with additive nondominant effect; that is, with multiple factors, later called polygenes. These crucial experiments assisted in the reconciliation of the biometricians' blending (continuous variation) hypotheses with the Mendelians' discrete units of heredity. Furthermore, all this provided an experimmental basis for the principles of quantitative genetics.

The theoretical basis for the accommodation of the biometricians and the Mendelians was provided by R. A. Fisher in a 1918 paper (Provine, 1971) that ascertained the biometric properties of a Mendelian population and demonstrated that correlations between relatives in man did not contradict a Mendelian scheme of inheritance. Further work in 1931 by Sewall Wright (Provine, 1971) confirmed Fisher's earlier mathematics by a different method and focused attention on the interactions of systems of genes that resulted in a general theory of the quantitative consequences of inbreeding. In 1930, J. B. S. Haldane (Provine, 1971) agreed with much of Fisher's work but was more oriented toward establishing a quantitative theory of natural selection with emphasis (unlike Wright) on single-gene effects. (The definitive reconciliation at the theoretical level is given by R. A. Fisher, 1930, in the first chapter of *The Genetical Theory of Natural Selection.*)

One of the peripheral issues that emerged from this dispute was the distinction between more easily traced major genes, which have massive effects on the phenotype of organisms, and minor genes with less striking phenotypic effects. In this latter instance, several minor genes (polygenes) may contribute to the formulation of a characteristic. For example, body weight, height, and intelligence in humans are influenced by many genes that act in more or less additive ways. And although it is the cumulative number of more or less contributory alleles that determines the individual phenotype, it is the arrangement of alleles on specific chromosome pairs that influences genetic transmission of the trait. By the time a gene is expressed, its capacities have been altered by its association with other genes as well as by environmental interactions.

To better understand the nature of continuous variation let us consider, as an example, height. In this instance it is not possible to place people into discrete classes such as tall or short because all degrees of height occur. The frequency distribution of many quantitative traits approximates the *normal distribution* (Fig. 1.6). It can be completely described in terms of two parameters. One is the mean or average value; the other is an expression of variability around the mean. In some cases the variability around the mean is small (Curve *A;* elsewhere, the population distribution may be more diffuse, and the variability around the mean is larger (Curve *B*). The term for the quantity measuring variability is *variance,* which is the sum of squares of deviations around the mean divided by $n - 1$ *(where n = the number of observations)*. The square root of the variance is the *standard deviation.* It can be shown that 95% of the population of a normally distributed trait lies within two standard deviations on either side of the mean. Much of the theory of quantitative genetics is based on the assumption of a

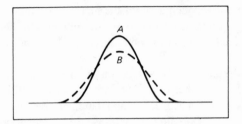

FIG. 1.6. Normal distribution. Curves A and B have the same mean, but the variance of B is larger than that of A.

normal distribution. If it is impossible to assume a normal distribution, a suitable algebraic transformation may be found that will convert the data to an approximately normal distribution.

Components of Phenotypic Variance

If we assume that a trait such as height is partially genetically controlled we must ask how the intrinsically discontinuous variation caused by genetic segregation is translated into the continuous variation of quantitative traits. Suppose two individuals *Aa* and *Bb* are crossed, where *Aa* and *Bb* signify gene pairs at two unlinked loci. Further, if genes *A* and *B* act to increase the measurement of a quantitative trait by one unit and genes *a* and *b* act to decrease the trait by one unit, we can arrive at a quantitative value for a genotype by counting the number of plus (+) and minus (−) genes. The cross would yield five genotypes, the most common genotype +/− +/− (*Aa Bb*) having a genotype value of zero; this is the *mean genotypic value*. The least common genotypes are the two extremes +/+ +/+ (*AA BB*) and −/− −/− (*aa bb*) with values of +4 and −4 respectively. If there is a third locus with two similar alleles, then a cross between multiple heterozygotes would raise the number of genotypic classes to seven, a fourth locus to nine, and so on. As more genes are channeled into the expression of a trait, the more smoothly continuous the distribution appears. This is true because the differences among the classes become progressively smaller as the number of segregating loci rises. At the stage when the differences among classes become about as small as errors of measurement, the distribution becomes continuous as in Fig. 1.6. Then too, any variation due to nongenetic causes, (i.e., to environmental influences) blurs the underlying discontinuity of the segregating alleles still further so that the variation is seen as continuous irrespective of the accuracy of the measurement. Because of this, many genes each with small effects on a trait plus the variability due to nongenetic or environmental causes lead us to expect a continuous distribution of a given trait—the normal distribution or bell-shaped curve already cited.

As mentioned previously, genes that contribute to a quantitative trait but are not directly identifiable by classic Mendelian segregation (i.e., cannot be studied individually) are polygenes. Genes whose effects can be studied individually are referred to as major genes. The distinction does not imply any fundamental difference between the two categories of genes. It is merely a matter of convenience of nomenclature because the breeding methods used to study the effects of major genes cannot in general be used in the study of polygenes. Quantitative or biometrical genetics has been developed to deal with the variation caused by the simultaneous segregation of many polygenes.

Heritability

Behavioral traits such as activity scores in mice, duration of copulation in *Drosophila*, or IQ in humans are essentially quantitative. To analyze these traits an appreciation of the aims and methods of quantitative genetics is essential. One of the main goals of biometrical genetics is to make known the apportionment of the total phenotypic variance (V_p) associated with a given trait into components due to the genotype (V_g) and to the environment (V_e). If we assume no interaction between genotype and environment—the simplest assumption possible—then phenotypic variance is the sum of the genotypic and environmental variances: $V_p = V_g + V_e$. It is reasonable then, to compute the proportion of the total phenotypic variance that is genotypic:

$$\frac{V_g}{V_g + V_e} = \frac{V_g}{V_p}. \quad \leftarrow cut$$

This ratio is referred to as the *heritability* of the trait *in the broad sense*. However, the important measure in plant and animal breeding is *heritability in the narrow sense*.

Heritability varies from zero when the genotype is unimportant to one when the trait's determination is entirely genotypic, as with blood groups. It should be noted here that the estimated heritability refers to the population under analysis and not to the trait itself. A different population may give a different heritability for even the same trait.

There are several ways of calculating heritability. One method is based on the use of laboratory animal strains that have been inbred for many generations (usually via sister × brother matings) and are, by the end of the breeding regime, essentially homozygous. This means that *within* each of these strains all individuals are "genetically identical," so that any variation that occurs will be due to nongenetic causes. However, *between* strains there will, in general, be variation due to the different strains as well as to environmental variation. Using a statistical technique known as "analysis of variance," (which was devised by R. A. Fisher) the variation within and between strains can be separated. This method enables us to calculate estimates of V_G and V_E and therefore to compute heritability.

The average effect that an allele has, in all genotypes in the population that carries it, is its *additive genetic value*. Deviations in any one individual may be due to dominance effects, epistasis (an interaction where one gene modifies the effect of another at a different locus), and the environment. Certain breeding procedures allow division of the genotypic variance, V_G into variance due to additive genes, V_A, dominance, V_D and epistasis, V_I, so that we can formulate $V_G = V_A + V_D + V_I$. This permits a more refined definition of heritability as the ratio of the additive genetic variance to total variance V_a/V_p, which is referred to as *heritability in the narrow sense* and written as h^2 in contrast to heritability in the broad sense (V_g/V_p), or degree of genetic determination. The parameter h^2 is a measure of the proportion of variation due to additive genes and as such is a more useful concept than degree of genetic determination. (Heritability in the narrow sense provides a measure of genetic variability transmitted from one generation to the next because it takes into account only additive genes and provides a basis for selective breeding.) It has been widely applied in the study of quantitative behavioral traits including intelligence, temperament, emotional behavior, specific abilities, and some other aspects of personality (e.g., neuroticism). The evidence strongly supports polygenic influences on these traits.

Parents pass on fractions of their genes via gametes to succeeding generations, not their entire genotypes. This means that dominance and epistatic relationships in parental genotype are, in general, broken down and not transmitted from one generation to the next in an outbreeding population. These altered gene interactions have important intergenerational implications for families in search of genetic counseling.

Stabilizing, Directional and Disruptive Selection

There are three primary types of natural and artificial selection to which a quantitative trait may be subjected. The form whereby intermediate phenotypes are favored at the expense of the extremes is referred to as stabilizing selection. For a trait such as height in humans, it seems reasonable to expect a normal distribution, as the extremes of very short and very tall are disadvantageous because they are less likely to reproduce than the median individual and so, on the average, phenotypically extreme individuals will contribute fewer genes to subsequent generations. This type of selection is common in natural populations. Evidence for stabilizing selection has been found for many traits (e.g., for hatchability of duck's eggs of assorted sizes and for birth weight of humans in relation to longevity). Thus, in natural populations, selection will tend to reduce the variability of many quantitative traits by favoring intermediate phenotypes. Because quantitative traits are under the control of many genes, unique genotypes, and hence extreme phenotypes, will be produced continuously by genetic segregation and recombination but will usually be at a selective disadvantage and are likely to be eliminated through natural selection. Yet because environments can never be constant, some extreme variants may be advantageous in part of the geographic range of a population or at certain times or both. Further, some of these variations may prove useful if the organism is challenged to adapt to an environment that is continuously changing in one direction.

The converse of stabilizing (or canalizing) selection is diversifying (disruptive) *selection,* wherein the two extremes of a distribution are favored. This can occur when an interbreeding population occupies different ecological niches, perhaps seasonally. In one of these, a particular phenotype may be favored because of specific niche qualities to which it is better adapted, but in the other niche a different phenotype may prevail. The importance of this type of selection is still speculative (Fig. 1.7).

Finally, there is *directional selection,* in which one extreme is favored generation after generation. This is the form of selection employed by many plant and animal breeders favoring traits of economic importance (high yield and disease resistance in wheat, fertility in poultry, size and body weight in cattle, etc.) In behavior genetics, one of the main experimental methods often employed involves selection of individuals at the extremes of a distribution and by directional selection, forming separate high or low lines in subsequent generations. Using special breeding techniques, it is even possible in some cases to localize genes controlling a quantitative trait to specific chromosomes or even to regions of chromosomes.

In predicting a response to directional selection, the mean value of the populations selected must be considered. This will depend on the size of the population available at the outset, which, in animals, will be finite. The likelihood of inbreeding will be increased if the selected population is too small, which will lead to an excess of homozygotes and a reduction of the level of genetic variability. Frequently, fertility and general fitness decrease during selection because these more homozygous genotypes with their lowered variability have not been previously exposed to the severe action of natural selection. Indeed their response must diminish rapidly over generations.

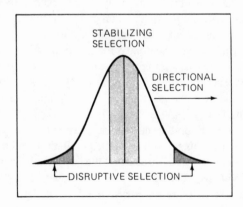

FIG. 1.7. Directional, disruptive, and stabilizing selection. Sections of normal population distribution favored under these three selection regimens.

Examples of successful selection for behavioral traits include ferocity or gentleness in dogs, egg production in chickens, speed and strength in horses, and mating speed in *Drosophila*.

POPULATION GENETICS

Basic to the work of Fisher, Haldane, and Wright was the notion of the mathematical consequences of Mendelian heredity. This is embodied in the Hardy–Weinberg equilibrium principle (derived independently in 1908 by Wilhelm Weinberg, a German physician and by G. H. Hardy, a British mathematician). It resolved the question of what proportions of alleles at a given locus would be propagated in a large mixed population. Indeed, the study of population genetics is based on this principle. Whereas Mendelian genetics is concerned with gene segregation and action in progenies at the familial level, consideration was now extended to a population (a randomly breeding group of individuals). Much of population genetic theory is based on the assumption of random mating (panmixia), but this does not necessarily always apply. One important modification occurs as a result of inbreeding—the mating of individuals related to each other by ancestry. Individuals with a common ancestor are likely to carry replicas of an allele present in that ancestor and, if they mate, both replicate alleles may be passed on to their offspring. This process increases the proportion of homozygotes compared to that produced by panmixia.

Phenotypic assortative mating is the mating of individuals based on phenotypic resemblance. This occurs when like phenotypes mate more frequently than under random mating. Because like phenotypes may be under the control of like genotypes, this will lead, as will inbreeding, to more homozygotes than expected via panmixia. Positive assortative mating has been found for many traits in humans (e.g., for height, educational attainment, and various behavioral traits).

The geneticists' infinitely large, randomly interbreeding population constitutes a gene pool—the total genetic information of the breeding members of a population of sexually reproducing organisms. The Hardy–Weinberg law states that in a large, randomly mating population, gene frequencies and genotype frequencies will remain constant from generation to generation unless outside forces act to change them. In the "simplest" situation, where the frequency of the alleles A and a are designated as p and q respectively, the basic equilibrium formula is $p^2(AA) + 2pq(Aa) + q^2(aa) = 1$. Such population equilibrium is based on several assumptions: each individual's genotype must be as reproductively fit as that of others in the population; and random (panmictic) mating must consistently occur throughout the population, which must have a large number of individuals, bearing no newly arisen genetic variation (mutation). Essentially it is a description of a static Mendelian population—one not evolving. Mathemati-

cal formulations explaining population changes in allele frequencies were provided by Fisher, Wright, Haldane, and others who conceived the theoretical
models and investigated the mechanisms of change that allow the exploration of
the genetic architecture of different populations.

Population dynamics are a complex interaction of changes in allele frequency
caused by selection, mutation, chance (random genetic drift), and differential
migration. Also important are the frequency of chromosomal changes—duplications, deletions, inversions, and translocations and that of numbers (ploidy)—
which provide the basis for inquiry into the relations of individual genotypes and
their ever-changing environment.

A by now classic illustration of the changes in allele frequency in a population
as a result of a changed environment may be seen in the studies conducted by
Kettlewell (1955). He collected elsewhere and released in grimy industrial Birmingham 447 black moths *Biston betularia*. At the same time 137 white moths
(also not of the vicinity) were released. After a period of time, moths in the area
were caught and classified and the black moths (who blended best with their
surroundings) had a 2:1 survival advantage over the white ones. Note that cam-
ouflaged insects are obliged to manifest suitable behaviors (i.e., they must settle
on an appropriate background). The converse of this experiment was performed
in nonindustrialized, rural Dorset where the survival rate of the white moths was
three times that of the black. The implication is that certain genotypes in a
species provide their carriers with fitness traits (in this case, color) that are
advantageous for survival and reproduction, but these can be altered dramatically
in value with correlated environmental alterations. And what natural environment is constant?

ORGANIC AND CULTURAL EVOLUTION

Environmental variability is one of the conditions of organic evolution. Dobzhansky (1973) stated that evolutionary theory provides a basic paradigm for the
natural sciences: "Nothing in biology makes sense except in the light of evolution [p. 125]." Humans are animal in origin and in biologic characteristics—
anatomically similar to the great apes, particularly the chimpanzee and gorilla.
Also the chemical structure of our proteins (especially hemoglobin) is very
similar in humans and in apes. These similarities suggest that human ancestors
were apes along an evolutionary continuum that produced gorillas, chimpanzees,
and humans—a sequence that is supported by the fossil record.

But we are much more than merely biologic, as no other species has the
capacity to control its destiny as we do, nor the ability to remember and profit
from the past, communicate this experience to others of our species and look
forward to and plan for the future. These abilities have enabled humans to change
their way of life at a dazzling pace and in formidable fashion—faster than any

other species and in a qualitatively different way; and this has occurred without our undergoing any major change in anatomical features or brain structure for the last 35,000 years. Thus an understanding of human evolution involves the recognition of both its biologic and its cultural aspects and the interrelations between them.

Though human anatomic traits have not varied much in 35,000 years, human society has changed so rapidly and in so many different ways that our species has established control over the inanimate environment and other animals. What we do with this dominance is the subject of other texts and disciplines (philosophy, social science, ethics, etc.) but it has been gained not by organic but by cultural evolution. The latter resembles organic evolution in that both processes result in changes, often directional, involving both a population and its environment.

Like organic evolution, the cultural variety can proceed in many directions and at differing speeds but the stored variability that makes cultural evolution possible is not genetic in nature. It consist of ideas, inventions, laws, traditions, and all the other learned responses by which society is regulated. Individuals acquire their share of it not by heredity but through conditioning, training, and learning; and the stored variability of culture increases not genetically, but by additions to and synthesis with the ongoing collection of inventions, ideas, and customs that connote our culture. Another profound difference between organic and cultural evolution is the way that changes in the relationship between population and environment occur. Alfred Russell Wallace (1858), whose evolutionary theory strongly paralleled that of Darwin, noted that animals evolve by adjusting themselves to their environment; man by adjusting the environment to himself.

These considerations exist on a more restricted level with other species. Bonner (1980) notes that: "Many birds and mammals have a primitive form of culture in which there is a transmission of behavioral information and there is a maintenance of accumulation in the form of tradition. . . . The only difference between their culture and that of man is that the transmission of the information is by imitation rather than by true teaching [p. 171]." He cites as examples the "termiting" of chimpanzees, where a twig is poked into a termite nest opening, revolved, and extracted with a snack of termites adhering to it. Another celebrated case is the opening of milk bottles by titmice in Britain. A particularly bright and/or hungry bird pecked through the aluminum cap on the milk bottle and this skill spread so rapidly that soon most of the bird population of the British Isles was obtaining its breakfast cream in this fashion. An equally impressive behavioral innovation was that of a female macaque who washed the sand from sweet potatoes before eating them. All these creative inventions were passed on to other members of the species and are prime examples of the importance of behavior in cultural evolution.

But what of the ubiquitous mode of transmission—organic evolution? Recently, Dobzhansky, Ayala, Stebbins, and Valentine (1977) stated that:

Organic evolution is a series of partial or complete and irreversible transformations of the genetic composition of populations, based principally upon altered interactions with their environment. It consists chiefly of adaptive radiations into new environments, adjustments to environmental changes that take place in a particular habitat, and the origin of new ways for exploiting existing habitats. These adaptive changes occasionally give rise to greater complexity of development pattern, of physiological reactions, and of interactions between populations and their environment [p. 8].

The principle of evolution provides a unifying theme for contemporary biologic sciences but the link that makes the "modern synthesis" of evolutionary theory an invaluable paradigm was the contribution of Charles Darwin in his concept of *natural selection* noted in the beginning of this chapter as one of the mainstreams of thought from which the discipline of behavior genetics emerged. Darwin was the first to recognize that the connection between environmental variability and evolutionary change was provided by this phenomenon. In Darwin's words—from the Introduction to *Origin of Species* (1859):

As many more individuals of each species are born than can possibly survive; and as, consequently, there is a frequently recurrent struggle for existence, it follows that any being, if it vary however slightly in any manner profitable to itself, under the complex and sometimes varying conditions of life, will have a better chance of surviving, and thus be *naturally selected*. From the strong principle of inheritance, any selected variety will tend to propagate its new and modified form.

One inference is that through natural selection, some individuals in a population survive and some do not. The notion that populations are culled in this fashion came initially from Thomas Malthus' *Essay on Population,* which Darwin read in 1838 when he was formulating his ideas about natural selection.

Darwin deduced that because population growth in a species is checked by some means, not all individuals are equally susceptible to being eliminated. This idea of the variability of individuals within populations led to the hypothesis that the factors checking the increase in numbers of individuals, within a species, act more strongly on those that are relatively poorly adapted and favor those best fitted to their environments. Individuals thus favored will leave more offspring over generations than their less well-adapted conspecifics.

This concept that Darwin recognized and documented—that among diverse individuals in a population some have a higher probability of survival than do others—lacked one fundamental. He was not aware of the nature or mechanisms of hereditary variation. It was the contribution of Gregor Mendel and the combination of Darwinism and Mendelism that provided the basis for the synthetic theory of evolution.

Natural selection favors certain individuals in a population and as a result acts on populations of genes controlling traits; so, slowly the genetic constitution of the population changes in response to selection. As an evolutionary force, natural selection acts at the population rather than at the individual level. Artificial selection, whether for exotic breeds of animals or for increased food yield, acts in the same way.

It should be evident that the behavior of individuals within a population (mating patterns, habitat selection, resource utilization) is both a result of selection and of the factors that mediate it. Behavioral phenotypes are manifestations of the underlying genotypic variability of populations. Those behavioral traits that are relevant to population continuity need to be isolated and studied under a multiplicity of environments because these behaviors are among the most important components in evolutionary processes.

Why should these topics be subjects of continuing and intense study? Because they provide clues to the answers to questions that humans have been posing since the evolution of consciousness: Where do we come from? Why are we as we are? Where are we going?

REFERENCES

Ambrus, C. M., Ambrus, J., Horvath, E., Pederson, H., Sharma, S., Kant, C., Mirand, E., Guthrie, R., Paul, T., Phenylalanine depletion for the management of phenylketonuria: Use of enzyme reactors with immobilized enzymes. *Science,* 1978, *201,* 837–839.

Beach, F. A. The snark was a boojum. *American Psychologist,* 1950, *5,* 115–124.

Bonner, J. T. *The evolution of culture in animals.* Princeton, N.J.: Princeton University Press, 1980.

Boring, E. G. *A history of experimental psychology.* New York: Appleton–Century–Crofts, 1950.

Darwin, C. *The origin of species by means of natural selection.* London: Murray, 1859.

Dobzhansky, T. Nothing in biology makes sense except in the light of evolution. *American Biology Teacher,* 1973, *35,* 125–129.

Dobzhansky, T., Ayala, F., Stebbins, G. I., Valentine, J. W. *Evolution.* San Francisco: W. H. Freeman, 1977.

Fisher, R. A.. *The genetical theory of natural selection.* Oxford: Clarendon Press, 1930.

Fox, S. W. (Ed.). *The origins of prebiological systems.* New York: Academic Press, 1965.

Galton, F. *Hereditary genius.* London: Macmillan, 1869.

Haldane, J. B. S. The origin of life, *The Rationalist Annual.* London, 1929.

Hartl, D. L. *Our uncertain heritage: Genetics and human diversity.* Philadelphia: Lippincott, 1977.

Kettlewell, H. B. D. Selection experiments on industrial melanism in the Lepidoptera. *Heredity,* 1955, *9,* 323–342.

McKusick, V., & Ruddle, F. The status of the gene map of the human chromosome. *Science,* 1977, *196,* 390–405.

Miller, S. L., & Orgel, L. E. *The origins of life on earth.* Englewood Cliffs, N.J.: Prentice–Hall, 1974.

Oparin, A. I. *Genesis and evolutionary development of life.* New York: Academic Press, 1968.

Probber, J., & Ehrman, L. Pertinent genetics for understanding gender. In E. Tobach & B. Rosoff (Eds.), *Genes and gender.* New York: Gordian Press, 1978.

Provine, W. B. *The origins of theoretical population genetics.* Chicago: University of Chicago Press, 1971.

Urey, H. C. *The planets, their origin and development.* New Haven: Yale University Press, 1952.

Wachtel, S. H–Y antigen and the genetics of sex determination. *Science,* 1977, *198,* 797–799.

Wallace, A. R. On the tendency of varieties to depart indefinitely from the original type. *Proceedings of the Linnean Society,* 1858.

Watson, J. D., & Crick, F. H. C. Molecular structure of nucleic acids. *Nature,* 1953, *171,* 737–738.

2 Genetics of Human Nervous System Functioning

Gordon Claridge
Gordon Mangan

University of Oxford

It is appropriate to open this chapter by defining the scope of the material it covers and the orientation from which it is written. This is especially necessary in the present case because the material potentially available for review, perhaps more than most, overlaps considerably with that considered in several other chapters. It is, after all, through the central nervous system that many of the genetically determined influences on behavior discussed elsewhere are mediated. This means that data about the genetic determination of nervous system functioning considered relevant here may also bear closely on questions discussed in a different context by other contributors. Where we place emphasis and how we have selected studies for inclusion therefore reflect, more than usually, our personal bias.

Our interpretation of the subject matter of the chapter is that the research reviewed should have, as its common theme, the aim of trying to unravel the biological basis of human psychological differences; it is evidence relating to this problem with which we therefore are mainly concerned. In order to arrive at a more exact set of guidelines for collating such evidence we begin by sketching in the historical and theoretical background from which many of the genetics studies reviewed here have either explicitly proceeded, or against which they can be usefully evaluated.

Three research areas are historically important. The first is psychophysiology, which has established a set of experimental techniques for examining, and a body of knowledge about the physiological correlates of psychological states in the intact organism. On the theoretical side it has given rise to various models of the "conceptual nervous system," enabling inferences to be made about the actual brain mechanisms that may be involved in observed psychophysiological phenomena. Interpreted in a narrow sense, psychophysiology confines itself to the measurement of strictly physiological phenomena, such as EEG and autonomic nervous system (ANS) responses, but it can be said also to embrace a wider range of experimental techniques that allow inferences to be drawn about central nervous function. Included here are conditioning techniques, drug manipulations, and sensory threshold procedures. We use the term *psychophysiology* in this broader sense.

A second research area of relevance is the special application of psychophysiology—in both its broad and narrow sense—to the problem of individual differences. Particularly important is the work of the Eysenck school, and that carried out under the rubric of the Pavlovian theory of "nervous types." These two approaches to the biology of individual differences have proceeded to some extent in parallel, but have slightly different emphases. Those working within the framework of, or influenced by Eysenck's theory have used psychophysiological measures to try to establish the biological basis of statistically derived or clinically recognized personality characteristics, the latter themselves often having been subjected to independent genetical analysis. In Eastern Europe, however, differential psychophysiologists have been less concerned to relate their findings to personality traits. Instead, they have concentrated more on developing techniques for measuring the "fundamental properties of the nervous system" along the lines initiated by Pavlov. Although not exclusively, they have also relied more on procedures that, as noted earlier, extend the scope of psychophysiology beyond the measurement of purely physiological responses.

The third research area of relevance here is experimental psychopathology, part of which has been concerned specifically with the biological basis of abnormal mental states, either as an extension of the studies of normal variation just referred to, or, in more empirical vein, as a search for objective descriptors of different psychiatric disorders. In both cases work in this area is now taking on special significance for human nervous system genetics, through its contribution to an understanding of "high-risk" factors that predispose to serious mental illness such as schizophrenia.

Of course, investigators working in all three research areas just delineated have always been interested in how far hereditary influences determine the phenomena they have observed, and biological explanations of individual differences have often contained an assumption of partial genetic control. It is the results of their inquiries, looked at from a genetic viewpoint, which form the bulk of the studies considered here.

In selecting material for review, three guidelines have been followed. First, preference has been given to studies carried out from, or the results of which bear on some defined theoretical viewpoint; it is hoped thereby to give the chapter a more coherent theme and make it easier to draw general conclusions and identify areas for future research. Second, in keeping with the title of the chapter, attention has been focused mainly on research in human behavior genetics. And third, emphasis has been placed on measures of the type investigated within the framework of psychophysiology, as we have defined it. Needless to say, none of these guidelines is followed inflexibly, studies not meeting these criteria being discussed where context demands.

Because of the overlapping nature of the material, any subdivision for review purposes is inevitably arbitrary and has a somewhat awkward appearance. We decided, however, that it was most convenient to organize the chapter as follows: We start by reviewing a number of studies in which psychophysiological measures—mainly of EEG and of autonomic nervous system function—have been examined from a genetic viewpoint. That part of the review is confined to studies of Western origin, parallel work carried out under the umbrella of Soviet psychology then being considered in a separate section. These two approaches are then brought together through their common link with models of personality, and some conclusions drawn about the light they shed on the genetics of human nervous system function. Up to that point we restrict ourselves to those studies that have focused on "normal" variations, in the sense that they have not been primarily concerned with elucidating the genetics of serious mental disease. Evidence pertaining to the latter that we consider within the scope of the chapter is discussed later. We then close the chapter by identifying some areas for future research.

Before undertaking the review, brief comment is necessary about the validity of the data presented as evidence of genetic influences in human nervous system functioning. Research in human behavior genetics faces many methodological problems and difficulties of interpretation. The results of the studies described must therefore be evaluated with such limitations in mind.

GENETICS OF PSYCHOLOGICAL VARIATION

Studies of "Western" origin

The studies of interest here have been of several kinds, differing in emphasis according to the empirical or theoretical background that has inspired them. They all have in common, however, the fact that, directly or indirectly, they have been thought to shed light on the genetic basis of psychological variation, usually personality variation.

The particular measures we consider fall into three types: measures of EEG activity, indices of autonomic nervous system function, and other measures that, although not strictly psychophysiological, have been examined because of the inferences that may be drawn from them about CNS differences. Each of these three types of measure is considered in turn.

Studies of EEG Activity

For obvious reasons, psychophysiologists seeking evidence for a biological correlate of human psychological variation have frequently turned to the EEG; a logical extension of that interest has been to examine the possible influence of genetic factors on the differences in EEG activity commonly observed across individuals. Indeed, not long after Berger, in the late 1920s, first described human brain waves, several investigators reported that the EEGs of monozygotic (MZ) twins were very similar (Davis & Davis, 1936; Raney, 1934). Slightly later, Lennox, Gibbs, and Gibbs (1945) confirmed that conclusion, describing what they considered to be practically identical EEGs in the majority of a group of MZ twins they studied; by comparison, dizygotic (DZ) twins, they reported, were mostly dissimiliar in brain-wave pattern. Two other studies stand out in that early literature. One, by Juel-Nielsen and Harvald (1958), examined the EEGs of a small group of MZ twins who had been separated shortly after birth, brought up apart, and reunited as adults. Their EEGs were evaluated for three parameters: the frequency and amplitude of dominant occipital activity, and the percentage time of that activity. Although the authors attempted no statistical analysis of their data, they were convinced that, for all three parameters, there was strong evidence for inheritance of EEG pattern. The second study of note, important because of its use of a very large twin sample, was reported by Vogel (1958), whose more recent work we discuss in detail shortly. In the study in question, Vogel examined 208 twin pairs, 110 of whom were monozygotic. EEGs were recorded under conditions of rest, hyperventilation, oxygen lack, and during sleep. Using a blind classification procedure, it was demonstrated that the individuality of the EEG and its similarity in MZ twins was such that, according to Vogel, it could be considered, under normal conditions, to be exclusively determined by heredity.

Building on these early observations, later studies have examined the EEG in more detail. In doing so they have gradually taken advantage of two developments—the more sophisticated methodology for analyzing the human electroencephalogram, and progress in thinking about its psychological and behavioral significance. Furthermore, paralleling work on the EEG in general, behavioral genetic studies have used two approaches: the analysis of the background EEG and the investigation of the cortical-evoked potential.

Background EEG.   Probably the most comprehensive genetic studies of the human background EEG have been carried out by Vogel and his colleagues. Following the twin study just referred to, they have extended their work in several ways, a central feature being the detailed examination of individuals showing different EEG "types," recognized as variants on the resting EEG pattern. In a paper describing these variants, Vogel (1970) reported on their population frequencies and the results of family investigations in which the relatives of probands of particular EEG types were themselves examined and genetic analyses attempted. The results of these studies are summarised in Table 2.1. Of the variants recognised, Vogel suggested that a fairly straightforward autosomal dominant mode of inheritance can be claimed for two EEG patterns— low voltage EEG and quick (16–19 Hz) alpha variants. This may also be true, he suggested, for what he called "monotonous alpha waves," that is, unusually regular alpha rhythm of high amplitude found not only occipitally but also over frontal areas of the cortex. Also shown in Table 2.1 are the results of a cross-cultural comparison of the incidence of the different EEG variants in German and Japanese samples, as reported by Vogel and Fujiya (1969). According to the latter authors, and as seen in the table, population frequencies seem to be very similar for the alpha variants, but the beta variants were observed somewhat less commonly in the Japanese.

Not included in Table 2.1 is another, rare variant on the alpha rhythm described by Vogel in his 1970 paper, and given special attention by Kuhlo, Heintel, and Vogel (1969). This is a regular 4–5 Hz rhythm seen in some adult individuals under resting conditions in the occipital or occipitotemporal areas. The waves in question are said to differ from superficially similar waves observed in children in that they do not reappear immediately after closing of the eyes. Instead, normal alpha waves first appear, to be replaced, after a few

TABLE 2.1
Resting EEG Variants, According to Vogel, with
German and Japanese Population Frequencies
(Adapted from Vogel and Fujiya, 1969)

		Population Frequency	
		German	Japanese
Low voltage EEG	Typical	4.20%	4.57%
	Borderline	2.25%	2.14%
Monotonous α waves		3.85%	4.27%
Fronto-precentral β groups	Numerous	1.47%	0.68%
	Rare	0.45%	0.12%
Occipital slow (16-19 cps) β (α-variant)		0.60%	0.42%
Diffuse		5.54%	4.03%

seconds, by 4–5 Hz activity; furthermore, the latter can be easily abolished by slight stimulation, which causes a reappearance of alpha rhythm. The characteristic is of some interest because of its possible association with various forms of behavioral abnormality, such as emotional instability (Dongier, Tournadre, Naquet, & Gastaut, 1965; Petersén & Sörbye, 1962; Pitot & Gastaut, 1956). Kuhlo et al. (1969) themselves confirmed this association and also comment on its possible genetic basis following their study of 40 individuals showing the trait, and some of their relatives. They found complete concordance in two sets of MZ twins, but the rhythm was rarely found among relatives, and, even then, only in siblings. It was, however, more common in males than in females, and appeared to become rarer with advancing age. Kuhlo et al. conclude that their results may suggest a genetic basis, but not of a simple form, and that in some cases the picture may be complicated by exogenous causation, such as early brain damage, giving rise to phenocopies.

Returning to the main body of Vogel's work, in a very recent series of papers (Vogel & Schalt, 1979; Vogel, Schalt, & Krüger, 1979; Vogel, Schalt, Krüger, Propping, & Lehnert, 1979) he extends his discussion to consider how far the EEG variants he has recognized might account for, or be related to psychological differences in personality and cognitive performance. The empirical data he is mainly concerned with are the psychological characteristics of individuals, selected from a group of 298 healthy adult males, showing one or other of the EEG variants he describes. The psychological characteristics in question were those also studied in a parallel twin study carried out by his colleague, Propping, who had administered the same battery of personality, cognitive, psychomotor, and perceptual tests to 26 MZ and 26 DZ adult male twins. The logic of the strategy is therefore obvious. It was hoped that the twin study would reveal systematic evidence about the heritability of certain important psychological traits. By comparing individuals selected on the basis of their allegedly inherited EEG pattern, Vogel sought to build a genetic bridge between the psychological and physiological modes of description. Unfortunately, the results of the twin study were disappointing, in that the psychological tests presented a confused picture with regard to heritability estimates. Furthermore, surprisingly, Vogel and his colleague failed to follow through the logic of their procedure; namely, they did not apply Vogel's typology to the EEGs of the *twins*. Had they done so, it might have been possible to shed more direct light on the genetic contribution to the EEG variants he describes, as well as to examine in more depth the reasons for any discrepancy between psychological and physiological parameters.

Nevertheless, Vogel, Schalt, and Krüger (1979) did attempt some description of the psychological test characteristics of individuals showing different EEG variants. Thus, they describe the "low-voltage" group as high in intelligence, especially spatial orientation, and normally extraverted in personality; individuals with "diffuse beta rhythm," on the other hand, emerged as low in intelligence and poor in emotional control. The authors admit, however, that many of the differences among groups were small, and that the whole range of test scores could be found in all the EEG types. Despite these limitations, Vogel's studies represent a unique attempt to extend the genetic analysis of the background EEG beyond mere physiological description and into the realm of psychological interpretation; we return to some of his theoretical speculations later.

A study by Hume (1973) had a similar end in view, though it started from a different point and, with respect to the EEG at least, was more limited in scope. Hume's study formed part of an investigation of adult twins carried out by Claridge, Canter and Hume (1973). The aim of the whole investigation was to examine possible genetic determinants of some personality, cognitive, and psychophysiological variables that previous studies of individual differences (Claridge, 1967) had proved to be of theoretical interest. Hume's contribution was to study psychophysiological variations; these were mainly indices of autonomic function and perceptual response, but also included five EEG measures. All five measures concerned different features of occipital alpha rhythm activity. One was alpha frequency (eyes closed), and a second the duration of alpha blocking following closing of the eyes after inspection of a rotating Archimedes spiral. The other three were all measures of percentage time alpha (alpha index), two with eyes closed while resting, and the third during the spiral inspection period. In this part of the investigation 39 MZ and 43 DZ twins were tested and compared using intraclass correlations and F-ratios of within-pair variances. The measures of both alpha blocking and alpha index (eyes open) showed quite negative results, the DZ correlations in fact being somewhat higher than the MZ correlations, though the difference between the two twin types was not significant. A significant difference was found, however, on one, but not on the other measure of alpha index determined with eyes closed; in both cases MZ twins showed higher intraclass correlations than DZ twins (+.69 and +.60 compared with +.42 and +.23). The most striking finding for the data in this form concerned alpha frequency, where the MZ and DZ correlations were +.75 and +.40, respectively; the within-pair variances were also significantly different.

As part of a further analysis of the data from that twin study, the psychophysiological measures, including those from the EEG, were subjected to principal components analysis. In the larger of two such analyses, involving all the twins on whom EEG data had collected, Hume found that of four interpretable factors, one was clearly composed of the alpha rhythm measures, which appeared to define a distinct dimension of cortical activity.

In the same study, Claridge (1973) carried out a second principal components analysis that included part of Hume's data. Here the sample of subjects was reduced to 10 MZ and 10 DZ twin pairs whose data on another measure of interest—the sedation threshold—were considered important to include in that particular analysis. Three of Hume's EEG measures were used—alpha frequency, alpha blocking response, and alpha index (eyes closed). The result of the analysis was slightly different from that found by Hume in that it partly separated the alpha blocking response from the other two EEG measures. However, a major EEG factor did again emerge with very high loadings on alpha index and alpha frequency; the loadings were, predictably, opposite in sign, being +.79 and −.82, respectively. The further step was then taken of calculating, for each subject, a factor score on that component, using the measure as a basis for comparing the two twin types. It was found that the intraclass correlation for MZ twins was +.96 ($p < .001$) and for DZ twins +.54 ($p < .05$); in other words, the values closely approached those expected from the hypothesis of complete genetic determination of this EEG factor.

In a study similar to that just described, though confined to raw EEG measures, Young, Lader, and Fenton (1972) also examined various characteristics of the spontaneous EEG in 17 MZ and 15 DZ twins, as well as in a group of unrelated subjects. They scored the EEG for alpha index and alpha amplitude, and also measured activity in four frequency bands: 2.3–4.0 Hz, 4.0–7.5 Hz, 7.5–13.5 Hz, and 13.5–26.0 Hz. In all but one case, MZ twins showed higher intraclass correlations than DZ twins, whose correlations in turn exceeded those of unrelated individuals. The exception was the measure for the 2.3–4.0 frequency band, where the correlations for all three samples were in any case very low. The most convincing finding in this study was the result for the 13.5–26.0 frequency band; here the MZ correlation was +.90 and the DZ correlation +.56, the value for unrelated subjects being +.05.

In the same investigation Young et al. also measured what they called *alpha attenuation*, that is the amount of alpha rhythm observed following presentation of a series of 60 paired flash stimuli. For the purpose of analysis they considered, not only the 60 trials as a whole, but also two phases—the first 25 trials (prehabituation) and the second 35 trials (posthabituation). For all three sets of data, unrelated individuals showed small negative correlations, but in twins the values were always positive, and higher in MZ than DZ pairs. The greatest difference between the two twin types was observed in the posthabituation phase, where the MZ and DZ correlations were, respectively, +.62 and +.21. The latter values, apart from the two exceptions already mentioned, were typical of those obtained in the study as a whole. However, Young et al. were of the opinion that their results provided rather conservative estimates of ''heritability,'' and concluded that genetic factors do seem to be important in the determination of background EEG.

Surwillo (1977) was also of the opinion that the basic frequency of the brain's electrical activity may be genetically determined, following a rather different kind of study from those considered so far. He measured half-waves taken from the left parietal–occipital EEG recorded while his subjects performed a reaction-time task. Interval histograms of EEG period, based on 780 readings, were constructed and compared for a number of distribution parameters. His experiment was rather limited in scope, in that it involved only seven pairs of MZ twins, all children, and seven pairs of unrelated individuals matched with the twins for age. However, he reported that the intrapair variance was significantly greater in the latter group than in the twins on five of six of his measures. MZ twins also showed significant intraclass correlations on his three measures of central tendency—mean, mode, and median—though not for the other distribution characteristics he studied, namely dispersion, kurtosis, and skewness. None of the intraclass correlations for unrelated subjects was significant.

A more guarded conclusion about the heritability of the EEG was reached by Lykken, Tellegen, and Thorkelson (1974), after studying 39 MZ and 27 DZ twin pairs. The source of their data was rather unusual, in that they were obtained during an investigation of the genetic influence on hypnotic susceptibility. The experimental conditions were so arranged that the twins sat side by side as a pair listening to a tape intended to induce hypnosis, their EEGs being recorded at the same time. The EEG recordings were analyzed for six spectrum parameters. Three of these were fairly standard, namely delta (.1–2.9 Hz), theta (3.0–7.9 Hz), and beta (13.0–19.9 Hz). A fourth measure, phi, was alpha rhythm, defined as the median frequency in a 3Hz band centred on the peak ordinate from 7.0–13.9 Hz. The proportion of the total spectrum centred on phi was itself designated alpha; finally a measure named kappa described the kurtosis of the alpha peak. For the MZ twins all these measures gave very high intraclass correlations—mostly in excess of .80. Indeed, the correlations were sufficiently high, and consistently so, for Lykken et al. to conclude that the similarities on EEG were "remarkable," approaching those for morphological traits.

However, the authors then go on to question the force of their own conclusion in the light of their finding that in the DZ sample the twin correlations ranged between only −.20 and +.23. These, they argue, are "too low"; one would expect values in the region of +.40–.50 if a straightforward genetic hypothesis were to be sought. Lykken and his coauthors suggest that one reason for the low correlations found could have been that the particular DZ sample tested may have had less than the expected 50% of their genes in common; they reject this explanation on what might be considered the somewhat dubious grounds that their DZ twins did show correlation values around .50 on morphological traits, such as height and weight.

Another possibility they consider is that environmental factors may have been operating to influence the EEG measures, but in opposite directions in the two kinds of twin, causing MZ twins to become more alike and DZ twins less alike. To support their argument they quote evidence that similar effects may influence the measurement of other phenotypic characteristics in twins. They are referring here to Shields' (1962) findings that twins reared apart were actually more alike than those reared together when compared on a measure of extraversion, suggesting that the interaction between members of a twin pair living together suppresses some of their genetic similarity due, in this case, to the adoption of mutually dominant and submissive roles. The effect certainly seems to be a genuine one, Canter (1973) having obtained a similar result for personality traits when comparing intraclass correlations for twins divided according to whether they had been living apart for more than or less than 5 years. However, she does point out that the effect of separation—whether it increases or decreases twin similarity—depends on the personality trait being examined. Futhermore, the effect seems to operate regardless of the zygosity of the twins and is therefore unlikely to account for the differential distortion of MZ and DZ correlations that Lykken et al. try to explain. In any case, results such as those reported by Shields and Canter refer, not to short-lived phenomena such as psychophysiological changes, but to long-term developmental influences on personality, where a plausible mechanism through which they are mediated can be discerned. Only in a very general sense, therefore, can a parallel be drawn between the personality findings and the EEG data described by Lykken et al.

Nevertheless, their result does serve to remind us that estimates of heritability or twin similarity, especially in the case of psychophysiological measures, are always dependent on the conditions under which they are made; in this respect it is of interest that the use of hypnosis in the experiment just described was a significant departure from the procedure used by most previous investigators of the EEG in twins. Given this rider, that the actual estimate of the genetic determination will depend partly on the conditions under which recordings are taken, there does seem to be consistent evidence that hereditary factors play a major role in individual differences in the pattern of background EEG activity. Paricularly under genetic influences seem to be certain parameters underlying the frequency of the alpha rhythm, and some aspects of activity in the fast frequency (beta) range.

Evoked Potentials.   As an alternative approach to the study of EEG genetics, some workers have chosen to investigate the cortical-evoked potential. This was made possible by the development of rapid and accurate averaging methods for extracting from the background EEG the typical wave form following a train of sensory stimuli. The technique has many pitfalls, and interpretation of the evoked potential is still problematical (Regan, 1972). Nevertheless, the procedure has a certain attractiveness because of the possibility it offers of describing more exactly than the gross EEG those features of brain activity that might be of biological significance and of psychological importance. Certainly, wide individual variations in the wave form of the evoked response have frequently been noted, as has also, however, its stability when recorded on different occasions in the same person. Evoked potentials have been studied also in relation to a large number of psychological and behavioral characteristics (Callaway, 1975).

Application of the technique to the study of twins was pioneered in an investigation by Dustman and Beck (1965), who examined the visual evoked potential in 12 MZ and 11 DZ pairs. In addition, they tested one set of triplets and 12 unrelated pairs of individuals matched with the MZ twins. Results were analyzed by calculating a correlation between the wave forms for each pair of individuals being compared, that is, for each set of twins or unrelated subjects; in some cases test–retest correlations were also determined when, as on some occasions, subjects were examined twice. It was found that in all three types of subject the range of correlations was fairly wide, indicating that even in some MZ twins the evoked potentials were not similar. However, this appeared to be the exception rather than the rule, because the *average* correlation for the MZ sample was +.82 compared with a value of +.58 for the DZ sample and +.61 for unrelated pairs. In discussing their results Dustman and Beck note that the evoked potentials of some twin pairs were so similar that the correlation found between a twin and his or her sibling was sometimes greater than that for the same twin tested as long as 6 months later. However, they also comment on the finding, described previously, that some MZ twins showed remarkably low intrapair correlations, and report that in some cases where such pairs were retested the correlation often increased markedly—for example, from +.50 or below to +.70. They suggest the possiblity that variation in the stability of the evoked potential may itself be a source of individual difference. The idea is an interesting one and accords with the view put forward by Callaway (1975), who presents evidence relating evoked response variability to cognitive performance, especially those cognitive style differences that may be associated with creativity.

In a second study of the evoked potential in twins (Lewis, Dustman, & Beck, 1972) Dustman and his colleagues expanded their original investigation to include auditory and somatosensory, as well as visual stimulation. They examined a larger group of subjects: 44 MZ twins, 46 DZ twins, and 46 unrelated pairs. The sample also covered a wide age range, from 4–40 years. The results were very much in line with those of the earlier study, MZ twins again showing a consistently greater degree of similarity on all measures than either DZ twins or unrelated individuals. This was true for all of several scalp areas from which they recorded, and for all three modalities, although responses to somatosensory stimuli were slightly less consistent than those for audition or vision.

Other workers who have looked at wave-form similarities have tended to confirm these findings. In the study referred to in the previous section, Young et al. (1972), as well as examining background EEGs in their twin sample, also recorded auditory evoked potentials. Wave forms were analysed for two time periods after stimulus onset—latency 20–300 msec and latency 20–498 msec. For the former time period the mean MZ and DZ correlations were, respectively, +.71 and +.39; unrelated controls showed a mean correlation of +.07. Sampling over the longer time period produced essentially similar results, the differences among the three groups being only slightly reduced. The evidence for a genetic influence on the evoked potential, therefore, appears to hold up for relatively late components, at least of the auditory response.

Osborne (1970), however, suggested that in the case of the visual system the components that are most stable and therefore perhaps more likely to be genetically determined are those occurring up to about 250 msec after the stimulus.

Osborne himself added to the twin data on evoked potential wave forms by studying visual responses in 13 MZ and 16 DZ pairs; he also included 38 unrelated controls. His study was unusual in comparing not only twin types and unrelated individuals, but also hemispheric differences in each subject. His results were remarkably similar to those of other investigators. As in Dustman and Beck's original experiment, the correlations between pairs in the twin groups showed a wide range of values, but averaged +.77 for the MZ and +.53 for the DZ pairs; the mean correlation for the control pairs was +.11. The intra-hemispheric comparisons naturally produced high correlations, but Osborne notes that in four of his MZ twins the correlation with the cotwin was greater than that between different sides of the proband's head!

A notable departure in the method of analyzing twin data on the evoked response is found in the most recent study, that by Rust (1975), who is so far the only investigator to apply to his results the model-fitting techniques of biometrical genetics (Eaves, Last, Young, & Martin, 1978). His data consisted of latency and amplitude measurements for several major peaks, identified by inspection, in the auditory evoked potentials of 20 MZ and 20 DZ twin pairs. Rust tested these measures against the simplest biometrical genetic model, which assumes a single additive genetic component, with simple within and between environment effects, and no genotype/environment interaction. He reported a statistically significant fit for the model, concluding that genetic effects were strong for evoked response amplitude; calculated heritabilities were over 80% for all three amplitude measures he took. The evidence was weak for a genetic influence on latency variations, which, he considered, were mainly due to between-family environmental effects. This latter result is intuitively surprising and may have something to do with Rust's failure to take account of what is known to be a very systematic regression of latency on age (Callaway, 1975).

In biometrical genetic terms, according to Eaves and Eysenck (1976), age can, at least formally, be regarded as part of the environmental variances, so that Rust's neglect of its contribution to latency differences leaves his analysis, in our opinion, incomplete and his interpretation of his data misleading. There is, however, a more serious problem—the very small sample of twins involved in the experiment. In a recent paper entitled "The power of the classical twin study" Martin, Eaves, Kearsey, and Davies (1978) discuss the question of what constitutes an adequate twin sample needed to reject or accept particular models of variation. They calculate that for several frequently encountered cases, including the model tested by Rust, more than 600 pairs of twins would be required to reject inappropriate alternatives. Even to achieve much more modest power would require 200 pairs. It is therefore doubtful whether, despite the apparent sophistication of the statistical analysis, Rust's experiment takes our understanding of the origins of evoked potential variation very much further. Nevertheless, it makes a useful addition to existing evidence that genetic influences appear to play a significant role.

All the studies considered so far in this section were concerned with the *shape* of the evoked potential wave form, in the sense that they have examined characteristic compoentnts of the response itself. Furthermore, they were all carried out from an essentially empirical viewpoint, the aim simply being to establish similarities between genetically related individuals. A slightly different kind of twin investigation, and one having more theoretical underpinning, was that carried out by Buchsbaum (1974). His interest in the evoked potential arose out of its use in the study of the so-called "augmenting–reducing" response (Buchsbaum & Pfefferbaum, 1971; Buchsbaum & Silverman, 1968). This refers to the fact that the amplitude of the evoked potential increases, though in some individuals paradoxically decreases, with increasing intensity of stimulation. According to Buchsbaum and his colleagues, augmenting–reducig reflects the operation of a stimulus control or regulating mechanism that modulates the cortical response to sensory input. The wide individual variations observed in the response appear to represent differences in central nervous "style"; as an experimental technique it also seems closely to resemble that used by Soviet scientists to measure nervous system "strength," discussed in a later section.

As part of a long series of studies of the augmenting–reducing response Buchsbaum (1974) conducted the twin investigation referred to previously. The purpose of the experiment was to try to establish biological indicators of vulnerability to psychiatric disorder; such indicators should ideally show evidence of stability and genetic determination when assessed using normal twin strategies.

The twin study in question involved 30 MZ and 30 DZ twins. The experimental procedure and data analysis were both extremely comprehensive, but essentially fell into three parts. The first concerned the evoked response to flash stimuli at four different light intensities, an individual's augmenting–reducing response being expressed as the slope of the regression line relating stimulus strength to evoked potential amplitude. A number of different ways of determining the latter was examined, including the one most commonly used in augmenting–reducing research, namely a visually assessed peak-to-trough measure from the positive component, occurring between 80 and 100 msec (P100), and the following negative peak, occurring around 140 msec (N140). In addition, slopes based on statistically derived 'point-by-point' amplitudes determined at 4-msec intervals in the evoked response curve were also studied.

In a second part of his procedure, Buchsbaum (1974) investigated the response to a 10-Hz sinusoidally modulated light presented at four depths of modulation. Augmenting–reducing was again calculated as a regression line slope, relating, in this case, modulation depth to the amplitude of the EEG response for the first harmonic (10 Hz) and the second harmonic (20 Hz) determined by Fourier analysis of the EEG.

A third source of data was not concerned with augmenting–reducing, as such, but with the variability of the evoked response. In order to study variability, auditory stimuli were used. A long series of 60-dB 500-Hz tones was presented, eight averaged evoked potentials then being derived for sets of stimuli randomly selected from the tone series. The results were analyzed by calculating, for each subject, measures of variability based on split-half correlation coefficients for replicates of the whole evoked response and standard deviations of amplitudes determined at each of a number of different time intervals after stimulus onset.

The variability procedure produced equivocal results and is not considered further here. However, all the direct evoked potential measures demonstrated uniformly greater similarity between MZ than between DZ twins, as assessed by intraclass correlations. This was true, not only for the various slope indices of augmenting–reducing, but also for the individual amplitude parameters on which they were based, and which Buchsbaum also examined. For the amplitude measures themselves, the actual values of the correlations were, however, somewhat lower than those reported by other workers. For flash stimuli they never exceeded +.59 in MZ twins and often fell as low as −.03 in DZ twins; for sine-wave stimulation the equivalent values were +.65 and −.04. The same was true, though to a lesser extent, for the augmenting–reducing slopes. For flash stimuli the twin correlations varied according to the amplitude criterion on which the slope was based; 25 different criteria are considered in the paper. Among these, the MZ correlations reanged between −.04 and +.76; the corresponding range for DZ twins was from −.01 to +.39. The best discrimination was achieved for slopes based on the point-by-point amplitude analysis, especially for positive waves occurring around a latency of 100 msec, and then later at around 200 msec. Thus, at 100 ec the slopes were correlated +.76 in MZ and +.09 in DZ twins; at 228 msec tne correlations were, respectively, +.76 and +.14.

The augmenting–reducing data derived from the sine-wave procedure produced very similar results to those found using flash stimuli; again MZ correlations were almost always higher than those for DZ twins, but the latter had uniformly low values. Indeed, as Buchsbaum himself comments, a general feature of his results was that the differences found between the two types of twin largely reflected the low intraclass correlations observed in the DZ sample.

This makes it difficult to draw any firm conclusions about the genetic influence on the particular feature of the evoked response—augmenting–reducing—on which Buchsbaum has placed so much emphasis. Certainly, a strong heritable component seems to be indicated when his data are judged against the *consistency* with which high MZ correlations were found using different methods of calculating evoked potential amplitude and, hence, augmenting–reducing slopes. However, it must be borne in mind that many of these were variants on a basically similar procedure and were almost certainly themselves highly correlated. Furthermore, the uniformly low correlations found in DZ twins militate against a simple genetic interpretation; at the very least, they suggest that transient situational or other individual difference variables were again influencing or interacting with the EEG parameters.

In this respect, a recent study by Birchall and Claridge (1979) may be relevant. Taking repeated amplitude/intensity slope measurements in the course of a single recording session, they showed that the amount and sometimes the direction of augmenting–reducing was not a fixed characteristic of the individual. Instead, it varied with the subject's concomitant state of arousal as measured by skin conductance level. However, it did so in a highly systematic fashion, suggesting that it may be possible, using appropriate research strategies, to find stable underlying determinants of evoked response augmenting and reducing that *are* genetic in origin. Certainly, on the evidence available to date, evoked response techniques seem to be promising tools for investigating genetic influences on central nervous functioning.

Studies of Autonomic and Somatic Activity

Probably the most common form of measurement adopted by psychophysiologists has been the recording of some aspect of autonomic or somatic function—electrodermal activity, heart rate, blood pressure, muscle tension, and so on. The inferences that can be drawn from such peripheral indicators about the action of the central nervous system are necessarily more indirect than in the case of data derived from the EEG. However, they have been popular for two historical reasons—one methodological, the other theoretical. First, compared with the EEG, until recent years they have had the advantage of being more easily measurable. And, secondly, starting with the early speculations of Duffy (1934), intepretation of such data has been comfortably subsumed under psychological concepts like "activation" and "arousal," notions that imply a major central nervous influence on the peripheral signs of emotion, consciousness, and behavioral vigor. These arousal explanations of autonomic and somatic activity have stimulated a good deal of research, some of it from a genetic viewpoint. They have also generated considerable controversy, a dispute that does not concern us at this stage in the discussion, except for one observation, namely that many workers using such techniques often overlook the fact that the systems from which they are recording, in addition to reflecting central influences of some possible psychological interest, also subserve a purely physiological function. Thus, alterations in heart rate may reflect changes in posture, as well as changes in emotional excitement, and shifts in vasomotor tone a response to cold as well as a reaction to psychological stress. This dual function of peripherally measured systems has important implications for behavioral genetic studies. The conditions under which measurements are taken, for example, are especially critical. Furthermore, the proportion of variance, either genetic or environmental, that can be attributed to factors of psychological, as distinct from purely physiological significance may be difficult to establish.

A good example of this latter problem occurs in the case of blood pressure. That genetic factors contribute to individual variations in blood pressure is beyond reasonable doubt (Hamilton, Pickering, Roberts, & Sowry, 1954; Miall &

Oldham, 1955). However, much of the evidence has been obtained, not within the context of psychophysiological research, but as part of the enquiry into the causes of cardiovascular disease. Admittedly, insofar as the latter can be said to be partly "psychosomatic" in origin, the results of such studies do bear on the questions at issue here. Nevertheless, complexities of interpretation are introduced and must constantly be borne in mind when evaluating findings from this area of psychophysiology.

This being said, it is comforting that such difficulties are minimal in the case of one autonomic function that has long been of special interest to psychophysiologists, namely electrodermal activity. To the best of our knowledge, this has no significant biological function apart from its role in arousal, and indeed has been of very little interest to pure physiologists. Electrodermal indices, however, are robust and easily measured. In genetic studies the most usual index of electrodermal activity chosen has been some aspect of skin resistance (or its reciprocal, conductance), either its basal level or a feature of the galvanic skin response (GSR) itself. However, one of the earliest investigations, by Siemens in 1924, looked at the sweating response, which, by consensus, lies at the basis of variations in electrodermal activity. Siemens classified the members of twin pairs into one of three categories according to the amount of palmar sweating observed, and recorded that all but one of a group of 35 MZ twins were concordant for the category to which they could be assigned; only 12 of a sample of 23 DZ twins were concordant.

The indication from this experiment that electrodermal activity might be partly genetically determined received some support from two early studies of skin resistance by Carmena (1934) and by Jost and Sontag (1944). The former examined the GSR in 36 MZ and 16 DZ twins and concluded that both the magnitude and duration of the response were highly similar in 26 of the MZ, but in only one of the DZ pairs. The study by Jost and Sontag was part of a larger investigation involving a battery of autonomic measures and a substantial sample of subjects, all children. Most of these pairs (1009) were unrelated, but 16 pairs of MZ twins and 54 pairs of sibling children were also included. The twins proved to be more alike in skin resistance level than the other two groups, though siblings were also very similar.

More recent investigators of electrodermal activity have naturally used more sophisticated methods of recording and data analysis, and some, at least, have examined more specific features of the response. All have reported evidence for a genetic influence on *some* aspect of electrodermal activity, with the exception of Vandenberg, Clark, and Samuels (1965). They compared 34 MZ and 26 DZ twins for their GSRs to a series of startle stimuli, and found that in no case was there a significant difference in within-pair variance. However, Block (1967) found a very high intraclass correlation for skin resistance *level* in MZ twins, as did Rachman (1960) for the *latency* of the GSR; but the latter, who studied only seven pairs, reported that they were not significantly alike in *habituation* of the GSR.

Perhaps the most elegant investigations of electrodermal activity in twins were those conducted by Lader and Wing (1966) and by Hume (1973), to whose work we have already referred. Both studies included autonomic measures other than electrodermal indices, but only the latter are discussed at this point. Lader and Wing's investigation formed part of a larger study, the aim of which was to try to establish autonomic parameters for describing anxiety in neurotic patients and for monitoring the effects of psychotropic drugs. Having successfully done so, they then introduced their measures into a study of normal twins, 11 monozygotic and 11 dizygotic. Each individual underwent a standard habituation procedure—the presentation of 20 consecutive pure tones, with randomly distributed intervals between them. In the case of the electrodermal data (log skin conductance) six scores were derived. One of these was of skin conductance *level* (its change from the beginning to the end of the habituation procedure) and two were *response* measures, namely the size of the GSR to the first tone, and the number of spontaneous fluctuations in GSR occurring during the experimental procedure. The other three measures described the habituation data, which, for each individual, were analyzed by fitting regression lines to GSR amplitudes plotted against stimulus number (the first stimulus was excluded). This procedure yielded two measures—intercept and slope (habituation rate). Finally, as these two latter measures were themselves highly correlated (subjects having a larger initial response showed a steeper decline) an additional slope measure, called the H-score, was derived by partialling out the regression of slope on intercept.

Comparison of the twin samples on these measures showed that in every case MZ pairs were more alike than DZ pairs, as judged by intraclass correlation. The largest, and only significant differences, however, were found on two of the measures. These were the H-score and the number of spontaneous fluctuations in GSR. In the latter case the MZ and DZ correlations were, respectively, $+.68$ and $-.02$; the H-score gave similar values of $+.75$ and $+.13$. In both instances, therefore, the differences between the two types of twin reflected the relatively low values found in the DZ sample. Nevertheless, the results as a whole provide striking confirmation of earlier evidence that some aspects of electrodermal activity, at least, seem to be under genetic control. They are especially convincing because the two measures showing clearest indication of hereditary influence—the H-score and spontaneous fluctuations—were precisely those that, elsewhere in the research program, Lader and Wing had shown to be the best discriminators of normal subjects and neurotic patients suffering from pathological anxiety where, according to clinical studies, a genetic component in etiology has been implicated (Shields & Slater, 1960).

The other detailed twin study of electrodermal activity, by Hume (1973), differed from that of Lader and Wing in using, not the exogenously derived index of skin resistance, but a measure of the endogenous response, expressed as skin potential. The same group of twins was studied as that involved in the investigation of background EEG described in a previous section, the skin potential measurements being taken during the same recording session. For technical reasons, however, the sample sizes were different, skin potential data being available for 40 MZ and 48 DZ twins. Two main stimulation procedures provided information on skin potential. One was a habituation regime, exactly the same as that used by Lader and Wing. From it Hume derived three scores, one of which—the size of the GSR to the first stimulus—was identical to that of Lader and Wing. The other, a linear regression slope relating GSR amplitudes to stimulus number, was slightly different, being based on the first 10 tone stimuli only. Like Lader and Wing, Hume also counted the number of spontaneous fluctuations in skin potential occurring during the habituation procedure. The stimulation procedure was a conventional cold pressor test, in which the subject kept his or her hand in cold water for one minute. The skin potential measures taken here were the change in level over the period of immersion, and the time to recover to the prestimulus level. In addition to these five measures, Hume also derived three others: two indices of resting skin potential level and an additional measure of spontaneous fluctuations, including those occurring during sensory stimulation with an Archimedes spiral.

Comparison of the two twin samples on these raw measures revealed that in all cases the intraclass correlations were higher for MZ than for DZ pairs. Six of the eight MZ, and only one of the DZ correlations reached significance, though none reached a high value, the largest being +.54. Two measures showed a significant difference between the intrapair variances of the twin samples; these were the cold pressor response and the habituation slope, the latter finding therefore confirming that of Lader and Wing.

It will be recalled that Hume also subjected his data to principal components analysis in which he included his skin potential scores. Because he also included other autonomic measures, it is convenient at this point to consider his analysis as a whole. The additional measures were both cardiovascular, namely heart rate and finger pulse volume. The scores derived for each of these were obtained during the same cold pressor and habituation procedures that yielded the skin potential measures. However, they were fewer in number—five in each case—and included two measures of resting level, two cold pressor measures (a change score during immersion and the time to recover), and the size of response to the first habituation tone.

In the case of the raw scores for these two cardiovascular variables, several measures showed a significant difference in the intrapair variance of the two types of twin, the best discrimination being achieved for the two measures of resting heart rate. Here the intraclass correlations were +.67 and +.65 for MZ twins, and +.33 and +.44 for DZ twins. Results for finger pulse volume were generally poorer, with the exception of the cold pressor response, where the intrapair variances were significantly different. Taking the ten cardiovascular measures as a whole, in seven cases the MZ correlations were greater than the DZ correlations, although often the values were quite low.

Turning to Hume's principal components analysis, this yielded five factors, four of which could be interpreted. One, as we have described previously, was an EEG factor. The other three were defined by combinations of the autonomic measures, a particular feature being the division into two separate factors of spontaneous skin potential responses, on the one hand, and the remaining electrodermal indices, on the other. The major autonomic component was what Hume described as an "autonomic level" factor, made up mainly of the cardiovascular, particularly the finger pulse volume measures, but also including two skin potential measures. When factor scores were derived for this component, and the MZ and DZ samples compared, the intrapair variances were found to be significantly different; intraclass correlations were, respectively, +.57 and +.16. Similar correlation values were also found using factor scores for the component of "spontaneous skin potential responses," the intrapair variances again being significantly different.

It is instructive at this point to return to the study by Lader and Wing, who also included finger pulse volume and heart rate in their investigations, and to compare their results with those of Hume. The actual measures used by Lader and Wing in their twin analysis were the levels on each cardiovascular variable at the end of the experiment. Pulse volume produced disappointing results, in the sense that, although the intraclass correlation for MZ twins was high and significant, it was even higher in DZ twins. The result for heart rate confirmed the significant difference between MZ and DZ samples found by Hume, though this was partly because the DZ correlation was substantially negative ($-.38$) compared with a value of $+.78$ in MZ twins.

Comparing the two studies, there is clearly considerable agreement over the greater similarity of MZ twins on electrodermal measures such as spontaneous activity and habituation rate; this seems to be true whether an exogenous or endogenous index of electrodermal activity is used. Hume's MZ correlations, however, were generally lower than the figure of $+.75$ found by Lader and Wing for the H-score, though both seem to agree that some statistically "derived" measure, such as the latter, or those arrived at from factor analysis, give better MZ/DZ discrimination than simple raw scores. This may be because such manipulations "iron out" the transient features that are characteristic of most autonomic responses. An example of this is seen in the results for finger pulse volume found in the two studies. Although Hume's data were more clear-cut than those of Lader and Wing, in its raw score form pulse volume was the worst of his three measures. However, in his factor analysis it emerged as a major defining parameter of his "autonomic level" component, significantly differentiating MZ and DZ twins.

The other cardiovascular variable common to both studies, namely heart rate, seems more robust in this respect, and the evidence for a genetic influence here confirms that found by most other workers. Thus, Shapiro, Nicotero, Sapira, and Scheib (1968) found MZ twins to be more alike than DZ twins on measures of both resting and stimulus-induced change in heart rate. Mathers, Osborne, and De-George (1961) confirmed this for resting heart rate, whereas Block (1967), although he did not examine DZ twins, reported a significant intraclass correlation in MZ twins under resting and stimulation conditions. Furthermore, Vandenberg et al. (1965), in a study already referred to, despite their negative results for electrodermal activity, found that four of the five "startle" stimuli they used produced significant differences in the intrapair variances of their MZ and DZ samples. Finally, and sounding a tempering note, in one of the earliest investigations of cardiovascular function, Weitz (1924), although confirming the aforementioned results, reported that occasionally MZ twins could be quite discordant for resting pulse rate where one member of the pair had a history of engaging in sports or heavy physical labor. His observation serves to remind us of the dual function of the autonomic system referred to earlier, and of the limitations this imposes on peripheral psychophysiological measures as indicators of central nervous variation.

Turning now to blood pressure, as we have already seen, there is good evidence from family studies carried out under the rubric of cardiology rather than psychophysiology for genetic control over this aspect of cardiovascular function. Twin studies have added to and frequently formed part of that evidence. Generally speaking, studies relying on single casual readings of blood pressure have shown poor discrimination between MZ and DZ twins (Barcal, Simon, & Sova, 1969; Downie, Boyce, Greig, Buchanan, & Alepa, 1969); but this is not always the case, Osborne, DeGeorge, and Mathers (1963) finding a significant difference for casual blood pressure in 34 MZ and 19 DZ twin pairs. MZ twins were shown to be more alike than DZ twins in their response to pain by Shapiro et al. (1968) and to cold stimulus by Kryshova, Beliaeva, Dmitrieva, Zhilinskaia, and Perrov (1962), although the latter included only two DZ pairs and presented no statistical analysis. Another study also limited by sample size (18 MZ and 13 DZ) and by the unusual age range of the twins—51–74—was that recently reported by Theorell, DeFaire, and Fagrell (1978). Their investigation was of interest, however, in that they looked at cardiovascular changes occurring during a stressful psychiatric interview. They concluded that there was evidence for a genetic influence on blood pressure—and peripheral pulse volume—as measured at the end of the interview.

One of the most comprehensive studies of blood pressure, involving 75 MZ and 84 DZ twins, was that carried out by Takkunen (1964). He not only investigated blood pressure, but also made a detailed analysis of the electrocardiogram in his subjects, as well as studying numerous anthropometric variables thought to relate to cardiovascular function. Blood pressure was recorded after a long rest period in order to approach a true basal level. The findings led Takkunen to conclude that the systolic blood pressure, at least, is under genetic control. He found a significant difference between his MZ and DZ samples, and significant correlations of +.67 and +.34, respectively. Pulse pressure showed similar values, as did several of the electrocardiographic parameters; pulse rate itself, however, did not significantly differentiate the two types of twin, although the correlation for MZ pairs was higher than that for DZ pairs. Apart from one anthropometric measure, diastolic blood pressure was the poorest discriminator of all the variables he studied, the twin difference failing significance, and the intraclass correlations being low—+.44 in MZ and +.26 in DZ twins. Takkunen concluded that there is no evidence for an hereditary influence on diastolic blood pressure, though systolic blood pressure and pulse pressure do seem to be under genetic control. However, he also reported correlations between these cardiovascular measures and body type, a covariation that itself seemed to be partly genetically determined. Given, too, the additional influence of such factors as age on the cardiovascular system, these results illustrate more than most the difficulties of imposing psychophysiological interpretations on autonomic measures.

Genetic studies of other somatic variables have been sparse. Abe and Pérez de Francisco (1972) described some unpublished data which, they considered, suggest that the salivary response may be a dominant hereditary trait. The study in question was rather unusual, involving a comparison of salivary secretion in infants who were cutting their teeth with that of their parents at the same period of infancy. Information on parents was obtained by questionnaire from their own mothers, and it was claimed that excessive salivation was found more frequently in children, one or both of whose parents had shown the same characteristic during childhood. The data on which this study was based, of course, must be considered of questionable relability, though it is of interest that in their early investigation of MZ twins and siblings Jost and Sontag (1944) included salivation in a battery of autonomic measures and decided that there was evidence for a genetic influence. Jost and Sontag also included a measure of respiration, but reported no difference between twins and siblings. However, in their comparisons of MZ and DZ twins, both Block (1967) and Vandenberg et al. (1965) reported results consistent with a genetic hypothesis. Finally, in what appears to be the only twin study of electromyographic response, Lader and Wing (1966), in their investigation, included an EMG measure—the mean forearm extensor muscle activity occurring at the end of their habituation procedure. However, the results were quite negative, the intraclass correlations in both MZ and DZ twins being very low.

It is evident from the studies reviewed here that in examining peripheral psychophysiological variables from a genetic viewpoint most attention has been paid to electrodermal indices, on the one hand, and cardiovascular measures, on the other. In both cases it can be claimed that there is consistent evidence for part of the variance at least being genetic in origin. What this tells us about the direct influence of heredity on central nervous function can be decided only by inference from data about the psychological significance of variations on these peripheral measures. In this respect electrodermal activity presents little difficulty, to the extent that a large body of background research indicates that it is extremely sensitive to and perhaps mainly determined by psychological, and therefore higher central nervous influences; neurophysiological control of the response is also well established (see Edelberg, 1972).

Cardiovascular function, although also under central nervous influence and subject to change during emotional and other psychological states, offers more problems of interpretation from both a genetic and a physiological point of view. Despite these difficulties, it still seems notionally possible to separate those aspects of cardiovascular function that, looked at genetically, are of psychological significance, and those that subserve a purely physiological role. Several of the studies reviewed, particularly of heart rate, examined twin similarities, usually in young healthy subjects, in which the psychological state was manipulated. In those cases a genetic hypothesis at least as strong as that for electrodermal activity was usually upheld, suggesting that both types of measure reflect a common change in central nervous arousal. Explicit support for this conclusion was evident in the study by Hume, whose application of principal components analysis to his data allowed general factors of autonomic activity to be identified. Indeed, Hume comments with surprise that, given the transient nature and difficulty of interpreting autonomic data, so few studies have used such procedures to arrive at more stable composite indices. Where such techniques have been employed, results have generally been confirmative. Thus, Jost and Sontag (1944) derived from their measures a component of autonomic balance based on the factor analytic studies of Wenger (Wenger & Ellington, 1943) and showed that MZ twins were more alike than siblings; Hume considered this factor to be very similar to his first principal component of autonomic level. Another study, an early one by Eysenck (1956), also applied factor analysis to a battery of tests that included autonomic measures. The autonomic factor that emerged, when used to derive factor scores, gave a high MZ correlation of $+.93$, the DZ value being $+.72$. Admittedly, as Eysenck himself notes, both values were probably somewhat inflated by a high loading of sex on the factor, but the results, nevertheless, are in general accord with the conclusion reached by Hume.

A different approach, though one that has the same aim of sharpening up the analysis of autonomic data, is exemplified in the work of Lader and Wing. Many studies, particularly those of more chronologically distant origin, have been methodologically poor; they have used inadequate recording techniques, paid little regard to the instability of peripheral autonomic measures, and have rarely deliberately examined different parameters of the physiological system in which they have been interested. Furthermore, there has usually been no attempt to select measures, or to interpret them, with reference to any theoretical model, however low-level. In contrast, Lader and Wing's work on electrodermal activity showed an awareness of methodological issues in general psychophysiology, leading them, for example, to take account of the "law of initial values" (Wilder, 1957) in deriving their H-score. Their use—and indeed that by Hume—of a habituation procedure also allowed them to take the interpretation of their findings somewhat beyond that of a mere description of twin similarity.

Considered in those terms—of MZ/DZ differences—the findings for peripheral psychophysiological measures are perhaps less impressive than those for the EEG. As we have just implied, however, this may be because they are more complexly determined and require greater care to extract relevant parameters. They are certainly less direct indicators of central nervous function. They may, of course, be under less genetic control or, at least, the hereditary component may be less easy to disentangle in adults because of the continuous involvement of autonomic conditioning processes during the development of emotional responsiveness and even personality trait characteristics like anxiety. For this reason different research strategies may be required to establish the true genetic influence on the temperamental characteristics toward the understanding of which many of the twin studies of autonomic function have been directed. In this respect it is interesting to note the very wide variations in autonomic responsiveness that can be observed in the first few days of life (Lipton, Steinschneider, & Richmond, 1961); more detailed examination of these differences within a genetic context might be extremely fruitful.

Finally, a complication of psychophysiological research on peripheral indices of "arousal" that has not yet been mentioned, but which may also prove to be important, is that of response patterning or "response specificity", the tendency for individuals to differ in the autonomic system through which they optimally react to stress (Lacey, Bateman, & Van Leyn, 1953; Lacey and Lacey, 1958). As far as we are aware, there is no published evidence on the possible origin— whether genetic or environmental—of these response profiles, although Hume has relevant twin data that, according to a personal communication, he is currently analyzing. The results of his investigation should prove extremely interesting in helping to answer a question that he himself recognised in his original report; namely, the extent to which autonomic patterning influences observed twin similarities and differences, especially when assessed by scores derived from factor analysis, which, of course, by its very nature ignores variations in response specificity.

Studies Using Drug Strategies

In this section we consider studies that have in common the use of drug response techniques in order to examine genetic influences on central nervous system activity. The studies in question are few in number, and in no sense represent the field of pharmacogenetics, which, in any case, is fully reviewed elsewhere (Chapter 4). Indeed, they are less derived from classic pharmacogenetics than from psychophysiology, being carried out within the same context as that of studies described already, the offshoot into psychopharmacology being a logical extension of strategy. Although not as yet giving a detailed enough explanation of the results of the studies to be described, the *kind* of theory proposed many years ago by Eysenck (1957) has argued that biological variations, largely of genetic origin, determine individual differences in personality. Inasmuch as it is claimed that the neurophysiological mechanisms underlying personality are precisely those that are affected by centrally acting drugs, using the latter as a research strategy has a twofold purpose: to examine the biological basis of personality and, when applied in a genetic context, to help explore the hereditary influence on those CNS variations that may account for behavioral differences. Russian nervous typological research, from which Eysenck's ideas in any case were partly derived, has similarly incorporated drug procedures into its research methodology, though to our knowledge these have not yet been employed in human genetic studies.

Very little of the limited psychopharmacogenetic work we are concerned with here has actually proceeded from a theoretical viewpoint of the kind just described; nevertheless it is included either for completeness, or because the study in question bears directly on the central theme of this chapter.

One of the earliest attempts to investigate the genetic influence on the behavioral response to a centrally acting drug in human subjects was a study on the effects of caffeine, reported by Glass (1954). By contemporary standards his methodology was crude, one might even say quaint. He used a single pair of monozygotic twins, and in the first part of his experiment compared the effects of five cups of cold coffee with a "no-drug" condition. Under each condition the response curves of the twins, plotted by performance on a target-aiming task, were remarkably similar. Glass then carried out a more controlled experiment using caffeine citrate and a proper placebo, and found similar results that, he suggested, contrasted markedly with the great variability in response curves seen in randomly selected subjects.

Another twin study, also methodologically rather dubious, and also concerned with caffeine—as well as alcohol—was reported by Abe (1968). He confined his investigation to an assessment of subjective reactions, and his sample to MZ twins only, a small group of 11 pairs, who were also atypical in that one or both members of each pair had been diagnosed as suffering from an affective illness. Each twin was asked about his reactions to coffee and alcohol, particular emphasis being placed on changes in activity, mood, sleep, and autonomic effects. The member of each pair was also asked about the reactions of his co-twin, additional information being obtained from relatives. The most notable finding concerned the effect of coffee drinking in causing sleeplessness. There was a significant intrapair similarity on this variable, only one of the 10 twin pairs being discordant. The effects of alcohol were less clear-cut, though there was a significant tendency for the twin to be concordant for "overtalkativeness," and for sleeplessness after coffee to be correlated, negatively, with reddening of the face after alcohol. The results of the study might perhaps be said to offer a weak pointer toward some constitutional influence on response to these two commonly imbibed drugs.

A much more complete, and in every respect more sophisticated investigation—of alcohol—was recently carried out by Propping (1977). Working under the auspices of Vogel, whose research on the EEG we have already described, Propping undertook a study of the effects of alcohol on the EEG in 52 twin pairs, 26 MZ and 26 DZ. EEGs were taken at rest, and then after a single oral dose of 1.2 ml/kg 100% ethanol, recordings under the drug condition being made at hourly intervals for 4 hours after ingestion. The EEG measurements used in the analysis were the mean frequency and the number of waves in the alpha, beta, and theta bands; in all cases the results reported were for occipital leads, both right and left recordings being taken.

Propping's findings for his twin comparisons under resting conditions were very similar to those previously reported for background EEG. That is to say, MZ twins were consistently and significantly more alike than DZ twins on all parameters, though he notes that the differences between the two kinds of twin were more evident in the right than in the left hemisphere. The overall effect of alcohol—irrespective of twin status—was the well-documented one of producing better synchronisation of the EEG, as reflected in an increase in the amount of alpha and theta activity and a decrease in beta activity. The MZ/DZ differences observed at rest were, if anything, exaggerated under alcohol. Thus, in the MZ sample the mean intrapair differences on all frequency parameters remained nearly constant during the drug phase of the experiment, whereas those in the DZ group steadily increased, at least for the first 120 minutes; in fact, at around that point, many of the intrapair correlations for DZ twins were near zero. The latter finding is of interest in view of our earlier discussion concerning the possible effect of recording conditions on twin similarities and differences, this particular feature of Propping's results being reminiscent of that reported by Lykken et al. (1974), who, it will be recalled, were puzzled by the very low correlations found in DZ twins, given the very high values seen in MZ twins. Although not entirely comparable, the hypnosis conditions under which Lykken et al. obtained their EEG recordings are perhaps functionally somewhat similar to those induced by Propping with alcohol. Both can be seen as examples of the experimental conditions having a further differentiating effect on individuals who are constitutionally dissimilar, an effect that of course would be expected in DZ, but not in MZ twins.

In conducting his experiment Propping also took care to investigate the possibility that the results he obtained could have been due to kinetic factors concerned with the absorption and elimination rates of alcohol in the two kinds of twins. To this end he determined blood alcohol concentrations at various points during the experiment. He reported that during the time of the main effect on the EEG the blood alcohol curves of MZ twins were no more similar than those of

DZ twins and concludes that the genetic influences he inferred from his data were primarily mediated through the central nervous system. He also concludes that of the macroscopic techniques available for studying these influences the EEG is the most powerful; a series of other psychophysiological procedures, such as CFF, which he also used in the experiment, failed to reveal the same effects of alcohol. It should also be recalled in passing that Propping's study was the source of personality data that, as we saw in a previous section, Vogel attempted to relate to EEG differences. Neither author explicitly linked the results of the alcohol experiment with personality, though, as we shall see shortly, Vogel did attempt a theoretical alignment with a biological model within which the final study described here was undertaken.

The experiment in question was one reported by Claridge and Ross (1973), whose strategy was somewhat different from that of Propping. They used the technique of sedation threshold, a procedure for determining an individual's tolerance of intravenously administered sedative drugs. First introduced by Shagass (1954), the sedation threshold has been the subject of considerable research in relation to both psychiatric status and personality (see Claridge, 1970). In principle, therefore, it would seem to be an ideal tool for exploring the areas of overlap among personality, genetics, and pharmacology. However, it appears that only one study, that by Claridge and Ross (1973), has used the technique as a tool in behavior genetics.

The experiment in question formed part of the larger twin study referred to earlier (Claridge, Canter, & Hume, 1973). The sedation threshold for intravenous amylobarbitone sodium was determined in twins using the procedure devised by Claridge and Herrington (1960). Briefly, this requires the subject to double a series of digits, played over a tape recorder, while receiving a continuous infusion of the drug. The procedure is continued until the subject is making 50% errors, the amount of drug injected up to that point, divided by the body weight, being taken as the sedation threshold.

Because of the inconvenience of the technique (its one major disadvantage) only a small number of twins could be tested—10 MZ and 11 DZ. Despite this limitation, the results were of some interest. They were analyzed by calculating intrapair difference scores and comparing the two twin samples using a relatively rigorous cut-off point of 1 mg/kg. Divided in this way, it was found that MZ twins were significantly more alike than DZ twins, nine of the latter differing by more than 1 mg/kg, whereas only two of the former differed by as much as that amount. A striking feature of the results, in fact, was that where MZ twins were alike the similarity was remarkable, suggesting a major genetic influence on this measure of drug tolerance. However, the two MZ pairs who were discordant differed to an extreme degree, in one case greater than that seen in any of the DZ pairs. The reasons for this were difficult to determine, but in one of the pairs it was thought to be due to the fact that one member of the pair, but not the other, had acquired an unusual tolerance for barbiturates following a long series of surgical operations.

Part of the rationale for the study of the sedation threshold just described was the central importance of the measure in a previous program of research conducted by Claridge (1967) into the psychophysiology of normal personality and psychiatric disorder. There it was shown that when entered with a number of other autonomic and EEG variables into factor analysis it had helped to define a major factor that was named "tonic arousal" and that appeared to account for some variations in normal personality and psychiatric state related, according to Eysenckian theory, to dimensions of neuroticism and introversion–extraversion. In order to investigate those results further, Claridge (1973) took advantage of the availability of the twin data on sedation threshold, as well as of the other psychophysiological measures collected on twins by Hume, and carried out the principal components analysis referred to in the earlier section on background EEG. It will be recalled that one of the components that emerged was a clear EEG factor. Another, confirming the results of the earlier analysis, was a factor loading highly on sedation threshold and on a number of autonomic measures, especially heart rate. The factor seemed similar to that originally identified as tonic arousal.

The further step was then taken of calculating factor scores for this component, as a basis for making twin comparisons. However, unlike the EEG factor, where evidence of high heritability had been observed, no difference between intraclass correlations for MZ and DZ twins was found; indeed the DZ correlation was somewhat higher. It is probable that this discrepancy between the results obtained using the raw measure of sedation threshold and that found when the latter formed part of a statistically derived factor was due to the very marked discordance of the two MZ pairs, whose data were included in the principal components analysis. Looked at another way, the results based on the latter can, in a sense, be said to reflect the true state of affairs, because the existence of the two very discordant pairs demonstrates that environmental factors can occasionally have a marked influence on the sedation threshold. A major problem, of course, was the very small sample of twins whom Claridge and Ross were able to test; the data are thus very vulnerable to discordant pairs, however exceptional.

The fact that the studies reviewed in this section are so few in number, taken together with their results, suggests that this is a neglected area of research in human behavior genetics, and one that deserves more attention. It is, of course, natural that most work in pharmacogenetics should address itself to different problems, more concerned with the biochemistry of drug action; this, if nothing else, brings it nearer to the genotype and provides a closer interface between genetics and pharmacology. At the other end of the spectrum, however, there is considerable evidence for systematic variations in the response to drugs when assessed by more gross behavioral or psychophysiological measures. More research using such measures could help to bridge the vital gap between macroscopic and microscopic descriptions of the nervous system and help to elucidate the common genetic link between them.

Intermediate between the two approaches, and deserving particular mention here, are attempts to relate drug response variations to strain differences in animals. A theoretically important example of this approach is the long program of research undertaken by Broadhurst on the Maudsley reactive and nonreactive rat strains, partly bred originally in an attempt to provide an animal equivalent of human personality dimensions being investigated by the Eysenck school. A very large number of studies, reviewed by Broadhurst (1975), has investigated the two animal strains with respect to several drugs, such as alcohol, barbiturates, morphine, nicotine, reserpine, and caffeine. The results of these experiments have sometimes been conflicting and are difficult to summarize coherently, but Broadhurst's own conclusion is that the strain differences do tap a generalized trait of emotional reactivity. Given that this is so, the continued use of pharmacological strategies might help to throw light on the genetically determined differences in drug response suggested by human studies.

Soviet Studies of Nervous System Properties

Proceeding largely independently of the studies just reviewed, workers in Eastern Europe have addressed themselves to similar questions, albeit within a different theoretical framework, and occasionally using different methodology. Some genetic studies of mental processes and nervous functioning were actually carried out as early as the 1920s at the Institute of Medical Genetics in Moscow (Luria, 1978) but in the mid-1930s such work was proscribed by the Central Committee of the Communist Party. It is only very recently that human genetic studies have again become respectable in the Soviet Union. The data are still relatively limited, but are of considerable interest in complementing those collected in the West. Before considering these findings, however, it will be useful to describe briefly their theoretical and historical origins, as the "conceptual nervous system" adopted by Soviet workers, and the language in which this is couched, will be unfamiliar to most Western readers.

Theoretical Background

The starting point for Soviet differential psychophysiologists is still the typological or temperamental model introduced by Pavlov. However, this has undergone a number of revisions since Pavlov's own final statement on human typology in 1935. The modern theory postulates the existence, in a highly organized nervous system, of a number of orthogonal properties or parameters describing the dynamics of excitatory and inhibitory processes that, in various combinations, are considered to form the neurophysiological bases of psychological differences (Nebylitsyn, 1966). Typological profiles, derived from measurements of these nervous system properties, are assumed to provide a complete description of human temperament. The various properties recognized are six in number. The first is excitatory strength, usually measured by resistance to extinction with reinforcement, by photic driving, and, more recently, by a reaction time (RT) method describing the function relating RT to stimulus intensity. Strength of excitation is a consequence of low reactivity, and weakness a consequence of high reactivity or sensitivity of receptors and cortical projection areas. Strength of the inhibitory process is usually measured by a discrimination procedure in which the interstimulus interval is progressively reduced until discrimination breaks down.

The second property is the readiness with which the nervous system generates processes of excitation and inhibition in the formation of conditional responses and is referred to as the dynamism of the nervous system. Dynamism of excitation is defined by the rate of formation of conditional reflexes, dynamism of inhibition by the rate of formation of inhibitory reflexes (i.e., of conditioned inhibition).

A third property, mobility, which has recently attracted a good deal of interest in Russian laboratories, initially described all aspects of nervous activity involving speed. However, the term is now reserved for nervous system activity concerned with the alteration of stimulus signs, that is, the speed and ease with which the subject can inhibit a response to a previously excitatory stimulus (CS+) and disinhibit response to a previously inhibitory stimulus (CS−) (transformation mobility).

Fourthly, the speed with which the nervous system generates processes of excitation and inhibition is now referred to as lability, a property similar to that described by Vvendenskii as functional lability. The latter is defined by the maximum number of impulses the nervous tissue is able to produce per second, in conformity with the rhythm of stimulation. Ukhtomskii regarded lability as the speed with which a physiological structure is able to pass from a state of rest to a state of excitation, and vice versa, and this is the sense in which lability is construed in typological theory. A number of authors have suggested that the capacity to arrest excitatory processes quickly is directly proportional to the subject's capacity to process information.

A fifth property—of concentration—is an extremely tentative concept, referring to the speed with which process of excitation and inhibition irradiate and concentrate. Few data concerning this property, however, have been reported.

Finally, equilibrium, or balance of nervous processes, is considered to be a secondary property derived from comparisons of levels of excitatory and inhibitory processes for any given property.

Twin Studies of Nervous System Properties

An obvious question that has arisen is the extent of genetic contribution to temperament, and hence to nervous system properties. Pavlov (1927) clearly assumed a high degree of involvement, for he commented that: "Temperament . . . is the general, basic character of the nervous system" (p. 85). Because the latter was, for him, represented in the complex of nervous system properties, which are genetically coded, temperament for him constituted the genotype. On the other hand, Teplov (1951) expressed some doubt about the simple equating of the two, considering that whereas nervous system properties might be "natural" properties, they are not necessarily inherited, because they might be influenced by prenatal and postnatal factors, particularly intrauterine influences and early social experience. The problem could, of course, be resolved by direct test for heritability, and the recent revival of interest in genetic studies in the Soviet Union is now beginning to throw some light on the question.

Perhaps the most informative studies to date are those reported in a series of investigations described in the biennial publication (1978), entitled *Problems in the Genetics of Psychophysiology,* of the Institute of Psychology, Academy of Sciences, Moscow. These extend some earlier work on heritability of excitatory strength and mobility reported by Shlyakhta (1972), Vasiletz (1974), Ravich-Shcherbo (1974), and Elkin and Khoruzheva (1975) and take account of a number of previously disregarded methodological and statistical problems (cf. Dubinin, 1971). Five studies are reported from Ravich-Shcherbo's group dealing with the properties of excitatory strength, mobility, lability, and dynamism, the nervous system properties that have been most carefully researched by the Moscow group.

The first study is that of Shlyakhta and Panteleyeva (1978), in which excitatory strength was measured by three classic indices—the EEG version of extinction with reinforcement, the reaction time method referred to earlier, and photic driving to low frequency stimuli. The first of these provided no evidence for genetic contribution and is not considered further here. However, the other two indices, compared in MZ and DZ twins, gave more interesting results. Summary data for the study are shown in Tables 2.2 and 2.3, where it can be seen that considerable differences between the two twin types were found for both the photic driving and reaction-time measures. This was especially so for the latter, though a comparison of different age groups showed a substantially higher correlation for the younger, relative to the older MZ twins. The greatest correlation was found in the oldest (adult) group. In this respect is worth noting that these twins had lived apart for at least 10 years.

The second study, reported by Vasiletz (1978) was concerned with mobility, measured by transformation, sensory aftereffects, and personal tempo. Transformation refers to the ease with which the signal values of CS+ and CS− can be reversed in a discrimination sequence, sensory aftereffects to the persistence of the sensory (visual, auditory) trace, and tempo to preferred and optimal speed in performing simple tasks, usually checking or easy cognitive tasks.

Twenty MZ and 20 DZ pairs were compared in each of two age groups, 7–11 years and 35–55 years. In all, 25 mobility indices were derived from the data.

TABLE 2.2
Intra-Class Correlations for Twin Pairs (14—16 Years)
Using the Photic Driving Index (** p < 0.025, ***p < 0.01)

Pairs	N	4 imp/sec	5 imp/sec	6 imp/sec
MZ	19	+ 0.69	+ 0.64	+ 0.53
DZ	22	+ 0.24	- 0.11	+ 0.24
NR	19	- 0.05	- 0.24	+ 0.01
F		2.45**	3.08***	1.62

These are, however, somewhat difficult to interpret unequivocally. Of the 50 possible MZ/DZ comparisons only 10 (five at each age level) were statistically significant; furthermore, Vasiletz reports little correlation among the different indices of mobility. This seriously raises the question of whether mobility can be assumed to be a unitary dimension. In fact, one of the measures used by Vasiletz—"personal tempo"—is now regarded more as a measure of lability than of mobility. It is therefore of some interest that this particular index did show clear-cut differences between the two twin types; this was true for both age levels, the MZ and DZ correlations being +.68 and +.03, and +.90 and −.16 for the younger and older groups respectively.

Interpreted as a measure of lability, this finding would support the results of a third study, that by Panteleyeva and Shlyakhta (1978). They report a twin study of lability, measured by critical flicker fusion (CFF), photic driving to high frequency stimuli, and measures of optimal and maximal tempo (Tenning test). The results for their study, which examined both MZ and DZ twins as well as unrelated individuals, are shown in Table 2.4. It can be seen that CFF appears to have a strong genetic component in all three age groups studied. The same is true for the 13- to 16-year age group on the photic driving and tempo measures.

TABLE 2.3
Intra-Class Correlations for Twin Pairs in Three
Age Ranges Using An RT Index (***p < .01)

Age	Pairs	N	r	F
8–11	MZ	20	+ 0.91	
	DZ	20	- 0.14	12.67***
	NR	24	- 0.13	
13–16	MZ	20	+ 0.67	
	DZ	20	+ 0.41	1.79
	NR	22	- 0.28	
33-56	MZ	23	+ 0.94	
	DZ	20	+ 0.09	15.17***
	NR	20	+ 0.11	

TABLE 2.4
(a) Intra-Class Correlations and F Ratios for CFF
in Three Age Groups (***p < 0.01)

Age	Pairs	N	CFF	F
9–11	MZ	20	+ 0.78	
	DZ	20	+ 0.35	2.95**
	NR	24	+ 0.12	
13–16	MZ	20	+ 0.84	
	DZ	22	- 0.01	6.3***
	NR	22	- 0.14	
35–86	MZ	23	+ 0.72	
	DZ	20	- 0.30	4.64***
	NR	20	+ 0.15	

(b) Intra-Class Correlations and F Ratios for Photic
Driving and Tempo Scores in the 13–16 Year Group
(***p < 0.01, **p < 0.025, *p < 0.05)

			Photic Driving imp/sec			Tempo	
Age	Pairs	N	18	25	30	$T_{opt.}$	$T_{max.}$
13–16	MZ	20	+ 0.81	+ 0.60	+ 0.30	+ 0.78	+ 0.69
	DZ	22	+ 0.24	+ 0.29	+ 0.23	+ 0.15	+ 0.25
F			4.0***	1.78	1.1	3.86***	2.42*

The fourth study, reported by Shibarovska (1978), concerned dynamism of excitation and inhibition, that is, the speed of formation and extinction of conditional responses, as measured using a wide range of EEG indices. The results of the study are summarised in Tables 2.5, 2.6, and 2.7. Table 2.5 concerns resting EEG measures, and it can be seen that many of the MZ/DZ differences reach significance, particularly those for alpha rhythm. Tables 2.6 and 2.7 show orienting and conditioning data for the alpha-blocking response, and again reveal a number of significant MZ/DZ differences, particularly for the speed of extinction and persistence of the alpha-blocking response during conditioning.

The finding by Shibarovska, together with similar results from other Soviet studies (Mikhevey, 1979), that certain features of the EEG may be genetically determined to a significant degree, confirms the results of those Western studies we have already reviewed. It is also in line with a further Soviet twin study of the EEG reported by Meshkova and Smirnov (1978). They examined a number of EEG parameters, using a sample of 20 MZ and 20 DZ pairs, aged 18–26 years. The EEG measure employed was the coefficient of local nonstationariness (CLN) devised by Klyagin and Kovalev (1974). CLN, which is based on a

TABLE 2.5
Intra-Class Correlations for 30 MZ and 26 DZ Pairs, Aged
11–12 Years, on Four EEG Indices
($*p < 0.05$, $**p < 0.025$, $***p < 0.01$)

Index		MZ	DZ	F
Alpha amplitude		+0.80	+0.33	3.35***
Alpha index	(Beginning	- 0.04	+0.03	
— (	(End	+0.56	+0.11	2.02*
Frequency of rhythms	(delta	+0.65	+0.07	2.66***
	(theta	+0.85	+0.82	
	(alpha	+0.87	+0.48	4.0***
	($beta_1$	+0.24	+0.38	
	($beta_2$	+0.30	+0.51	
Total energy of rhythms	(delta	+0.66	+0.57	1.26
	(theta	+0.78	+0.54	2.09*
— (	(alpha	+0.81	+0.53	2.47**
	($beta_1$	+0.89	+0.79	1.91
	($beta_2$	+0.71	+0.18	2.83***

TABLE 2.6
Intra-Class Correlations for 30 MZ and 26 DZ Pairs, Aged 11–12
Years, for Alpha-Blocking Responses to Auditory Stimulation
($*p < 0.05$)

		MZ	DZ	F
1st stimulus Presentation	Latency	- 0.06	+ 0.26	
	Persistence	+ 0.55	+ 0.37	1.4
Mean Values	Latency	- 0.41	+ 0.33	
	Persistence	+ 0.35	+0.01	1.52
Speed of Extinction		+ 0.50	+ 0.04	1.92*

TABLE 2.7
Intra-Class Correlations for EEG Indices of Conditioning
(***p < 0.01) in 30 MZ and 26 DZ Pairs, Aged 11—12 Years

		MZ	DZ	F
Trials to criterion CR		+ 0.15	- 0.26	
Conditioning Index		+ 0.30	- 0.05	
	(Latency	+ 0.35	- 0.04	
Alpha-blocking	− (Persistence	+ 0.76	+ 0.03	4.04***
	(Index	+ 0.69	+ 0.25	2.42***

complex frequency analysis of the EEG, is, generally speaking, regarded as an index of level of alertness and has been used in a number of Soviet studies. For example, Klyagin and Kovalev report a correlation between CLN of the resting EEG and strength of the nervous system, whereas Kovalev, Smirnov, and Rabinovich (1976) describe a relationship with the intensity of certain emotions, such as joy and rage.

Meshkova and Smirnov (1978) used multiple electrode placements, a particular feature of their study being a comparison of the EEG from different sites. Taking all leads together, a significant difference in the MZ and DZ correlations was found, the values being + .66 and − .10, respectively. Similar values—+ .61 and − .04—were observed for right hemisphere recordings. However, consistent with the similar result from Propping's study reported in a previous section, recordings from the left hemisphere showed no significant difference between the two twin types, nor were there any differences found for anterior and posterior sites of recording.

These data, as well as confirming that certain general features of the EEG may be under strong hereditary control, also suggest that genetic effects may not be uniform over the whole brain, or at least that activity in certain areas may be differentially modified by environmental factors, perhaps during early development. Considering the greater heritability of right, compared with left EEG indices, Meshkova (1976) herself in an earlier paper reached this conclusion; she suggested that certain left hemisphere characteristics may develop in postnatal ontogeny under the influence of environmental and social factors, largely speech and specific motor activity associated with the right hand. Her conclusion is in line with a study reported by Panashchenko (cited by Kol'tsova, 1973), who reports that training the fingers of a child to execute precise movements stimulates speech development and, at the same time, considerably alters resting EEG parameters in the appropriate hemisphere.

Meshkova's suggestion raises an important point about the relative influence, on nervous system properties, of strictly genetic effects on the one hand and, on the other, of environmentally modifiable developmental or maturational processes that may lead to a progressive alteration in phenotypic variation with age. It is these latter effects that are perhaps being revealed in the different heritabilities found at different age levels in some of the studies already reviewed. Similar age-related effects have been found in other Soviet twin research using measures that, though of a more "psychological" nature, are considered to reflect the nervous system properties postulated by Russian typological theory. Of particular interest have been measures of voluntary attention, as assessed by "proofreading" ability, which has been shown to be related to excitatory strength (Ermolayeva-Tomina, 1960) and mobility (Akhimova, 1971). Several such studies of voluntary attention have shown evidence of a genetic influence on performance. Thus, Glukhova and Vorob'eva (1974) found heritability coefficients of $+.70$ for number of errors and $+.75$ for time required to complete a proofreading task. These findings replicate earlier work by Shvartz (1970) and were themselves confirmed by Mosgovoy (1974), who reported significantly higher MZ correlations for several measures of proofreading performance—number of errors, efficiency under distracting conditions, time taken, and speed and accuracy. In that study Mosgovoy used a relatively small twin sample (20 MZ and 14 DZ pairs) within the age range 9–11 years. However, in a later, more extensive, investigation, which is the fifth study appearing in the 1978 publication from the Moscow Institute, Mosgovoy (1978) examined 130 pairs covering three age ranges—10–11, 14–15, and 20–50 years. The proofreading material was Landolt's rings, the parameters of voluntary attention being distribution (efficiency in identifying rings with breaks at two points), steadiness (efficiency in identifying rings with a break at one point only), and shifts (efficiency in identifying rings where the response criterion changes every two rows). Number of errors, time taken, and the ratio of errors over time were measured for each of these parameters.

The results of the study are shown in Table 2.8, where it can be seen that in the youngest group (10–11 years) the MZ twin similarity is significantly greater on all the experimental measures. However, this was maintained in the older groups only for the steadiness measure, twin diffences and heritabilities for the other indices being considerably reduced, especially in adults. It was concluded that variations in voluntary attention seem to depend on the subject's developmental level, as well as on a genetic contribution. The latter appears to be maximal in childhood, but later to influence only certain aspects of attentional performance.

The diminishing role for genetic factors noted by Mosgovoy for certain aspects of voluntary attention agrees with earlier data reported by Luria (1962) who, in a study of memory processes with preschool and school-age children, reports a progressive reduction in genetic contribution. This reduction was more pronounced in tasks involving what Luria describes as "voluntary verbal regulation of behavior." The greater the child's command of language, the smaller the genetic contribution. In Mosgovoy's study, however, this reduction did not

TABLE 2.8
F Ratios and Heritability Coefficients for Three Parameters of Voluntary Attention

Age of twins (years)	Statistical parameters	Distribution			Steadiness			Shifts		
		N	T	N/T	N	T	N/T	N	T	N/T
10–11	F	3.42**	2.95**	3.24**	3.25**	4.27**	4.20**	5.08**	4.01**	4.89**
	H	0.71	0.66	0.69	0.69	0.76	0.76	0.80	0.75	0.80
14–15	F	1.38	1.56	1.36	4.00**	1.96*	2.84**	1.71	1.94	1.86
	H	0.27	0.35	0.26	0.75	0.49	0.64	0.42	0.50	0.46
20–50	F	0.65	0.30	0.95	2.45*	0.82	4.66**	1.84	0.61	1.66
	H	–	–	–	0.68	–	0.76	0.43	–	0.38

$*p < 0.05$
$**p < 0.01$

apply equally to all aspects of voluntary attention. Steadiness, which involves concentration on a single cue, was found to have a strong genetic determination even in the oldest group. Why this should be so is a matter of conjecture. It has been claimed that this particular attentional parameter reflects mobility of nervous processes (Akhimova, 1971), but would now be regarded more as a measure of *lability*. That being so, the finding would be consistent with other data quoted earlier suggesting that quite different measures of that property, personal tempo (Vasiletz, 1978) and CFF (Panteleyeva & Shlyakhta, 1978), also show a strong genetic influence at all age levels.

Clearly the particular concern of Soviet research with developmental as well as strictly genetic influences on central nervous functioning raises many interesting questions. The recency with which behavior genetics research has been revived in the Soviet Union, together with the uncertainty surrounding the measurement and interpretation of parts of the theory of CNS activity that guides their work, must, for the moment, leave many of these questions unanswered. Nevertheless, the data that have emerged so far do, at an empirical level, add further important evidence for a genetic contribution to the nervous system characteristics underlying individual psychological differences.

Towards a Synthesis

In this section we try to bring together briefly the results of the studies reviewed so far and draw some preliminary conclusions from them. But first it is necessary to make some remarks that serve to emphasize the tentative nature of any synthesis that may be attempted.

First, we should reinforce a point made at the beginning of this chapter, namely that from a genetics point of view the studies considered here must be judged against the limitations imposed by the methodology almost all researchers have used. It is now generally recognized that the classic twin method can do little more than act as a pointer to those parameters of function that may have a significant genetic component, and that may therefore be worth examining in more detail using other analytic techniques, such as those of biometrical genetics. Furthermore, in the area reviewed, there has invariably been the additional restriction imposed by small sample sizes, which makes anything other than simple twin comparisons rather dubious, if not impossible. The reason the studies considered here have been confined to small samples is obvious. Many of the techniques are time-consuming, inconvenient for the subjects, and usually involve a visit to a laboratory. By comparison, procedures involving, say, questionnaires, often sent by post, can generate large quantities of data, more suitable for detailed genetical analysis. Against this advantage, however, can be offset the fact that notionally, at least, psychophysiological and other biological procedures can be said to get nearer to the hard substrate of the organism, and therefore be more likely to reveal stable dimensions of central nervous function.

Although the results of individual studies, each based on a small twin sample, cannot therefore be taken as definitive, where several studies using comparable techniques are in broad agreement it seems reasonable to reach firm, if cautious conclusions.

A second difficulty facing the reviewer in this area is the varied choice of measures used by different investigators to tap central nervous functioning. Analysis of the EEG, has, for many, provided the mainstay of measurement, whereas others have felt able to make inferences about the central nervous system from more peripheral autonomic indices or, even less directly, from psychological and psychophysical measures such as CFF and reaction time.

Over and above, though partly accounting for, these differences there have also been variations in the theoretical framework, or lack of it, that has dictated the choice of measure or measures used in particular experiments. Many studies have been of a "look and see" variety, often involving a single measure or a set of measures from a single domain, such as EEG. In contrast, other studies have been undertaken from a very particular theoretical viewpoint, and relevant concepts may be difficult to translate into other theoretical domains. Even where some convergence of theory is evident, experimental strategies may have differed, and here it is instructive to compare Western and Soviet work. As an example, some Western workers, Hume for instance, have sampled a variety of psychophysiological measures in the hope of identifying, by factor analysis, broad parameters that might describe genetically determined properties of the nervous system. Search for the latter has also, of course, motivated all Soviet research in the area, but workers there have preferred a different approach, namely the detailed and idiographic analysis of single measures that are themselves selected and interpreted with close reference to neo-Pavlovian theory.

Nevertheless, despite these differences in experiment and in presence, absence, or sophistication of theory, some coherence in the studies reviewed so far can be observed. There seems little doubt that major features of the EEG are substantially inherited, whether this is judged in terms of certain parameters of the resting rhythms, on the basis of the wave form of the evoked response, or as reflected in the response to drugs such as alcohol. For reasons inherent in their nature, there is less certainty about the results obtained with autonomic and somatic measures. Even if subject to genetic control, we can only make inferences about what this reveals of higher nervous activity. However, a good deal of surrounding research in psychophysiology is convincing enough to suggest that where genetic influences appear, as they often do, these relate to central nervous mechanisms, possibly different from or only partly overlapping those tapped by EEG measures.

The data considered in the previous sections, of course, represent only part of a larger body of research into the biological basis of individual variation. Indeed, they can be said to be buttressed on either side by two other kinds of evidence. One concerns the relationship between measures of CNS function of the kind discussed here and psychological variables, especially personality. The other concerns genetic studies of the latter, as represented, for example, in questionnaire studies of twins. Together these three sources of evidence can be seen, potentially at least, as being mutually supportive of one another and have given rise to the main theoretical models that, among them, might make it possible to interpret data on the genetics of the normal nervous system.

The most vigorous exponent of this three-pronged approach is undoubtedly Eysenck. Directly, or indirectly through his influence on other workers, Eysenck has contributed extensively to the data and theory in all three areas, evolving in the process a conceptual nervous system model that he claims can account for what he perceives as genetically determined personality variations. Eysenck's own model has so far been most highly developed with respect to the two personality dimensions of introversion–extraversion and neuroticism. The personality variations described by these two dimensions are, he suggests, due to inherited differences in the activity of two brain circuits, one involving the ascending reticular formation and the other the limbic system. According to his theory, high reticular arousal and high limbic system activation are associated, respectively, with increased introversion and neuroticism. In support of the theory, Eysenck (1967) has quoted evidence of the three kinds demanded for a test of its validity: data on the biological correlates of personality, on the genetics of its physiological basis, and on twin similarities as assessed by questionnaire.

A detailed critique of Eysenck's own position is beyond the scope of this chapter. Suffice it to say that the picture is less straightforward than Eysenck originally claimed. It is true that a good deal of early evidence suggested strong genetic determination of extraversion (E) and neuroticism (N), a finding supported by recent studies using more powerful biometrical–genetical techniques; an example is that carried out by Young, Eaves, and Eysenck (1980), who report significant genetic variation for both E and N in large adult and juvenile samples. Nevertheless, the relationship between central nervous differences and such descriptive personality characteristics is still uncertain. One problem is that, taken individually, introversion and neuroticism do not relate in any systematic way to relevant biological parameters but, instead, interact in a complex fashion. Thus, using the method of "zone analysis," several studies have shown that it is only when different combinations of E and N are considered that a systematic relationship emerges with such variables as EEG response (Savage, 1964), autonomic activity (Sadler, Mefferd, & Houck, 1971), and drug tolerance (Claridge, Donald, & Birchall, 1981).

This being said, Eysenck's theory has provided a rough template of CNS differences from which other investigators have worked. Claridge (1967), as we have seen, used a similar strategy of examining the psychophysiological correlates of personality and then followed this up with a twin study (Claridge, Canter and Hume (1973), in order to see how far the biological parameters isolated in the original experiments were genetically determined. The first study suggested that it was possible to identify a major descriptive dimension of personality, "dysthymia–hysteria," which to some extent cuts across Eysenck's dimensions, and which, at the biological level, seems to be determined by a combination of two factors; one was associated with EEG activity and the other with autonomic measures and drug response variations. Conceptually, these two factors were visualized as homeostatically related brain systems of arousal modulation and tonic arousal. The subsequent twin study suggested that the level of activity in each of these systems, and possibly the balance between them, is to some extent inherited.

Vogel and Schalt (1979), in interpreting their own data on the genetics of the EEG, have also arrived at a model that, they note, is very similar to that just described. They have speculated that the different EEG variants identified in their EEG studies may reflect differential activity from three sources in the brain: cortical, thalamic, and reticular. The cortex is seen as a "battery," responsible, through the asynchronous activity of cortical elements, for the fast frequency beta rhythm. Vogel and Schalt then recognize the pacemaker functions of the thalamus and the general activating influence of the ascending reticular formation as being in a feedback relationship with the cortex. The relative balance between these systems is considered to account for genetically determined differences in EEG and the behavioral variations associated with them. Although Vogel and his colleagues have had difficulty finding statistically significant relationships between the EEG and psychological tests, they do claim, as we saw earlier, that the various EEG types recognized show broad differences in personality and cognitive function.

Turning to the work of Soviet scientists, historically the nervous typological theory that has guided their research is the precursor of many Western models. Indeed, Eysenck (1955) originally adopted a simplistic version of Pavlovian theory in postulating a cortical "excitation–inhibition" balance as the basis of introversion. Subsequently, Gray (1964, 1967) has attempted a more sophisticated translation of the Russian theory into Western concepts, suggesting, convincingly, that there is a similarity between the notion of arousal and that of strength of nervous system. There is, indirectly, some support for this proposition from studies reporting a relationship, admittedly weak, between extraversion and strength–sensitivity of the nervous system, measured by lower absolute threshold in a number of sensory modalities (cf. reviews by Eysenck, 1967; Mangan, 1972), and between extraversion and the threshold of transmarginal inhibition, which describes the limit of working capacity of cortical cells, and thus strength of nervous processes (White & Mangan, 1972).

Other nervous system properties hypothesized by Pavlovian theory are more difficult to align with Western concepts, partly because the Pavlovian properties are themselves less highly developed within the theory, and partly because relevant experimental work reported in the Western literature is limited and generally unsophisticated. Nevertheless, there are now sufficient grounds for suggesting, tentatively, some relationship between extraversion and neuroticism and dynamism of excitation and of inhibition, measured by autonomic responses (i.e., "conditionability") (Eysenck & Levy, 1972; Mangan, 1974, 1978a), and between extraversion and transformation mobility (Mangan, 1967b, 1978b) and lability (Mangan, 1967a). Also relevant here is the reported relationship between perceptual flexibility and lability (Mangan, 1967a), and between cognitive flexibility (creativity), as measured by Guilford's tests, and mobility (Klonowicz, 1979; Mangan, 1967b, 1978b; Strelau, 1977).

In this general context we might also note two recent studies from our own laboratory that have shown a close alignment of Western and Soviet concepts. One is a study by Paisey (1980) who reports a considerable overlap between Eysenck's extraversion measure and nervous system properties of excitatory strength and mobility, derived from a questionnaire that Strelau (1972) claims to have validated against experimentally derived indices of these properties. The other, by Robinson (1982), has demonstrated correlations between certain features of the EEG and the Eysenck dimensions of extraversion and neuroticism. Deriving his CNS parameters from a systems analysis of the EEG, Robinson has argued that the covariation observed with personality reflects individual differences in thalamocortical relationships that correspond to and provide a physiological substrate for the Pavlovian notions of excitatory and inhibitory strength.

Returning to genetic studies per se, the results of Soviet twin studies, by Shibarovska (1978) and others, would support similar Western findings in suggesting a genetic effect on certain EEG parameters, as measured by indices of alpha rhythm activity such as frequency, amplitude, and responsiveness to stimulation. In Pavlovian terminology these data would support the idea of a partly inherited CNS characteristic separate from strength, related to excitatory and inhibitory nervous influences. There is some similarity here, therefore, with the distinction made in Claridge's model between general arousability (strength) and arousal modulation, identified empirically with the same alpha rhythm indices of EEG activity. In this respect it is also of interest to note again Gray's comparison of Russian and Western theory (Gray, 1967). Gray quotes evidence reported by Nebylitsyn (1963) for a central nervous factor that is defined partly by alpha rhythm measures and is considered, according to its direction, to reflect the predominance of the two opposite features of dynamism, namely excitation or inhibition. In his paper Gray was concerned with the question of which of the two separable nervous properties of strength and dynamism can account for variations in introversion–extraversion. That he comes down in favor of the former is not of concern here, not least because neither the data nor the theories in question are, in our view, robust enough for a firm conclusion to be reached. The more relevant point is the similarity between the psychophysiological data themselves and the form their interpretation has taken.

Clearly there is some convergence of theory and evidence in the studies reviewed so far, suggesting that important parameters of central nervous functioning are subject to genetic influence. Precisely what these parameters are, and how they relate to psychological differences, are still in doubt, though it seems very likely that they will ultimately prove to be the biological substrate of certain descriptive personality characteristics. It is significant that some of the most coherent theorizing and systematic data collection have occurred in those studies that, explicitly or implicitly, have been guided by models of individual psychological differences, especially personality. So far, work in that direction has been dominated by attempts to account for differences that, at a psychological level, can be referred to dimensions like introversion–extraversion and, to a lesser extent, neuroticism and anxiety. In other words, it is the study of these "normal" dimensions of individual variation that has inspired most of the research in the field so far. More recently, however, research starting originally from a quite different direction—namely the attempt to understand the etiology of disorders like schizophrenia—has begun to suggest that it may be able to throw light, not just on the genetics of serious mental disease, but also on the inheritance of some aspects of normal nervous system function. The state of theory and experimentation in that instance is even more precarious than in the case of the evidence reviewed already. Nonetheless, we feel that some reference to it should be made, if only as a pointer to future research possibilities.

PSYCHOSIS AND NORMAL CNS GENETICS

The data reviewed in this section represent two separate, though recently converging lines of research undertaken against somewhat different theoretical backgrounds. One set of data comes from work attempting to identify characteristics, especially central nervous characteristics, that describe individuals at high risk for schizophrenia. The other arises from attempts to extend personality analysis in order to encompass the idea that psychotic traits make up part of normal personality variation. Because the studies concerned with the latter have found their stimulus mainly from biological models of personality, they can be seen as an elaboration, as it were, of the research discussed in earlier parts of this chapter; in other words, they have involved use of the same experimental strategies and manipulations of the same concepts about the nervous system, albeit in an even less highly developed form. They have also made implicit assumptions about the genetic determination of the traits in question.

Such assumptions are, of course, more explicit in the first, high-risk kind of research mentioned. The very rationale for many of these studies lies in the surrounding evidence that the predisposition to schizophrenia is partly under genetic control (see Chapter 6) and that appropriately selected individuals ought therefore to show central nervous characteristics that describe their increased risk of psychotic breakdown. Part of the background literature that has given particular impetus to high-risk research (but has also added credence to the personality dimensional studies with which they converge) is the current popularity of the polygenic hypothesis, offered by workers like Gottesman and Shields (1976), as an interpretation of the genetic evidence on schizophrenia. Such a hypothesis clearly leaves room for the possibility that certain features of central nervous function, as well as defining the disposition to serious mental breakdown, might also, perhaps in lesser degree, or in still healthy people, describe sources of normal individuality. Because it is this possibility that is our sole concern here we do not attempt a review of high-risk research per se; that in any case would be outside the terms of reference of the chapter. Instead we concentrate on a few studies that help to illustrate what, in our view, is an exciting potential contribution from research on the genetics of schizophrenia to our understanding of normal nervous system functioning.

It is more appropriate to start, however, by discussing work of the other kind we mentioned, which has approached the question from the different direction of personality theory. Again Eysenck has been prominent here, suggesting many years ago (1952) that in order to give a complete description of personality it is necessary to postulate a third dimension of "psychoticism," additional to those of extraversion and neuroticism. It is only recently, however, that he has begun actively to pursue the idea, following his development of a questionnaire scale that he claims measures psychotic traits (Eysenck & Eysenck, 1975). Because of the relative immaturity of psychoticism within Eysenck's framework of personality description, research from his own laboratory has not yet made full use of all three strategies that, as noted earlier, help to give theories of this kind their internal strength as biological models of variation. On the genetic side he has so far confined himself to several biometrical genetical studies of his psychoticism questionnaire measure (Eaves, 1973; Eaves & Eysenck, 1977). He has also assembled the evidence and arguments for psychoticism as a valid concept in normal personality description (Eysenck & Eysenck, 1976). So far, however, he himself has not suggested a "conceptual nervous system" for psychoticism, a fact that has limited his own investigations of the biological correlates of the dimension, including its genetical analysis using physiological measures.

Nevertheless, work along similar lines has tended to support the general conclusion that one variant of normal nervous system organization may be of a kind seen in its extreme pathological form as schizophrenia. This idea was discussed some years ago by Claridge (1972) under the title "The Schizophrenias as Nervous Types," the suggestion being that the personalities and cognitive styles of some otherwise normal individuals have a similar biological basis to that underlying schizophrenia. The theory was derived from the research program previously referred to (Claridge, 1967), in which it had been possible to identify two hypothetical brain systems of "tonic arousal" and "arousal modulation." As discussed earlier, it was considered that the level of activity in these two systems accounted for variations along a continuum of "dysthymia– hysteria," encompassing the personality domain of Eysenck's extraversion and neuroticism dimensions. However, it was argued that this is so only where the systems are in a state of equilibrium. Evidence was also cited that in schizophrenics, and in some normal people, the two systems could be in a state of disequilibrium or relative imbalance, which, it was thought, might define a further major personality dimension comparable, perhaps, to what Eysenck subsequently attempted to measure with his psychoticism questionnaire.

Support for the latter possibility has been found in some recent studies that have looked at the psychophysiological correlates of the Eysenck scale. Briefly, those studies have focused on a form of analysis of psychophysiological data suggested in the original research by Claridge (1967); this involves examining, not absolute *levels* of psychophysiological parameters, but the *covariation* between them. It appears that relative imbalance between the two hypothetical systems referred to here is reflected in an inversion of the usually expected covariation between measures of arousal, on the one hand, and measures of sensory responsiveness, on the other. Such inversion of correlation has been shown when subjects high and low in psychoticism have been compared using measures of, respectively, skin conductance level and two-flash threshold discrimination (Claridge & Birchall, 1978; Claridge & Chappa, 1973; Robinson & Zahn, 1979). Thus, subjects high in psychoticism appear to show a counterintuitive negative relationship between autonomic arousal and sensory sensitivity, a pattern similar to that found in schizophrenics (Claridge & Clark, 1982) and in normal subjects under LSD–25 (Claridge, 1972). Such results therefore support the hypothesis that a psychotic form of central nervous organization may indeed be found naturally in some normal individuals, and that, at a descriptive level, this may correspond to Eysenck's notion of psychoticism as a personality dimension.

Although the possible genetic influences on these particular psychophysiological variations have not been examined, some data relevant to the model from which they were derived were obtained in the twin study reported by Claridge, Canter and Hume (1973). As part of that investigation Claridge carried out a "nervous typological" analysis of some of the psychophysiological data collected on the twins. It will be recalled from a previous section that principal components analysis of the data yielded two factors that appeared to correspond to "arousal modulation" and "tonic arousal"; one was an EEG factor, whereas the other had loadings on sedation threshold and autonomic measures. According to the direction of their scores on these factors, subjects were classified into two groups—individuals who showed "congruent" scores (high tonic arousal and high EEG activation) and those in whom the factors were related in the opposite direction. The two samples were designated the neuroticisim (*N*) and psychoticism (*P*) groups, respectively, to indicate that, psychophysiologically, they resemble individuals who, in Claridge's original 1967 analysis, had served to define the dimensions of dysthymia—hysteria and psychoticism. The *P*-group subjects were therefore considered to be of a hypothetical "psychotic nervous type."

Comparison between the two groups on psychological variables revealed theoretically coherent differences between them. On personality measures, in line with previous findings, relationships with psychophysiological variations differed in the two types of individual. Thus, in the N group, higher tonic arousal was, as expected, associated with increasing introversion. In the P group, however, it was associated with the opposite—increasing extraversion and impulsivity—a pattern previously considered to be indicative of the relative imbalance between tonic arousal and arousal modulation found in psychosis. The groups were also found to differ on measures of cognitive function, P subjects being significantly more divergent on some measures of creative thinking. The latter finding was considered to be of special interest because, on theoretical grounds, it would be expected that if the notion of a psychotic nervous type were viable, then it should particularly manifest itself as a variation in cognitive style, given that cognitive disturbance is a prominent feature of schizophrenia itself.

In fact, other studies of both creative individuals (Dykes & McGhie, 1976) and subjects high on Eysenck's psychoticism dimension (Woody & Claridge, 1977; see also Eysenck & Eysenck, 1976) do indeed suggest that differences in attentional and thinking style might form the phenotypic expression of one type of central nervous organization found in some normal people. That the latter may have a partly genetic basis, related to the schizophrenic process, is further suggested by the results of studies of a more deliberately high-risk nature; for example, some data suggest that "divergent," "allusive," or "loosened" thinking styles are characteristically found in the healthy relatives of schizophrenic patients (Chapman, 1979; Lidz, Wild, Schafer, Rosman, & Fleck, 1963; McConaghy, 1960; Phillips, Jacobsen, & Turner, 1965).

In the psychophysiological evidence, too, there is some sign of a sensible convergence between high-risk research and personality dimensional studies. One strategy used by high-risk researchers has been the long-term follow-up of children considered vulnerable because of their genetic relatedness to a schizophrenic parent. The classic pioneer study of this type was the investigation begun by Mednick in the 1960s. The research subsequently revealed that a salient feature of the psychophysiology of high-risk children was a particular feature of the GSR—its rapid recovery after stress (Mednick & Schulsinger, 1973; Mednick, Schulsinger, Higgins, & Bell, 1974).

Since then the recovery limb of the GSR has become a focus of interest for some workers in the field, both high-risk and personality researchers. Thus, Nielsen and Petersen (1976), in a personality study, showed that similar rapid recovery of the GSR could be found in individuals selected on the basis of a questionnaire measure of "schizophrenism." Of greater import from a genetic point of view, however, is the fact that Mednick, in conjunction with Venables, used GSR recovery rate as an index with which to select children for their recent large-scale follow-up study in Mauritius. The complete results of that study are, of course, not yet available, though Venables, Mednick, Schulsinger, Raman, Bell, Dalais, and Fletcher (1978) have reported some preliminary findings, one feature of which is of interest here.

Part of the Mauritian study has involved examining the play activities of children supposed, because of their rapid GSR recovery, to be at high risk for schizophrenia. Venables et al. (1978) report that the so-called high-risk children show significantly greater amounts of constructive play, a quality that is seen as possible evidence of greater creativity. The interpretation put by Venables et al. on their Mauritian finding derives from Venables' (1974) model of the possible neurophysiological basis of schizophrenia. He has argued that rapid GSR recovery is a psychophysiological sign of one mode of attention deployment in which there is relative "openness to environmental stimuli," this in turn reflecting the sensory response gating function of the hippocampus. Venables et al. (1978) speculate that, although such openness to the environment might, in an extreme form, lead to schizophrenic disorder, it might also result in more extensive sampling of stimuli, and hence the greater creativity apparent in the play activity of their high-risk children.

Psychophysiological findings of a different kind, but also of interest because of their relevance here, have emerged from the work of Buchsbaum and his colleagues, whose research on the evoked potential augmenting–reducing response we have already mentioned in our review of twin studies of the EEG. As part of their research program, Buchsbaum and his collaborators have used a combined high-risk/personality strategy to examine the augmenting–reducing response, sometimes in conjunction with other, biochemical, measures of CNS function, such as platelet MAO levels. In one such study (Haier, Buchsbaum, Murphy, Gottesman, & Coursey, 1980) they analysed their data using the covariation approach adopted by Claridge, classifying their subjects according to which of the four combinations of augmenting or reducing or high or low platelet MAO level they showed. Using the MMPI, Haier and his co-workers found more elevated scores in individuals showing what they considered to be unbalanced combinations of their two experimenal measures; namely augmenting in association with low MAO (interpreted as indicative of high CNS arousal) or reducing in association with high MAO. In discussing their findings, Haier et al. arrive at a model similar to that proposed by Claridge (1967), arguing that some personality types are characterized by an apparent imbalance between arousal level (as measured in their case by MAO) and the degree of sensory regulation that Buchsbaum considers responsible for EEG augmenting–reducing variations. Elsewhere in their writings Buchsbaum and his coauthors have considered the genetic implication of their results (Buchsbaum, Murphy, Coursey, Lake, & Zeigler, 1978; Murphy & Buchsbaum, 1978). The jigsaw is clearly a complicated one, but parts of it are beginning to piece together, such as the evidence from twin research (Winter, Herschel, Propping, Friedl, & Vogel, 1978) that platelet MAO levels are under strong genetic control, and the possibility that this biochemical index of CNS activity may be a useful marker for detecting vulnerability to serious mental disorder (Wyatt & Murphy, 1976).

In summary then, although a definitive picture has yet to emerge, there does seem to be a certain coherence in the research that has tried to extend our understanding of the genetics of normal nervous system functioning by seeking inspiration from studies of psychotic variations in behavior. Two themes run through the work reviewed in this section. First, there is increasing convergence on the idea that psychotic forms of central nervous organization, already known to be partly under genetic control, may be widely represented in the general population, manifest in behavior as particular forms of personality structure and cognitive style. Secondly, there is some resemblance in the conceptual nervous system models proposed by different writers to explain these variations in behavior. Thus, Venables' hippocampal filtering model, Claridge's notion of arousal modulation, and Haier and Buchsbaum's concept of sensory regulation are all very much alike, the latter two theories being especially similar in proposing that the critical feature of psychoticism may be a relative imbalance, or a particular kind of feedback relationship, between CNS arousal and the facilitatory and inhibitory controls over sensory response.

FUTURE DIRECTIONS

The account given here of very disparate kinds of research into the genetics of the human nervous system provides, in a sense, its own evidence of where future work should lie. Many of the advances that could be made must undoubtedly await developments in other fields, especially, of course, in our understanding of the nervous system itself, and in the refinement of exact techniques for measuring its functioning. Here we have concentrated particularly on psychophysiological measurements, for two reasons. First, they have provided the bulk of the evidence, and, secondly, being close to behavior, some, albeit scrappy alignment can be attempted with descriptions of the psychology of the organism, especially personality. There is no doubt that psychophysiology will continue to make its impact, perhaps not so much in the development of technology, but rather in disentangling the significance for behavior of different psychophysiological parameters. Here more sophisticated models of the conceptual nervous system must surely prove to be important, and these in turn will benefit from a wider perspective on the individual psychological differences they seek to explain. The field has to date been to constrained by its concentration on limited sources of variation, like introversion–extraversion; it has also paid too little attention to what we can learn about the normal nervous system from examining the severely abnormal, which in genetics, as elsewhere, has in the past been considered to belong to a different universe of discourse.

If we were to pick out a common growth point for all of these suggested developments it would be the investigation of genetic influences on individual differences in hemisphere function. Asymmetries of hemisphere function have already been firmly implicated in several forms of psychopathology, especially schizophrenia (Gruzelier & Flor-Henry, 1979); while evidence is beginning to emerge (Claridge, in press) that some normal personality traits, associated with "psychoticism", may have a similar physiological basis. Apart from a few studies cited earlier, this perspective on the nervous system has so far received scant attention from a genetics point of view. In our opinion it would be extremely profitable to make more detailed twin comparisons, utilizing the techniques developed in hemisphere function research and guided by the conceptual nervous system models to which such research is now giving rise.

Despite its importance, psychophysiology can be seen as an intermediate level of description of the nervous system, and we can anticipate that major advances in technique and understanding will come from developments in biochemistry. These should make it possible to get nearer to the genetic substrate though, in our view, progress will be more rapid if it occurs in an orderly fashion, the results of biochemical studies being interpreted within, and constantly referred back to psychophysiological and psychological models of the nervous system and behavior.

Given more exact and better understood measures of CNS function, a question remains concerning their application in genetic studies. We have commented several times in this chapter on the methodology of most of the research we have reviewed. Almost all of it has made use of the classic twin method, usually with small samples, a fact that has stimulated criticism, particularly from the biometrical geneticists. To some extent we feel that their case is overstated, reflecting their own frequent concentration on relatively soft data, such as questionnaires, which contain many sources of variation, including high error variance. It is significant, for example, that for some biological variables very clear-cut results can be obtained using quite small twin samples, and this is likely to become increasingly the case as more exact biological parameters of variation are identified. Nevertheless, from a genetics viewpoint, there are other advantages to be gained from larger samples and from samples that include, not just twins, but other forms of kinship. For it is unlikely that the parameters of interest in this field follow a simple mode of inheritance; they are more likely to be continuous sources of variation, and subject to the complex interaction of genetic and environmental influences that the techniques of biometrical genetics are designed to disentangle. Collection of sufficient amounts of the right kind of data still, of course, presents a practical problem, but one that will have to be faced if progress in the field is to be made.

Finally, perhaps the most important growth point for future research in the area we have reviewed is its elaboration within a framework of human development. The human nervous system is extremely plastic, and although in a book of this kind it undoubtedly sounds a truism, it is worth repeating that, whatever the genetic effects on the CNS, they only find expression in behavior—and thereby achieve significance—as a result of their interplay with environmental influences. Most of the research we have reviewed, with the exception of a few Soviet studies, has paid scant attention to that question. A good deal of research needs to be done to establish how such factors as age and life experience modify the phenotypic expression of genetically controlled sources of CNS variation. In practical terms this means adapting many of the techniques and ideas we have discussed for use in studies employing other research strategies, such as the investigation of infants and their long-term follow-up.

REFERENCES

Abe, K. Reactions to coffee and alcohol in monozygotic twins. *J. Psychosom. Res.*, 1968, *12*, 199–203.

Abe, K., & Pérez de Francisco, C. Genetic determinants of psychophysiological variables. *Rev. Latinoamer. de Psi.*, 1972, *4*, 75–88.

Akhimova, M. N. Individual characteristics of attention and basic properties of the nervous system. Unpublished dissertation, Kuibyshev, 1971,.

Barcal, R., Simon, J., & Sova, J. Blood pressure in twins. *Lancet*, 1969, *1*, 1321.

Birchall, P. M. A., & Claridge, G. S. Augmenting–reducing of the visually evoked potential as a function of changes in skin conductance level. *Psychophysiology*, 1979, *16*, 482–490.

Block, J. D.. Monozygotic twin similarity in multiple psychophysiologic parameters and measures. *Recent Advances in Biological Psychiatry*, 1967, *9*, 105–118.

Broadhurst, P. L. The Maudsley reactive and non-reactive strains of rats: A survey. *Behav. Genet.*, 1975, *5*, 299–319.

Buchsbaum, M. Average evoked response and stimulus intensity in identical and fraternal twins. *Physiol. Psychol.*, 1974, *2*, 365–370.

Buchsbaum, M. S., Murphy, D. L., Coursey, R. D., Lake, C. R., & Zeigler, M. G. Platelet monoamine oxidase, plasma dopamine-beta-hydroxylase and attention in a "biochemical high-risk" sample. *J. Psychiat. Res.*, 1978, *14*, 215–224.

Buchsbaum, M., & Pfefferbaum, A. Individual differences in stimulus intensity response. *Psychophysiology*, 1971, *8*, 600–611.

Buchsbaum, M., & Silverman, J. Stimulus intensity control and the cortical evoked response. *Psychosom. Med.*, 1968, *30*, 12–22.

Callaway, E. *Brain electrical potentials and individual psychological differences.* New York: Greene & Stratham, 1975.

Canter, S. Personality traits in twins. In G. S. Claridge, S. Canter, & W. I. Hume (Eds.), *Personality differences and biological variations: A study of twins.* Oxford: Pergamon, 1973.

Carmena, M. Ist die persönliche Affektlage oder Nervosität eine ererbte Eigenschaft. *Z. Neurol. Psychiat.*, 1934, *150*, 434.

Chapman, L. Recent advances in the study of schizophrenic cognition. *Schiz. Bull.*, 1979, *5*, 568–580.

Claridge, G. S. *Personality and arousal.* Oxford: Pergamon, 1967.

Claridge, G. S. *Drugs and human behaviour.* Harmondsworth: Penguin, 1970.

Claridge, G. S. The schizophrenias as nervous types. *Brit. J. Psychiat.,* 1972, *121,* 1–17.

Claridge, G. S. A nervous typological analysis of personality variation in normal twins. In G. S. Claridge, et al. (Eds.), *Personality differences and biological variations: A study of twins.* Oxford: Pergamon, 1973.

Claridge, G. S. The Eysenck Psychoticism Scale, In: J. N. Butcher & C. D. Spielberger (Eds.). *Advances in Personality Assessment.* Vol. II. Lawrence Erlbaum. New Jersey (in press).

Claridge, G. S., & Birchall, P. Bishop, Eysenck, Block and psychoticism. *J. Abn. Psychol.,* 1978, *87,* 664–668.

Claridge, G. S., Canter, S., & Hume, W. I. *Personality differences and biological variations: A study of twins.* Oxford: Pergamon, 1973.

Claridge, G. S., & Chappa, H. J. Psychoticism: a study of its biological basis in normal subjects. *Brit. J. Soc. Clin. Psychol.,* 1973, *12,* 175–187.

Claridge, G. S., & Clark, K. H. Covariation between two-flash threshold and skin conductance level in first-breakdown schizophrenics. Relationships in drug-free patients and effects of treatment. *Psychiatry Research,* 1982, *6,* 371–380.

Claridge, G. S., Donald, J. R., & Birchall, P. M. A. Drug tolerances and personality: Some implications for Eysenck's theory. *Person. Ind. Diffs.* 1981, *2,* 153–166.

Claridge, G. S., & Herrington, R. N. Sedation threshold, personality and the theory of neurosis. *J. Ment. Sci.,* 1960, *106,* 1568–1583.

Claridge, G. S., & Ross, E. Sedative drug tolerance in twins. In G. S. Claridge et al. (Eds.), *Personality differences and biological variations: A study of twins.* Oxford: Pergamon, 1973.

Davis, H., & Davis P. A. Action potentials of the brain in normal persons and in normal states of cerebral activity. *Arch. Neurol. Psychiat.,* 1936, *36,* 1214–1224.

Dongier, S., Tournadre, R. de, Naquet, R., & Gastaut, H. A. Psychological study of 34 subjects presenting a 4 c/sec rhythm. *EEG Clin. Neurophysiol.,* 1965, *18,* 722p.

Downie, W. W., Boyce, J. A., Greig, W. R., Buchanan, W. W., & Alepa, F. P. Relative roles of genetic and environmental factors in control of blood pressure in normotensive subjects. *Brit. Heart J.,* 1969, *31,* 21–25.

Dubinin, N. P. Philosophical and sociological aspects of human genetics. *Vop. Filosof.,* 1971, No. 1.

Duffy, E. Emotion: An example of the need for reorientation in psychology. *Psychol. Rev.,* 1934, *41,* 239–243.

Dustman, R. E., & Beck, E. C. The visually evoked potential in twins. *EEG Clin. Neurophysiol.,* 1965, *19,* 570–575.

Dykes, M., & McGhie, A. A comparative study of attentional strategies of schizophrenic and highly creative normal people. *Brit. J. Psychiat.,* 1976, *128,* 50–56.

Eaves, L. J. The structure of genotypic and environmental covariation for personality measurements: An analysis of the PEN. *Brit. J. Soc. Clin. Psychol.,* 1973, *12,* 275–282.

Eaves, L. J., & Eysenck, H. J. Genotype × age interaction for neuroticism. *Behav. Genet.,* 1976, *6,* 359–362.

Eaves, L. J., & Eysenck, H. J. A genotype–environmental model for psychoticism. *Adv. Behav. Res. Therap.,* 1977, *1,* 5–26.

Eaves, L. J., Last, K. A., Young, P. A., & Martin, N. G. Model-fitting approaches to the analysis of human behaviour. *Heredity, 41,* 1978, 249–320.

Edelberg, R. Electrical activity of the skin. In N. Greenfield & R. Sternbach (Eds.), *Handbook of psychophysiology.* New York: Holt, 1972.

Elkin, V. L., & Khoruzheva, S. A. Major properties of nervous processes in twins. *Zh. Vyssh. Nervn. Deyat. im I. P. Pavlova,* 1975, *25,* 1.

Ermolayeva-Tomina, L. B. Individual differences in level of concentration of attention and strength of the nervous system. *Vopr. Psikhol.,* 1960, No. 2.

Eysenck, H. J. *The scientific study of personality.* London: Routledge & Kegan Paul, 1952.

Eysenck, H. J. A dynamic theory of anxiety and hysteria. *J. Ment. Sci.,* 1955, *101,* 28–51.

Eysenck, H. J. The inheritance of extraversion–introversion. *Acta Psychol.,* 1956, *12,* 95–110.

Eysenck, H. J. Drugs and personality. I. Theory and methodology. *J. Ment. Sci.,* 1957, *103,* 119–131.

Eysenck, H. J. *The biological basis of personality.* Springfield, Ill.: Charles C. Thomas, 1967.

Eysenck, H. J., and Eysenck, S. B. G. *Manual of the Eysenck Personality Questionnaire. London: Hodder & Stoughton, 1975.*

Eysenck, H. J., & Levy, A. Conditioning, introversion–extraversion and the strength of the nervous system. In V. D. Nebylitsyn & J. A. Gray (Eds.), *Biological bases of individual behaviour.* New York and London: Academic Press, 1972.

Eysenck, H. J., & Eysenck, S. B. G. *Psychoticism as a dimension of personality.* London: Hodder & Stoughton, 1976.

Glass, A. B. Genetic aspects of adaptability. *Genetics and the inheritance of integrated neurological and psychiatric patterns. Proceedings of the association for research in nervous and mental diseases* (Vol. 33). Baltimore: Williams & Wilkins, 1954.

Glukhova, R. C., & Vorob'eva, A. L. The study of certain psychophysiological parameters of intellectual work capacity in school-age twins. In *New studies in developmental physiology* (No. 9). Moscow: Pedagogika Publishers, 1974.

Gottesman, I. I., & Shields, J. A critical review of recent adoption, twin and family studies of schizophrenia: Behavioural genetics perspectives. *Schiz. Bull.,* 1976, *2,* 360–398.

Gray, J. A. *Pavlov's typology.* Oxford: Pergamon, 1964.

Gray, J. A. Strength of the nervous system, introversion–extraversion, conditionability and arousal. *Behav. Res. Therap.,* 1967, *5,* 151–169.

Gruzelier, J. & Flor-Henry, P. (Eds.) *Hemisphere assymetries of function in psychopathology* Elsevier/North Holland, Amsterdam, 1979.

Haier, R. J., Buchsbaum, M. S., Murphy, D. L., Gottesman, I. I., & Coursey, R. D. Psychiatric vulnerability, monoamine oxidase and the average evoked potential. *Arch. Gen. Psychiat.* 1980, *26,* 340–345.

Hamilton, M., Pickering, G. W., Roberts, J. A. F., & Sowry, G. S.C. The aetiology of essential hypertension. 4. The role of inheritance. *Clin. Sci.,* 1954, *13,* 273–304.

Hume, W. I. Physiological measures in twins. In G. S. Claridge, Canter, S. & Hume, W. I. (Eds.), *Personality differences and biological variations: A study of twins.* Oxford: Pergamon, 1973.

Jost, H., & Sontag, L. W. The genetic factor in autonomic nervous system function. *Psychosom. Med.,* 1944, *6,* 308–310.

Juel-Nielsen, N., & Harvald, B. The electroencephalogram in twins brought up apart. *Acta Genetica,* 1958, *8,* 57–64.

Klonowicz, T. Transformation ability, temperament traits and individual experience. *Polish Psychological Bulletin,* 1979, *10,* 215–223.

Klyagin, V. S., & Kovalev, A. N. Statistical analysis of EEGs and diagnosis of nervous system strength. In *Diagnostic problems of mental development.* Tallin, Moscow 1974.

Kol'tsova, M. M. *Motor activity and the development of cerebral functions in the child.* Moscow: 1973.

Kovalev, A. N., Smirnov, L. M., & Rabinovich, L. A. The relationship between the internal structure of EEGs and emotionality. *Novye Issledovaniya v Psikhologii,* 1976, No. 1.

Kryshova, N. A., Beliaeva, Z. V., Dmitrieva, A. F., Zhilinskaia, M. A., & Pervov, L. G. Investigation of the higher nervous activity and of certain vegetative features in twins. *Soviet Psychol. and Psychiat.,* 1962, *1,* 36–41.

Kuhlo, W., Heintel, H., & Vogel, F. The 4–5 c/sec rhythm. *EEG. Clin. Neurophysiol.,* 1969, *26,* 613–618.

Lacey, J. I., Bateman, D. E., & Van Leyn, R. Autonomic response specificity. An experimental study. *Psychosom. Med.*, 1953, *15*, 8–21.

Lacey, J. I., & Lacey, B. C. Verification and extension of the principle of autonomic response–stereotypy. *Amer. J. Psychol.*, 1958, *71*, 50–73.

Lader, M. H., & Wing, L. *Physiological measures, sedative drugs and morbid anxiety.* Maudsley Monographs No. 14. London: Oxford University Press, 1966.

Lennox, W. G., Gibbs, F. A., & Gibbs, E. C. The brain wave pattern, an hereditary trait. Evidence from "normal" pairs of twins. *J. Hered.*, 1945, *36*, 233–243.

Lewis, E. G., Dustman, R. F., & Beck, E. C. Evoked response similarity in monozygotic, dizygotic and unrelated individuals: A comparative study. *EEG clin. Neurophysiol.*, 1972, *32*, 309–316.

Lidz, T., Wild, C., Schafer, S., Rosman, B., & Fleck, S. Thought disorders in the parents of schizophrenic patients: A study utilizing the Object Sorting Test. *J. Psychiat. Res.*, 1963, *1* 193–200.

Lipton, E. L., Steinschneider, A., & Richmond, J. B. Autonomic function in the neonate. IV Individual differences in cardiac reactivity. *Psychosom, Med.*, 1961, *23*, 472–484.

Luria, A. R. Ontogenetic variation in psychological functions. *Vopr. Psikhol.*, 1962, No. 3.

Luria, A. R. *Selected writings of A. R. Luria* (Michael Cole, Ed.) White Plains, N.Y.: M. E. Sharpe Inc., 1978.

Lykken, D., Tellegen, A., & Thorkelson, K. Genetic determination of EEG frequency spectrum. *Biol. Psychol., 1,* 1974, 245–259.

Mangan, G. L. Studies of the relationship between neo-Pavlovian properties of higher nervous activity and Western personality dimensions: II. The relation of mobility to perceptual flexibility. *J. Exp. Res. Pers.*, 1967, 2, 107–116. (a)

Mangan, G. L. Studies of the relationship between neo-Pavlovian properties of higher nervous activity and Western personality dimensions: III. The relation of transformation mobility to thinking flexibility. *J. Exp. Res. Pers.*, 1967, 2, 117–123. (b)

Mangan, G. L. The relationship of strength–sensitivity of the visual system to extraversion. V. D. Nebylitsyn & J. A. Gray (Eds.), *Biological bases of individual behaviour.* New York: Academic Press, 1972.

Mangan, G. L. Personality and conditioning: Some personality, cognitive and psychophysiological parameters of classical appetitive (sexual) GSR conditioning. *Pav. J. Biol. Sc.*, 1974, *9*, 125–135.

Mangan, G. L. Factors of conditionability and their relationship to personality types. *Pav. J. Biol. Sc.*, 1978, *13*, 226–235. (a)

Mangan, G. L. The relationship of mobility of inhibition to rate of inhibitory growth, and measures of flexibility, extraversion and neuroticism. *J. Gen. Psychol.*, 1978, *99*, 271–279. (b)

Martin, N. G., Eaves, L. J., Kearsey, M. J., & Davies, P. The power of the classical twin study. *Heredity*, 1978, *40*, 97–116.

Mathers, J. A. L., Osborne, R. H., & DeGeorge, F. V. Studies of blood pressure, heart rate and the electrocardiogram in adult twins. *Amer. Heart J.*, 1961, *62*, 634–642.

McConaghy, N. Modes of abstract thinking and psychosis. *Amer. J. Psychiat.*, 1960, *117*, 106–110.

Mednick, S. A., & Schulsinger, F. A learning theory of schizophrenia: Thirteen years later. In M. Hammer, R. Salzinger, S. Sutton (Eds.), *Psychopathology: Contributions from the social, behavioural, and biological sciences,* New York: Wiley, 1973.

Mednick, S. A., Schulsinger, F., Higgins, J., & Bell, B. *Genetics, environment and psychopathology.* Amsterdam: North–Holland, 1974.

Meshkova, T. A. Study of the genetic basis of various resting EEG parameters in humans using twin methodology. Unpublished dissertation, Moscow, 1976.

Meshkova, T. A., & Smirnov, L. M. The genetic bases of individual differences in the human EEG at rest. *Vopr. Psikhol.*, 1978, No. 5, 66–75.

Miall, W. E., & Oldham, P. D. A study of arterial blood pressure and its inheritance in a sample of the general population. *Clin. Sci.*, 1955, *14*, 459–488.

Mikhevey, V. F. The role of genetic factors in the formation of human trace reactions. *Zh. Vyssh. Nervn. Deyat. im I. P. Pavlova*, 1979, *29*, 510–517.

Mosgovoy, V. D. Study of genetic determination of attentional parameters. Proceedings of a Symposium; *Relationship between the Biological and the Social in Human Development*. Moscow: Vil'nyus, 1974.

Mosgovoy, V. D. Genetic determination of voluntary attention. In B. M. Lomov & I. V. Ravich-Shcherbo (Eds.), *Problems in the genetics of psychophysiology*. Institute of Psychology, Academy of Sciences SSSR Nauka, 1978, 244–253.

Murphy, D. L., & Buchsbaum, M. S. Neurotransmitter-related enzymes and psychiatric diagnostic entities. In R. L. Spitzer & D. F. Klein (Eds.), *Critical issues in psychiatric diagnosis*. New York: Raven, 1978.

Nebylitsyn, V. D. An electroencephalographic investigation of the properties of strength of the nervous system and equilibrium of the nervous processes in man using factor analysis. In B. M. Teplov, (Ed.), *Typological features of higher nervous activity in man* (Vol. 3). Moscow: Akad.Pedagog. Nauk. RSFSR, 1963.

Nebylitsyn, V. D. *Fundamental properties of the human nervous system*. Moscow: Prosveshchenie, 1966.

Neilsen, T. C. & Petersen, N. E. Electrodermal correlates of extraversion, trait anxiety and schizophrenism. *Scand. J. Psychol.*, 1976, *17*, 73–80.

Osborne, R. T. Heritability estimates for the visual evoked response. *Life Sciences*, 1970, *9*, 481–490.

Osborne, R. H., DeGeorge, F. V., & Mathers, J. A. L. The variability of blood pressure: Basal and casual measurements in adult twins. *Amer. Heart J.*, 1963, *66*, 176–183.

Paisey, T. J. H. Individual differences in psychophysiological response. Doctoral dissertation Oxford University, England, 1980.

Panteleyeva, T. A., & Shlyakhta, N. F. Genetic components of certain indices of lability of nervous processes. In B. M. Lomov & I. V. Ravich-Shcherbo (Eds.), *Problems in the genetics of psychophysiology*. Moscow: Institute of Psychology, Academy of Sciences SSSR. Nauka, 1978.

Pavlov, I. P. In: *Complete Works*, Vol. III, No 2 (1951–52) Moscow, Acad. Nauk. SSSR, 1927.

Petersén, I., and Sörbye, R. Slow posterior rhythm in adults. *EEG Clin. Neurophysiol.*, 1962, *14*, 161–170.

Phillips, J. E., Jacobsen, N., & Turner, W. M. Conceptual thinking in schizophrenics and their relatives. *Brit. J. Psychiat.*, 1965, *111*, 823–890.

Pitot, M., & Gastaut, Y. Aspects électroencéphalographiques inhabituels des séquelles des traumatismes crâniens: II Les rhythmes postérieurs à 4 cycles-seconde. *Rev. Neurol.*, 1956, *94*, 189–191.

Propping, P. Genetic control of ethanol action on the central nervous system. *Hum. Genet.*, 1977, *35*, 309–334.

Rachman, S. Galvanic skin response in identical twins. *Psychol. Rep.*, 1960, *6*, 298.

Raney, E. T. Brain potentials and lateral dominance in identical twins. *J. Exp. Psychol.*, 1934, *24*, 21–39.

Ravich-Shcherbo, I. V. Genetic factors underlying nervous system characteristics and their stability. *Symposium*, Tallin, 1974.

Regan, D. *Evoked potentials in psychology, sensory physiology and clinical medicine*. London: Chapman and Hall, 1972.

Robinson, D. L. Properties of the diffuse thalamocortical system and human personality: A direct test of Pavlovian/Eysenckian theory. *Person. Individ. Diff.*, 1982, *3*, 1–16.

Robinson, T. N., & Zahn, T. P. Covariation of two-flash threshold and autonomic arousal for high and low scorers on a measure of psychoticism. *Brit. J. Soc. Clin. Psychol.*, 1979, *18*, 431–441.

Rust, J. Genetic effects in the cortical auditory evoked potential: A twin study. *EEG Clin. Neurophysiol.*, 1975, *39*, 321–327.

Sadler, T. G., Mefferd, R. B., & Houck, R. L. The interaction of extraversion and neuroticism in orienting response habituation. *Psychophysiology*, 1971, *8*, 312–318.

Savage, R. D. Electro-cerebral activity, extraversion and neuroticism. *Brit. J. Psychiat.*, 1964, *110*, 98–100.

Shagass, C. The sedation threshold. A method for estimating tension in psychiatric patients. *EEG Clin. Neurophysiol.*, 1954, *6*, 221–333.

Shapiro, A. P., Nicotero, J., Sapira, J., & Scheib, E. T. Analysis of the variability of blood pressure, pulse rate and catecholamine responsivity in identical and fraternal twins. *Psychosom. Med.*, 1968, *30*, 506–520.

Shibarovska, G. A. Genetic basis of dynamism of nervous processes. In B. M. Lomov & I. V. Ravich-Shcherbo (Eds.), *Problems in the genetics of psychophysiology*. Institute of Psychology, Academy of Sciences SSSR. Nauka, 1978, 137–144.

Shields, J. *Monozygotic twins brought up apart and brought up together*. London: Oxford University Press, 1962.

Shields, J., & Slater, E. Heredity and psychological abnormality. In H. J. Eysenck, (Ed.), *Handbook of abnormal psychology* (1st Ed.). London: Pitman, 1960.

Shlyakhta, N. F. Twin study of nervous system properties. In V. D. Nebylitsyn (Ed.), *Problems of differential psychophysiology* (Vol. 7). Moscow: Pedagogika Publishers, 1972.

Shlyakhta, N. F., & Panteleyeva, T. A. Study of genetic factors underlying excitatory strength of the nervous system. In B. M. Lomov & I. V. Ravich-Shcherbo (Eds.), *Genetics of psychophysiology*. Institute of Psychology, Academy of Sciences SSSR. Nauka, 1978, 94–110.

Shvartz, V. G. Some genetic aspects of physical work capacity in children. Proceedings of a Conference: *The Organization of Public Helath, the History of Medicine and Social Hygiene*. Tallin, 1970.

Siemens, H. *Die Zwillingspathologie*. Berlin: Springer, 1924.

Strelau, J. A diagnosis of temperament by non-experimental techniques. *Polish Psychological Bulletin*, 1972, *3*, 97–105.

Strelau, J. Behavioural mobility versus flexibility and fluency of thinking: An empirical test of the relationship between temperament and abilities. *Polish Psychological Bulletin*, 1977, *8*, 75–82.

Surwillo, W. W. Interval histograms of period of the electroencephalogram and the reaction time in twins. *Behav. Genetics*, 1977, *7*, 161–170.

Takkunen, J. Anthropometric, electrocardiographic and blood pressure studies on adult male twins. *Ann. Acad. Scient. Fenn.* Series A., V. Medica 107. Helsinki: Suomalainen Tiedeakatemia, 1964.

Teplov, B. M. Typological properties of the nervous system and their psychological manifestations. *Vopr. Psikhol.*, 1951, *5*, 108–130.

Theorell, T., De Faire, U., & Fagrell, B. Cardiovascular reactions during psychiatric interview: A non-invasive study on a twin sample. *J. Human Stress*, 1978, *4*, 27–31.

Vandenberg, S. G., Clark, P. J., Samuels, I. Psychophysiological reactions of twins: Hereditary factors in galvanic skin resistance, heartbeat, and breathing rates. *Eugen. Quart.*, 1965, *12*, 7–10.

Vasiletz, T. V. Genetic determination of mobility of nervous processes in motor reactions. *Vopr. Psikhol*, 1974, No. 5.

Vasiletz, T. V. Mobility as a nervous system characteristic: Genetic aspects. In B. M. Lomov & I. V. Ravich-Shcherbo (Eds.), *Problems in the genetics of psychophysiology*, Institute of Psychology, Academy of Sciences SSSR. Nauka, 1978, 111–126.

Venables, P. H. The recovery limb of the skin conductance response in "high-risk" research. In S. A. Mednick, F. Schulsinger, J. Higgins, & B. Bell (Eds.), *Genetics, environment and psychopathology*. Amsterdam: North-Holland, 1974.

Venables, P. H., Mednick, S. A., Schulsinger, F., Raman, A. C., Bell, B., Dalais, J. C., & Fletcher, R. P. Screening for risk of mental illness. In G. Serban (Ed.), *Cognitive defects in the development of mental illness*. New York: Brunner/Mazel, 1978.

Vogel, F. Über die Erblichkeit des normalen Elektroencephalogramms. Stuttgart: Thieme, 1958.

Vogel, F. The genetic basis of the normal human electroencephalogram (EEG). *Humangenetik*, 1970, *10*, 91–114.

Vogel, F., & Fujiya, Y. The incidence of some inherited EEG variants in normal Japanese and German males. *Humangenetik*, 1969, *7*, 38–42.

Vogel, F., & Schalt, E. The electroencephalogram (EEG) as a research tool in human behaviour genetics: Psychological examinations in healthy males with various inherited EEG variants. III Interpretation of the results. *Hum. Genet.*, 1979, *47*, 81–111.

Vogel, F., Schalt, E., & Krüger, J. The electroencephalogram (EEG) as a research tool in human behaviour genetics: Psychological examinations in healthy males with various inherited EEG variants. II Results. *Hum. Genet.*, 1979, *47*, 47–80.

Vogel, F., Schalt, E., Krüger, J., Propping, P., & Lehnert, R. F. The electroencephalogram (EEG) as a research tool in human behaviour genetics: Psychological examinations in healthy males with various inherited EEG variants. I. Rationale of the Study, Materials, Methods, Heritability of test parameters. *Hum. Genet.*, 1979, *47*, 1–45.

Weitz. W. Studien an eineiige Zwillingen. *Z. Klin. Med.*, 1924, *101*, 115.

Wenger, M. A., & Ellington, M. The measurement of autonomic balance in children: Method and normative data. *Psychosom. Med.*, 1943, *1*, 365–371.

White, K. D., & Mangan, G. L. Strength of the nervous system as a function of personality type and level of arousal. *Behav. Res. Ther.*, 1972, *10*, 139–146.

Wilder, J. The law of initial values in neurology and psychiatry. *J. Nerv. Ment. Dis.*, 1957, *125*, 73–86.

Winter, H., Herschel, M., Propping, P., Friedl, W., & Vogel, F. A twin study on three enzymes (DBH, COMT, MAO) of catecholamine metabolism. *Psychopharmacology*, 1978, *57*, 63–69.

Woody, E. Z., & Claridge, G. S. Psychoticism and thinking. *Brit. J. Soc. Clin. Psychol.*, 1977, *16*, 241–248.

Wyatt, R. J., & Murphy, D. L. Low platelet monoamine oxidase activity and schizophrenia. *Schiz. Bull.*, 1976, *2*, 77–89.

Young, J. P. R., Lader, M. H., & Fenton, G. W. A twin study of the genetic influences on the electroencephalogram. *J. Med. Genet.*, 1972, *9*, 13–16.

Young, P. A., Eaves, L. J., & Eysenck, H. J. Intergenerational stability and change in the causes of variation in personality. *Person. Ind. Diffs.*, 1980, *1*, 35–55.

3 Genetics of Exploratory Behavior and Activity

Edward C. Simmel
Marjorie Bagwell

Miami University

How behaviors are labeled, defined, combined, and measured have a considerable amount of impact on generalizations that can be made from them and, consequently, on our understanding of their importance in the life and evolution of organisms. Although problems of definition and measurement are important in any area of behavioral investigation, they are especially critical in behavior-genetic analyses, for unless a behavioral phenotype is consistently and unambiguously measured, there cannot be consistency and unambiguity regarding its genetic determinants or its relationship to other behavioral, somatic, or morphological phenotypes. There are few areas of behavioral investigation in which this "phenotype problem" is so clearly shown as in the genetic analysis of activity and exploratory behavior, as we illustrate repeatedly, directly and indirectly, throughout this chapter.

Activity and exploratory behavior are fundamental, yet surprisingly complex behavioral phenotypes. Important in their own right for the organism, activity and exploration are also of importance to researchers in that each of these behaviors has been used as an index for more abstract behavioral concepts, such as emotionality and curiosity.

In this chapter, we have dealt solely with the genetic determinants of activity and exploratory behavior. To do so adequately we needed to go into some detail regarding definitional problems and techniques of measurement. We did not intend to provide a thorough review of the voluminous literature on these topics, but rather have emphasized studies that provide clear statements of genetic effects or of a genetic model, or that serve as definitive examples of a particular technique. An additional criterion in the selection of a major reference was to select, where possible, the study that would be most accessible to the greatest number of readers.

ACTIVITY

Meanings

Any movement of any part of the body of an animal is, according to a broad definition of the term, *activity*. However, as we use the term, and as it is most commonly used in the literature, "activity" refers to movement of the whole organism. In order to distinguish whole-organism movement from such behaviors as orienting response (head-turning responses, movement of the pinnae, or movement of the eyes), the term *locomotor activity* is often used. Fuller and Thompson (1978) recommend *ambulation* as a more unambiguous term. The problems with activity as a phenotype, however, derive not so much from nomenclature as from what this form of behavior represents, and how it is measured.

Activity as a Trait.   For several decades prior to the 1960s, activity *qua* activity was outside the mainstream of American psychology. During this era of the major behavioristic learning theories, the concept of "behavioral homeostasis" was in vogue. This concept held that the "normal" state of an organism was not behaving (i.e., being inactive), and that activity occurred only as the result of a "drive," that resulted from an imbalance in a physiological system, such as a tissue deficit or increase in hormonal activity. According to the assumption of behavioral homeostasis, then, one does not (indeed, cannot) account for activity as a general trait, because activity occurs only as a consequence of other internal and external events.

In recent years, the waning influence of the major learning theories, together with a number of disparate empirical findings (for example, exploration of their surroundings by satiated animals; the neural and bodily activity seen in many species during sleep), has led to a decline in the acceptance of the behavioral homeostasis assumption. There is an increasing degree of acceptance of its opposite: The normal state of an organism is behaving (i.e., being active).

When comparing the amount of movement of different individuals within similar settings, in the absence of specific physiological states, external goals, or noxious stimuli, one is investigating activity as a trait. If individual organisms do indeed possess specific "general activity levels" that yield strong predictions of other sorts of behavior or that are highly consistent across environmental settings and resistant to alteration, the theoretical and practical implications would indeed be great. Knowledge of the extent and nature of the genetic determinants of activity as a trait would magnify the importance of these implications.

Activity as an Index.   Activity has been of interest not only as a trait in its own right, but also as an index, or means of measurement, of more complex and difficult to define forms of behavior. The most frequent uses for activity measures in this regard are exploratory behavior, whiich we discuss at length in later sections of this chapter, and emotionality. As an index of emotionality, activity is often combined with other measures, especially urination and defecation. Stemming largely from the selection studies in rats by Hall in the early 1930s (Hall, 1951), emotionality is, in turn, an indication of the autonomic and behavioral responsiveness of organisms to potentially dangerous or noxious situations. Organisms are assumed to be either more active or much less active (e.g., "freezing") in these situations, depending on the species or the genotype within a species.

The advantages of using activity as an index are obvious: (1) activity is relatively easy to measure, with a number of measurement procedures yielding reliable, readily interpretable scores; (2) a considerable body of activity data exist for a variety of species of laboratory animals; and (3) a number of successful selection and strain-comparison studies have been completed, providing an abundance of subjects more or less "genetically standardized" on this phenotype.

The problems associated with using activity as an index are more subtle, but no less real. If activity serves as an index of a more complex characteristic, such as emotionality, there should be a strong correlation between activity and any other index measures. Such a correlation would indicate that the various phenotypes bear some causal relationship to one another; or that they are due to pleiotropic or linkage effects of the same genes; or, on the other hand, that the correlation between indices is a spurious, chance occurrence, perhaps furthered by measurement artifacts. When the correlation between indices is high for some selected lines but not for others, or for some inbred strains but not others, one must give the greatest credence to the last of these explanations (spurious correlation). We deal with this problem, a major factor in phenotype confusability, later in this chapter.

When activity is used as the sole index of a more complex phenotype, it is the same as operationally defining the more complex phenotype in terms of the measure of activity (for example, defining exploratory behavior *as* "squares crossed in an open field"). It then becomes the obligation of the "consumers" of such information to decide for themselves whether this sort of measure is appropriately definitive and adequately representative of the complex phenotype.

An example of the problems that can arise from the use of activity as an index can be illustrated by a study performed by Whimbey and Denenberg (1967), and further discussed by Denenberg in 1969. In the portions of this study relevant to our discussion, they placed rats in an open field briefly for 5 days and obtained a variety of measures that were then intercorrelated and factor analyzed. On Day 1, activity (squares crossed in the open field) correlated positively with defecation scores and could be considered an index of "emotionality." However, on Day 2 and thereafter this same measure of activity, taken under presumably the same conditions from the same subjects, produced a *negative* correlation with defecation and yielded different factor loadings. Then, the authors imply, activity becomes an index of exploratory behavior. Is activity a useful index for more complex behavioral phenotypes? As used here, hardly.

Measurement

The Activity Wheel.   This simple, time-honored device is still available from a number of laboratory equipment suppliers in sizes suitable for rats, mice, and other small animals. These devices usually operate a counter that provides an easily understood and interpreted measure of activity expressed as the number of revolutions per unit of time. (Most commercial activity wheels will run both clockwise and counterclockwise, and will operate the counter when turned in either direction.) In addition to obvious extraneous variables such as room temperature, illumination level, and the like, the major variables to be controlled with this type of apparatus are: (1) the degree of effort needed from the subject to overcome friction in order to turn the wheel; and (2) the degree of choice available to the animal to move freely from an adjoining cage into the wheel. Assuming the greatest possible control, the activity wheel provides what is, perhaps, the purest measure of ambulation, at least in terms of energy output. However, the very conditions providing the control and "purity" of this measurement technique also make it the most artificial; ordinarily, the movement of an organism results in a changing environment for the animal as it progresses. Inasmuch as running in the wheel does not lead to changes in the animal's environment and does not therefore result in exploratory or neophobic excursions, it is not surprising that the "distance" run in such a device is often greater than the distance traversed under more natural conditions. Perhaps this type of activity, unlike that measured by other means, is primarily exercise.

Maze Running.   Where the activity wheel allows only totally circumscribed locomotion, the maze (an alley Y maze is most commonly used) allows some degree of directionality of movement. The maze thus measures ambulation in a relatively more naturalistic fashion than does the activity wheel, especially for thigmotaxic and burrowing species, such as rats. However, it is possible that an animal moving through an unfamiliar maze is also exploring its environment, or seeking an escape route, or is engaged in some other form of behavior that involves ambulation. Therefore, attempting to measure *activity* in a more realistic manner can create questions as to just what is being measured.

Activity Meters.   These are automated measuring devices that unobtrusively detect and count movements of animals. Newer versions consist of a platform containing a series of electromagnetic sensors upon which any nonmetallic container with one or more animals inside can be placed. Any movement by the animal operates a counter, with the degree of movement required for detection controllable by setting the sensitivity of the sensors. The activity meter has the unique property of permitting measurement of movement in any sort of box or container that will fit on its surface, including home cages. Another unique property of this type of device is that it measures all movement by the subject, rather than merely ambulation from one location to another. Although this has the advantage of being a meaningful measure of activity as a trait, it dictates caution when comparing scores with those obtained in other types of activity-measuring devices. The unique and specific conditions that must be controlled when using an activity meter involve the placement of the animal container over the sensors and the sensitivity setting of the meter. Although the raw scores taken from the counter activated by movement over the sensors are difficult to interpret, meaningful and statistically reliable ordinal data have been obtained with this device in studies comparing genotypes and experimental conditions.

The Open Field. This is the most frequently used technique for the measurement of activity, a fact that should have resulted in a major advantage for its use: the establishment of reliable standardized scores to serve as a basis for comparison across genotypes and experimental conditions. There is also a high degree of face validity for its use in the measurement of activity as a trait—if ambulation within an open arena is not activity, what is? The open field can be constructed in a variety of shapes (round, square, and rectangular being the most popular) and made in sizes appropriate for almost any species. More limited versions of this apparatus are the straight runway and the shuttle box, which share many of the advantages and disadvantages found with the open field.

Unfortunately, despite its apparent advantages, the open-field test has not turned out to be free of severe problems. As Walsh and Cummins (1976) point out in their thorough and critical review of the many studies using this technique: "Apparatus, techniques, subjects, parameters, analyses, and interpretations have diversified enormously, yet in the literature, sweeping generalizations and conflicting interpretations continue to be made on the basis of univariate studies [p. 182]." Table 3.1 lists the variety of behavioral and autonomic measures that have been taken, although number of squares crossed seems to be the most common measure in most behavior-genetic studies. One of the difficulties referred to by Walsh and Cummins results from the attempt to compare studies that have used different measures.

Many independent and dependent variables have been tested, and there are numerous and sometimes inconsistent effects due to several conditions inherent in the testing situation, such as size of the open field, level of illumination, or prior experience in the apparatus. When genetic studies are reported in which such conditions are not standardized or systematically varied, it is difficult to see how valid comparisons and generalizations can be made.

TABLE 3.1
Classification of Open-Field-Dependent Parameters
(Adapted from Walsh & Cummins, 1976)

I. Behavior

  A. *Whole or major body movement*

    1. Type of movement

      a. Distance covered per unit time
      b. Time spent in ambulation
      c. Rearing frequency
      d. Escape attempts
      e. Latency (usually time taken to leave start area)
      f. Time spent without movement

    2. Locations

      a. Field area visited (inner or peripheral areas, corners, etc.)
      b. Affiliation (distance from partner subject)
      c. Stimulus interaction (e.g., distance from stimulus object)

  B. *Part body movement*

      a. Manipulation of objects
      b. Sniffing
      c. Scratching
      d. Digging
      e. Teeth chattering
      f. Grooming
      g. Vocalization
      h. Visual exploration

II. Autonomic nervous system

      a. Defecation
      b. Digestive transit time
      c. Urination
      d. Heart rate and rhythm
      e. Respiratory rate

Despite its high face validity, ease of scoring, and long history of usage, the sensitivity of subjects to changes in experimental conditions in open-field testing suggest that this technique is not the *sine qua non* for the determination of "general" activity levels. Because there are a number of questions as to just what is being measured in the open field, we suggest that studies using open-field measures (especially those using a single dependent variable) be interpreted with caution.

Genetic Differences

The presence of a genetic effect was originally established for activity through selective breeding of animals showing high activity scores, low activity scores, or both on a standardized measuring technique. One of the earliest attempts to establish the hereditary basis of activity using this method was a selection study by Rundquist (1933), in which he selected rats for high and low activity following testing in an activity wheel. His results (high-line versus low-line differences) were disappointing initially, because of an error in the method of selection he employed: the mating of high-active and low-active offspring from nonselected parents. That is, rats showing high activity scores were selected not only from parents that also had high activity scores, but from those yielding low activity scores as well, with the reverse holding true for rats with low activity scores. As a result, there was no differentiation between groups from generations F_1 to F_4. When this error was corrected, differentiation into high and low active groups did occur and laid the foundation for further research in this area.

Hall (1938, 1951) performed selection experiments in the early 1930s to establish the genetic basis of emotionality in rats using as phenotypes the behavior of the rats in an open field (urination and defecation, primarily). Although he was not concerned with activity per se, his selection procedures and his use of the open-field test have served as the prototype for many of the modern studies on this problem.

In the 1950s, it became possible to use another method for determining hereditary basis for activity and for any other phenotype. This method involves finding differences in activity between strains of genetically controlled animals, usually highly inbred strains of mice and rats, but also established breeds of dogs and inbred strains of various species of *Drosophila*.

Thompson (1956) tested five strains of inbred mice (C57BR/a, C57BL/6, AKR, BALB/c, A/J) on two activity measures, the open field and the Y maze. His results showed marked differences in the activity levels of the five strains, and the two C57 strains showed much higher activity levels in both open-field test and the Y maze than did the BALB/c mice.

In mouse studies, McClearn (1961) has shown that there is a clear relationship between locomotor activity and the percentage of C57BL/Crgl genes present (from backcrosses with A/Jax mice). Thompson (1953) illustrated that there is considerable variability between strains in exploratory activity, with C57BR/a, C57BL/6, and C57BL/10 having the highest mean score (for activity) and A, AK/e, and BALB/c strains having the lowest. These orderings were fairly invariant with another testing apparatus (enclosed arena versus Y maze). McClearn (1960) noted that there are differences in the ordering of mice on exploratory activity with different environmental conditions. For example, he found that in bright white light, A/Crgl mice showed less exploratory activity than in dim red light. In the C57BL/Crgl mice, activity increased in white light and decreased in dim red light. Although he found no change in the ordering of the strains, McClearn did find a significant interaction between strain and illumination.

Differences have also been found among strains of inbred rats. In a study by Harrington (1972), the Iowa Nonreactive (INR) strain was found to be the most active and Tryon Maze-bright (TS1) rats the least active in open-field ambulation among the 12 rat strains tested: ACI, August 990 (A990), A35322, Fischer 344

(F344), Iowa Nonreactive or Hall Nonreactive (INR), Iowa Reactive (IR), Maudsley Nonreactive (MNR), Maudsley Nonreactive Albino (MNR-a), Maudsley Reactive (MR), Tryon Maze-bright (TS1), Tryon Maze-dull (TS3), Wistar Albino (WAG). A replication was completed recently (Harrington, 1979b) with an additional strain (B) included that had high ambulation scores. Otherwise, the results were the same as those found in the 1972 study: INRs were highly active, followed closely by WAG and MNR-a rats. Again, the TS1s were the least active. In home-cage movement, TS3s were most active, but the INRs are highly active on this measure also. The MNR/Hars are least active in the home-cage environment, but are highly active in running-wheel activity (Harrington, 1971a).

Of particular historic and current interest are the MR and MNR rats originated by Broadhurst (1958, 1969). Using a procedure essentially similar to that used by Hall more than two decades earlier for selection for emotionality, Broadhurst made several important improvements; the open field was more uniformly and brightly lighted and initial responses of the subjects were weighted more heavily, giving less emphasis to the possible adaptation of the rats to a familiar environment. MR and MNR rats clearly differ on open-field defecation scores, an index of emotionality, and are often regarded as "benchmark" strains for this category of behavior. However, on several *activity* measures, these strains do not differ greatly.

Selection studies illustrating genetic differences in activity have also been carried out in species other than rats and mice. The comparison of the normal behavior of different established breeds of dogs provides an excellent example; for "everyday" differences in activity levels, one need only compare a bassett hound and a fox terrier. Dogs have also been used as subjects in more systematic genetic studies, with different breeds treated analogously to inbred strains of rodents. A prime example of this sort of research is the 19-year study conducted by Scott and Fuller (1965) that dealt with, among other things, the emotional behavior of several breeds of dogs, together with their hybrids and backcrosses. Clear genetic differences were found in the five breeds (Basenji, beagle, cocker spaniel, Shetland sheepdog, and fox terrier) on a variety of reactivity tests, including open-field behavior.

Some work on activity and its implications has also been done with the fruit fly (especially *Drosophila melanogaster*). In one study (Manning, 1961) flies were selected for mating speed. Activity differences for these two selected strains (fast mating speed and slow mating speed) were measured by counting the number of squares a fly entered in an open field arena over a given period of time. It was found that slow mating lines exhibited more open field activity than did fast mating lines. In contrast, Ewing (1963), in selecting for spontaneous activity, found that the less active lines displayed *more* sexual behavior than did the more active lines. Although these results appear to contradict Manning's findings, the discrepancy appears to have been due to an "apparatus effect," as he used an apparatus other than the open field used by Manning. When Ewing replicated his study using the type of open field apparatus used by Manning, no significant differences between the selected lines were found.

Besides being alterable by apparatus effects and other environmental changes, activity is readily changed, often differentially for different genotypes, by internal events, especially pharmacological agents. A thorough review of drug effects on activity, and the genetic factors involved, has recently been prepared by Broadhurst (1978).

The studies we have cited are typical of many others to be found in the literature, in that many of the findings are contradictory with one another, with varying differences among genotypes (whether inbred strains or selected lines) depending on procedural details, types of apparatus, and so on. Therefore, generalizations on activity levels of specific genetically controlled animals should be treated cautiously.

An excellent example illustrating this point can be found in an article by Lassalle and Le Pape (1978). They demonstrated that the environmental conditions associated with testing of two inbred mouse strains (C57BL/6 and BALB/c) can reverse the general finding we referred to earlier that C57s are more active than BALB/cs. Under seminatural and breeding cage conditions, the BALB/cs showed a higher level of activity than did the C57s, and in the seminatural condition, the BALB/cs were more active in the dark (Table 3.2).

Their concluding point is worth repeating: "A clear implication of the current study is simple and obvious, although often overlooked and neglected. Strains cannot be said to possess a given behavioral trait. In each instance it is necessary to specify precisely the environmental conditions in which the trait was measured [p. 376]."

TABLE 3.2
Locomotor Activity: Amount and Distribution[a]

	BALB/c		C57BL/6
Seminatural environment			
Diurnal activity	323	<	881
Nocturnal activity	7111	>	3462
Total activity	7325	>	4828
Percent of total activity occurring at night	94	>	75
Breeding cages			
Diurnal activity	1006	>	589
Nocturnal activity	2732	>	2153
Total activity	3734	>	2662
Percent of total activity occurring at night	72.5	≃	78

[a]The values are medians of the different groups. Where indicated, differences are significant at $p = 0.039$ (Fisher test). (From Lassalle & Le Pape, 1978, p. 373)

In an issue related to the genetic determination of activity, several studies have attempted to determine heritabilities and degrees of genetic determination (see DeFries & Hegmann, 1970). However, as these findings are limited to the specific genetic population measured within a specific type of apparatus and a specific set of environmental conditions, these data are of rather limited use in the generalizable information they provide for an understanding of the genetic basis of activity.

Genetic Models

There are various avenues of research that have suggested that a polygenic model can best account for activity. One form of evidence for this type of genetic determination can be found in score distributions; typically, a continuous, unimodal distribution of scores has been found for most types of activity measurement when genetically heterogeneous animals were tested. Tryon (1942) suggests a polygenic model on the basis of this form of evidence alone.

An excellent example of a more thorough analysis supporting a polygenic model can be found in the series of studies reported by DeFries and Hegmann (1970). They tested large numbers of two inbred strains of mice (C57BL/6J and BALB/cJ), as well as their reciprocal F_1 and F_2 hybrids and backcrosses. The mice were tested for 3 minutes on each of 2 successive days in an automated, white Plexiglas open field, 90 cm^2. The means and variances of the transformed scores are shown in Table 3.3.

The degree of increase in variance from the F_1 to the F_2 generation is inversely related to the number of loci affecting a trait. It can be seen in Table 3.3 that the

TABLE 3.3
Open-Field Activity Scores of C57BL/6J, BALB/cJ, and
F_1 and F_2 Hybrid Mice
(Data from DeFries & Hegmann, 1970, p. 33)

Generation	Genotype[a]	Means[b]		Variances	
		Males	Females	Males	Females
Parental	C	4.67	4.25	6.33	4.27
	B6	15.73	16.40	4.86	6.56
F_1	B6C	15.09	15.04	6.70	10.37
	CB6	13.98	13.66	15.80	21.06
F_2	CB6 x CB6	11.88	12.63	17.96	9.23
	CB6 x B6C	11.76	12.30	16.56	13.68
	B6C x CB6	12.66	12.86	15.87	16.44
	B6C x B6C	11.58	11.84	14.22	16.28

[a]B6 = C57BL/6J; C = BALB/cJ; female parent listed first among hybrids.
[b]Square root of the sum of squares crossed over two-dat test period.

TABLE 3.4

Mean[a] and Variance of Activity and Defecation Scores of
Albino and Pigmented Animals in Generations F_2–F_5,
Pooled Across Generations
(From DeFries & Hegmann, 1970, p. 49)

	Activity		Defecation		
	Mean ± S.E.	Variance	Mean ± S.E.	Variance	N
Pigmented	12.72 ± 0.18	18.65	1.78 ± 0.03	0.55	1729
Albino	10.44 ± 0.10	17.71	2.13 ± 0.02	0.49	557

[a]Unweighted mean of scores for ♀s and ♂s in each generation and line (generations F_4 and F_5 are generations S_1 and S_2 of the selection experiment).

variance between the F_1 and F_2 generations is very slight, especially when compared with the difference in variances between the parental strains and the F_1 generation. (One can discount the variance differences between the reciprocal F_2s as a maternal effect, especially as it diminishes in later generations.) DeFries and Hegmann conclude that activity is controlled by genes at several loci, suggesting *at least* 3.2 loci from an earlier analysis by Hegmann.

In the same paper, DeFries and Hegmann report on an activity selection study they completed. As a foundation population, they used the F_3 generation from the C57BL/6J and BALB/cJ mice used in the study reported earlier. Using an open-field activity measure similar to that used in their earlier study, they were able to obtain two highly active, two low-active, and two control lines, with each selection (and its replication) being successful within six generations. Of particular interest for a genetic model for activity, however, was their finding that albino rats were less active than those with any other coat coloring (agouti, cinnamon, black, and brown). These data (see Table 3.4), together with their previous findings that the albino BALB/c mice are less active than the pigmented C57BL/6 mice, led DeFries and Hegmann to suggest that: "Although open-field behavior is clearly influenced by genes at many loci, a major gene effect was found: albino Ss have lower activity and higher defecation scores than pigmented animals [1970, p. 53]." The influence of the albinism (c) locus has also been suggested by several other authors.

This is plausible from the data presented, although it is possible that the effect found may be due to genes linked to the c locus in certain strains, with activity not actually moderated by the c locus itself. The finding that some albino (cc) mice do not exhibit a reduction in activity warrants caution in making generalizations about the "effects" of albinism. It has been found, for example, that albino LG/J mice are highly active (Simmel, Haber, & Harshfield, 1976; Simmel & Walker, 1972), and that albino BALB/cBy mice do not show reduced activity when tested on an activity meter (Simmel & Eleftheriou, 1977). DeFries and Hegmann used BALB/cJ mice, as did several of the other studies supporting the "albino effect." Perhaps there is linkage of the c locus with other genes that reduces activity in this strain (at least under some conditions of measurement— see the study by Lassalle and Le Pape referred to earlier). There could also be a different sort of linkage with the reverse effect in highly active albino strains such as the LGs mentioned previously. At this point, however, attributing reduced activity to the lack of pigmentation seems premature.

EXPLORATORY BEHAVIOR

Measurement and the Phenotype Problem

Exploratory behavior is frequently confounded with, measured by, or operationally defined as ambulation (i.e., activity), most often in an open field. Defining and then measuring exploratory behavior in this fashion creates obvious problems for any attempt to determine the nature of the genetic factors underlying exploration. These problems are especially severe for the determination of specific genetic models. If one accepts the polygenic systems most often found for activity (as discussed in the previous section of this chapter), and if one accepts the definition of exploratory behavior as ambulation within an open field (as many do), then the same polygenic models established as underlying activity also hold true for exploratory behavior. Following this rationale, the phenotypes—activity and exploratory behavior—are logically and empirically identical.

If activity and exploratory behavior have been shown to be identical—merely two terms for the same phenotype—our chapter could end at this point. Clearly, howevever, this is not the case. Although rather similar behavioral measures and techniques have been used, the nature of the results obtained and especially the genetic models that these results have suggested point to the conclusion that activity and exploratory behavior are distinguishable phenotypes.

Even though ambulation through an open field or movements back and forth in a shuttle box are the most frequently employed measures, there is a subtle difference when these behaviors are used to investigate exploration rather than activity. One way to separate *exploratory* activity from general activity is to measure ambulation for brief (usually 5-minute or 10-minute periods) single trials, or to count only those movements occurring when the subject is first placed in the apparatus. The assumption here is that exploratory behavior is the activity that occurs in a novel environment. Another method is to measure different *kinds* of movement (e.g., rearing, as opposed to ambulation), especially when barriers or other objects are placed in the novel environment. Birke (1979) provides a good recent discussion of the aforementioned problem, and offers a similar method (using specific object exploration) for improving the measurement of exploratory activity.

The likelihood that these changes in procedure have resulted in measuring something other than "mere" activity is supported by the nature of the genetic models suggested. Although polygenic systems seem to be the rule for activity (DeFries & Hegmann, 1970), it is not uncommon to find that exploration is controlled by a single locus, as illustrated by the two research programs described next.

Van Abeelen (1975, 1977a, 1977b). Two inbred strains of mice (SRH and SRL) were developed from two lines originally selected for high and low frequency of rearing behavior in a novel environment. The novel environment consisted of an open field that sometimes contained an empty food hopper. Van Abeelen (1977b) created two backcross populations by crossing the high-rearing-frequency SRH and the low-rearing-frequency SRL mice, respectively, with mice from an inbred strain (DBA/2; one of the progenitor strains for the original selection study). When he then tested the intercrosses (offspring of the two backcross generations) in the novel enviornment, the rearing scores formed a clearly bimodal distribution, indicating the likelihood that a single locus controls this phenotype. There was also a bimodal distribution of open field activity scores, although not as striking as that obtained for rearing frequency. This, together with the high correlations between rearing frequency (vertical activity) and open-field movement (horizontal activity), led van Abeelen (1977b) to conclude that the two phenotypes are both under monogenic control, and that: "The genetic factor might be allelic with the gene *Exa* [affecting "exploratory activity level"] discovered by Oliverio, Eleftheriou, and Bailey [1973] [p. 403]."

It does not seem farfetched to consider rearing up to investigate even a familiar stimulus (an empty food hopper) in a novel environment as a form of exploration. The argument in favor of a single-locus controlling activity, and a necessary causal connection between activity and exploration, is considerably less compelling. First, mice ambulating in a novel enviornment are not necessarily being active for the same reasons as those ambulating through a more familiar environment. Second, if mice of two genotypes differ to the extent to which they will rear up to investigate an object, they would also be apt to differ in the horizontal activity needed to approach these objects initially. (We assume they do not walk on their hind legs.) Thus, the "causal relationship" between exploration (rearing) and activity (open field ambulation) could well be an artifact of the experimental setting. Nonetheless, van Abeelen's studies do provide a distinctive approach to the genetic analysis of one type of exploratory behavior.

Oliverio, Eleftherious, and Bailey (1973). Using a completely different technique for genetic analysis, and also a different behavioral testing procedure, Oliverio et al. (1973) also reported that what they termed "exploratory activity" is controlled by a single locus. The genetic technique they used involved the use of recombinant inbred (RI) strains and congenic histocompatibility lines of mice to establish a precise genetic model. (The use of RI strains is discussed in an earlier chapter of this volume and reviewed later in this chapter.) In the Oliverio et al. study, mice of the seven CXB (Bailey) RI strains, together with their C57BL/6By (B6) and BALB/cBy (C) progenitors and reciprocal F_1 hybrids were tested in a small shuttle box for 10 minutes. Shortly before testing, each subject received an intraperitoneal injection of saline. The scores on "short-term exploratory activity" were measured by crossings from one end of the shuttle box to the other. The results yielded a strain distribution pattern, confirmed by the testing of a congenic histocompatibility line, showing that short-term exploratory activity is controlled by a gene (*Exa*) linked to the *H-26* locus on chromosome 4 (Linkage Group VIII). In this same study, they found that exploratory activity is changed (for several strains, reversed) by the effects of increasing dosages of scopolamine. A single gene (*Sco,* linked to the *H-2* locus on chromosome 17, LG IX) controls the changes in exploratory activity due to the administration of scopolamine. Changes in exploratory activity following injection of amphetamine were found to be due to a polygenic system. In a later study (Oliverio & Eleftheriou, 1976), the genetic model for short term exploratory activity (following a saline injection) was replicated, and the locus for its modification by ethanol was found: *Eam,* linked to the $H-16^c$ locus on chromosome 4 (LG VIII).

These studies are important for several reasons. The 1973 study was one of the first to make use of recombinant inbred strains (and the related techniques) to establish a specific genetic model for a *behavioral* phenotype. Furthermore, these investigations have demonstrated an important method for comparing the genetic determinants of various pharmacological agents on behavior.

Unfortunately, the influence of these same studies on our understanding of exploratory behavior has been largely negative, due to the misapplication of the phenotype. The 1973 paper has been widely cited, and the gene discussed in that paper was named *Exa,* which has led to the conclusion by some that the locus controlling exploratory behavior (or exploratory activity) has been determined. But were Oliverio et al. measuring exploration?

It seems unlikely that either exploratory behavior or general activity is involved to any great extent in whatever behavior is controlled by *Exa.* It hardly seems credible that a small shuttle box, hardly longer than the lengths of three mice, could serve as a novel environment for as long as 10 minutes, or even a fraction of that time. Evidence against considering shuttle-box activity to be a measure of exploratory behavior is found later in this chapter in the discussion on stimulus reactivity. Satiation to much more complex settings has been shown to occur within that time span in several species, including mice, as is also discussed in the following section. It would also be going beyond the data to term shuttle-box crossings general activity, as Oliverio et al. do, in the absence of correlated measures.

What Oliverio et al. could well have been measuring instead of exploratory activity might have been an activity response to stress or painful stimulation.

Remember that shortly before testing, each subject received an intraperitoneal (IP) injection of saline. We attempted to replicate this study in our laboratory, using the same strains of mice and the identical apparatus and procedure, but *without an IP injection*. The results we obtained were vastly different, with not only many differences in the rankings of the strains, but also a strong indication of a polygenic effect (unpublished data). There is other evidence that IP injection can have major effects on behavior. For example, blank IP insertions can greatly mitigate the severity of audiogenic seizures in DBA/2 mice (Richardson & Simmel, 1973). There seems to be sufficient contrary evidence to dispute that *Exa* is directly concerned with either exploratory behavior or exploratory activity.

Exploration of Novel Stimuli

In an earlier part of the chapter we wrote of an animal's activity—its movement in a particular apparatus—but no mention was made as to why the animal moves about. It could be a spontaneous behavior as a result of physiological changes in the animal—one theory (drive-reduction) includes this presumption. Perhaps there is something of "interest" to the animal that motivates it to explore its surroundings. Before discussing exploratory behavior studies, it is important to understand the various components that are involved. Stimuli have several properties that can be varied to affect behavior: novelty, complexity, and the interaction of these two. The novelty of a stimulus is quite important in the initiation of exploratory behavior of that stimulus. The more novel a stimulus, the greater the likelihood that it will be approached and examined by the animal. Given a choice between two stimuli, the subject generally will choose to examine the more novel of the two (Berlyne, 1955; Thompson & Solomon, 1954). There appears to be a bell-shaped curve in the effect novelty has on exploratory behavior. There is a degree of intermediate novelty that produces maximum exploration; too much or too little and the animal becomes indifferent to the stimulus or spends less time examining it (Berlyne, 1960). One characteristic of novelty is that of change; the changing of a familiar stimulus makes it more novel. This ties in with the concept of satiation or familiarization. If the animal is familiar or satiated with the stimulus, it will normally spend less time exploring it. If, however, some aspect(s) of the stimulus is/are changed, the animal will examine it for a longer period of time. The satiation of the subject to a stimulus or environment poses a problem in the measurement of exploratory behavior, as mentioned previously. As the trial or observation continues, the stimuli or environment involved in the experiment becomes more familiar and less novel until it loses its novelty almost completely. So the behavior recorded at the start of the trial is more likely to be indicative of exploratory behavior than is the behavior at the end of the trial. Studies by Berlyne (1960), Montgomery (1952, 1953), and Thompson and Solomon (1954) have shown that exploration declines with time in a trial where the subject is exposed to the same object or environment for some specified period of time. Furthermore, Thompson and Solomon (1954) also found that when rats were confined in a limited space (to which they were not habituated) for a period of time, exploration initially increased before it began the ultimate decline. This could be due to an interaction between fear and exploration and makes studies using such techniques (such as shuttle boxes) more vague in the measurement of exploratory behavior.

Another study by Berlyne (1958) illustrated the attraction of novel stimuli over familiar stimuli. Using rats as subjects, he projected pairs of pictures of animals side by side on a screen for 10 seconds, for 10 trials. One animal in the pair reappeared constantly on one side, whereas the other side had a different animal projected each time. Berlyne found that the subjects spent more of the 10 seconds fixating on the novel pictures on the varying side than on the recurring picture.

Once the organism approaches the stimulus, the complexity of the stimulus becomes a major determinant of attention and exploration by the subject. If two stimuli are both novel, the subject will generally react more to the more complex of the two. A complex stimulus contains more variety than simple stimulus, and thus provides more ways for the subject to group the various components and come up with different responses. Several components seem to comprise the attribute of complexity (all other things being equal): the number of distinguishable elements; an increasing dissimilarity between those elements; and an inverse relationship between complexity and the degree to which several elements are responded to as a unit. That is, the perceptual organization of the stimuli by the subject has a bearing on its degree of complexity (Berlyne, 1960).

Clearly, the novelty and the complexity of a stimulus will integrate with each other to give each stimulus a particular characteristic. Both are variables that influence exploratory behavior, but the methods for manipulating each are different. The modification of a previous stimulus results in novelty and implies that the organism has an expectancy about the stimulus that is not fulfilled (when it views the modified novel stimulus). A succession of events (stimuli) containing item(s) with short-term novelty will probably be more complex during the presentation than a succession that is purely repetitious. In addition, "a pattern with a high degree of synchronous complexity will probably have a high degree of relative novelty [Berlyne,1960,p.43]." Assessing the involvement of novelty and complexity in experiments is quite difficult, as these two variables generally are confounded. Dember, Earl, and Paradise (1957) offer a nice illustration of both variables. In this experiment, two circular pathways were joined together to form a figure-eight apparatus. One of the pathways had its walls lined in black and white horizontal stripes, and the other walls of the second pathway were either homogeneously black or white. Sixteen rats were run for 60 minutes per day for 5 consecutive days. They found that on the first day, eight rats preferred the more complex pathway (the one with the horizontal stripes). On Day 2, four more rats preferred the complex pathway, and the last four rats shifted preferences on the third day. None of the subjects regressed to a preference for the plain pathway. These results indicate that "any change in preference will be from the less to the more complex path [Dember et al. 1957, p. 514]."

It can be seen that even though novelty and complexity are operationally different, they can be considered to be similar in their effects.

The environment in which an organism responds is as much a complex of stimuli as are the stimuli to which it is presumably responding. If the environment is unfamiliar, the organism may be responding to its novelty rather than the properties of the stimulus. Novel environments may unwittingly be created by the experimenter, as in the study by Kivy, Earl, and Walker (1956). They introduced rats into a T-maze with the goal arms blocked by glass partitions at the choice point. The rats were satiated over varying exposure times and the goal arms were either both black or both white. On the test trial the color of one of the arms was changed and the glass partition removed, allowing the subject to choose between a black or white alley. The results were as expected; the subjects chose the novel stimulus (the changed arm), indicating a "curiosity motive"—as Kivy et al. stated it. Or it could be that the animals chose the changed arm as a result of being satiated by the other arm. Dember (1956) gave support to the novelty of environment hypothesis by reversing the satiation and the test trials of the Kivy et al. experiment. What this points out is that a change in the environment, such as brightness in this case, can result in a different response by the subject. Dember's experiment predicted and found the same behaviors as the Kivy et al. experiment, but provided a different interpretation to those findings. This shows that there can be a novel environment or novel stimuli in a familiar environment. Granted, the brightness change in the aforementioned experiment is a stimulus as well as part of the environment. This illustrates the problem in controlling as many aspects as possible so that a particular response can be accurately attributed to the stimulus or to the environment, depending on the situation.

Responsiveness to Novel and Complex Stimuli

Exploratory behavior can be viewed from several different perspectives: approach situations (exploration) and avoidance. Avoidance to novel stimuli generally taps into fear responses and degree of conflict. In these cases, the exploratory behavior exhibited by the subject is such that the animal avoids a particular stimulus or environment. Normally, these studies are confounded, as it is difficult to tell if a subject is not moving because of fear or because of lack of interest. This situation is not as commonly used as approach measurement, but has been used in several studies (Harrington, 1979c, 1979e). Regarding exploratory behavior as an approach situation is more common. In most exploratory behavioral studies, the animal can either approach a novel stimulus or not, and the resulting behavior is taken to be an indication of the degree of novelty and complexity of the stimulus.

Genetic Differences in Exploratory Behavior. As in activity, genetic differences have been found in exploratory behavior. Harrington (1979a–e) has done many studies on various measures of exploratory behavior in the rat. In runway learning, he has shown that MNRA, INR, TS3, and WAG strains were the fastest (see the section of this chapter on Activity). The F344 had very poor runway performance (Harrington, 1979d). He also found strain differences in rats in the exploration of different environments, with MNR, ACI, and MNR-a lines having the highest latencies of emergence from a home cage into a novel environment (Harrington, 1971b). Harrington brings up the dilemma of confusing a "novel environment" with novel stimuli within a familiar environment and admits that this problem still exists in the literature (Harrington, 1979a).

McClearn (1959), also using six inbred strains of mice (C57BR/cd, C57BL/10, LP, AKR, BALB/c, A/J), employed four measures of activity: arena, hole-in-wall, open-field, and barrier, to determine the generality of genetic differences. Again, both strains of C57s were found to be more active than BALB/c mice on all but the hole-in-wall measures. In 1972, Simmel and Walker demonstrated that exploration of novel stimuli can be differentially enhanced in different strains by the presence of a companion. For other exploratory activity studies with inbred mice, see Wimer and Sterns (1964) and Wimer and Fuller (1965).

We have shown that the investigation of exploratory behavior as responsiveness to novel/complex stimuli is certainly not new. Some of the earlier studies, including those discussed in the preceding section, have been widely cited over the years in a number of theoretical contexts. (See also the review by Barnett, 1975; and the critical comments by Henderson, 1980.) Despite this, most behavior genetic studies have used activity measures for the investigation of exploratory behavior, with the resulting confusion regarding the appropriate phenotype.

Stimulus Reactivity

We devote the remainder of this chapter to a discussion of a technique that permits the direct measurement of the responsiveness of an animal to novel stimuli (*stimulus reactivity*) separately from "general" locomotor activity.

Procedure. The apparatus for the measurement of stimulus reactivity is shown in Fig. 3.1. It consists of a 20-cm square arena constructed of black Plexiglas. The arena is diagonally bisected by a partition, also of black Plexiglas, with a 1-cm^2 opening in the bottom center. This allows the subject free access between the two halves of the arena. One-half of the arena (*the novel side*) has murals consisting of black and white diagonal stripes, vertical stripes, and check-

FIG. 3.1.   Arena used for measurement of stimulus reactivity and activity. Arena is placed on activity meter for automatic measurement of gross activity.

erboard designs. The other half (*the plain side*) has no stimuli on the walls. The entire arena is placed atop the sensors of an automated activity meter in such a fashion that identical movements on either half of the arena will yield identical responses by the meter's counter.

A 10-minute trial begins when a subject is placed on the plain side of the arena. Two experimenters observe the subject by viewing a mirror angled atop the apparatus. Operating a counter and a stopwatch, the following measures are taken: Latency (LAT), the time between the start of the trial and the subject's initial entry into the novel side; novel-side time (NOV), the number of seconds (of the 600-second trial length) the subject spends on the novel side of the arena; arena crossings (ARX), the number of crossings through the hole in the center partition between the novel and plain sides; and activity (ACT), the number of movements by the subject, as recorded on the counter of the activity meter. Under this procedure, it is *possible* to obtain almost any combination of scores on the four measures taken. For example, a subject can be highly active on the plain side without ever venturing into the novel side; or can spend nearly all its time, inactively, on the novel side; or dart quickly back and forth between the two sides, and so on.

Multivariate Analysis of Response Measures.   The first study using this procedure determined the relationship among the several dependent measures (Simmel, 1976; Simmel & Eleftheriou, 1977). A total of 66 male mice, 50–70 days old, were tested. The mice represented 11 different genotypes (six mice from each genotype): C57BL/6By, BALB/cBy, B6Cf₁, CB6CF₁, and the seven CXB RI strains: CXBD, CXBE, CXBG, CXBH, CXBI, CXBJ, and CXBK. All subjects were tested using the procedure described and, in addition, immediately after this testing, were placed in a shuttle box similar to that used by Oliverio et al. (1973) for 10 minutes. The four arena measures (LAT, NOV, ARX, and ACT) and shuttle-box crossings (SHX) were intercorrelated and subjected to a principal components factor analysis. The factor loadings are shown in Table 3.5 It can be seen that two factors emerged. Two of the arena measures, latency and novel-side time, had high loadings on Factor I (stimulus reactivity) and loadings of nearly zero on Factor II (undifferentiated activity). One arena measure, gross motor activity, had a high loading on Factor II and the lowest loading among the four arena measures on Factor I. Arena crossings had relatively high loadings on both factors. Both measures having high loadings *only* on Factor I (latency and novel-side time) are behaviors dependent on the subjects' responsiveness to novel stimuli and are thus considered indices of stimulus reactivity. They are also relatively independent of gross motor activity. Arena crossings appears to measure *both* stimulus reactivity and activity. Notice that shuttle-box crossings appears to be independent of stimulus reactivity and to be primarily a measure of motor activity.

Genetic Model.   The factor structure found in the study just described clearly showed that stimulus reactivity (response to novel stimuli) can be distinguished from activity. Next, it was necessary to establish as precise a genetic model as possible, in order to determine the "fit" between the nature of the genetic determinants of the four measures and their loadings on the two factors obtained. The technique used (strain distribution pattern—SDP) for this purpose is described in an earlier chapter in this volume and is summarized in the following section.

The technique referred to was developed by Bailey (1971) and involves the use of recombinant inbred (RI) strains for the determination of strain distribution

TABLE 3.5
Factor Loadings of Stimulus Reactivity and Activity Measures
(Data from Simmel & Eleftheriou, 1977)

Variable	Factor I	Factor II
LAT (latency to first novel entry)	-.641	-.011
NOV (novel-side time)	.506	-.122
ARX (crossings from plain to novel side)	.586	.302
ACT (gross locomotor activity)	-.253	.414
SHX (shuttle box crossings)	.206	.616

patterns (SDPs). The Bailey RI strains were derived from a cross of two unrelated but highly inbred progenitor strains, BALB/cBy (C) and C57BL/6By (B6). They are then maintained from the F_2 generation onward by strict inbreeding. Inbreeding of the independent RI lines obtained from this cross results in genetically fixing, as full homozygosity is approached, the chance recombination of genes that occurred after the F_1 generation. Unlike conventional crosses from inbred strains, this technique provides a continuing (and thus, replicable) supply of genotypically identical animals for the investigation of complex phenotypes, enabling one to establish genetic models through use of SDPs, as described in the following section (Swank & Bailey, 1973).

The seven Bailey RI strains (designated CXBD, CXBE, CXBG, CXBH, CXBI, CXBJ, and CXBK) are tested together with the progenitor strains (BALB/cBy and C57BL/6By) and the reciprocal hybrids (B6CF$_1$ and CB6F$_1$). If the progenitor strains differ significantly on the phenotype under investigation, it is possible to compile a strain distribution pattern denoting which of the RI strains resemble BALB/c (C-like), and which resemble C57BL/6 (B6-like). It is then frequently possible to match the SDP obtained for the phenotype under investigation with one previously established for distinctive loci, such as histocompatibility. Matching SDPs indicate the possible identity or close linkage of the genes. Because it is possible for two SDPs to match by chance [$p = (\frac{1}{2})^7 = .0078$], it is advisable to confirm linkage by testing candidate congenic lines.

The congenic (histocompatibility) lines were developed independently from an initial cross of B6 to C by a regimen of skin graft testing and backcrossing to B6 for at least 12 generations. This procedure resulted in a battery of congenic B6 lines, each line being distinguished from B6 solely by a chromosomal segment that includes a C-strain allele at a distinctive histocompatibility locus. Each of these congenic lines has been tested against the seven Bailey RI strains to determine which strains carry the B6 allele and which the C allele at a given histocompatibility locus (Bailey, 1975).

Thus, matching an SDP obtained for any given phenotype with an SDP for histocompatibility indicates the likelihood of pleiotropism or linkage of controlling genes at the locus involved. Additional confirmation regarding the specification of the locus in question is derived from testing the congenic line that carries the histocompatibility gene of the matching SDP. If the scores of the congenic-line mice are C-like, this would confirm the control of the phenotype under investigation by a gene at this locus, as the mice of this congenic line are identical with B6 at all other loci. The probability of an unlinked gene outside this chromosomal segment after 12 backcross generations is $\frac{1}{2}^{12}$ or 24×10^{-5} (Eleftheriou & Bailey, 1972).

Whether or not an SDP can be established, it is possible to obtain other sorts of information through the use of RI strains, progenitors, and hybrids (referred to as an "RI set"). For example, a continuous distribution of scores on a phenotype among the RI strains, together with a lack of a difference between the progenitors, indicates that a polygenic system underlies that phenotype. It is possible to determine dominance by the conventional method of backcrossing hybrids to the progenitors, which differ from each other; sex linkage can be established beginning with the analysis of a difference between the reciprocal hybrids, etc. In addition to the Bailey RIs, several sets of RI lines using various progenitors are currently being developed (Haber & Simmel, 1978; Taylor, 1976).

The study to determine the genetic analysis of stimulus reactivity (Simmel & Eleftheriou, 1977) tested 132 mice (which, as in the first study, were 60-day-old males), using the same testing procedure as was used for the earlier study, except that the shuttle box was omitted. There were 12 mice of each genotype of the Bailey RI set. On the two measures that had been found in the first experiment to have high loadings on stimulus reactivity and low loadings on activity (latency and novel-side time), it was possible to establish a strain distribution pattern

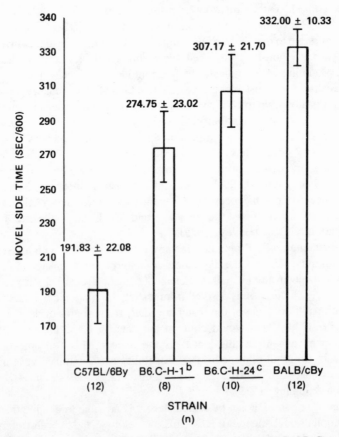

FIG. 3.2.   Comparison of congenic histocompatibility lines with BALB/cBy and C57BL/6By on (a) latency and (b) novel-side time. For both measures, congenic

(SDP), with the same SDP being found for both measures. When confirmed by testing mice from two congenic histocompatibility strains—B6.C-H-24^c and B6.C-H-1^b—it was confirmed that both stimulus reactivity measures (LAT and NOV) are controlled by genes at chromosomal segments of H-24^c and H-1^b on chromosome 7 (Linkage Group I). The results of these tests on the congenic lines and progenitor strains are shown in Fig. 3.2. A polygenic model was found for the two measures with loadings on Factor II (activity and arena crossings), with the progenitor C57BL/6By and BALB/cBy strains not differing from each other on either activity measure.

The results of the study show that the genetic model is supportive of the factor structure previously obtained: Exploratory behavior measured in terms of reactivity to novel stimuli can be determined separately from general activity and is

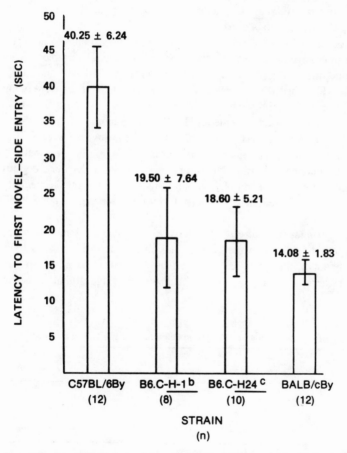

lines differ from BALB/c but not from C57BL/6, confirming the involvement of these loci on stimulus reactivity.

under distinctively different genetic control. *We suggest that this general strategy (seeking agreement between multivariate and genetic analyses) might be useful in other areas where there have been problems with confused or confounded phenotypes.*

Generality and Stability of Stimulus Reactivity. The measures obtained in the stimulus reactivity arena appear to be quite reliable over time. Table 3.6 presents data obtained 3 and 4 years apart on mice of identical age, sex, and genotype, but using many different experimenters, at different seasons of the year, and, in some cases, different locations (Maine and Ohio).

The procedures and the general factor structure have been found to be basically similar when species other than mice are tested, with appropriate modifications of the apparatus. Stimulus reactivity and activity measures have been obtained for rats (Haber & Friedman, 1979), gerbils (Rosenfeld, Lasko, & Simmel, 1978), and dogs (Wright, 1980).

In mice, the stimulus reactivity measures have been found to be surprisingly resistant to change due to either subject or environmental variables. When male and female mice, either 60 or 125 days of age (of C57BL/10J and LG/J strains and B10LGF$_1$ hybrids) were tested, only genotype differences were obtained on the two stimulus reactivity measures (LAT and NOV), whereas activity differed according to age and sex (Simmel et al., 1976). In another study (Haber, 1976), C57BL/6By and BALB/cBy mice were reared for several weeks in complex three-tiered cages or were placed in such complex cages one week before testing or were maintained in normal laboratory cages. Even under these conditions of early and prior enhanced stimulation, stimulus reactivity scores (especially LAT) were largely unaffected, whereas activity scores were altered considerably.

The Importance of Exploratory Behavior. It has been demonstrated, several decades ago as well as in the current literature, that exploratory behavior is a viable phenotype and can be measured as a response to novel and complex stimuli, rather than as a vague form of activity. We suggest that further research

TABLE 3.6
Replication of Mean Scores on Stimulus Reactivity and Activity

Variable	74-75[a] (N = 132)	77-78[b] (N = 139)
LAT	22.15	23.30
NOV	275.96	270.96
ARX	37.97	37.69
ACT	412.97	401.82

[a]Data from Simmel & Eleftheriou (1977).

[b]Data from Simmel, Bagwell & Slattery (1982)

on the genetic aspects and ramifications of exploratory behavior would be useful, especially in two general areas.

One research area of potential importance is the relationship between exploratory behavior and other complex forms of behavior. For example, several years ago, Hayes (1962) suggested that general intelligence (in humans and in other animals) is an acquired characteristic, resulting from an individual's experience with the environment. Individual differences between those in similar environments are due to largely inherited differences in "experience-producing drives," or the tendency to explore.

Another, and perhaps more fundamental topic relates to the adaptive significance of exploratory behavior. Exploration is the form of behavior by which an organism is most likely to obtain information about changes in its environment. Humans, rats, and mice, among other species, seem to adapt well to rapidly changing conditions in their environments, and much could be learned about the adaptive processes through further investigation of exploratory behavior.

We have tried to make clear in this chapter that advances in any of these important areas of research in exploratory behavior can occur only if the phenotypes are unambiguously defined and consistently measured.

REFERENCES

van Abeelen, J. H. F. Genetic analysis of behavioural responses to novelty in mice. *Nature*, 1975, *254*, 239–241.

van Abeelen, J. H. F. Biometrical genetic analysis of exploratory behavior in mice. *Behavior Genetics*, 1977, *1*, 90–91. (Abstract) (a)

van Abeelen, J. H. F. Rearing responses and locomotor activity in mice: Single-locus control. *Behavioral Biology*, 1977, *19*, 401–404. (b)

Bailey, D. W. Recombinant inbred strains. *Transplantation*, 1971, *11*, 325–327.

Bailey, D. W. Genetics of histocompatability in mice. I. New loci and congenic lines. *Immunogenetics*, 1975, *2*, 249–256.

Barnett, S. A. *The rat: A study in behavior* (Rev. Ed.). Chicago: University of Chicago Press, 1975.

Berlyne, D. E. The arousal and satiation of perceptual curiosity in the rat. *Journal of Comparative and Physiological Psychology*, 1955, *48*, 238–246.

Berlyne, D. E. The influence of complexity and change in visual figures in orienting response. *Journal of Experimental Psychology*, 1958, *55*, 289–296.

Berlyne, D. E. *Conflict, arousal, and curiosity.* New York: McGraw–Hill, 1960.

Birke, L. I. A. Object investigation by the oestrous rat and guinea-pig: The oestrous cycle and the effects of oestrogen and progesterone. *Animal Behaviour*, 1979, *27*, 350–358.

Broadhurst, P. L. Determinants of emotionality in the rat: III. Strain differences. *Journal of Comparative and Physiological Psychology*, 1958, *51*, 55–59.

Broadhurst, P. L. Psychogenetics of emotionality in the rat. *Annals of the New York Academy of Science*, 1969, *159*, 806–824.

Broadhurst, P. L. *Drugs and the inheritance of behavior.* New York: Plenum, 1978.

DeFries, J. C., & Hegmann, J. P. Genetic analysis of open-field behavior. In G. Lindzey & D. D. Thiessen (Eds.), *Contributions to behavior-genetic analysis: The mouse as a prototype.* New York: Appleton–Century–Crofts, 1970.

Dember, W. N. Response to environmental change. *Journal of Comparative and Physiological Psychology*, 1956, *49*, 93–95.

Dember, W. N., Earl, W., & Paradise, N. Response by rats to differential stimulus complexity. *Journal of Comparative and Physiological Psychology*, 1957, *50*, 514–518.

Denenberg, V. H. Open-field behavior in the rat: What does it mean? *Annals of the New York Academy of Science*, 1969, *159*, 852–859.

Eleftheriou, B. E., & Bailey, D. W. A gene controlling plasma serotonin levels in mice. *Journal of Endocrinology*, 1972, *55*, 225–226.

Ewing, A. W. Attempts to select for spontaneous activity in *Drosophila melanogaster*. *Animal Behaviour*, 1963, *11*, 369–378.

Fuller, J. L., & Thompson, W. R. *Foundations of behavior genetics*. St. Louis: C. V. Mosby, 1978.

Haber, S. B. *Synergistic effects of genotype and experience on the development of exploratory behavior in Mus musculus*. Doctoral dissertation, Miami University, Oxford, Ohio, 1976.

Haber, S. B., & Friedman, R. Psychobiology of experimental hypertension: Evaluation of the Dahl selected rat lines. *Behavior Genetics*, 1979, *9*, 454. (Abstract)

Haber, S. B., & Simmel, E. C. Behavioral variability in developing recombinant inbred strains of mice: A progress report. *Behavior Genetics*, 1978, *8*, 548–549. (Abstract)

Hall, C. S. The inheritance of emotionality. *Sigma Xi Quarterly*, 1938, *26*, 17–27.

Hall, C. S. The genetics of behavior. In S. S. Stevens (Ed.), *Handbook of experimental psychology*. New York: Wiley, 1951.

Harrington, G. M. Strain differences among rats initiating exploration of differing environments. *Psychonomic Science*, 1971, *23*, 348–349. (a)

Harrington, G. M. Strain differences in rotating wheel activity of the rat. *Psychonomic Science*, 1971, *23*, 363–364. (b)

Harrington, G. M. Strain differences in the open-field behavior of the rat. *Psychonomic Science*, 1972, *27*, 51–53.

Harrington, G. M. Strain differences in light-contingent barpress behavior of the rat. *Bulletin of the Psychonomic Society*, 1979, *13*, 155–156. (a)

Harrington, G. M. Strain differences in open-field behavior of the rat. II. *Bulletin of the Psychonomic Society*, 1979, *13*, 85–86. (b)

Harrington, G. M. Strain differences in passive avoidance conditioning in the rat. *Bulletin of the Psychonomic Society*, 1979, *13*, 157–158. (c)

Harrington, G. M. Strain differences in runway learning in the rat. *Bulletin of the Psychonomic Society*, 1979, *13*, 159–160. (d)

Harrington, G. M. Strain differences in shuttle avoidance conditioning in the rat. *Bulletin of the Psychonomic Society*, 1979, *13*, 161–162. (e)

Hayes, K. J. Genes, drives, and intellect. *Psychological Reports*, 1962, *10*, 299–342.

Henderson, N. D. Effects of early experience on the behavior of rodents: The second 25 years of research. In E. C. Simmel (Ed.), *Early experiences and early behavior*. New York: Academic Press, 1980.

Kivy, P. N., Earl, R. W., & Walker, E. L. Stimulus context and satiation. *Journal of Comparative and Physiological Psychology*, 1956, *49*, 90–92.

Lassalle, J. M., & Le Pape, G. Locomotor activity of two inbred strains of mice in a seminatural and a breeding cage environment. *Behavior Genetics*, 1978, *8*, 371–376.

McClearn, G. E. The genetics of mouse behavior in novel situations. *Journal of Comparative and Physiological Psychology*, 1959, *52*, 62–67.

McClearn, G. E. Strain differences in activity in mice: Influence of illumination. *Journal of Comparative and Physiological Psychology*, 1960, *53*, 142–143.

McClearn, G. E. Genotype and mouse activity. *Journal of Comparative and Physiological Psychology*, 1961, *54*, 674–676.

Manning, A. The effects of artificial selection for mating speed in *Drosophila melanogaster*. *Animal Behaviour*, 1961, *9*, 82–92.

Montgomery, K. C. A test of two explanations of spontaneous alternation. *Journal of Comparative and Physiological Psychology*, 1952, *45*, 287–293.

Montgomery, K. C. Exploratory behavior as a function of "similarity" of stimulus situations. *Journal of Comparative and Physiological Psychology*, 1953, *46*, 129–133.

Oliverio, A., & Eleftheriou, B. E. Motor activity and alcohol: Genetic analysis in the mouse, *Physiology and Behavior*, 1976, *16*, 577–581.

Oliverio, A., Eleftheriou, B. E., & Bailey, D. W. Exploratory activity: Genetic analysis of its modification by scopolamine and amphetamine. *Physiology and Behavior*, 1973, *10*, 893–899.

Richardson, E. J., & Simmel, E. C. Moderation of the severity of audiogenic seizures in DBA/2 mice following intraperitoneal insertion. *Bulletin of the Psychonomic Society*, 1973, *1*, 429–430.

Rosenfeld, J., Lasko, L. A., & Simmel, E. C. Multivariate analysis of exploratory behavior in gerbils. *Bulletin of the Psychonomic Society*, 1978, *12*, 239–241.

Rundquist, E. A. The inheritance of spontaneous activity in rats. *Journal of Comparative Psychology*, 1933, *16*, 415–438.

Scott, J. P., & Fuller, J. L. *Genetics and the social behavior of the dog.* Chicago: University of Chicago Press, 1965.

Simmel, E. C. Behavior genetic analysis of exploratory behavior and activity through use of recombinant inbred strains of mice: A progress report. *Behavior Genetics*, 1976, *6*, 117–118. (Abstract)

Simmel, E. C., Bagwell, M., & Slattery, J. Age and experience effects on stimulus reactivity in the CXB recombinant inbred mice. Paper presented at Behavior Genetics Association, 1982.

Simmel, E. C., & Eleftheriou, B. E. Multivariate and behavior genetic analysis of avoidance of complex visual stimuli and activity in recombinant inbred strains of mice. *Behavior Genetics*, 1977, *7*, 239–250.

Simmel, E. C., Haber, S. B., & Harshfield, G. Age, sex, and genotype effects on stimulus exploration and locomotor activity in young mice. *Experimental Aging Research*, 1976, *2*, 253–259.

Simmel, E. C., & Walker, D. A. The effects of a companion on exploratory behavior in two strains of inbred laboratory mice. *Behavior Genetics*, 1972, *2*, 249–254.

Swank, R. T., & Bailey, D. W. Recombinant inbred lines: Value in the genetic analysis of biochemical variants. *Science*, 1973, *181*, 1249–1251.

Taylor, B. A. Development of recombinant inbred lines of mice. *Behavior Genetics*, 1976, *6*, 118. (Abstract)

Thompson, W. R. The inheritance of behavior: Behavioral differences in fifteen mouse strains. *Canadian Journal of Psychology*, 1953, *7*, 145–155.

Thompson, W. R. The inheritance of behavior (activity differences in five inbred mouse strains). *Journal of Heredity*, 1956, *47*, 147–148.

Thompson, W. R., & Solomon, L. M. Spontaneous pattern discrimination in the rat. *Journal of Comparative and Physiological Psychology*, 1954, *47*, 104–107.

Tryon, R. C. Individual differences. In F. A. Moss (Ed.), *Comparative psychology.* New York: Prentice–Hall, 1942.

Walsh, R. N., & Cummins, R. A. The open-field test: A critical review. *Psychological Bulletin*, 1976, *83*, 482–504.

Whimbey, A. E., & Denenberg, V. H. Two independent behavioral dimensions in open-field performance. *Journal of Comparative and Physiological Psychology*, 1967, *63*, 500–504.

Wimer, R. E., & Fuller, J. L. The effects of d-amphetamine sulphate on three exploratory behaviors. *Canadian Journal of Psychology*, 1965, *19*, 94–103.

Wimer, R. E., & Sterns, H. Controlled visual input and exploratory activity in C57BL/6J mice. *Perceptual Motor Skills*, 1964, *18*, 299–307.

Wright, J. C. Early development of exploratory behavior and dominance in three litters of German shepherds. In E. E. Simmel (Ed.), *Early experiences and early behavior.* New York: Academic Press, 1980.

4 Behavioral Pharmacogenetics

Gary P. Horowitz
State University of New York at Binghamton

Bruce C. Dudek
State University of New York at Albany

Individual differences in both physiological and behavioral responses to pharmacologic agents have been known for many years; the sources of these differences are often assumed to be a complex constellation of biological, environmental, experiential and for humans, sociological factors. In one sense, drug responsivity can be viewed as following a continuum; a given dose of a given drug may elicit little or no response, a moderate (often desirable) level of the same response, or an extreme response in various individuals of a patient or subject population. These variations might be considered quantitative differences in the magnitude of a given response following drug administration. Descriptive statistics, such as ED50 (the dose of a drug that will be effective in producing the desired response in 50% of the subject population) and LD50 (the median lethal dose), are sometimes useful indices of the average response to a particular drug. However, the necessity of using such indices of central tendency is evidence for the inherent variability in drug responsiveness among individuals in a given subject population.

A second category of individual differences in response to drugs is evident when a proportion of individuals in a population display a qualitatively unique response to even moderate doses of the drug. This second category has often been called adverse or idiosyncratic drug reactions (Goldstein, Aronow, & Kalman, 1974). An example of an adverse reaction is the hemolytic anemia produced by primaquine and related drugs in certain individuals with glucose 6-phosphate dehydrogenase deficiency (Beutler, 1978). Such idiosyncratic adverse drug reactions can come about through abnormal ways in which the body acts on drugs or via abnormal states of the body that are revealed only upon drug challenge (Vesell, 1979). Documentation of idiosyncratic drug reactions is probably greatest for drugs used in therapeutic settings, examples of which are discussed later, and more complete reviews of which are presented elsewhere (Goldstein et al., 1974; LaDu, 1973; Vesell, 1979).

Pharmacogenetics is an emergent discipline that seeks to elucidate the contribution of genetic factors to the types of individual differences in drug responses presented in the preceding paragraphs. The related discipline of behavioral pharmacogenetics stems from the realization that behavioral responses to drugs are no different in principle from physiological or biochemical responses, and further that altered biochemical processes often can be reflected in variations in observable behaviors. The principles and methods of behavior-genetic analysis apply equally for evaluating differences among genotypes of mice in an open field, regardless of whether or not the mice have been previously treated with a psychoactive substance. Similarly, the methods used to attempt to describe the relative importance of genetic and nongenetic factors for individual differences in specific cognitive abilities among humans can also be applied when the phenotype in question is one related to drug responsivity (e.g., differences in responses to acute or chronic administration of alcohol). That is, the roots of behavioral pharmacogenetics are based in the older disciplines of psychology, genetics, and pharmacology. In its evolution, the field passed through the interdisciplinary fields of behavior genetics, behavioral pharmacology, and pharmacogenetics.

In describing the nature of the then emerging discipline of psychopharmacogenetics, Fuller and Hansult (1975) and Eleftheriou (1975) referred to its hybrid nature. Fuller and Hansult went on to ask how far we might extend the metaphor by asking the relevant question: "Is this simply an area of overlap between three established areas or does it have characteristics of its own, even a touch of heterosis or hybrid vigor [p. 11–12]?" Some years later, Horowitz (1981) claimed that heterosis had indeed occurred. However, he was referring to heterosis in the context of the rapid development and increasing numbers of researchers with interest in the area. Surely this increase in published articles in the field has begun to be realized, as evidenced by the summaries provided by Broadhurst (1978) and others. However, it has yet to be established if heterosis, in the sense that Fuller and Hansult (1975) intended, has been realized in the field of behavioral pharmacogenetics. Does the field, in fact, have characteristics that are unique and independent of its progenitor disciplines? In the following discussions of critical issues and future directions, we hope to offer suggestions for directions that might result in this latter type of heterosis.

Until we address those issues, however, we should like to be bold enough to introduce a new metaphor to describe the progress of behavioral pharmacogenetics. We refer to it, not only as a hybrid science, but also as a fully segregating science. The metaphorical question thus becomes whether or not the elements derived from the dihybrid cross follow Mendel's law of independent assortment. We think that they have, and that this independent assortment is evident in at least two divergent strategies that have emerged within the broader goal of behavioral pharmacogenetics. This broader goal is the demonstration that individual differences occur in behavioral responses to drugs, and that genetic factors must be shown to contribute to these individual differences. This primary goal has been realized by behavioral geneticists and pharmacologists alike. As Mc-Clearn (personal communication, 1978) has stated, we could probably find strain differences in virtually any trait of interest to behavioral pharmacology. The two strategies referred to differ as to how to proceed from this initial goal (Horowitz, 1979). One approach is more concerned with elucidating the genetic architecture of the behavioral response to a given drug. Its advocates suggest the application of such tools of behavior-genetic analysis as classical genetic breeding designs, diallel crosses, and family studies, to learn more about such genetic parameters as mode of inheritance, heritability, and phenotypic and genetic correlations. The second approach uses the tools of genetics to produce (e.g., via selection) or to find (e.g., via strain surveys) populations of subjects that differ in their behavioral responses to drugs, and then to look for underlying physiological or neurochemical mechanisms that might mediate these differences. It should be noted that the differences between the two strategies is not so much in the initial methods employed, but rather in the interest and subsequent course of further investigation.

The former approach, here termed a biometrical approach, was the thesis for Broadhurst's (1978) recent survey of behavioral pharmacogenetics. In this work, Broadhurst offered an admirable survey of the research to date in this field. He further suggested that careful examinations of the genetic architecture of the traits in question have rarely and only recently been conducted. Therefore, in the present chapter, we concentrate on the contributions made to the field by re-searchers employing the second approach, that of attempting to define possible physiological and neurochemical substrates of genetic differences in behavioral responsiveness to psychoactive agents. It should become apparent, however, that the two approaches are not always separable in practice, and we discuss the relationship between these strategies in a following section.

Human Pharmacogenetics

Ecogenetics is a field of research defined by Omenn and Motulsky (1978) as the study of genetically determined differences in susceptibility to environmental agents. Although this broad definition includes such interesting topics as nutri-tional requirements and carcinogens, more related to the present chapter is the field of pharmacogenetics. Both therapeutic and recreational drugs have effects that show considerable individual variation. A thorough review is provided by Omenn and Motulsky (1975). We present a few examples here for the purpose of describing the kinds of pharmacological agents and genetic mechanisms that might be involved in these individual differences.

Important therapeutic drugs such as antipyrine, phenylbutazone, and halo-thane are known to demonstrate marked individual differences in the attained levels in blood and in metabolism rate. These pharmacokinetic factors are known to have a considerable basis in hereditary factors, as demonstrated by twin and family studies (Vesell, 1974). Genetic control of pharmacokinetics of drugs used in psychiatric therapy is well demonstrated by the polygenic nature of inheritance of nortriptyline kinetics (Asberg, Evans, & Sjoqvist, 1971). Response to two classes of antidepressants, MAO inhibitors and tricyclics, is thought to reveal two genetically specific types of depression (Pare & Mack, 1971). Responders to MAO inhibitors generally respond to all MAO inhibitors, but not necessarily to tricyclics and vice versa. The mode of inheritance was not determined in this work, but the mechanism of action is probably not one of dispositional influences, but more related to primary sites of drug action. Omenn and Motulsky (1976) describe a dramatic condition where the widely used muscle relaxant, succinylcholine, can paralyze breathing for several hours. One in 2500 Caucasians shows this response due to an abnormal variant of plasma pseudocholinesterase, the enzyme responsible for inactivation of the drug. This condition is nearly as dramatic as another in which a single autosomal dominant gene produces susceptibility to malignant hyperthermia induced by inhalation anesthetics such as halothane or ether, and also succinylcholine. This rapid rise in temperature often results in death (Omenn & Motulsky, 1976).

Another question that can be addressed is the issue of a hereditary predisposition for abuse of psychoactive drugs. The answer to this question may lead us on a search for "addictable" physiologies or, perhaps, personalities (Nichols, 1972). Clearly, ethanol is a drug with widely varying individual effects. Vesell, Page, and Passananti (1971) demonstrated high heritability for rate of ethanol elimination in twins ($H = .98$). Yet the fact that alcoholism runs in families and probably has a genetic component is not likely to be explained on the basis of metabolism rates alone (Goodwin, 1979). Recent work by Schuckit, Goodwin, and Winokur (1972) not only demonstrates a reliable genetic component in alcoholism, but may be interpreted as indicating that environmental influences of sharing a household with an alcoholic are of less importance than sharing common genes. The question of etiology of this as well as other addictions may center on the aforementioned "addictable" physiologies and/or personalities.

Nonhuman Pharmacogenetics

The majority of studies using nonhuman subjects reviewed herein were designed to examine at least possible differences among groups of genetically defined subjects in behavioral responses to various psychoactive drugs. However, it should be noted that individual differences are almost always observed, and have often been reported with interest by some investigators using random-bred populations. For example, Goldstein and Sheehan (1969) found large variations among Swiss–Webster mice in levels of locomotor activity following activating doses of the opiate agonist, levorphanol. These authors speculated that genetic factors contributed to the observed differences. Perhaps these early observations were the impetus for a selective breeding study conducted in the same laboratory several years later. Using Swiss–Webster mice as the founding population, Judson and Goldstein (1978) successfully selected for quantitative differences in changes in locomotor activity induced by an acute dose of this same opiate agonist.

Cappell and LeBlanc (1971) investigated the role of morphine pretreatment on later ingestion of a morphine-adulterated solution in randomly bred Wistar rats. Individual differences in acceptance of the morphine adulterated solutions were observed in both premedicated and control groups. Those rats in the pretreatment group who were classified as "nondrinkers" could be induced to drink morphine solutions by subsequent injections of morphine and finally reached a level of morphine ingestion equal to the amount consumed by "drinkers" in the control group. Other studies using different populations of rats have shown clear genetic differences in acceptance of morphine-adulterated solutions (Meade, Amit, Pachter, & Corcoran, 1973). Although the differences in acceptance of morphine solutions observed by Cappell and LeBlanc (1971) were not established to be of genetic origin, their results are compatible with a possible genotype by treatment interaction.

Individual differences in drug responsivity have been reported also in similar populations of randomly bred mice. Ho, Loh, and Way (1977) reported significant differences in morphine-induced analgesia, lethality, tolerance, and physical dependence in Swiss–Webster and ICR mice. Furthermore, on several measures, there were significant differences among different populations of Swiss–Webster (or ICR) mice when these populations were purchased from different sources. Thus, although individual differences in drug responsiveness among animals of randomly bred populations are often sources of hypotheses that can be tested via genetic or pharmacologic methods, these differences may also contribute to uncertainty when attempts are made to replicate results from different laboratories.

Perspectives and Organization

We have previously alluded to the prolific growth of research in the area of behavioral pharmacogenetics. In nonhuman subjects, genetic factors have been shown to influence responsiveness to virtually every class of psychoactive agents, including (but not limited to) stimulants, mild tranquilizers, barbiturates, alcohol, anesthetics, nicotine, opiates, and neuroleptics. Much of this work has been reviewed or collected previously by Broadhurst (1978) and Eleftheriou (1975). Similarly, reviews of human psychopharmacogenetics have also been compiled (Omenn & Motulsky, 1976; Vesell, 1979). It is beyond the scope of the present chapter to provide an exhaustive review of all literature in this field. In keeping with the title of this volume, we have concentrated on the principles and applications of the methods of behavior-genetic analysis with respect to their relevance to the study of individual differences in behavioral responses to drugs. Specifically, we have concentrated on research using alcohol and opiates. It is hoped, however, that the principles, applications, and implications that we discuss will extend to the study of the behavioral effects of any drug of choice by a given researcher.

Our decision to limit much of our discussion to the pharmacogenetics of alcohol and opiate responsiveness goes beyond our own research interest in these two classes of drugs. First, these two classes of drugs are important from a sociomedical perspective; both alcohol and the opiates (primarily, heroin) remain major drugs of abuse. Secondly, alcohol is probably the most widely studied drug in behavioral pharmacogenetics, and research in the pharmacogenetics of opiates is steadily growing (Crabbe & Belknap, 1980). Finally, there has been an increased interest in recent years in possible commonalities and interactions between alcohol and opiates (see Blum, Hamilton, & Wallace, 1977, for a review).

The remainder of the chapter is divided into four major organizational topics. First, we present the major techniques of genetic analysis of drug responsiveness, for both humans and nonhumans. Secondly, we define phenomena of interest from a pharmacologic perspective and give examples that illustrate each from a pharmacogenetic perspective. In the third section, issues of evaluation and interpretation are presented and discussed. Finally, we offer some personal impressions of several directions in which the field of behavioral pharmacogenetics might be fruitfully extended by future research.

GENETIC TECHNIQUES

Human Studies

The genetic methodologies in human pharmacology are the typical approaches in human behavior genetics. They may range from pedigree analysis to adoption studies and segregation analysis. We emphasize one approach that has been useful in the study of alcoholism. Detailed treatment of other methodologies may be found in standard sources (Fuller & Thompson, 1978). Recent work by Schuckit (1980) has emphasized a search for biological markers that might be identifiable in individuals genetically at risk for alcoholism. Toward this end, individuals from families with a history of alcoholism respond differently on subjective reports of degree of intoxication in that they report feeling less intoxicated than controls. Physiological response to alcohol shows similar differentiation. More interesting, though, is the report that men with familial history of alcoholism exhibit higher levels of circulating acetaldehyde after ethanol consumption than controls. It is possible that some of the addictive properties of alcohol are actually attributable to its metabolite acetaldehyde (see later section on locus of effect). If so, the finding of higher blood levels may be more than just a marker. It may be a prime factor in the etiology of alcoholism. At any rate, the need for such markers is obvious in order that individuals at risk for alcoholism can be closely studied. In this way those environmental factors that precipitate alcoholism may be uncovered with the use of a genetic tool.

Nonhuman Studies

Strain Surveys

The major source of nonhuman subjects in behavioral pharmacogenetic research, as is the case in behavioral genetics in general, has been inbred strains of mice and, to a lesser extent, rats. Over half of Broadhurst's (1978) recent book reviews the literature in which strain differences have been reported in studies using a wide variety of drugs, including alcohol and opiates.

The process of inbreeding leads to individuals who are more closely related than are members of a population at large. In the development of inbred strains, inbreeding is secured by sequential and continuous brother–sister mating. After 20 generations of brother–sister mating, virtually all loci will be fixed for alleles in the homozygous state. Thus, all members within an inbred strain are virtually uniformly homozygous and furthermore, for practical purposes, are genetically identical to each other (Plomin, DeFries, & McClearn, 1980). However, the specific alleles that are fixed through the process of inbreeding are randomly determined (with the exception of those combinations that are eliminated by selective pressures). Therefore, although individual members within an inbred strain are (within limits) genetically identical to other members of the same strain, in all likelihood they will differ from members of a different inbred strain in the alleles fixed at many loci.

In the simplest model of quantitative genetic theory, the observable phenotype of an individual is the additive effects of its genotype and its environment (Plomin et al., 1980). Given that two different inbred strains are reared and tested in the same (within practical limits) environment, then differences between strains on some observable trait (e.g., alcohol preference) must be determined in large part by the different alleles fixed within each. Thus, differences between two inbred strains that are significantly greater than differences within each strain (which may be due to such factors as a lack of homozygosity, microenvironmental differences, and errors of measurement) are usually construed to represent a genetic substrate for the observed phenotypic difference.

Differences between inbred strains is a useful initial demonstration for both strategies of behavioral pharmacogenetic research. Strain comparisons have yielded significant differences in, and hence a genetic contribution to, such alcohol-related responses as in preference for alcohol in a choice situation (McClearn & Rodgers, 1961; Rodgers, 1966; Thomas, 1969), in acceptance of alcohol using a forced-ingestion procedure (McClearn, 1968), in the hepatic metabolism of ethanol (Belknap, MacInnes, and &McClearn, 1972; Schlesinger, Bennett, & Hebert, 1967; Sheppard, Albersheim, & McClearn, 1970), and in neural sensitivity to ethanol (Collins, Lebsack, & Yeager, 1976), as well as in numerous other behavioral and biochemical phenotypes. Similarly, genetic factors have been shown to contribute to differences among mice in preference for morphine-adulterated drinking solutions (Horowitz, Whitney, Smith, & Stephan, 1977), in acceptance of morphine solutions using a forced-ingestion procedure (Meade et al., 1973), in readdiction susceptibility (Eriksson & Kiianmaa, 1971), in opiate-induced analgesia and hyperactivity (Castellano & Oliverio, 1975; Collins & Whitney, 1978), in neurochemical responses to opiates (Castellano, Espinet-Llovera, & Oliverio, 1975; Trabucchi, Spano, Racagni, & Oliverio, 1976), and in the development of physical and behavioral dependence on opiates (Brase, Loh, & Way, 1977; Whitney, Horowitz, & Collins, 1977).

For the investigator interested in further elucidating the genetic architecture of the alcohol or opiate-related phenotype, these initial strain comparisons are usually extended via the tools of behavior-genetic analysis, such as classical Mendelian breeding designs and analysis, diallele crosses and subsequent analysis, or the use of recombinant inbred strains (discussed later) to gain further information regarding the genetic parameters of the trait in question. Although these analytical designs and tools have been applied infrequently in traits of pharmacogenetic interest, the reader is referred to Broadhurst (1978) for a survey of these attempts and their results.

With respect to the second general strategy of behavioral pharmacogenetics, the demonstration of strain differences often leads to hypotheses regarding the relationship between a trait and an underlying physiological or biochemical mechanism, or a relationship between two behavioral traits. For example, C57BL and DBA mice are known to differ in their preference for 10% ethanol solutions when it is offered versus tap water in a two-bottle choice situation (McClearn & Rodgers, 1961; Rodgers, 1966). These two strains also differ in the activity of hepatic enzymes of ethanol metabolism (Sheppard et al., 1970). The question now becomes whether or not these traits are physiologically or genetically related. Historically, attempts to answer such questions have relied principally on three different approaches.

The first approach might be considered one of pharmacologic intervention. For example, if the higher preference for ethanol in C57BL mice is casually related to increased ability to metabolize the drug, then interfering with ethanol metabolism should reduce the alcohol preference of these mice. Schlesinger, Kakihana, and Bennett (1966) demonstrated that pretreatment with disulfuram (an aldehyde dehydrogenase inhibitor) significantly lowered the alcohol preference of C57BL mice. However, the magnitude of the decrease was slight. Similarly Ho, Tsai, and Kissin (1975) reported that C57BL mice significantly incorporate higher levels of acetylcholine and greater uptake of radiolabeled choline than did DBA mice. Pretreatment with a choline acetyltransferase inhibitor, which would interfere with acetylcholine synthesis, significantly but transiently reduced the voluntary alcohol ingestion of C57BL mice. Although these two pharmacological intervention studies may contribute to our understanding of why C57BLs prefer ethanol, they may tell us little of why DBA mice avoid it (see later discussion on issues of locus of effect).

A second direction that may be taken to elucidate further a difference in two traits in two different inbred strains is to expand the number of genotypes tested. For example, Castellano and Oliverio (1975) have reported that C57BL mice respond to morphine injections with increases in locomotor activity, whereas DBA mice do not. Horowitz et al. (1977) have further shown that C57BL mice prefer morphine–saccharin solutions over water, whereas DBA mice in the same situation drink almost exclusively water. Subsequently, Whitney et al. (1977) have reported a positive correlation of about .92 between preference for morphine–saccharin solutions and morphine-induced hyperactivity in six different inbred strains. This strong positive correlation was not due to differences in baseline preference for saccharin, nor to differences in level of activity following saline injections. Similarly, we have recently derived a positive relationship between ethanol preference (data from Rodgers, 1966) and preference for morphine–saccharin solutions (data from Whitney et al., 1977). With respect to morphine responsivity, Brase et al. (1977) tested four inbred strains and two randomly bred lines for hyperactivity following acute morphine administration and the degree of jumping observed following withdrawal from chronic morphine exposure. The correlation between morphine-induced hyperactivity and jumping elicited by either abrupt or naloxone-precipitated withdrawal was .93–.96.

Collectively, these correlation strain surveys suggest a possible genetic relationship between the two traits of interest. However, a more stringent test would be to see if whether or not the two traits of interest are covariates in individual mice of a heterogeneous genetic population (disussed later). However, many of the techniques of pharmacogenetics are invasive, if not destructive. For example, if one were interested in the possible genetic relationship between morphine-induced analgesia and the density of central opiate receptors, the use of heterogeneous mice would require measuring both traits in an individual animal. Thus, the researcher is forced to accept the possible confounding effect of order of testing. Counterbalancing in this case is not practical, as it is as yet impossible to assay for central opiate receptors without first removing the subject's brain. Ideally, then, a researcher would like the best of both worlds: the genetic uniformity of inbred strains and the variability of a segregating population.

One population of inbred mice that comes close to meeting these criteria consists of recombinant inbred (RI) lines. A fuller description of these mice is presented elsewhere (Eleftheriou & Elias, 1975). Basically, two inbred strains are crossed to produce an F_1 generation, which in turn are mated *inter se* to produce the F_2 generation. Beginning with the F_2 generation, a number of brother–sister mating pairs are begun, and brother–sister mating is continued for at least 20 generations. Collectively, the surviving lines represent unique recombinations of the alleles at which the parental types were different. However, within each RI line, all members are virtually genetically identical and all alleles are fixed once again in the homozygous state.

The use of recombinant inbred lines in pharmacogenetic research has recently been reviewed by Broadhurst (1978) and by Crabbe and Belknap (1980). The hypothetical example of the relationship between morphine-induced analgesia and density of opiate receptors was actually the subject of an interesting application of recombinant inbred lines by Baran, Shuster, Eleftheriou, and Bailey (1975). These researchers measured morphine-induced analgesia and the density of opiate receptors in the brain for 11 isogenic populations of mice: C57BL and BALB (the progenitor strains), the two reciprocal F_1 generations, and seven RI strains derived as described previously. Their study amounted to a strain survey for each trait in question, as well as a strain ranking correlation between the two traits. Genetic differences were apparent for both morphine-induced analgesia, as measured by the tail flick method, and for density of opiate receptors, as measured by stereospecific binding of radiolabeled naloxone (an opiate antagonist). Finally, although there was a positive strain ranking correlation between binding and analgesia, the differences in opiate receptor density were not great enough to explain a significant proportion of the variation in morphine-induced analgesia. Clearly, properties of opiate receptors are one possible source of genetic differences in the ability of morphine to produce analgesia. However, other factors, including possible differences in neurotransmitter systems modulated by these receptors, are also possible contributors to the phenotypic differences in morphine-induced analgesia previously reported among a number of inbred strains (Brase et al., 1977; Castellano & Oliverio, 1975; Collins & Whitney, 1978; Whitney et al., 1977).

Selection Studies

With inbred strains, the combinations of alleles that are fixed by the inbreeding procedure are random. Thus, the combinations of phenotypic expressions of these genes are also random. By selective breeding procedures, a researcher can select a phenotype of interest and selectively breed for quantitative differences in that phenotype. In research, unlike agricultural applications, it is usually desirable to select for both high and low levels of the phenotype. Different types of selective breeding techniques are available, depending in part on the demands dictated by the desired phenotype (Falconer, 1960). The basic scheme is to start with a heterogeneous founding population and to breed individuals with higher levels to others with higher levels; conversely, individuals with low phenotypic levels are bred with other low individuals. The response to selection is based, in part, on the variability in the founding population, the heritability of the trait in question, the system of selective breeding, and the selective pressure exerted by the experimenter. Given heritable variance, the appropriate selective pressure, and enough generations of selective breeding, the procedure will produce two lines of subjects that differ markedly in the phenotype undergoing selection.

Historically, a number of selective breeding programs have been developed for various phenotypes of interest in behavioral genetics, using both mice and rats as subjects. Many programs have selected for phenotypes initially not related to pharmacologic interests, but later evaluated for differences in response to various psychoactive agents. These studies are reviewed by Broadhurst (1978; see esp. pp. 33–73). In addition, several selective breeding studies have been conducted in which the selection criterion was a response to a psychoactive agent; the majority of these studies used some response to alcohol as the phenotype in question. These studies, including the Colorado sleep lines (McClearn & Kakihana, 1973), the most and least affected rat lines (Lester, Lin, Anandam, Riley, Worsham, & Freed, 1977), the alcohol nontolerant and tolerant lines (Rusi, Eriksson, & Maki, 1977), the Finnish Alko Alcohol and Nonalcohol lines (Eriksson, 1975), and the Indiana alcohol preferring and nonpreferring rats (Lumeng, Hawkins, & Li, 1977) have recently been reviewed, by workers involved in each, in a research monograph published by NIAAA (McClearn, Deitrich, & Erwin, 1981). Selection studies using opiate-related phenotypes are reviewed in a separate chapter of the same volume (Horowitz, 1981).

A successful selective breeding design serves at least two valuable functions. First, the demonstration that selective breeding can produce systematic quantitative changes in a drug-related phenotype gives definitive proof of the contribution of genetic factors to individual differences in that particular response. More importantly, such studies often generate hypotheses regarding the selected trait and other traits thought to be related to it. Ideally, the selection procedure should effect changes in the frequency of only those genes that are related to the response undergoing selection (Crabbe & Belknap, 1980). Therefore, differences among selected lines in some response other than that being selected may be indicative of a physiological or genetic relationship between the two traits in question. For example, rats that have been selectively bred for susceptibility or resistance to readdiction liability to morphine also show predictable differences in their susceptibility to alcohol readdiction (Nichols & Hsiao, 1967). These data are suggestive of a genetically based commonality in the processes underlying addiction to alcohol and opiates. It is our contention that such demonstrations are valuable in suggesting hypotheses of commonality, but these must be tested by more rigorous experiments.

By way of further example, we would like to discuss some research generated by the Colorado sleep lines. McClearn and Kakihana (1973; 1981) have selectively bred two lines of mice for differential sensitivity to the hypnotic effects of ethanol. The selection index was duration of the loss of righting reflex ("sleep time") following 3.4 g/kg ethanol, although this dose was increased in subsequent generations of selection. By the fourteenth generation, there was virtually no overlap in the sleep time distribution of mice forming the long sleep and short sleep lines. Subsequent research has shown that these lines of mice differ on a number of other response measures to both alcohol and other agents (see Collins, 1981, for a review). These responses may or may not be genetically correlated to the selection index, but they furnish useful hypotheses regarding the mechanisms of action of behavioral and physiological responses to alcohol.

We would like to illustrate such hypothesis testing by three examples, each involving a different concept. First, traits for which an *a priori* reason exists to suggest a relationship with the selected trait, may play no realizable role in the product of such selection. Previous research by Belknap et al. (1972) had demonstrated an inverse relationship between ethanol-induced sleep time and aldehyde dehydrogenase activity following alcohol challenge in the heterogeneous population from which LS and SS mice were eventually derived. This relationship would lead to the intuitive hypothesis that part of the differential response to selection in LS and SS mice might be due to differences in rates of ethanol metabolism. Yet Heston, Erwin, Anderson, and Robbins (1974) reported similar rates of ethanol metabolism in LS and SS mice.

A second phenomenon is concerned with the demonstration of concomitant responses in selected lines. LS mice have a longer duration of the loss of righting reflex following hypnotic doses of ethanol than do SS mice. On the other hand, Goldstein and Kakihana (1975) demonstrated that SS mice showed a greater severity of seizures, relative to LS mice, following withdrawal from 3 days of chronic exposure to ethanol vapors, even though blood alcohol levels were maintained at approximately the same level in both lines. These results might be interpretable as demonstrating a relationship between acute tolerance to, and acquired dependence on ethanol. However, in the same paper, these authors further demonstrated the independence of ethanol-induced narcosis and severity of seizures induced by withdrawal from chronic alcohol exposure by showing that the two traits were not covariates among individual animals of the original heterogeneous founding population from which LS and SS mice were derived. Thus, the concomitant responses observed in the first part of their study may have been due to a stochastic association of genes subserving each response.

The final phenomenon regarding research subsequent to the development of selected lines is that, even if hypotheses derived from them are tested and and not confirmed, further hypotheses might evolve from them. Ethanol-induced sleep time and severity of seizures following withdrawal from chronic ethanol exposure may not be genetically correlated characters; at least no evidence to support this hypothesis can be derieved from the results of the latter study by Goldstein and Kakihana (1975). However, LS and SS mice still may represent different genetic populations for the withdrawal phenotype. Given the previous suggestion of common responses to alcohol and opiates reviewed by Blum et al. (1977), Horowitz and Allan (1982) investigated withdrawal from chronic morphine exposure in LS and SS mice. Briefly, mice of both lines were implanted with sustained-release morphine pellets and were withdrawn 3 days later via naloxone challenge. On two classic measures of opiate withdrawal (wet-dog shakes and jumping), SS mice were more sensitive to withdrawal from chronic morphine administration. These results were consistent with those reported for severity of seizures following chronic withdrawal from alcohol exposure, where SS mice again showed the more severe score on one index of alcohol withdrawal.

Two cautions must be taken regarding the interpretation not only of this latter study, but of all tests of concomitant responses in selected lines. First, the results may be specific to the methods and behavioral assays employed. SS and LS mice may or may not differ with respect to their general level of physical dependence on alcohol or morphine following these drug regimens. Secondly, as Satinder (1977) and others have pointed out, the true test of a genetic relationship between two traits is to see whether or not the traits are covariates in individual animals of a population of heterogeneous genetic stock. Thus, the relationship suggested by the results of the studies by Goldstein and Kakihana (1975) and Horowitz and Allan (1982) remain to be evaluated.

Heterogeneous Stocks

Genetically heterogeneous stocks are found in several different forms, each of which has its peculiar advantages and disadvantages. The common characteristic of these stocks is a lack of genetic uniformity from animal to animal. Within that definition, however, is a wide range of genetic diversity. The presence of this diversity provides several tools for the pharmacologist interested in the control of but not specific investigation of genetic variables. For the geneticist, these stocks have a wide variety of uses. First, we distinguish several types of stocks available.

"Random-Bred" Stocks.   Many commercial suppliers of laboratory animals, particularly rodents, characterize their populations as random bred. The extent of genetic variability in most of these stocks is not known. In-house breeding colonies maintained by investigators are likely to be rather inbred owing to founder effects. This fact and general lack of information on ancestry of these types of stocks make them less suitable for genetic work than the other types of stocks described here. However, Rice and O'Brien (1980) have characterized the often used Swiss mouse. This work suggests more genetic variability (enzyme polymorphisms) in this stock than is suggested by the foregoing discussion. The usual lack of controlled breeding schemes and the many different populations of Swiss type mice call into question the generality of this finding.

Inbred Strain Crosses.   The production of an F_2 generation from two inbred strains is relatively inexpensive monetarily, as well as in time, and provides a heterogeneous stock with an amount of genetic variability determined by the number of allelic differences in the two parental stocks. Maintenance of such a stock by production of further segregating generations could be accomplished by a random mating scheme with some care taken to avoid inbreeding. Such a stock with controlled breeding would be suitable for a variety of genetic analyses. The degree of usefulness is limited by the number and character of differences between the two parental stocks. More than two alleles at a locus would not be possible. Choice of progenitor stocks should attempt to maximize genetic diversity for the phenotype(s) of interest or in general characteristics if genes underlying such phenotypes are not identified. The latter approach would simply suggest using maximally unrelated strains.

The Eight- or Four-Way Cross.   The limitations suggested for the F_2 type stocks may be overcome by complex crosses involving more than two parental stocks. Such crosses have several times served as the foundation stock for selective breeding experiments. Their usefulness can, however, extend beyond this role. McClearn, among others, has suggested the use of the eight-way cross (McClearn, Wilson, & Meredith, 1970). Beginning with eight inbred strains of diverse origins it is possible to construct a stock with an extremely high degree of genetic diversity, expressed in a variety of phenotypes and specifically identifiable in numerous enzyme polymorphisms. It is conceivable that within such a stock, allelic combinations might arise that would not be possible in natural populations. This could occur if some of the progenitor strains were derived from completely separate populations. For example, recently inbred strains from *Mus musculus castaneus* and *M.m.molossinus* both show allelic variants at some loci that do not exist in any of the more commonly used strains (T. H. Roderick, personal communication, 1981). Such uniqueness might very well provide useful tools for the analysis of drug action. This notion is analogous to the argument that single-gene mutants may provide information on "normal" nonmutant physiology. In this case however, the degree of phenotypic aberration may provide either qualitative or quantitative uniqueness of response.

A method for construction of such a stock has been suggested by McClearn et al. (1970). Several such eight-way crosses exist in mice and have been useful in a number of pharmacological experiments (McClearn & Anderson, 1979). The size of the breeding colony required to avoid inbreeding is the major drawback of use of such a stock. However, for certain uses, as described later, such an investment may be worthwhile.

Experimental Advantages.   For the pharmacogeneticist, the major advantage of a genetically heterogeneous stock may be the appropriateness for investigations of genetic covariance or correlations. As is discussed later, these correlations may involve several behaviors affected by a single pharmacological agent or single phenotype effects of several agents. Although the number of subjects required for such analyses may be large, and vary in a negative fashion with character heritabilities (Klein, DeFries, & Finkbener, 1973), the benefits of such work may be large. In the present context, heterogeneous stocks provide a clear advantage over the use of inbred strains in such analyses. The fact of complete homozygosity in such inbred strains produces a physiologically aberrant situation. It is conceivable that pharmacological effects might depend not only on allelic genotype but also on the fact of homozygosity. More importantly, the random fixation of allelic pairs as a result of the inbreeding process may have produced unique polyallelic combinations. To the extent that these genotypic profiles might produce correlated behavioral/pharmacological effects, potentially large genetic correlations could result. These relationships might however be transient effects that could disappear or diminish in a random breeding population where they would occur with a smaller frequency, as those genotypic combinations would be rare. The presence, then, of genetic correlations in heterogeneous stocks indicates that they are probably more general phenomena.

Another experimental approach that finds powerful application has been termed the genotypic method in behavior genetics (Fuller & Thompson, 1978). The identification of an allelic variant can allow assessment of the pleiotropic effects of such on behavioral response to drugs. This approach is most often used with coisogenic stocks, but could be accomplished in heterogeneous stocks as well. For example one might identify electrophoretic variants of mouse liver alcohol dehydrogenase, which are known to be under Mendelian control (Holmes, 1977, 1979), and proceed to assess alcohol responsiveness in mice of differing genotype. If the phenotype examined were known to be influenced by dispositional variables, then one might expect an influence of ADH. Such pleiotropic effects could be examined for a wide variety of identified allelic variants (i.e., coat colors or enzymes). The advantage of the heterogeneous stock in this case is the ability to examine effects of one genetic variable (the marker phenotype), while maintaining a more or less random genetic background on which it is found. Again, observed effects might be more general than those seen with coisogenic stocks.

The utility of heterogeneous stocks for nongenetic research also deserves comment. Although the advantages of using inbred strains derives from their reproducibility and isogenicity, these variables are also potentially disadvantageous for the researcher not interested in genetic mechanisms. Aside from the generality issue raised by pharmacological work with one or few strains, the fact of genetic homogeneity at all loci may be less than desirable. Simply stated, a completely homozygous mouse is clearly an aberrant type of animal. Inbreeding depression effects may very well extend to pharmacological responsiveness, particularly if the behavioral phenotype is related to fitness. The heterogeneous stock provides a physiologically more normal animal. In stocks produced by careful crossing of inbred strains (i.e., 2-, 4-, or 8-way crosses), the genetic ancestry of any particular subject can be as well known as that of any inbred animal. Contrary to popular belief, the variability associated with measurement of a phenotype is not necessarily larger and can be smaller in heterogeneous stocks than in inbreds (McLaren & Michie, 1956). The good general health and large reproductive capacity of heterogeneous stocks is also a consideration.

An interesting use of the Colorado HS mice was reported by Reed (1978). In a population of 1055 mice, heart rate was measured before and after administration of ethanol. What occurred was a "normalizing" response. Mice that had deviant baselines scores tended toward the mean on the postalcohol test. Thus a substantial negative correlation between preethanol and postethanol scores existed. This normalizing response was also shown to exist in humans. In another study with HS mice (Reed, 1977), it was suggested that effects of ethanol on body temperature, open-field activity and heart rate were not genetically correlated.

Mutants and Coisogenic Stocks

The genotypic approach in the preceding section has received relatively little attention from the experimental pharmacogeneticist. This is not for a lack of genetic material to work with. Numerous allelic variants of enzymes have been identified in a variety of species. The mouse, as a mammal, is probably the species of choice here as well. As has been discussed, a variety of enzymes could be examined for allelic variant effects on pharmacologic responsiveness. In addition, the presence of numerous neurological mutants provides an obvious subject matter for neuropharmacological if not behavioral investigation.

Two examples of mutants are pertinent. First, the single-gene neurological mutant, Jimpy, is at least 10 times less sensitive to the analgesic effects of morphine (Law, Harris, Loh, & Way, 1978). These mice also have fewer opiate receptors (Loh, Cho, Wu, Harris, & Way, 1975). Whether these effects are direct or executed during development is unknown. A second example is not an identified mutant, but a plausible one. Poley (1972) reported that an Alberta subline of the C57BL/6 mouse would not consume ethanol in a choice situation the way that other closely related substrains would. Thus, although both C57BL/6J and C57BL/10J mice consumed considerable ethanol, the C57BL/-6UAE mice did not. Whitney and Horowitz (1978) replicated that work, but also showed that the Alberta subline would consume as much sweetened morphine as would the other sublines. The difference in ethanol consumption is certainly suggestive of a single-gene difference.

PHARMACOLOGICAL PHENOMENA

Acute Responses

Initial Sensitivity

Naive sensitivity to alcohol may be assessed in a number of ways depending on the behavior or process of interest. Loss of the righting reflex in the mouse has been a useful phenotype for genetic study. Although some variation in procedure occurs from lab to lab, the basic method is that of Kakihana, Brown, McClearn, and Tabershaw (1966). After injection, the mouse may or may not be assessed for the length of time needed to fall from a suspended grid (fall time). The primary measurement is the time it takes the subject to right itself spontaneously three times in 30 seconds. This measure gives a general estimate of hypnotic potency of a depressant drug. When combined with assessment of blood levels of the drug at time of waking, an estimate of neurosensitivity is obtained. This, of course, was the criterion phenotype for the selective breeding program for the LS and SS lines. Extensive study of strain variation in this response to alcohol also exists (Crabbe & Belknap, 1980).

Measures of intoxication are numerous. The tilting plane test, the grid test, and the Rota-Rod have all been utilized in pharmacogenetic work. The grid test (Belknap, 1975) involves suspension of a hardware cloth grid over a metal plate. Placement of a paw through the grid onto the plate results in a circuit activation and can be registered as a count. The more intoxicated a mouse, the greater number of "step-throughs." When combined with simultaneous assessment of locomotor activity, a relatively sensitive measure of intoxication can be obtained (Church, Fuller, & Dudek, 1977). That approach determined LS mice to be relatively more intoxicated by subhypnotic doses of ethanol than were SS mice. Dose-dependent effects may be obtained but, at times, only if a correction for general activity levels is made.

A third measure of initial sensitivity examines the purported biphasic shape of the ethanol dose-response curve. Locomotor activity is generally stimulated by low (1–2 g/kg) doses of ethanol. Higher doses of ethanol produce depression. The stimulant properties of low doses are of interest in that they seem to present a paradox: stimulant actions by a depressant drug. The genetic studies of this behavior at times indicate quantitative genotypic differences and in other cases qualitative differences. For example, both LS and SS mice are activated by 1.0 g/kg, but higher doses depress LS mice and continue to activate SS mice (Church et al., 1977). Alternatively, BALB/c mice show activation to several low doses and C57BL/6 mice apparently cannot be stimulated at all (Randall, Carpenter, Lester, & Friedman, 1975).

These examples were chosen to indicate that if an understanding of underlying mechanism is of interest, then it is probably best to study simple behaviors. Study of complex behaviors, although important, is difficult from a genetic viewpoint because more genes, and thus more physiological systems, are probably involved.

Toxicity

Both alcohol and opiates should be seen as toxic agents. Cirrhotic livers and heroin overdoses are *prima facie* evidence. As such, standard toxicological techniques can be applied. A study by Whitney and Whitney (1968) showed mortality rate among 10 genotypes to be negatively related to their free choice preference for ethanol. Also, the LD50 values for LS and SS mice were equivalent in mice from the 14th generation (Erwin, Heston, McClearn, & Deitrich, 1976).

Another type of toxicity is studied in the conditioned taste aversion experiment where ethanol can serve as a UCS or poison to produce flavor aversions. Toward this end, Horowitz and Whitney (1975) showed DBA/2 mice to form much more severe conditioned aversions, when ethanol was the UCS and saccharin flavored water the CS, than did C57BL/6 mice. Of similar interest is a recent report that BALB/c mice do not form conditioned taste aversions when doses of ethanol up to 6.0 g/kg are used as the UCS (MacPhail & Elsmore, 1980).

Chronic Drug Administration

Tolerance

This refers to the waning of responsiveness with repeated administration of a drug. It can be expressed in two ways. First, tolerance is evident if the magnitude of the response decreases with repeated administration of the same dose of the drug. Alternatively, tolerance can be demonstrated if, with repeated exposure, greater doses of the drug are required to maintain a given magnitude of the response being evaluated. Furthermore, tolerance observed at the behavioral level may be a result of alteration of drug availability (e.g., metabolic tolerance), adjustments made by the organism to compensate for the effects of the drug (e.g., behavioral tolerance), or changes that occur at the site of action due to continued presence of the drug (e.g., pharmacodynamic tolerance). Genetic factors at any of these levels could contribute to individual differences in the degree of tolerance to either alcohol or the opiates in both humans and nonhumans.

Pharmacogenetic aspects of drug tolerance and dependence, including both alcohol and the opiates, have been reviewed extensively in an excellent recent summary article by Crabbe and Belknap (1980). Thus, in the present chapter. one example is presented for genetic differences in the development of tolerance to alcohol and to the opiates. A similar strategy is employed in the following discussion of drug dependence.

Moore and Kakihana (1978) examined the development of tolerance to the hypothermic effects of ethanol in three inbred strains of mice. Animals were injected with 3 g/kg ethanol daily for a period of 8 days. Rectal body temperatures were recorded on the first, third, and last day of drug administration. As a general summary, tolerance developed to the hypothermic effects of alcohol in mice of the BALB and C57BL strains, but not in DBA mice. In fact, DBA mice may have become more sensitive to the hypothermic effects of ethanol with repeated administration. Although interpretation of the degree of tolerance is complicated by the magnitude of hypothermia induced by the initial exposure to ethanol in mice of these three genotypes, it is clear that genetic factors affected the rate and direction of changes in body temperature resulting from repeated ethanol administration. Furthermore, although differences in alterations in alcohol elimination rates were apparent in this study, these metabolic differences were not great enough or consistent with the genetic differences in the pattern of tolerance to the hypothermic effects.

Tolerance to the analgesic effects of morphine was evaluated by Oliverio, Castellano, and Eleftheriou (1975). Both initial sensitivity and tolerance were evaluated using the hot-plate technique in the Bailey BALB, C57BL, F_1 and recombinant inbred lines. Mice were initially tested following three doses of morphine and then made tolerant by four additional administrations of morphine. Tolerance evaluation followed the fifth morphine injection. In general, genetic differences were apparent in both initial sensitivity and acquired tolerance to the analgesic effects of morphine as measured by response to thermal stimulation. However, for both the initial sensitivity and the tolerance test, analgesia was evaluated at several postinjection times. Thus, interpretation of the results is complicated by dose by postinjection trial, which clearly could be influenced by the repeated-measures design (Collins & Whitney, 1978; Horowitz, 1981).

Dependence

Dependence is a consequence of chronic drug administration in which the drug is needed in order for the organism to maintain normal functioning. Both alcohol and opiates can lead to dependence following repeated administration. As Crabbe and Belknap (1980) have pointed out, dependence is inferred by the consequences of terminating drug administration (i.e., withdrawal or the abstinence syndrome). Genetic differences in the degree of dependence on alcohol and morphine have been alluded to in our discussion of selected lines. Goldstein and Kakihana (1975) reported greater seizure severity following withdrawal from chronic alcohol exposure in SS than LS mice, whereas Horowitz and Allan (1982) found analogous results in mice of these lines in both jumping and wet-dog shakes induced by withdrawal from chronic morphine administration. Similarly, the demonstration by Brase et al. (1977) of genotypic differences in severity of opiate withdrawal has been mentioned previously in our discussion of the use of inbred strains.

Self-Administration

This is another phenomenon associated with chronic drug administration. It can be demonstrated in a number of different paradigms, but in each case, the organism determines the amount of drug to be administered. Due to the number of subjects required in pharmacogenetic studies using nonhuman subjects, the most common techniques involve some variation of oral ingestion (see Horowitz, 1981, for a review). When the procedure employs an alternative to consuming a drug-adulterated solution or food source, drug preference or drug seeking can be examined. Genetic differences in preference for alcohol (Rodgers, 1966) and morphine-adulterated solutions (Whitney et al., 1977) among geneticaly defined subjects were discussed in the previous section on the use of inbred strains.

A cautionary note should be made regarding the importance of task specificity when evaluations are made regarding the genetic contributions to differences in tolerance, dependence, and self-administration. These cautions will be expanded in subsequent discussion. First, degree of tolerance for a psychoactive drug is often dependent on the choice of response used as a criterion of strength of effect. Secondly, inasmuch as dependence is inferred from the consequences of drug termination, genetic differences may reflect differences in the response being measured rather than differences in the absolute degree of dependence (see Goldstein, 1981; Horowitz & Allan, 1982). Finally, oral ingestion regimens of self-administration may be confounded by factors unrelated to the neurochemical effects of the drug, such as preabsorptive orosensory cues (Belknap, Belknap, Berg, & Coleman, 1977). Although all these task-related factors may be of interest from a behavioral pharmacogenetic perspective, such factors must be considered in making statements regarding genetic differences in the more general phenomena under investigation.

GENERAL ISSUES

Locus of Effect

Absorption and Distribution

When psychoactive agents such as alcohol and opiates are administered systemically and when genetic differences are observed in response to such administration, it is tempting to attribute the observed differences to genetically determined differences at the site of action, usually in the central nervous system. However, such conclusions are predicated on the assumption that equal amounts of the drug reach the site of action. The assumption may not be warranted, if genetic factors contribute to different amounts of the drug reaching the central nervous system via such mechanisms as differential absorption, metabolism, and permeability of the blood–brain interface.

Many careful attempts to evaluate such possible differences in drug distribution have been conducted in pharmacogenetic research involving alcohol and the opiates. For example, Brase et al. (1977) measured morphine-induced hyperactivity and analgesia (using an abdominal constriction technique) in four inbred and two randomly bred lines of mice. These researchers also measured brain morphine levels at the time of each test. Strain differences were apparent for all measures, including brain morphine levels. However, brain morphine levels did not correlate significantly with either analgesia or locomotor activity, indicating that the observed differences in these latter two phenotypes could not be explained by differences in distribution of the drug. Similarly, Moore and Kakihana (1978) demonstrated differences in blood alcohol levels following ethanol administration in three inbred strains of mice, but further showed that these differences could not totally account for differences among these strains in ethanol-induced hypothermia. Finally, Goldstein and Kakihana (1975) equated blood alcohol levels in LS and SS mice, but still found differences in the severity of seizures following withdrawal from chronic alcohol exposures.

In a slightly different sense, issues of drug absorption may be of primary interest in the investigation of behavioral phenotypes in pharmacogenetic research. For example, the substrate for the previously discussed difference in alcohol preference between C57BL and DBA mice has been the subject of considerable research. Recently, Belknap and his associates (Belknap et al., 1977; Belknap, Coleman, & Foster, 1978) have suggested that different mechanisms may underlie the avoidance of alcohol by DBA mice, and the preference that has been consistently demonstrated by most sublines of C57 mice. Specifically, DBA mice may avoid alcohol on the basis of preabsorptive (e.g., taste and olfactory) cues, rather than on the pharmacological consequences of ethanol. C57BL mice, on the other hand, may be less sensitive to these preabsorptive cues, and thus postabsorptive cues may be more relevant in explaining alcohol preference in mice of this genotype. Thus, it might be of interest to see whether a specific subline of C57BL mice (C57BL/Uae), which have been reported (Poley, 1972; Whitney & Horowitz, 1978) to avoid 10% ethanol solutions, show DBA-like responses to the preabsorptive cures of ethanol demonstrated by Belknap and associates (Belknap et al., 1977, 1978).

Peripheral Effects

Although drug action on behavioral phenotypes is often assumed to occur via effects on the central nervous system, it should be apparent from the preceding discussion that genetic influences on these drug responses may operate through peripheral mechanisms. Dispositional and pharmacokinetic factors are, however, not the only characters of this type. Cardiovascular as well as autonomic nervous system effects of pharmacological agents are well known. It is conceivable that alteration of such characters as heart rate, blood pressure, and sympathetic tone may play a large role in the general behavioral response to a drug. In other words, subjective evaluation of drug response may depend on information about changes in activity of these peripheral systems. A probable genetic component in variability of peripheral response to alcohol is well known. Racial differences in such variables as heart rate response, blood pressure effects and facial flush response are documented (Ewing, Rouse, & Pellizzeri, 1974). The greater sensitivity of Orientals to alcohol may be due in part to metabolic factors (Stamatoyannopoulos, Chen, & Fukui, 1975; von Wartburg & Schurch, 1968). However, the marked differences in peripheral response to alcohol may be related to differential drinking patterns as well (Wilson, McClearn, & Johnson, 1978). The issue may be even more complex, as suggested by Zeiner, Peredes, and Christensen (1979), who investigated the role of the primary metabolite of ethanol, acetaldehyde, in these racial differences in peripheral response. If acetaldehyde is responsible for ethanol effects, then examination of its direct effects is required (Dudek & Fuller, 1978).

Central Nervous System

Within the central nervous system (CNS) there are a number of physiological/ biochemical loci where genetic influences and drug effects might converge. One important way in which such genetic influences might occur is not directly discussed here. Genetic control over CNS development and the general issue of epigenetic processes are discussed in a later section. There remain several ways in which the genome can exert an influence over functional states of the CNS and, both by demonstration and inference, behavior.

Membranes and Channels. The unique attribute of the neuron, its excitable membrane, depends on the regulated passage of several kinds of ions through specialized protein channels. The action potential is in reality the generation of sodium and potassium currents as they traverse the neural membrane through these channels. The release of several neurotransmitters is dependent on similar actions of calcium. Modification of the number of these channels or their willingness to pass respective ions would affect neural function at its most basic level. Genetic control of affinity might be direct, given a single amino acid change due to mutation. Channel number control would seem to depend on gene regulation or posttranscriptional/epigenetic influences.

Several channel mutations have been isolated and mapped in paramecium species. Although this is not an example of a neural phenomenon, it remains an impressive fact that calcium and potassium channel mutants have behavioral sequelae in these single-celled organisms (see Kung, 1979, for a review). A particularly interesting use of these mutants was reported by Schein (1976), who took advantage of asexual reproduction and induced autogamy to follow alterations in membrane excitability during changes of paramecia from heterozygote to homozygous mutant form. Such a procedure allowed estimation of the calcium channel half-life at 5–8 days. Although no such demonstrations are, to our knowledge, available in neural tissue, the importance of such a possibility cannot be overlooked.

It has been long known that the narcotic effectiveness of depressant agents such as alcohol is related to the degree of lipid solubility. Thus the degree to which a compound can distribute itself into neural membranes is a direct predictor of its anesthetic potency (McCreery & Hunt, 1978). Because it is well accepted that anesthetics and many sedative hypnotic drugs can act in this way, it is a parsimonious assumption that genetic variation in membrane construction could result in differing lipid solubility of the same drug in different genotypes.

Although such genetic variation has, to our knowledge, not yet been demonstrated, the demonstrable hereditary variation in neurosensitivity to alcohols seems likely to involve these membrane mechanisms. The direct action of anesthetics on membranes is thought to occur via alteration of the proteins found there; these proteins are likely to be those of the channels discussed previously.

One investigation in this area has attempted to use genetic variation as a tool for studies of possible commonality of action of ethanol and halothane, a gaseous anesthetic. The LS and SS mice described earlier were shown to be similarly affected by halothane, even though their ethanol response is so strikingly different. It was concluded that the two drugs act via different mechanisms (Baker, Melchior, & Deitrich, 1980). In the present context, it would appear that different classes of membrane proteins are being affected by the two drugs. Alternatively, the differential LS and SS response to ethanol may not take its origin in membrane effects.

Receptors. In recent years it has become increasingly apparent that the functioning of the brain is critically dependent on highly specialized receptor molecules, protein in nature, and thus potentially subject to genetically controlled variation. These receptors are usually thought of as specific for particular neurotransmitters, neuromodulators, or hormones. Several clear examples of the involvement of such receptors exist in the psychopharmacogenetic literature. We deal here with drug effects and receptor studies of three types: benzodiazepines, opiates, and barbiturates.

Benzodiazepines have recently been shown to bind to specific sties in CNS membrane fragments. Thus it appears that the nervous system may possess endogenous neurotransmitter systems that are similar in function to the functions affected by this class of tranquilizers. The clinical efficacy of benzodiazepines (i.e., their anxiolytic effects) tends to correlate well with their affinity for these specific receptors. It has been suggested that characteristics of such endogenous benzodiazepine systems might be altered in pathological disorders of affect that are usually treated with this drug class. In animal model research, demonstrable genetic variation in emotionality/reactivity seems a likely candidate for explanation by variation in these receptors. Toward this end, Robertson, Martin, and Candy (1978) have studied benzodiazepine receptors in the rat strains selectively bred by Broadhurst (1960) for reactivity and nonreactivity. The Maudsley nonreactive rats (MNR) had a higher specific binding of diazepam in every brain region examined than did the Maudsley reactive (MR) rats. The largest differences were found in limbic structures. The nature of the differences was primarily one of receptor number rather than affinity. In hypothalamic tissue the number of binding sites was 1000 fmol/mg protein in MR rats and 1385 fmol/mg protein in MNR rats. The affinity for the ligand was only slightly higher in the MNR strain: $K_D = 3.64$ nmol for MR and 3.24 nmol for MNR. These authors suggest this issue of receptor density as a plausible basis for the reactivity differences in the strains. Presumably the more reactive MR rats have a less extensive endogenous "tranquilizing" system. In another study from the same laboratory (Robertson, 1979) similar results were found in four mouse strains. "Emotional" BALB/c mice had a lower density of benzodiazepine receptors than did three "less emotional" strains. Both these studies reinforce the notion that endogenous benzodiazepine systems play a role in emotional behavior. A more complete analysis in the future will, we hope, combine this type of study with actual examination of benzodiazepine effects on "emotionality."

Another type of receptor, the opiate receptor, has been similarly studied. As discussed previously, considerable genetic diversity has been demonstrated for behavioral effects of opiate compounds in mice. Again, a parsimonious prediction might be that opiate receptor density or affinity could serve as the basis for varied behavioral effects of opiates. Two groups of investigators have directly addressed this issue. Baran et al. (1975) investigated naloxone binding and analgesic response to morphine in C57BL/6By, BALB/cBy, their reciprocal F_1 hybrids, and seven recombinant inbred strains derived from those parental stocks. Considerable strain differences for total stereospecific naloxone binding in whole brain and also analgesic response were reported. Even though the strain with the lowest naloxone binding (CxBK) also had the lowest analgesic response, the correlation between binding and analgesia across strains was a nonsignificant .48. The relatively low statistical power for this correlation suggests that a real, but minor relationship may exist. Crabbe and Belknap (1980) point out that genotypic variation in behavioral response may well arise in part from dispositional factors, and thus the relative importance of receptor characteristics would diminish. We would suggest two additional possibilities. First, the Baran et al. (1975) study did not examine specific brain regions. To the extent that analgesic effectiveness of morphine is mediated by specific brain regions (e.g., the periaqueductal grey), analysis of a heterogeneous whole brain could well obscure important predictability from specific regions. The notion of non-homogeneous opiate receptor system distributions raises a second issue. Present conceptions of opiate systems within the brain assume not only anatomical topographies, but biochemical heterogeneity of opiate receptor subtypes. It is conceivable then, in this context, that naloxone specific binding may not reflect a more specific receptor subtype, which holds importance for the genetic variation in analgesic potency. Use of another ligand might reveal a correlation with the genetic effects on analgesia.

A recent study (Reggiani, Battaini, Kobayashi, Spano, & Trabucchi, 1980) has examined the role of subpopulations of opiate receptors in genotype-influenced morphine sensitivity. This work with C57BL/6 and DBA/2 strains of mice found that strain differences in opiate receptor number depended on both ligand (and thus subpopulation) and brain region. The C57BL strain had considerably more striatal opiate receptors when D-Ala$_2$-Met-enkephalin or Leu-enkephalin were used as ligands. Such differences did not occur when naloxone or dihydromorphine were ligands, nor did they occur in three other regions examined. These results suggested that the locomotor response to morphine in C57BL mice could take its origin in this striatal subpopulation of opiate receptors.

It is interesting that neither of these studies on strain differences in opiate receptor characteristics reported differences among genotypes in affinity (K_D) of the ligand for the receptor. This suggests that the kind of genetic variation present may not be that of structural gene mutation but rather complex regulation of the number of receptors inserted into neural membranes, or perhaps cell number. In this context, it may well be that control of receptor number in neural tissue could serve as a prime research area for regulation of gene action in higher organisms. A developmental perspective on this issue would probably also be of value.

As a third example of receptor characteristics and genetic control of drug response, the work of Waddingham, Riffee, Belknap, and Sheppard (1978) is of interest. The regulation of membrane-bound β-adrenergic receptors by catecholamines and the ubiquitous nature of norepinephrine influences on a variety of behaviors suggest these receptors to be of import in neural function. This study examined genetic influence on these β receptors and their response to phenobarbital treatment. C57BL/6 mice had generally higher densities of β-adrenergic receptors in several brain regions. After phenobarbital treatment, receptor proliferation was apparent, but dependent both qualitatively and quantitatively on strain and brain region. Again here, regulatory gene activity may be of interest particularly in the framework of an inducible system—induction produced by phenobarbital treatment.

Enzymes. As major gene products, enzymes in the CNS are likely candidates for heritable variation. The range of enzymes active in neural tissue is wide and provides a wealth of material for pharmacological study. At a time when research on behavioral control by various neurotransmitter systems is so predominant, it is perhaps easy to overlook other kinds of enzyme systems that could markedly influence drug response. Not the least of these are those biochemical pathways involved in energy metabolism (Cohen, Omenn, Motulsky, Chen, & Giblett, 1973). Small perturbations in cerebral energy metabolism can have potentially major effects on a wide variety of behaviors. The example provided by hypoglycemia and its effects on cognitive and affective characters is clear. The brain is a high energy utilization organ. All specialized functions of neural tissue depend on the more basic biochemistry of energy sources. Although we are not aware of genetic studies of enzymes involved in such energy metabolism and pharmacological alterations of activity, the caveat is relevant in the context of the distinction between direct and indirect drug effects on behavior discussed elsewhere in this chapter. A drug such as alcohol may very well modify neurotransmitter activity and some related behavior by actions on synthesis, degradation, receptor binding characteristics, or channel alteration. Numerous other "side" effects might also occur. For example, ethanol consumption produces a marked alteration in general metabolism, producing among other things a general acidosis and increasing blood β-hydroxybutyrate levels. As β-hydroxybutyrate is a usable energy source for brain tissue, marked changes in cerebral energy metabolism are likely to follow. Such "indirect" effects on neural function, and perhaps behavior, would certainly appear subject to the same sort of heritable variation as other biochemical systems. Thus the degree of behavioral modification by indirect as well as direct action of the pharmacological agent might also be influenced by genetic factors. While hoping not to be redundant, we believe the conclusion is obivous. That is, gene action may affect pharamcological responsiveness in a variety of ways other than via the specific neural system controlling the phenotype of interest.

Keeping the foregoing caution in mind, numerous informative avenues of research on neurotransmitter system regulation by genetic variables are available. The most well-researched of these is probably the area of catecholamine neurotransmitters. Norepinephrine and dopamine systems are subject to functional alteration by numerous pharmacological agents; many of these agents are of use in a clinical setting precisely because of their behavioral effects. These functional alterations can occur as a result of inhibition or induction of the enzymes responsible for synthesis and degradation of these catecholamine transmitters. Early work by Ciaranello, Barchas, Kessler, and Barchas (1972) first established the fact of strain differences in brain activity of tyrosine hydroxylase (TH), the rate-limiting enzyme in catecholamine biosynthesis. The same report also demonstrated marked strain differences in adrenal phenylethanolamine-N-methyl transferase (PNMT), the enzyme responsible for epinephrine production from norepinephrine. We here describe further examination of these mouse strain differences in TH activity in a context that serves as an example for pharmacogenetic study as well as neural development.

In midbrain of CBA/J mice, the activity of TH is 20% less than in BALB/cJ mice (Baker, Joh, & Reis, 1980). A comparable difference in immunocytochemically detectable dopamine neurons in midbrain suggests that this difference in TH activity is due to more cells with the enzyme, rather than to greater specific activity per cell. A correspondingly more dense innervation of the striatum, by what are presumably the axons of these midbrain nigral cells, occurs in the BALB/c mice. It is likely that this difference is related in a general way to the fact that the BALB/c brain is simply a larger one. Nonetheless, this represents a clear example of genetic control over neuronal organization and neurochemical characteristics of a behaviorally important system. The differences in density of striatal innervation apparently predispose a differential response to two agents with dopaminergic activity, amphetamine and apomorphine (Baker, Fink Joh, Swerdloff, & Reis, 1979). BALB/c mice are more responsive to the locomotor stimulant effects of amphetamine and less sensitive to the stereotypy-producing effects of apomorphine, a dopamine receptor agonist. Interestingly, these characteristics follow a developmental time course that lags slightly behind the appearance of the TH activity difference. Apparently both pre- and postsynaptic functions of the dopamine system are different between the two strains as reflected in the actions of the two drugs. A more detailed understanding of the genetic control over dopamine cell proliferation, growth, and effects on target tissue in this model system will answer many questions not only of pharmacogenetic interest, but also of basic importance to developmental neurobiology.

In this example of TH activity differences, the enzyme activity itself is probably not the focal point of either gene or pharmacological action. As noted, a more basic developmental issue is central. Yet the enzyme is a clear marker for a dramatic genetic influence. In other work, TH is seen to be more directly involved. Gamma-butyrolactone (GBL), through its metabolite gamma-hydroxybutyric acid (GHB), produces an inhibition of impulse transmission in dopamine neurons. A compensatory activation of TH occurs in these cells and an accumulation of dopamine occurs. In the LS and SS mice discussed previously, GLB has hypnotic effects that differ in the same direction as ethanol effects (Dudek & Fanelli, 1980). Of interest here is the fact that dopamine accumulation is 50% greater in the LS mice after GBL treatment. Presumably the activation of TH was greater in LS mice. This activation occurs via kinetic rather than molecular genetic induction and apparently reflects a differential effectiveness of GHB at a subcellular site in LS and SS dopamine neurons.

Although several studies of neurotransmitter systems' enzymes exist, there is yet to be a complete characterization of all biosynthetic and degradatory enzymes, transmitter levels and turnover rate, and receptor kinetics for any single system where genetic variation is also of interest. The difficulties of such a massive undertaking are obvious, but the benefits accruing to all three fields, neuropharmacology of drug action, neurochemical organization and dynamics, and mechanisms of genetic action are equally impressive. The field must surely progress along these lines.

Task Specificity

Alcohol

A recurrent theme in this chapter is that the genetic effects on a given behavioral response do not necessarily imply similar genetic effects on other phenotypes. The degree to which a similar genetic pattern occurs for more than one phenotype is a measure of the genetic correlation of these phenotypes. Yet task- or situation-specific genetic effects are not rare. We illustrate with two examples. The LS and SS mice discussed previously are usually very different in their response to ethanol. Yet in one study, across a wide range of doses, a particular behavior was not differentially affected. When ethanol was used as a poison

(UCS) in a conditioned taste aversion experiment, the magnitude of aversion produced was similar in the two lines (Dudek, 1982). Thus, whatever systems underlie the loss of righting-relfex differences do not control response to ethanol in the taste aversion experiment.

A second example may be drawn from the literature on strain differences in ethanol response. Although BALB/c mice lose the righting reflex longer than C57BL/6 mice (Kakihana et al., 1966), they show a greater locomootor stimulation following low-dose ethanol administration (Randall et al., 1975). It may be of interest that, in these two strains, amphetamine shows exactly opposite effects on locomotor activity. BALB/c mice are not activated, whereas C57BL/10 mice are (Moisset & Welch, 1973). Thus a negative genetic correlation is suggested for these effects of amphetamine and ethanol.

Measures of Responses to Opiates

It is becoming increasingly apparent that strain differences are readily obtained using the standard measures of opiate-induced analgesia and locomotor activity. However, it is equally apparent that both the magnitude and the direction of these differences change with the particular measures and procedures employed. For example, C57BL mice have been characterized as being sensitive to morphine-induced hyperactivity in open-field tests and toggle boxes (e.g., Brase et al., 1977; Collins & Whitney, 1978; Oliverio & Castellano, 1974). However, Eidelberg, Erspamer, Kreinick, and Harris (1975) have reported that C57BL mice respond to several doses of morphine with decreased activity when the measure is recorded on an activity wheel. Similarly, measures of analgesia, such as the hot-plate technique, tail-flick procedures, and abdominal writhing induced by agents such as acetic acid often yield very different strain distribution patterns. The exact nature of these differences among measures has been reviewed elsewhere (Crabbe & Belknap, 1980; Horowitz, 1981). For the present, the disparate results of these procedures lead us to argue for careful descriptions of the methods employed when discussing general phenomena associated with opiate responsivity. However, we share the optimism of Crabbe and Belknap (1980) that consistent patterns among several dimensions of opiate responsiveness may yet emerge as the results of further investigations are known.

Genetic Issues

Genetic Correlations

The basis of many inferences in pharmacogenetic research is the comparison of effects of more than one drug on the same genetic material. This is implicitly an example of genetic correlation. Both correlated drug effects and correlated effects of a single drug on several behaviors are means for analyzing the domain of effects of that pharmacological agent, or the common physiological substrate of a variety of agents. The quantitative genetic methods for deriving genetic correlations are readily available (Falconer, 1960; Lande, 1979; Hegmann & Possidente, 1981) and are not presented in detail here. We briefly discuss, instead, the major experimental approaches to genetic correlation methods and the general usefulness of such correlations.

One of the most utilized approaches to genetic correlations has been the examination of correlated characters in selected lines. A generalized example would find a set of lines selected for response to drug X, characterized at a variety of physiological levels, and perhaps other types of responses to the same drug being assessed as well. The usual argument is that care must be taken in interpretation of any correlated characters especially if replicated lines and unselected control lines are not available. The basis of this assertion is that when only two lines are sampled and two characters are measured on those genotypes, then the degrees of freedom for assessing the correlation are $N - 2 = 0$. Two points are worthwhile in counter. A set of once replicated lines would produce a situation with $df = 2$, hardly a powerful statistical test. Only when the numbers of genotypes exceeds 20 will actual correlations below .40 be detectable. In addition, correlational degrees of freedom are based on the assumption that both variables generating the covariance are free to vary. This is clearly not the case after a selection has been accomplished. The selected phenotype is at that point fixed. Only the as yet unmeasured phenotype is "free" to vary. Perhaps a more Bayesian approach is desirable.

The most rational alternative to numerous replicates is a simultaneous measurement of several phenotypes during the course of selection. As selection progresses, those characters sharing a true genetic covariance with the selected phenotype will diverge in somewhat parallel fashion. As this approach forces one to identify relevant variables even before the selection begins, it is a less than perfect one. Obviously a more practical approach is required. We would advocate a rational approach that makes use of a priori knowledge of the pharmacological system being examined and physiological intervention if the question being asked is one of mechanism of drug action. For example, in the LS and SS mice it is not surprising that alcohols other than ethanol have a differential effect (Erwin et al., 1976). However, if a correlated character is, for example, an altered preference for saccharin-sweetened water, then one must wonder if drift is responsible. In sum, no single approach seems completely suitable. Therefore, only careful and objective scrutiny can avoid erroneous conclusions.

The use of inbred strains and heterogeneous stocks has already been discussed. An additional tool in genetic correlation analysis is the diallel cross and also the triple-test cross (Jinks & Broadhurst, 1974). The inclusion of non-homozygous animals in the diallel cross analysis is probably an advantage, but again, unless the number of genotypes is large, statistical power is low.

The use of genetic correlations to assess commonalities of drug action is a powerful approach. Only the use of cross-tolerance studies can so clearly demonstrate common substrates of action as can the presence of a genetic correlation. The aforementioned effects of higher order alcohols on LS and SS mice led to the conclusion that considerably different mechanisms of action underlie the effects of alcohols and other sedative hypnotics that did not differentially affect the two lines. It is conceivable that, in a genetic system with high heritabilities, an experimenter could test a very large range of drug classes and attempt to classify them via one of a number of classification techniques on the basis of the genetic correlations.

The alternate approach examines the effects of a single drug on numerous behaviors. Examination of the kinds of genotype patterns of behaviors affected similarly by a given drug should indicate similar substrates for those behaviors. It would also be of interest to examine the pattern of genetic correlations in drugged and undrugged mice separately. Changes in those correlations would suggest an ability of the drug to dissociate genetically based covariation of behaviors. Thus even at a strictly behavioral level of analysis, the genetic correlation could be a very useful tool.

Developmental Phenomena

As already mentioned, the examination of drug responses in the developing organism and concomitant study of neural development is a powerful approach to the study of drug action and neural development. The addition of the genetic variable as a unique tool to tease apart neuropharmacological mechanisms is a potentially fruitful tactic. Correlated biochemical and pharmacological changes are more easily linked if the correlation holds in genotypes that show different time courses of developmental events.

An example from the literature on alcohol preference in mice is instructive. Wood (1976) reviews this literature on age-related differences in response to and preference for alcohol by laboratory rodents. Two studies in particular are instructive both for purposes of the genetic variable and its interaction with an experiential variable (Kakihana & Butte, 1980; Kakihana & McClearn, 1963). BALB/c mice are, as adults, avoiders of ethanol. Yet, from 4–9 weeks of age, they will voluntarily consume a fair amount of ethanol. Consumption diminished to near zero by 9–10 weeks unless a choice between alcohol and water was provided during the 4 to 8-week period. Interestingly, mice forced to drink alcohol during this period would typically not consume the alcohol as adults. Thus the fact of choice in young mice appears to predispose alcohol consumption in adults. This plasticity of phenotype or lack of phenostability makes clear the complex nature of the neural and behavioral systems on which the genes can act only very indirectly.

A more general view of pharmacogenetic research must realize that a most critical variable in drug effectiveness is the functional organization of the complex target organ called the nervous system. It is ultimately the genetic influence on the assemblage of the brain and other physiological systems that provides the substrate for individual differences produced in drug responsivity. A given drug acting on similar receptors in two brains that differ in organization of the neural network in which these receptors reside might produce different behavioral outcomes. The mechanism of such gene effects on neural development may be as simple as cell proliferation and migration rates or as complex as regulation of the timing of critical and sensitive periods for behavioral and emotional development. The study of such biochemical characters as those already enumerated may yield important findings. Ultimately, however, a concern with developmental issues in pharmacogenetic work will yield much more information about the mechanisms of gene and drug action. The adoption of this developmental–epigenetic perspective can only help to understand better the behavior(s) in question as well.

SUMMARY AND CONCLUSIONS

Broadhurst (1978) has emphasized a strong quantitative genetic approach to behavioral pharmacogenetics. Thus, extensive breeding designs, such as diallel cross, are stressed. The information derived from such biometrical analyses may be of considerable use to the geneticist. However, we have emphasized the search for mechanisms of gene action at the neural level. The two approaches are not mutually exclusive. It could be the case that the quantitative genetic information stressed by Broadhurst is necessary before complete hypotheses about mechanism can be generated. There may be an analogy in the distinction between the fields of psychopharmacology and its behavioral emphasis and neuropharmacological emphasis on substrate. That distinction appears to be fading; perhaps the genetic architecture–mechanism distinction is also somewhat artificial.

The biometrical approach could have at least one important, but as yet unexplored application. The interpretation of biometrical genetic architecture in evolutionary terms could be of interest in pharmacology also. Presumably, the types of drug sensitivities and resistances have some relationship to the ecological/ dietary history of the species. An animal species that never encounters a particular plant species with a toxic alkaloid component would probably not have evolved either resistance mechanisms or consummatory avoidance mechanisms. This should be reflected in the genetic architecture of that population's response to that alkaloid. What we are suggesting is an extension of Omenn and Motulsky's (1978) concept of ecogenetics to actual studies of species or subspecies populations, their evolutionary history of toxic substances, and their pharmacology.

Both genes and drugs share several common features (Fuller & Hansult, 1975). Therefore, for at least two reasons, the problems of task specificity have intruded upon both behavioral genetics and behavioral pharmacogenetics alike. It is one thing to recognize that genetic influences on responses to drugs might be affected by the behaviors we choose to measure. However, this is simply a summary of the findings to date. It is perhaps one of the most interesting challenges facing researchers in the field to search for the dimensions on which certain genetic differences in behavioral responses to each drug cluster, and which measures illuminate different aspects of drug responsivity (Crabbe & Belknap, 1980). In order to advance along this path, we need a better understanding into which aspects of nervous system integration each measure is providing the mirror. This challenge applies to all areas of behavioral research.

It bears reemphasis that the reductionistic approach to pharmacogenetics suggested here is potentially of great benefit not only to the pharmacogeneticist, but also to the neuroscientist. Any study that sorts out a neural mechanism underlying some genetically variable drug response not only provides information about that response, but also uncovers a probable basic functional rule in the nervous system. The genes truly can be tools for study of the nervous system as well as behavior. As Fuller and Hansult (1975) have previously noted, both genes and drugs can act as modifying agents of behavioral responses. It is hoped that the integrative field of behavioral pharmacogenetics will contribute to our understanding of the ways in which genes, drugs, and underlying physiological substrates interact to affect individual differences in behavioral responses.

Finally, we would like to return to the metaphor of heterosis. We now have ample evidence that genetic factors contribute to individual differences in behavioral responses to many psychoactive agents. Clearly, more examples can and should be explored. However, it is equally clear that many pieces of the puzzle are already available. It is time perhaps to try to put the puzzle together. That is, in one sense the key to heterosis may lie in the integration of the ways in which genes, drugs, and behaviors interact. The heterosis might then be realized by the contributions of this integration to our understanding of, not only the progenitor disciplines, but also to the larger issue of brain–behavior relationships. It should not be surprising if there are actually a number of different puzzles. Nonetheless, the integration may allow us to see which pieces are missing and may fruitfully guide our future research efforts.

REFERENCES

Asberg, M., Evans, D. A. P., & Sjoqvist, F. Genetic control of nortriptyline kinetics in man: A study of relatives of propositi with high plasma concentrations. *Journal of Medical Genetics,* 1971, *8,* 129–135.

Baker, H., Joh, T. H., & Reis, D. J. Genetic control of number of midbrain dopaminergic neurons in inbred strains of mice: Relationship to size and neuronal density of the striatum. *Proceedings of the National Academy of Sciences, USA,* 1980, *77,* 4369–4373.

Baker, H., Fink, J. S., Joh, T. H., Swerdloff, A., & Reis, D. J. Ontogeny of strain differences of nigrostriatal tyrosine hydroxylase activity and drug-induced behaviors in mice. *Neuroscience Abstracts,* 1979, *5,* 641.

Baker, R., Melchior, C., & Deitrich, R. The effect of halothane on mice selectively bred for differential sensitivity to alcohol. *Pharmacology Biochemistry and Behavior*, 1980, *12*, 691–695.

Baran, A., Shuster, L., Eleftheriou, B. E., & Bailey, D. W. Opiate receptors in mice: Genetic differences. *Life Sciences*, *1975*, *17*, 663–640.

Belknap, J. K. The grid test: A measure of alcohol and barbiturate-induced impairment of behavior in mice. *Behavior Research Methods and Instrumentation*, 1975, *7*, 66–67.

Belknap, J. K., Belknap, N. D., Berg, J. H., & Coleman, R. Preabsorptive vs. postabsorptive control of ethanol intake in C57BL/6J and DBA/2J mice. *Behavior Genetics*, 1977, *7*, 413–425.

Belknap, J. K., Coleman, R. R., & Foster, K. Alcohol consumption and sensory threshold differences between C57BL/6J and DBA/2J mice. *Physiological Psychology*, 1978, *6*, 71–74.

Belknap, J. K., MacInnes, J. W., & McClearn, G. E. Ethanol sleep times and hepatic alcohol and aldehyde dehydrogenase activities in mice. *Physiology and Behavior*, 1972, *9*, 453–457.

Beutler, E. Glucose-6-phosphate dehydrogenase deficiency. In J. B. Stanburg, J. B. Wyngaarden, & D. S. Fredrickson (Eds.), *The metabolic basis of inherited disease* (4th ed.). New York: McGraw–Hill, 1978.

Blum, K., Hamilton, M. G., & Wallace, J. E. Alcohol and opiates: A review of common neurochemical and behavioral mechanisms. In K. Blum (Ed.), *Alcohol and opiates: Neurochemical and behavioral mechanisms*. New York: Academic Press, 1977.

Brase, D. A., Loh, H. H., & Way, E. L. Comparison of the effects of morphine on locomotor activity, analgesia and primary and protracted physical dependence in six mouse strains. *Journal of Pharmacology and Experimental Therapeutics*, 1977, *201*, 368–374.

Broadhurst, P. L. *Drugs and inheritance of behavior*. New York: Plenum Press, 1978.

Broadhurst, P. L. Experiments in psychogenetics: Applications of biometrical genetics to the inheritance of behaviour. In H. J. Eysenck (Ed.), *Experiments in personality*. Vol 1, *Psychogenetics and psychopharmacology*. London: Routledge and Kegan Paul, 1960.

Cappell, H., & LeBlanc, A. E. Some factors controlling oral morphine intake in rats. *Psychopharmacologia*, 1971, *21*, 192–201.

Castellano, C., Espinet-Llovera, B., & Oliverio, A. Morphine-induced running and analgesia in two strains of mice following septal lesions or modification of brain amines. *Naunyn-Schmiedeberg's Archives of Pharmacology*, 1975, *288*, 355–370.

Castellano, C., & Oliverio, A. A genetic analysis of morphine-induced running and analgesia in the mouse. *Psychopharmacologia* (Berl.), 1975, *41*, 197–200.

Church, A. C., Fuller, J. L., & Dudek, B. C. Behavioral effects of salsolinol and ethanol on mice selected for sensitivity to alcohol-induced sleep-time. *Drug and Alcohol Dependence*, 1977, *2*, 443–452.

Ciaranello, R. D., Barchas, R., Kessler, S., & Barchas, J. D. Catecholamines: Strain differences in biosynthetic enzyme activity in mice. *Life Sciences*, 1972, *11*, 565–572.

Cohen, P. T. W., Omenn, G. S., Motulsky, A. G., Chen, S. H., & Giblett, E. R. Restricted variation in the glycolytic enzymes of human brain and erythrocytes. *Nature New Biology*, 1973, *241*, 229–233.

Collins, A. C. A review of research using the short-sleep and long-sleep mice. In G. E. McClearn, R. A. Deitrich, & V. G. Erwin (Eds.), *The development of animal models as pharmacogenetic tools*. Washington, D.C.: NIAAA Monograph, 1981.

Collins, A. C., Lebsack, M. E., & Yeager, T. N. Mechanisms that underlie sex-linked and genotypically determined differences in the depressant actions of alcohol. *Annals of the New York Academy of Sciences*, 1976, *273*, 303–316.

Collins, R. L., & Whitney, G. Genotype and test experience determine responsiveness to morphine. *Psychopharmacology*, 1978, *56*, 57–60.

Crabbe, J. C., & Belknap, J. Pharmacogenetic tools in the study of drug tolerance and dependence. *Substance and Alcohol Actions/Misuse*, 1980, *1*, 385–413.

Dudek, B. C. Ethanol-induced conditioned taste aversions in mice that differ in neurosensitivity to ethanol. *Journal of Studies on Alcohol*, 1982, *43*, 129–136.

Dudek, B. C., & Fanelli, R. J. Effects of gamma-butyrolactone, amphetamine and haloperidol in mice differing in sensitivity to alcohol. *Psychopharmacology*, 1980, *68*, 89–97.

Dudek, B. C., & Fuller, J. L. Task dependent genetic influences on behavioral response of mice (*Mus musculus*) to acetaldehyde. *Journal of Comparative and Physiological Psychology*, 1978, *92*, 749–758.

Eidelberg, E., Erspamer, R., Kreinick, C. J., & Harris, J. Genetically determined differences in the effects of morphine on mice. *European Journal of Pharmacology*, 1975, *32*, 329–336.

Eleftheriou, B. E. (Ed.). *Psychopharmacogenetics*. New York: Plenum Press, 1975.

Eleftheriou, B. E., & Elias, P. K. Recombinant inbred strains: A novel approach for psychopharmacogeneticists. In B. E. Eleftheriou (Ed.), *Psychopharmacogenetics*. New York: Plenum Press, 1975.

Eriksson, K. Alcohol imbibition and behavior: A comparative genetic approach. In B. E. Eleftheriou (Ed.), *Psychopharmacogenetics*. New York: Plenum Press, 1975.

Eriksson, K., & Kiianmaa, K. Genetic analysis of susceptibility to morphine addiction in inbred mice. *Annales Medicinae Experimentalis et Biologiae Fenniae*, 1971, *49*, 73–78.

Erwin, V. G., Heston, W. D. W., McClearn, G. E., & Deitrich, R.   Effects of hypnotics on mice genetically selected for sensitivity to ethanol. *Pharmacology, Biochemistry and Behavior*, 1976, *4*, 679–683.

Ewing, J. A., Rouse, B. A., & Pellizzari, E. D. Alcohol sensitivity and ethnic background. *American Journal of Psychiatry*, 1974, *131*, 206–210.

Falconer, D. S. *Introduction to quantitative genetics*. New York: Ronald Press, 1960.

Fuller, J. L., & Hansult, C. D. Genes and drugs as behavior modifying agents. In B. E. Eleftheriou (Ed.), *Psychopharmacogenetics*. New York: Plenum Press, 1975.

Fuller, J. L., & Thompson, W. R. *Foundations of behavior genetics*. St. Louis: C. V. Mosby, 1978.

Goldstein, A., Aronow, L., & Kalman, S. M. *Principles of drug action* (2nd ed.). New York: Wiley, 1974.

Goldstein, A., & Sheehan, P. Tolerance to opioid narcotics. I. Tolerance to the "running fit" caused by levorphanol in the mouse. *Journal of Pharmacology and Experimental Therapeutics*, 1969, *169*, 175–184.

Goldstein, D. B. What traits to breed for, and how to measure them. In G. E. McClearn, R. A. Deitrich, & V. G. Erwin (Eds.), *The development of animal models as pharmacogenetic tools*. Washington, D.C.: NIAAA Monograph, 1981.

Goldstein, D. B., & Kakihana, R. Alcohol withdrawal reactions in mouse strains selectively bred for long or short sleep times. *Life Sciences*, 1975, *17*, 981–986.

Goodwin, D. W. Alcoholism and heredity: A review and hypothesis. *Archives of General Psychiatry*, 1979, *36*, 57–61.

Hegmann, J. P, & Possidente, B. Estimating genetic correlations from inbred strains. *Behavior Genetics*, 1981, *11*, 103–114.

Heston, W. D. W., Erwin, V. G., Anderson, S. M., & Robbins, H. A comparison of the effects of alcohol on mice selectively bred for differences in ethanol sleep-time. *Life Sciences*, 1974, *14*, 365–370.

Ho, A. K. S., Tsai, C. S., & Kissin, B. Neurochemical correlates of alcohol preference in inbred strains of mice. *Pharmacology Biochemistry and Behavior*, 1975, *3*, 1073–1076.

Ho, I. K., Loh, H. H., & Way, E. L. Morphine analgesia, tolerance and dependence in mice from different strains and vendors. *Journal of Pharmacy and Pharmacology*, 1977, *29*, 583–584.

Holmes, R. S. The genetics of alpha-hydroxyacid oxidase and alcohol dehydrogenase in the mouse: Evidence for multiple gene loci and linkage between Hao-2 and ADH-3. *Genetics*, 1977, *87*, 709.

Holmes, R. S. Genetics and ontogeny of alcohol dehydrogenase isoenzymes in the mouse: Evidence for a cis-acting regulator gene (Adt–I) controlling C_2 isoenzyme expression in reproductive tissues and close linkage of ADH–3 and ADH–1 on chromosome 3. *Biochemical Genetics*, 1979, *17*, 461–472.

Horowitz, G. P. Behavioral pharmacogenetics. (Review of *Drugs and the inheritance of behavior*, by P. L. Broadhurst). *Science*, 1979, *203*, 1000–1001.

Horowitz, G. P. Pharmacogenetic models and behavioral responses to opiates. In G. E. McClearn, R. A. Deitrich, & V. G. Erwin (Eds.), *The development of animal models as pharmacogenetic tools*, NIAAA, Research Monograph, 1981.

Horowitz, G. P., & Allan, A. M. Morphine withdrawal in mice selectively bred for differential sensitivity to ethanol. *Pharmacology Biochemistry and Behavior*, 1982, *16*, 35–39.

Horowitz, G. P., & Whitney, G. Alcohol-induced conditioned aversion: Genotypic specificity in mice (*Mus musculus*). *Journal of Comparative and Physiological Psychology*, 1975, *89*, 340–346.

Horowitz, G. P., Whitney, G., Smith, J. C., & Stephan, F. K. Morphine ingestion: Genetic control in mice. *Psychopharmacology*, 1977, *52*, 119–122.

Jinks, J. L., & Broadhurst, P. L. How to analyse the inheritance of behavior in animals—the biometrical approach. In J. H. F. van Abeelen (Ed.), *The genetics of behavior*. Amsterdam: North-Holland. 1974.

Judson, B. A., & Goldstein, A. Genetic control of opiate-induced locomotor activity in mice. *Journal of Pharmacology & Experimental Therapeutics*, 1978, *206*, 56–60.

Kakihana, R., & Butte, J. C. Behavioral correlates of inherited drinking in lab animals. In K. Eriksson, J. D. Sinclair, & K. Kiianmaa (Eds.), *Animal models in alcohol research*. New York: Academic Press, 1980.

Kakihana, R., & McClearn, G. E. Development of alcohol preference in BALB/c mice. *Nature*, 1963, *199*, 511–512.

Kakihana, R., Brown, G., McClearn, G. E., & Tabershaw, I. Brain sensitivity to alcohol in inbred mouse strains. *Science*, 1966, *154*, 1574–1575.

Klein, T. W., DeFries, J. C., & Finkbeiner, C. T. Heritability estimates and genetic correlation: Standard errors of estimate and sample size. *Behavior Genetics*, 1973, *3*, 355–364.

Kung, C. Neurobiology and neurogenetics of *Paramecium* behavior. In X. O. Breakfield (Ed.), *Neurogenetics: Genetic approaches to the nervous system*. New York: Elsevier, 1979.

LaDu, B. N., Jr. The genetics of drug reactions. In V. A. McKusick & R. Claiborne (Eds.), *Medical genetics*. New York: HP Publishing Co., 1973.

Lande, R. Quantitative genetic analysis of multivariate evolution applied to brain size allometry. *Evolution*, 1979, *33*, 402–416.

Law, E. Y., Harris R. A., Loh, H. H., & Way, E. L. Evidence for the involvement of cerebroside sulfate in opiate receptor binding: Studies with azure a and jimpy mutant mice. *Journal of Pharmacology and Experimental Therapeutics*, 1978, *207*, 458–468.

Lester, D., Lin, G., Anandam, N., Riley, E. P., Worsham, E. D., & Freed, E. X. Selective breeding of rats for differences in reactivity to alcohol: An approach to an animal model of alcoholism. IV. Some behavioral and chemical measures. In R. G. Thurman, H. Williamson, H. Drott, & B. Chance (Eds.), *Alcohol and aldehyde metabolizing systems*. New York: Academic Press, 1977.

Loh, H. H., Cho, T. M., Wu, Y. C., Harris, R. A., and Way, E. L. Opiate binding to cerebroside sulfate: A model system for opiate-receptor interaction. *Life Sciences*, 1975, *16*, 1811–1816.

Lumeng, L., Hawkins, T. D., & Li, T. K. New strains of rats with alcohol preference and non-preference. In R. G. Thurman, J. R. Williamson, H. R. Drott, & B. Chance (Eds.), *Alcohol and aldehyde metabolizing systems*. New York: Academic Press, 1977.

MacPhail, R. C., & Elsmore, T. F. Ethanol-induced flavor aversions in mice: A behavior–genetic analysis. *Neurotoxicology*, 1980, *1*, 625–634.

McLaren, A., & Michie, D. Variability of response in experimental animals: A comparison of the reactions of inbred, F_1 hybrid, and random-bred mice. *Journal of Genetics*, 1956, *54*, 440–455.

McClearn, G. E. The use of strain rank orders in assessing equivalence of technique. *Behavior Research Methods and Instrumentation*, 1968, *1*, 49–51.

McClearn, G. E., & Anderson, S. M. Genetics and ethanol tolerance. *Drug and Alcohol Dependence*, 1979, *4*, 61–76.

McClearn, G. E., Deitrich, R. A., & Erwin, V. G. (Eds.). *The development of animal models as pharmacogenetic tools*. Washington, D.C.: NIAAA Monograph, 1981.

McClearn, G. E., & Kakihana, R. Selective breeding for ethanol sensitivity in mice. *Behavior Genetics*, 1973, *3*, 409–410. (Abstract)

McClearn, G. E., & Kakihana, R. Selective breeding for ethanol sensitivity: SS and LS mice. In G. E. McClearn, R. A. Deitrich, & V. G. Erwin (Eds.), *The development of animal models as pharmacogenetic tools*. NIAAA Monograph, 1981.

McClearn, G. E., & Rodgers, D. A. Genetic factors in alcohol preference of laboratory mice. *Journal of Comparative and Physiological Psychology*, 1961, *54*, 116–119.

McClearn, G. E., Wilson, J. R., & Meredith, W. The use of isogenic and heterogenic mouse stocks in behavioral research. In G. Lindzey & D. D. Thiessen (Eds.), *Contributions to behavior–genetic analysis: The mouse as a prototype*. New York: Appleton–Century–Crofts, 1970.

McCreery, M. J., & Hunt, W. A. Physico-chemical correlates of alcohol intoxication. *Neuropharmacology*, 1978, *17*, 451–461.

Meade, R., Amit, Z., Pachter, W., & Corcoran, M. E. Differences in oral intake of morphine by two strains of rats. *Research Communications in Chemical Pathology and Pharmacology*, 1973, *6*, 1105–1108.

Moissett, B., & Welch, B. L. Effects of *d*–amphetamine upon open field behavior in two inbred strains of mice. *Experientia*, 1973, *29*, 625–626.

Moore, J. A., & Kakihana, R. Ethanol-induced hypothermia in mice: Influence of genotype on development of tolerance. *Life Sciences*, 1978, *23*, 2331–2338.

Nichols, J. R. Alcoholism and opiate addiction: Theory and evidence for a genetic link between the two. In O. Forsander & K. Eriksson (Eds.), *Biological aspects of alcohol consumption*. Helsinki: Finnish Foundation for Alcohol Studies, 1972.

Nichols, J. R., & Hsiao, S. Addiction liability of albino rats: Breeding for quantitative differences in morphine drinking. *Science*, 1967, *157*, 561–563.

Oliverio, A., & Castellano, C. Genotype-dependent sensitivity and tolerance to morphine and heroin: Dissociation between opiate-induced running and analgesia in the mouse. *Psychopharmacologia* (Berl.), 1974, *39*, 13–22.

Oliverio, A., Castellano, C., & Eleftheriou, B. E. Morphine sensitivity and tolerance: A genetic investigation in the mouse. *Psychopharmacologia* (Berl.), 1975, *42*, 219–225.

Omenn, G. S., & Motulsky, A. G. Pharmacogenetics: Clinical and experimental studies in man. In B. E. Eleftheriou (Ed.), *Psychopharmacogenetics*. New York: Plenum Press. 1975.

Omenn, G. S., & Motulsky, A. G. Psychopharmacogenetics. In A. R. Kaplan (Ed.), *Human behavior genetics*. Springfield, Ill.: Charles C. Thomas, 1976.

Omenn, G. S., & Motulsky, A. G. Ecogenetics: Genetic variation in susceptibility to environmental agents. In B. H. Cohen, A. M. Lilienfield, & P. C. Huang (Eds.), *Genetic issues in public health and medicine*. Springfield, Ill.: Charles C. Thomas, 1978.

Pare, C. M. B., & Mack, J. W. Differentiation of two genetically specific types of depression by the response to antidepressant drugs. *Journal of Medical Genetics*, 1971, *8*, 306–309.

Plomin, R., DeFries, J. C., & McClearn, G. E. *Behavioral genetics: A primer*. San Francisco: Freeman, 1980.

Poley, W. Alcohol-preferring and alcohol-avoiding C57BL mice. *Behavior Genetics*, 1972, *2*, 245–248.

Randall, C. L., Carpenter, J. A., Lester, D., & Friedman, H. J. Ethanol-induced mouse strain differences in locomotor activity. *Pharmacology, Biochemistry and Behavior*, 1975, *3*, 533–535.

Reed, T. E. Three heritable responses to alcohol in a heterogeneous randomly mating mouse strain. *Journal of Studies on Alcohol*. 1977, *38*, 618–632.

Reed, T. E. Marked individual variability in heart rate "normalization" by ethanol. *Psychopharmacology*, 1978, *58*, 95–98.

Reggiani, A., Battaini, F., Kobayashi, Y., Spano, P., & Trabucchi, M. Genotype-dependent sensitivity to morphine: Role of different opiate receptor populations. *Brain Research*, 1980, *189*, 289–294.

Rice, M. C., & O'Brian, S. J. Genetic variance of laboratory outbred Swiss mice. *Nature*, 1980, *283*, 157–161.

Robertson, H. A. Benzodiazepine receptors in "emotional" and "non-emotional" mice: Comparison of four strains. *European Journal of Pharmacology*, 1979, *56*, 163–166.

Robertson, H. A., Martin, I. L., & Candy, J. M. Differences in benzodiazepine receptor binding in Maudsley reactive and Maudsley non-reactive rats. *European Journal of Pharmacology*, 1978, *50*, 455–457.

Rodgers, D. A. Factors underlying differences in alcohol preference among inbred strains of mice. *Psychosomatic Medicine*, 1966, *28*, 498–513.

Rusi, M., Eriksson, K., & Maki, J. Genetic differences in the susceptibility to acute ethanol intoxication in selected rat strains. In M. M. Gross (Ed.), *Alcohol intoxication and withdrawal* (Vol. 3a). New York: Plenum Press, 1977.

Satinder, K. P. Oral intake of morphine in selectively bred rats. *Pharmacology Biochemistry and Behavior*, 1977, *7*, 43–49.

Schein, S. J. Calcium channel stability measured by gradual loss of excitability in pawn mutants of *Paramecium auerlia*. *Journal of Experimental Biology*, 1976, *65*, 725–736.

Schlesinger, K., Bennett, E. L., & Hebert, M. Effects of genotype and prior consumption of alcohol on rates of ethanol-1-^{14}C metabolism in mice. *Quarterly Journal of Studies on Alcohol*, 1967, *28*, 231–235.

Schlesinger, K., Kakihana, R., & Bennett, E. L. Effects of tetraethylthiuram disulfide (Antabuse) on the metabolism and consumption of ethanol on mice. *Psychosomatic Medicine*, 1966, *28*, 514–520.

Schuckit, M. A. Biological markers: Metabolism and acute reactions to alcohol in sons of alcoholics. *Pharmacology, Biochemistry and Behavior*, 1980 (Supplement 1), *13*, 9–16.

Schuckit, M. A., Goodwin, D., & Winokur, G.   A study of alcoholism in half siblings. *American Journal of Psychiatry*, 1972, *128*, 122–126.

Sheppard, J. R., Albersheim, P., & McClearn, G. E. Aldehyde dehydrogenase and ethanol preference in mice. *Journal of Biological Chemistry*, 1970, *245*, 2876–2882.

Stamatoyannopoulos, G., Chen, S. H., & Fukui, M. Liver alcohol dehydrogenase in Japanese: High population frequency of atypical form and its possible role in alcohol sensitivity. *American Journal of Human Genetics*, 1975, *27*, 787–796.

Thomas, K. Selection and avoidance of alcohol solutions by two strains of inbred mice and derived generations. *Quarterly Journal of Studies on Alcohol*, 1969, *30*, 849–861.

Trabucchi, M., Spano, P. F., Racagni, G., & Oliverio, A. Genotype-dependent sensitivity to morphine: Dopamine involvement in morphine-induced running in the mouse. *Brain Research*, 1976, *114*, 536–540.

Vesell, E. S. Factors causing interindividual variations of drug concentrations in blood. *Clinical Pharmacology and Therapeutics*, 1974, *16*, 135–148.

Vesell, E. S. Pharmacogenetics. In L. G. Jackson & R. N. Schimke (Eds.), *Clinical genetics: A source book for physicians*. New York: Wiley, 1979.

Vesell, E. S., Page, J. G., & Passananti, G. T. Genetic and environmental factors affecting ethanol metabolism in man. *Clinical Pharmacology and Therapeutics*, 1971, *12*, 192–201.

von Wartburg, J. P., & Schurch, P. M. Atypical human liver alcohol dehydrogenase. *Annals of the New York Academy of Sciences*, 1968, *151*, 936–946.

Waddingham, S., Riffee, J., Belknap, J. K., & Sheppard, J. R. Barbiturate dependence in mice: Evidence for Beta–adrenergic receptor proliferation in brain. *Research Communications in Chemical Pathology and Pharmacology*, 1978, *20*, 207–219.

Whitney, G., & Horowitz, G. P. Morphine preference of alcohol-avoiding and alcohol-preferring C57BL mice. *Behavior Genetics*, 1978, *8*, 177–182.

Whitney, G., Horowitz, G. P., & Collins, R. L. Relationship between morphine self-administration and the effect of morphine across strains of mice. *Behavior Genetics*, 1977, *7*, 92. (Abstract)

Whitney, G. D., & Whitney, Y. Ethanol toxicity in the mouse and its relationship to ethanol selection. *Quarterly Journal of Studies on Alcohol*, 1968, *29*, 44–48.

Wilson, J. R., McClearn, G. E., & Johnson, R. C. Ethnic variation in the use and effects of alcohol. *Drug and Alcohol Dependence*, 1978, *3*, 147–151.

Wood, W. G. Age-associated differences in response to alcohol in rats and mice: A biochemical and behavioral review. *Experimental Aging Research*, 1976, *2*, 543–562.

Zeiner, A. R., Paredes, A., & Christensen, H. D. The role of acetaldehyde in mediating reactivity to an acute dose of ethanol among different racial groups. *Alcoholism: Clinical and Experimental Research*, 1979, *3*, 11–18.

5

Medical Genetics, Genetic Counseling, and Behavior Genetics

Gilbert S. Omenn

University of Washington

Clinical genetics has blossomed as a field of medicine in the past 15 years as a result of remarkable scientific progress and numerous practical applications of genetic principles in diagnosis, treatment, and counseling of patients and their families. Earlier, genetics was viewed by most physicians as preoccupied with rare or exotic diseases, and genetic diseases were considered to be untreatable or inevitable. A much more activist approach to at least some of these diseases is feasible now, and treatments drawn from the full array of the medical armamentarium have been applied. A sampling is listed in Table 5.1. Many of these treatments depend on knowledge of the biochemistry or even enzymology of the disease. Others represent surgery in circumstances of early diagnosis. Still others are treated or prevented by manipulation of the diet or manipulation of the immune response.

This chapter presents clinically significant applications of behavior genetics to medicine and emphasizes some of the psychological content of clinical genetics and genetic counseling. Aspects of the evolution of human behaviors have been reviewed previously (Omenn and Motulsky, 1972).

Classification of Genetic Diseases

Genetic diseases are grouped into three main categories. First, *chromosomal diseases* are associated with microscopically detectable aberrations of the chromosomes, observed as a change from the normal human number of 46 or as a structural rearrangement or deletion. There are many well-defined clinical syndromes associated with specific alterations of the chromosome pattern or karyotype. The chromosomal aberrations typically arise during formation of

TABLE 5.1
Therapy of Genetic Diseases by Conventional Means
(Without Changing the Abnormal Gene)

Treatment Approach	Examples
1. Add missing substance (product)	Thyroid hormone, cortisol, insulin, anti-hemophilic globulin, gamma globulin, blood cells, vitamin B_{12}
2. Prevent accumulation of toxic precursor (substrate)	Phenylalanine in PKU; galactose-1- phosphate in galactosemia
3. Replace the defective enzyme	Plasma pseudocholinesterase deficiency; Fabry; metachromatic leukodystrophy
4. Administer co-factor for altered enzyme	Vitamin B_{12} for methylmalonic aciduria
5. Use drug therapy (inhibit enzymes)	Allopurinol to prevent gout
6. Induce enzyme by drug	Phenobarbital for jaundice in newborn
7. Remove toxic substance	Metal-removing drugs for copper (Wilson disease) or iron (hemochromatosis)
8. Surgically remove organ	Colon in congenital polyposis; spleen in spherocytosis; lens with cataracts
9. Transplant	Kidney (polycystic kidneys, cystinosis); bone marrow (thalassemia); cornea
10. Use artificial aids	Eyeglasses; hearing aids; kidney machine
11. Block physiological (immune) response	Rhogam treatment of Rh-negative mothers who deliver Rh-positive babies

gametes (eggs or sperm) or shortly after fertilization. Most aberrations of the autosomal chromosomes are highly deleterious and are a frequent cause of spontaneous miscarriages. Aberrations of the sex chromosomes (X or Y) are better tolerated. Deletion of a whole chromosome is incompatible with life, except in a small percentage of cases of deletion of one of the two X chromosomes in females. Chromosome abnormalities most frequently appear *de novo* in a family and are not transmitted at a high frequency to other members of the family; however, some aberrations, especially balanced translocations, can be transmitted. The factors that predispose to chromosome aberrations—autoimmunity, radiation, advanced maternal age—contribute to a slightly increased risk of recurrence in subsequent pregnancies (see Fialkow, 1970; Vogel and Motulsky, 1979).

The diagnosis of chromosome disorders is made by growing cells, usually blood lymphocytes, in cultures in the laboratory, treating them with colchicine to block cell division at a time when the chromosomes are most easily visualized, and then staining. A major advance in chromosome work or cytogenetics occurred during the past decade as special stains were applied that now differentiate every pair of chromosomes from every other pair and show much detail along the length of each chromosome (Jacobs, 1977). With these techniques small deletions and previously unrecognizable rearrangements can be demonstrated. These techniques have been a boon to the mapping of specific human genes to specific locations on various chromosomes (see McKusick, 1978).

Second, *Mendelian disorders* are those due to inheritance of abnormalities of individual genes. This group consists of a very large number of distinct, mostly uncommon or rare diseases or syndromes. McKusick's catalog (1978) lists over 2000 such disorders, according to the pattern of transmission: autosomal dominant, autosomal recessive, and X-linked. Autosomal dominant disorders are transmitted from generation to generation, with 50% risk for children; if the disease has a high mortality, many observed cases will be due to new mutations. A typical pedigree (diagram of the family history) may show siblings, children, and parents affected (50% probability for all such first-degree relatives). Autosomal recessive diseases require that the affected person receive an abnormal gene from each parent; in most pedigrees, the parents are unaffected (carriers) and additional affected family members, if any, will be found among the siblings. Thus, the pattern of affected members is horizontal, rather than vertical, as in the case of autosomal dominant traits. X-linked recessive disorders will be transmitted by unaffected carrier women to their sons; additional affected members of the family will be found among the male relatives of the mother.

A great deal has been learned in recent years about the mechanisms of many of these single-gene diseases. Most recessive diseases are due to deficiency of a particular enzyme activity, such as hexosaminidase A in Tay–Sachs disease or phenylalanine hydroxylase in phenylketonuria. Mechanisms of dominant diseases are being elucidated, as well, and fall into three types:

1. Abnormal plus normal protein polypeptide chains (from the abnormal and normal forms of the gene on the two homologous chromosomes), interacting in a multimeric protein to give an abnormal or unstable product; an example is the unstable hemoglobin M.

2. Abnormalities in cell receptors, so that hormones or other molecules have decreased effect, sufficient over time to lead to illness; an example is familial hypercholesterolemia in which cell surface receptors for low-density lipoproteins are reduced to half-normal function in the heterozygotes (and missing altogether in the rare individual who has a double dose of the dominant, abnormal gene).

3. Enzyme deficiency to levels of about half-normal activity, rather than complete or nearly complete deficiency as is typical for recessive diseases; an example is acute intermittent porphyria, in which uroporphyrinogen synthetase is partially deficient, reducing the biosynthesis of a critical precursor of heme.

The third category of genetic diseases is by far the most common and most complicated, the *polygenic or multifactorial* class. These cause birth defects, developmental disorders, predisposition to heart diseases, cancers, arthritis, ulcers, and many other common diseases. More than one gene, plus environmental factors, is involved. One must also be alert to the likelihood that common diseases are highly heterogeneous; thus, the risk for family members may vary strikingly with the underlying cause. If schizophrenia is diagnosed, but the cause is Huntington disease, the risk for children will be 50%, rather than the approximately 12% that has been found to apply empirically to large numbers of children having one schizophrenic parent (see following). In genetic counseling, precise clinical and laboratory diagnosis is essential before risks and coping strategies can be discussed intelligently.

Frequency of Genetic Disorders

For the general population, it has been estimated that about .6% of newborns have detectable chromosomal abnormalities. That figure is derived from six studies involving some 56,000 unselected newborns in the United States, Canada, Denmark, and the United Kingdom. There was no difference between racial groups. Details are given in Table 5.2. Another 1.0% of liveborns has monogenic disorders, with striking variation in specific disorders according to ethnic origin (Table 5.3). Finally, a rough estimate of the frequency of polygenic disorders would be at least 10% of the population.

For hospitalized persons, especially children, the prevalence of genetic disorders is so high that the impact of genetic disease on hospital admissions and hospital costs is substantial. Studies from Seattle, London, Birmingham, Baltimore, Boston, and Montreal attribute 4–5% of admissions to monogenic and chromosomal disorders and more than 20% to polygenic and multifactorial disorders in children's hospitals. Data from Seattle are summarized in Table 5.4.

TABLE 5.2
Population Frequencies of Genetic Disorders in Newborns

Chromosomal abnormalities (per 1000 newborns)		Monogenic Disorders (per 1000)	
Sex chromosomes (in males)	2.60	Autosomal dominant	7
47,XXY	0.93		
47,XYY	0.93	Autosomal recessive	2.5
other	0.74		
Sex chromosomes (in females)	1.51	X-linked recessive	0.5
45,X	0.10		
47,XXX	1.04		
other	0.37		
Autosomal trisomies	1.44		
trisomy 21	1.25		
trisomy D, E	0.17		
Structural rearrangements			
balanced	1.95		
unbalanced	0.60		

Source: Hamerton et al, 1975.

Prenatal Genetic Diagnosis

The capability to obtain fetal cells from the amniotic fluid during the second trimester of pregnancy and carry out an increasingly broad array of specific diagnostic tests on those cells has transformed genetic counseling. For testable disorders, the patient and the physician no longer need to rely on probabilities of

TABLE 5.3
Ethnic Distribution of Genetic Diseases and Traits

Ethnic Group	Relatively High Frequency	Relatively Low Frequency
Ashkenazi Jews	Abetalipoproteinemia Bloom syndrome Dystonia musculorum deformans Familial dysautonomia Gaucher disease Niemann-Pick disease Pentosuria Spongy degeneration of brain Tay-Sachs disease	Phenylketonuria
Mediterranean peoples (Italian, Greek, Sardinian, Sephardic Jews)	Beta-thalassemia Glucose-6-phosphate dehydrogenase deficiency, Mediterranean type Familial Mediterranean fever	Cystic fibrosis
Blacks/Africans	Hemoglobinopathies: sickle cell anemia (Hb S), Hb C; thalassemias; persistence of fetal hemoglobin Glucose-6-phosphate dehydrogenase deficiency, African type Adult lactase deficiency	Cystic fibrosis Hemophilia Phenylketonuria Wilson disease Pseudocholinesterase deficiency α_1-anti-trypsin deficiency
Japanese & Koreans	Acatalasia Dyschromatosis	Slow acetylator HLA-B27
Chinese	Alpha-thalassemia Glucose-6-phosphate dehydrogenase deficiency, Chinese type Adult lactase deficiency	
Armenians	Familial Mediterranean Fever	G6PD deficiency
Finns	Congenital nephrosis Familial hypercholesterolemia Lysinuric protein intolerance	Phenylketonuria Krabbe disease
Eskimos	Pseudocholinesterase deficiency	
French Canadians	Tyrosinemia Oculopharyngeal muscular dystrophy	

Source: McKusick (1978)

TABLE 5.4
Contribution of Genetic Disorders to Hospital Admissions

Class of Disorder	Percent of All Admissions	
Chromosomal	0.6	
Autosomal dominant	1.2	
Autosomal recessive	2.2	
X-Linked	0.5	
Total Monogenic		4.5
Polygenic-Multifactorial		22.1
Other "familial"		13.2
Other developmental		13.6
Not genetic (trauma, etc)		46.6

Hall JG: Studies at Seattle Children's Orthopedic Hospital

occurrence of some particular disease; the fetus can be tested and found to be unaffected (usually) or affected. Of course, the fetus is still at risk for every other disease not tested for, just as for fetuses not tested at all. The development of this powerful technology has been discussed elsewhere (Milunsky, 1975; Omenn, 1978a), and its usefulness for psychiatric practice was reviewed also (Omenn & Motulsky, 1975). It is sufficient here to list the primary indications for amniocentesis and prenatal diagnosis (Table 5.5).

HUMAN BEHAVIOR GENETICS: MULTIFACTORIAL DISORDERS

Most of the important psychiatric, psychological, and developmental disorders fall into this category. It is likely that such classification makes our present ignorance more evident. Schizophrenia, manic–depressive illness and other affective disorders, alcoholism, epilepsy, minimal brain dysfunction, and mental retardation are certain to be heterogeneous diagnostic categories. Already we recognize dozens of specific causes of severe mental retardation, yet at the turn of the century it was thought that "feeblemindedness" was not genetic at all. When all cases are lumped together, it is difficult, indeed, to identify causes or mechanisms. Advances in specific diagnosis of mental retardation came from application of chromosome methods, clinical delineation of syndromes, and various laboratory measures. Even so, the majority of cases of severe mental retardation are still undifferentiated as to cause or mechanism.

The importance of ferreting out the heterogeneity lies not only in better understanding of the causes and perhaps their detection and avoidance; the importance lies also in devising appropriate treatment. For example, of all the children with mental retardation, only those with PKU can benefit from a low-phenylalanine diet and then only if started early in life.

In the absence of specific diagnostic criteria, geneticists do have methods to detect genetic influence on the frequency of a disorder and to estimate the recurrence risks for other relatives and subsequent pregnancies. The twin method (comparing incidence and prevalence of diseases in MZ and DZ twins), the family method (determining incidence for various classes of relatives), and the adoption method (separating inherited effects from common familial environment) have been applied to most of the common behavioral disorders. Strong evidence for genetic factors has been obtained. The adoption methods pioneered by Kety and Rosenthal were the most important in overcoming the views of many behaviorists that family attitudes and family interpersonal experiences were the dominant, if not the only factors determining risk for schizophrenias.

For counseling purposes, it is valuable to have data on the risks to relatives, data obtained from straightforward counting of diagnosed relatives among pooled index cases. Table 5. 6 presents such data for manic–depressive illness, schizophrenias, and epilepsy. At the same time it is wise to keep in mind a short list of specific genetic disorders that are commonly misdiagnosed as garden-variety schizophrenias or affective disorders. These include Huntington disease, Wilson disease, acute intermittent porphyria, hyperparathyroidism with hypercalcemia, and homocystinuria; they are discussed under monogenic disorders.

The challenge of counseling for the major psychoses can be illustrated with a case from our Medical Genetics Clinic at the University of Washington. A 2-week-old normal-appearing boy was brought to the clinic by a social services agency and a prospective adopting couple came to the clinic independently for

TABLE 5.5
Indications for Mid-Trimester Amniocentesis for Prenatal Genetic Diagnosis

Cytogenetic studies on cultured amniotic fluid cells:

> Mother's age over 35
> Previous child with Down syndrome
> Family history or carrier status for other chromosomal disorders
> Determination of sex in X-linked disorders

Biochemical studies on amniotic fluid:

> Previous child with neural tube closure defect
> High maternal serum alpha-fetoprotein (screening test for neural tube defects)
> Linkage study for myotonic dystrophy

Enzymatic analyses on cultured amniotic fluid cells:

> Previous child with testable inborn error of metabolism
> Couple at risk for Tay-Sachs disease, detected by population screening

Gene studies on cultured amniotic fluid cells:

> Couple at risk for alpha-thalassemia (cDNA/DNA hybridization)
> Couple at risk for sickle cell anemia (restriction enzyme linkage analysis)

TABLE 5.6
Empirical Risks for Behavioral Disorders

Relation to Proband	Manic-Depressive Illness		Schizophrenias		Epilepsy	
	% Risk	(range)	% Risk	(range)	% Risk	(range)
Parent	8	(3-23)	4	(1-12)	4	(2-5)
Sibling	9	(3-23)	7	(3-14)	4	(2-5)
Child	11	(6-24)	12	(7-16)	4	(3-6)
MZ Twin	74	(50-92)	45	(15-69)	85	(52-91)
DZ Twin	23	(16-38)	8	(2-11)	13	(4-23)
General Population	1-2		1		0.6	

Compiled from studies summarized in Rosenthal (1970) and Slater & Cowie (1971).

advice. The couple was eager to have a child and had waited a long time. However, they were anxious about the family history, for which the pedigree follows:

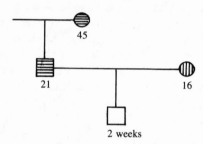

The parents of this boy were a 16-year-old woman hospitalized with a diagnosis of schizophrenia and a 21-year-old man hospitalized with a diagnosis of manic–depressive illness. Very little was known about their relatives, except that the father's mother had also been diagnosed as having a depressive illness. Working strictly from Table 5.6, we could tell the interested couple that the empirical risks for the boy would be about 12% for schizophrenia plus about 11% for manic–depressive illness. So far as we know, the risks are additive. The complementary probability, of course, is over 75% that the boy would develop neither schizophrenia nor depressive illness.

There were several subtleties. First, the knowledge that the grandmother was also affected increased his risk slightly. Second, a widely held hypothesis was that manic–depressive illness might be inherited, at least in some families, as an X-linked dominant trait. The rationale was that females are affected about twice as commonly as males, with vertical transmission in many families, which is compatible with X-linked dominant transmission. Evidence of linkage to X-chromosome genes (Xg blood group and color blindness) was published, satisfying many nongeneticists that X linkage of manic–depressive illness had been demonstrated. However, these studies were flawed and the results were implausible as well, because Xg and color blindness genes are located so far apart on the X chromosome that it would be extremely unlikely for any other gene to show linkage to both of them. Third, there was another proposal at the time that schizophrenia might be inherited as an autosomal dominant trait, manifested in some as the full-blown psychosis and in others as a schizoid personality (Heston, 1970). If true, this hypothesis would cause the adopting couple to worry about less severe psychiatric problems, on top of the risks already stated.

The counseling for so complicated a situation varies tremendously, depending on the questions the couple raise and depending on the counselor's view of the importance and reliability of the various kinds of information. In this case and many others, different couples hearing similar information make very different judgments about what course of action to take. For couples with many questions, the state of our knowledge, especially about multifactorial disorders, limits the capacity to respond.

HUMAN BEHAVIOR GENETICS: MONOGENIC DISORDERS

Throughout the history of genetics, analysis of rare mutants has yielded extraordinarily valuable information about normal processes and a better basis for investigating pathophysiology. The same approach is emphasized here. The metabolic disorders among which we may search for clues to human brain function and dysfunction can be differentiated roughly into those whose primary lesion is *extrinsic* to the brain and those *intrinsic* to the brain. Examples are listed in Table 5.7 and discussed in the following section.

In phenylketonuria and other aminoacidurias, the brain is damaged by normal metabolites in abnormal concentrations, a toxic mechanism, just as the brain may be damaged by lead poisoning from the more external environment. Hypothyroidism differs in representing a deficiency syndrome, in which thyroid hor-

TABLE 5.7
Origin of Brain Damage in Selected Inborn Errors of Metabolism

Extrinsic to the Brain	*Intrinsic to the Brain*
phenylketonuria	Lesch-Nyhan syndrome
urea cycle disorders	Wilson disease
maple syrup urine disease	homocystinuria
galactosemia	metachromatic leukodystrophy
hepatic porphyrias	mucopolysaccharidoses
hypothyroidism	Wernicke-Korsakoff syndrome
adrenogenital syndrome	

Source: Omenn (1981a); Omenn (1976)

mone, important to brain development and myelination, is lacking. The extrinsic disorders merit considerable attention because they are treatable or preventable by manipulation of the environment by special diets, administration of a specific hormone, or correction of fluid and electrolyte imbalance. Disorders that may be intrinsic to the brain provide clues to metabolic intermediates and metabolic pathways of significance in normal brain functioning, as well as to common patterns of psychiatric and neurologic dysfunction (see Omenn, 1981a; Robinson and Robinson, 1976. Chapters 3–6).

Extrinsic Disorders: Phenylketonuria (PKU)

Classical PKU results from a deficiency in the liver of the enzyme phenylalanine hydroxylase, with resultant deficiency of production of tyrosine from phenylalanine and with accumulation of phenylalanine itself and phenylpyruvic acid and other metabolites in blood and urine. PKU has been a leading model for study of mental retardation, because it is relatively common and because the dietary management of PKU is one of the most dramatic therapeutic developments in pediatrics. When infants are placed on a diet maintaining optimal levels of phenylalanine in the blood, the chance that nearly normal or normal IQ will result is vastly improved and normal growth can be achieved. Without treatment, children have a median IQ of about 40, with 95% having an IQ less than 50.

Følling identified phenylpyruvic acid in excess in the urine of patients with this condition in 1934; thereafter, his ferric chloride test became a simple means of detecting the 1% of patients with PKU among the large population with seemingly similar mental retardation. During the first year of life, there are usually few other clinical clues to the diagnosis. Some 25% have eczema of the skin and those who perspire profusely have a musty odor from phenylacetic acid during the first 2 or 3 months. Progressive dilution of skin pigment occurs over the first year because of inhibition of tyrosinase by excess phenylalanine. Nevertheless, the paucity of clinical clues makes biochemical diagnosis via a screening test in the newborn period imperative. Otherwise, the equivalent of an estimated 50 IQ points will be lost if treatment is not initiated until the end of the first year of life. The screening test, developed by Guthrie, utilizes a drop of blood from a heel stick of the infant and measures the extent of growth of phenylalanine-requiring bacteria. Analogous tests have been developed for a series of other rare inborn errors.

Heterogeneity. The prevalence of PKU was estimated at one case per 20,000 births on the basis of studies of prevalence of PKU among the mentally retarded. When the screening test was introduced widely and required by law in nearly all states, the apparent incidence nearly doubled. It was recognized only later that many of these additional cases represented hyperphenylalaninemia of heterogeneous origins that would not lead to classical PKU with mental retardation. For this reason, not all children placed on the phenylalanine-restricted diet benefited, and some suffered stunting of growth and even mental retardation. The heterogeneity includes a series of different mutations affecting phenylalanine hydroxylase, causing partial deficiency or altered affinity for the phenylalanine substrate. The former may lead to high enough blood phenylalanine levels to warrant the special diet. The latter leads to a requirement for higher than normal phenylalanine levels in order to make tyrosine, which is essential for protein synthesis, thyroid hormone and catecholamine synthesis, and skin pigmentation. Other babies give a positive test due simply to immaturity of liver function; on retest the phenylalanine hydroxylase activity is higher and the blood phenylalanine levels are within the normal range.

The detailed enzymology of phenylalanine hydroxylase is quite complicated. Tetrahydrobiopterin is a required cofactor, and the enzyme dihydrofolate reductase is coupled functionally to phenylalanine hydroxylase in order to shuttle the biopterin back and forth between the quinoid and the tetrahydro form. Deficiency of the dihydrofolate reductase, therefore, also can cause hyperphenylalaninemia. Such a defect was identified in three very unusual patients, who developed feeding difficulties shortly after birth, early evidence of developmental delay, and seizures, with no clinical response to a low-phenylalanine diet that controlled the blood phenylalanine level. By 18 months of age, voluntary movement and social awareness ceased, and the EEG was grossly abnormal. Dihydrofolate reductase deficiency was demonstrated in brain, liver, and cultured fibroblasts by Kaufman. Several patients have now been described, with more than one type of abnormality in the reductase enzyme likely. Dihydrofolate reductase deficiency clearly affects intrinsic pathways in the brain, as the same tetrahydrofolate cofactor system is used by tyrosine hydroxylase in the biosynthesis of dopamine and norepinephrine and by tryptophan hydroxylase in the biosynthesis of serotonin (5-hydroxytryptamine). Because the neurologic defects are probably due to secondary deficiency of formation of these monoamine neurotransmitters, one patient was treated with L–dopa, L–5–hydroxytryptophan, and carbidopa (inhibitor of the peripheral metabolism of L–dopa). The patient showed progressive improvement during 9 months; myoclonus, uncontrolled movements, tetraplegia, and some skin signs all disappeared.

Behavioral Features of PKU.   Untreated PKU children were described more than 20 years ago (Wright & Tarjan, 1957) as distinguishable from other mentally deficient children in the same institutions: "none could be described as friendly, placid, or happy," in striking contrast with Down syndrome children, in particular. The less retarded patients with PKU tend to be restless, jerky, and fearful; destructive and noisy psychotic episodes were observed in 10%. In another series (Paine, 1957), 32% had night terrors or uncontrollable temper tantrums. In fact, hyperactivity, irritability, and uncontrollable temper are often the reasons why these children are admitted to institutions. Even when dietary manipulations sustain development of IQ in the normal range, these children have a higher incidence of restless and hyperactive behavior.

Possible Effects of Heterozygotes.   The conventional wisdom that heterozygotes are entirely normal and that brain damage starts only after birth in homozygotes with PKU has been challenged by Bessman (1978). He postulated that tyrosine deficiency may be an important component of the mental retardation, in addition to the toxic effects of high phenylalanine and its metabolites on glycolytic enzymes, myelination, and various other vulnerable brain processes. Prenatal growth retardation and microcephaly does occur in some PKU newborns, indicating damage during fetal development. The distribution of IQs within the "normal range" (70–130) of apparently normal siblings of children with PKU has been reported to be bimodal, as might be predicted if the heterozygous sibs (two-thirds) were more vulnerable than the sibs who did not carry even one gene for PKU (one-third of sibs). Bessman hypothesized that a limited prenatal supply of tyrosine due to interaction of heterozygous mother and heterozygous fetus exerts a retarding effect on brain development. The therapeutic recommendation for such a situation is to supplement the diet with tyrosine late in pregnancy when the brain is growing most rapidly. This hypothesis has been challenged, however, on the grounds that the half-normal level of phenylalanine hydroxylase should be quite sufficient to support formation of tyrosine at more than half-normal levels (due to the kinetics of the reaction).

Maternal Phenylketonuria.   The extrinsic origin of brain damage in classical PKU is confirmed by the recognition of effects on the fetus of maternal PKU. These children are far more seriously affected than children with classical PKU, whose mothers are unaffected heterozygous carriers. Here the exposure to toxic levels of phenylalanine begins during fetal development of the brain, rather than postnatally. About 25% of children of PKU mothers have had congenital anomalies; and microcephaly, intrauterine growth retardation, and subsequent mental retardation have occurred in nearly all. The practical implications of this clinical phenomenon are staggering. First, those girls who have been treated successfully with dietary management to age 6 or 7 years will have to be restarted on a special diet during pregnancy, and as early as possible. The diets have generally been considered unpalatable, though more recent formulations are less objectionable. Also, breast feeding is contraindicated, because breast milk has high phenylalanine in these women. Second, without such obstetric management or contraception, the next generation may have as many or more severely retarded children due to PKU as were avoided by successful treatment after mass screening in the parental generation.

Extrinsic Disorders: Maple Syrup Urine Disease (MSUD)

Maple syrup urine disease, due to disorders of branched-chain amino acid metabolism, causes vomiting, muscular hypertonicity, convulsions, coma, and death in the first few weeks of life, unless the branched-chain amino acids leucine, isoleucine, and valine are restricted in the diet. A variant of MSUD, in which there is less than complete deficiency of the branched-chain amino acid decarboxylase, leads to symptoms only episodically, induced by dietary ingestion of the branched-chain amino acids. Initial manifestations have appeared months or years after birth, with attacks of ataxia associated with maple syrup odor and elevations of branched-chain amino acids and ketoacids in urine and blood. Mental and physical development can be normal in the partial deficiency. It is interesting that the primary symptom is ataxia (unstable gait), because leucine and to lesser extents isoleucine and valine are particularly toxic to cerebellum in organ explant systems in vitro. Leucine and its ketoacid alphaketoisocaproic acid inhibit myelination and inhibit pyruvate and glutamate decarboxylases. There is no information about the susceptibility to neurological or behavioral impairment in parents who are obligate carriers for the completely deficient infantile-onset form of MSUD. These parents should be investigated after high-protein meals.

Favorable responses to diets low in the branched-chain amino acids suggest that the deleterious effects result from abnormal accumulation of the metabolites proximal to the metabolic block, rather than from insufficiency of metabolites distal. These ataxic attacks are often induced by stress, particularly stress associated with infections, though some acute attacks have no apparent triggering causes. Ingestion of the amino acids simulates the clinical attacks.

Vitamin-Responsive Form. Yet another variant of MSUD involves a mutant branched-chain decarboxylase that requires higher than usual concentrations of the cofactor thiamine for its function. These rare patients have been responsive biochemically and clinically to administration of thiamine. They have been asymptomatic in the newborn period. With increased protein intake or intercurrent illness, producing catabolic stress, blood levels of branched-chain ketoacids rise and lethargy, ataxia, and vomiting ensue. A laboratory clue to the diagnosis is the combination of ketones without glucose in the urine.

Vitamin-responsive metabolic disorders may be considered somewhat analogous to vitamin-dependent endocrine disorders. Knowledge of the detailed biochemistry of the metabolic pathway, including the requisite cofactors and the metabolic interconversions of those cofactors, allows effective therapeutic interventions. Other examples are biotin-responsive and B_{12}-responsive ketotic hyperglycinemias, B_6-responsive homocystinuria, and biopterin-responsive hyperphenylalaninemia (described previously).

Extrinsic Disorders: Adrenogenital Syndrome

In adrenogenital syndrome there is a loss of feedback to the hypothalamus and pituitary resulting in hypersecretion of ACTH. The cause is a block (usually at the 21-hydroxylase step) in the enzymatic biosynthesis of cortisol in the adrenal cortex. ACTH stimulates the adrenal to make cortisol, but excess androgens are made instead, because of diversion of the precursors in the cortisol pathway. Exposure of female fetuses to the high circulating levels of androgens produces masculinization of the external genitalia, leading to hermaphroditism. Treatment involves administration of corticosteroids both as replacement therapy and to turn off the adrenal stimulation by ACTH. The gene for the 21-hydroxylase enzyme has been mapped to human chromosome 6 in close linkage to the histocompatibility (HLA) region.

Money and Lewis (1966) reported higher than normal IQ (mean 110) in the affected girls, as well as a shift toward "tomboyish" behaviors. Similar long-lasting behavioral effects of androgens on the developing brain were deduced in cases of progestin-induced hermaphroditism by Ehrhardt and Money (1967), supported by evidence from studies in animals. However, our family studies showed that patients' IQ was not higher than the IQ of unaffected siblings or the IQ expected from midparental values (McGuire & Omenn, 1975). Thus, the adrenogenital syndrome appears to confer no IQ advantage independent of family IQ level. As this analysis illustrates, it is useful to test family members, rather than relying on normalized population data for such parameters as IQ. On the other hand, because parents are obligate gene carriers and two-thirds of siblings would be heterozygotes also, it is conceivable that the putative IQ effects may be expressed in the heterozygote. The disorder is clearly extrinsic to the brain, however, and there is no evidence that 50% of normal activity for the adrenal 21-hydroxylase produces any significant changes in cortisol feedback and androgen levels. Tests that were expected to show male–female differences on cognitive and behavioral measures gave similar results for the patients and for a matched control group (McGuire, Ryan, and Omenn, 1975). The discrepancy between our findings and those of Money and his colleagues may reflect changes in the sexually dimorphic behaviors of the control groups over more than a decade between those studies.

Extrinsic Disorders: Hepatic Porphyrias

Disorders of hepatic biosynthesis of heme can be recognized by the excretion in urine or feces of excessive amounts of porphyrins and porphyrin precursors. These biochemical findings are associated primarily with attacks of abdominal pain, plus highly variable neurological and behavioral manifestations involving peripheral nerves, autonomic nervous system, cranial nerves, brain stem, and cerebral function (Stanbury, Wyngaarden, and Fredrickson, 1978). In the Swedish type of acute intermittent porphyria, the metabolic defect is a deficiency (about 50%) of uroporphyrinogen I synthetase, with resultant lack of heme biosynthesis and heme feedback on the production of porphyrin precursors. The defect lies in the liver, and the effects on the nervous system are assumed to represent the action of a yet unidentified toxic metabolite. It should be emphasized that cases of porphyria can masquerade as manic–depressive illness and occasionally as paranoid schizophrenia, providing an example of the heterogeneity in those common diagnostic categories. In addition, porphyric attacks are often triggered by drugs, including barbiturates, or sudden dietary changes; those triggering events may be due to behavioral dysfunction, leading to a very complicated clinical tangle.

Intrinsic Disorders: Lesch–Nyhan Syndrome

In 1964 a second-year medical student (Lesch) and his professor (Nyhan) reported a most remarkable metabolic/behavioral disorder that now bears their names and that has stimulated a vast array of investigations. Among patients with mental retardation due to inherited metabolic disorders, this syndrome may rank second only to PKU in frequency. The cause is a mutation on the X chromosome affecting the gene for the enzyme hypoxanthine-guanine phosphoribosyl transferase (HGPRT), involved in what was previously thought to be a minor pathway of purine metabolism. The clinical disorder focused attention on the underlying biochemistry. Boys with this X-linked recessive disorder have virtually complete deficiency of HGPRT activity in cells throughout the body. The highest activity of HGPRT normally is in the brain, particularly the basal ganglia, where many of the prominent neurological signs arise.

Affected males appear normal at birth and for up to 6 to 8 months postnatally. The earliest symptom is the occurrence of orange sand in the diapers, due to massive excretion of urates (hypoxanthine and uric acid). Screening diagnosis is made most reliably from an elevated uric acid/creatinine ratio. Hematuria or urinary tract stones may occur in the early months of life, and later these boys develop any or all of the manifestations of gout. Uric acid stones are radiolucent on Xray, so the cause of colicky abdominal pain may be overlooked in early episodes. Thus, the manifestations first noted are usually neurological abnormalities. Infants who had been sitting and holding up their heads lose those abilities. After a period of hypotonia, marked hypertonicity develops, with increased deep tendon reflexes and Babinski responses. Spasticity is so severe that bilateral dislocation of the hips may result. Dystonic and athetoid posturing, choreic movements, and athetoid dysarthria and dysphagia are characteristic and are worsened by tension or excitement. Motor disability is so great that none of the patients has walked, and they can sit in a chair only with trunk supports. They appear to have cerebral palsy. Mental development is retarded, with IQs less than 50, but most learn to speak. In fact, IQ testing in these children is highly suspect:

The motor defects make it difficult for them to perform adequately and writing is impossible. Nevertheless, they usually relate well to people and seem to understand what is said to them. They have bright, understanding eye contact, and they learn to communicate rather well, considering the dysarthric speech (Nyhan, 1978).

Aggressive self-mutilative behavior is an integral and most impressive feature of this syndrome. The behavior begins with the eruption of teeth, though in some patients it was delayed for 3 to 5 years. Most patients bite their lips or fingers in a stereotyped, highly individualized pattern and with such ferocity that loss of tissue almost always results. The hallmark of the syndrome is a distinct loss of tissue about the upper or lower lip. It is rare to see one of these patients without permanent damage to the lips, unless the primary teeth were removed early. They do not have a sensory defect; they scream in pain while they bite themselves, and they are really happy only when securely protected from themselves by physical restraints. They do seem to have insight into their propensity to hurt themselves. Sometimes they are engaging children when restrained, with a good sense of humor. However, when the restraints are removed, their personality changes abruptly and they appear terrified. As they get older, they may even call for help. The "switch" in their behavior is more dramatic than the switch described for manic–depressive patients.

As they get older, their behavior becomes more varied, picking with their fingers, scalding in hot water, catching themselves in braces employed for cerebral palsy or in the spokes of a wheelchair. Within their motor limitations, they can constitute a risk to others, including their favorite ward personnel. They may become verbally abusive, but they are remorseful about hurting others. As Nyhan has noted, these children are reminiscent of the principal character of the ballet "Petrouchka" with their admixture of good humor and tragedy, unusual posturing, and mittenlike coverings on the hands.

The metabolic disorder has been studied intensively, with many findings, but the link between purine metabolism and the self-mutilative behavior is still a mystery. Uric acid is not formed in the brain. The end products of purine metabolism in the brain are xanthine and hypoxanthine, with hypoxanthine levels in the cerebrospinal fluid four times control values. Treatment with the drug allopurinol is indicated to prevent the urinary tract complications, but it has no effect on the neurological or behavioral manifestations of the disease.

As little as 5% of normal HGPRT activity is sufficient to prevent development of the neuropsychiatric syndrome. Patients have been identified with gout due to other mutations affecting HGPRT, leaving 5 to 15% residual activity. These patients have no neurological abnormalities, or only mild spinocerebellar signs.

Much research has been directed at cyclic AMP, central adrenergic pathways, and serotonin functions, attempting to apply information on brain chemistry and aggressive behavior from animal studies. Administration of 5-hydroxytryptamine (serotonin) together with an inhibitor of peripheral decarboxylase activity (carbidopa) has been reported by Nyhan (1978) to have dramatic effects in reducing mutilative behavior. However, the effect could be sustained only for 3–4 weeks, due to pharmacologic tolerance. Attempts at behavior modification with mild aversive techniques of the sort employed with retarded patients with mild self-mutilative behavior have either had no effect on these boys, or have led to even worse behavior. Perhaps some combinations of pharmacologic and psychologic approaches will prove effective.

Neurochemical studies have been reported recently (Lloyd et al, 1981) on brain samples obtained post-mortem from three patients with Lesch-Nyhan syndrome. There was a striking decrease in dopamine and homovanillic acid levels, as well as DOPA decarboxylase and tyrosine hydroxylase activities, in regions rich in dopamine terminals (caudate nucleus, putamen, and nucleus accumbens). The deficit was less severe than in Parkinson disease, but corresponded to a 65 to 90 percent loss of nigrostriatal and mesolimbic dopamine terminals. The cell-body region in the substantia nigra had normal dopamine levels, in contrast with Parkinson disease. Thus, there seems to be a dearborization of dopamine terminals. Other measurements showed normal or near normal levels of norepinephrine, serotonin, dopamine beta-hydroxylase, and L-glutamic acid decarboxylase, with a moderate decrease in choline acetyltransferase. These observations should stimulate research on the effects of purines on development and function of dopaminergic neurons.

Hypoxanthine and other oxypurines which accumulate in the cerebrospinal fluid of patients with Lesch-Nyhan syndrome may interfere with binding of diazepam to receptors, as does caffeine, which can elicit self-mutilation in animals. Caffeine and theophylline also block high-affinity adenosine receptors. Perhaps some of these interactions are responsible for the bizarre behavioral manifestations of the syndrome (Kopin, 1981).

Intrinsic Disorders: Hepatolenticular Degeneration (Wilson Disease)

Wilson disease is an excellent example of a diagnosis that is seldom made unless the physician looks for it in the evaluation of patients with various neurological or psychiatric problems or with diagnoses of chronic active hepatitis, viral hepatitis, cirrhosis, acquired hemolytic anemia, or even renal disease. This rare autosomal recessive disease of young adults causes cirrhosis of the liver and degeneration of the brain, especially the basal ganglia. Copper deposition occurs in these organs and in greenish-brown Kayser-Fleischer rings at the limbus of the cornea of the eye. The copper-binding protein serum ceruloplasmin is deficient, and various tissue proteins are overloaded with loosely bound copper, but the primary lesion in copper metabolism is still not known. This disorder merits particular attention because treatment with a copper-chelating agent (D-PEN-ICILLAMINE) can reverse or prevent copper deposition and clinical signs of the disease. Untreated, on the other hand, the disease is invariably fatal. Psychiatric and behavioral changes are common, but vary in kind and degree. One series of patients included 60% with significant psychological manifestations as the first clinical indication of the disease; usually about 20% have such manifestations recognized before the neurological and liver disease signs emerge. Diagnoses include personality disturbances, hysteria, and schizophrenia. Intellectual capacities are maintained intact, though observers may be misled by childish personality changes, drooling, difficulties with verbal communication, and eventually a masklike expressionless face. Failure to consider Wilson disease in these patients allows the development of life-threatening complications and needless, ineffective psychiatric interventions.

A case history (Cartwright, 1978) demonstrates these points: A 24-year-old woman became nervous and developed a tremor. Her physician made a diagnosis of "nervous exhaustion" and prescribed tranquilizers. She became depressed and attempted suicide. A psychiatrist treated her with imipramine (antidepressant) and then chlorpromazine (tranquilizer), but she became psychotic. The next psychiatrist diagnosed acute schizophrenic reaction and ordered 11 electroconvulsive treatments, without benefit. An internist was called to see her because of abnormal liver function tests and attributed the abnormalities to chlorpromazine toxicity. The saga continued with a neurologist. By this time she had increased salivation, masklike facies, severe difficulty in swallowing, tremor, and dystonia; she was nearly completely unable to care for herself. His diagnosis also was chlorpromazine toxicity. Yet another psychiatrist was called, who diagnosed "conversion reaction" and prescribed doxepin and diazepam. At last she was admitted to a psychiatric ward because of dehydration; her diagnosis was "hysterical neurosis/conversion type," though "catatonic schizophrenia" could not be ruled out. Amazingly enough an alert on-call physician actually examined her carefully and recognized the distinctive Kayser-Fleischer rings in her eyes. He made the diagnosis of Wilson disease. It was 32 months after the onset of clinical symptoms.

The patient was told that her disorder was treatable and D-penicillamine was started. Her depression and psychiatric signs disappeared. Her neurologic signs improved slowly, but dystonia and spasticity persist and she walks with some difficulty. Nevertheless, she returned to college, graduated, remarried, and seemed well adjusted. She even wrote an account of her experiences in the American Journal of Nursing (Francone, 1976).

For so many neurological and psychiatric disorders, the physician has no really effective treatment to offer. Despite the relative rarity of this disease (1000 cases in the United States), the consequences of failure to diagnose and the results of proper diagnosis and treatment are sufficiently dramatic to warrant a watchful approach. The disorder can be diagnosed before symptoms appear, and administration of D-penicillamine can prevent the development of the whole disease. Siblings of patients are at 25% risk for the disease and must be tested.

Intrinsic Disorders: Homocystinuria

Homocystinuria is a fascinating autosomal recessive disorder of the metabolism of sulfur-containing amino acids. It illustrates well the matter of genetic heterogeneity and provides clues to potentially important aspects of brain metabolism. These patients have skeletal abnormalities similar to the gangling phenotype of Marfan syndrome, plus displacement of the lens of the eye, blotchy skin, and marked propensity for arterial and venous thromboses. Most interesting for this discussion, about half of the patients are described as mentally retarded (IQ below 70). However, it is not at all clear whether the half described as having normal IQ reflect heterogeneity of underlying disease mechanisms or whether they also have lost 20 or 30 points in IQ, yet remained in the "normal" range. Some patients have had quite high IQs (McKusick, 1972). To distinguish these possibilities, a study is needed that would compare the IQs of affected individuals with the IQs of their parents and siblings. Psychotic behavior has been noted in some patients, but no systematic evaluation has been done.

The primary biochemical defect is in the synthesis of the complex amino acid cystathionine, from homocysteine and serine, by the enzyme cystathionine synthase (CS). In affected persons, the enzyme has less than 5% of normal activity. Although the concentration of cystathionine is similar in the livers of rats and humans, the concentration is very much higher in the brains of humans and monkeys than in the brains of nonprimates. Its function is still unknown. It is conceivable that cystathionine acts as a neurotransmitter or neuromodulator, as many amino acids are now suspected of having such roles. In fact, the compounds widely regarded as neurotransmitters (acetylcholine, norepinephrine, dopamine, serotonin, GABA, and epinephrine) probably account for only 30% of the synapses of mammalian brain.

Two forms of cystathionine synthase deficiency can be distinguished by administration of vitamin B_6 (pyridoxine), the cofactor for this enzyme; such treatment leads to prompt metabolic correction in about half the patients. Most patients with the B_6-responsive form have normal intelligence, but retarded persons fall into both types. Nevertheless, it is important to recognize that administration of pyridoxine in the B_6-responsive patients may prevent or even reverse abnormalities in the brain. In several treated children, petit-mal seizures ceased and scholastic performance improved as the urine was cleared of homocystine. In another case, severe hyperactive behavior was controlled with pyridoxine.

A previously unrecognized cause of homocystinuria, due to deficiency of 5, 10-methylenetetrahydrofolate reductase (not cystathionine synthase), presented a clinical picture of recurrent episodes of mental deterioration and schizophrenic behavior. Folic acid was successful in reversing these episodes (Freeman et al, 1975). The deficient enzyme is necessary to form the folate compound that serves as methyl donor for methylation of homocysteine to methionine and possibly for methylation of some biogenic amines. Such rare metabolic disorders may provide a "handle" on the likely heterogeneity of causes of schizophrenia.

Intrinsic Disorders: Wernicke–Korsakoff Syndrome

This syndrome, which occurs in chronic alcoholics, comprises ophthalmoplegia (paralysis of eye movements), ataxia, and lethargy, plus a remarkable confabulatory psychosis. It is well known that deficiency of thiamine is responsible for the physical signs. Thiamine pyrophosphate (vitamin B_1) is a critical cofactor for transketolase in the pentose–phosphate shunt pathway and for glutamate decarboxylase in the Krebs cycle. Thiamine is normally found in high concentrations in the mammillary bodies and related brain regions. In alcoholics, who are likely to be B-vitamin depleted, thiamine should always be given by injection before intravenous administration of glucose, because metabolism of glucose also consumes thiamine and can precipitate the Wernicke complications.

The question here for medical geneticists is why only a few alcoholics with similar thiamine deficiency develop this organ-specific complication. Blass & Gibson (1977) have shown that such individuals have a variant of the thiamine-requiring enzyme transketolase (high K_m variant), which makes the enzyme need higher concentrations of thiamine in order to have the same enzymatic activity. This finding may be considered an example of what must be similar variation in target organ functions for such other organ-specific complications as cirrhosis of the liver, pancreatitis, cerebellar degeneration, cardiomyopathy, and fetal alcohol syndrome.

Intrinsic Disorders: Huntington Disease

This autosomal dominant disorder is a clear monogenic disease with awesome effects on cerebral and basal ganglia function, causing major psychiatric and neurological problems. The underlying biochemical abnormality is unknown, and numerous efforts to identify even a biochemical test for carriers of the abnormal gene or for diagnosis of those with clear signs of the disease have proved fruitless (Comings, 1981). The prevalence of the disease is about one in 10,000 people in the general population.

Those who have an affected parent must cope with the risk of 50% that they will come down with this disease. Typically, signs begin to appear only in the 30s or 40s; the median age of diagnosis is about 45. By then, these individuals have already had their own families, and they may have passed on the dread gene. The clinical signs often begin with personality changes, forgetfulness, improprieties, and other behavioral problems. All sorts of minor and major psychiatric diagnoses are made. Gradually neurologic signs appear, especially loss of control over movements. The pathologic signs and changes in the brain are most severe in the basal ganglia, stimulating speculation about the possible role of dopamine. Drugs that are dopamine antagonists do ameliorate some of the neurologic abnormalities, but they have no effect on psychological abnormalities.

Huntington disease is one of the most trying problems in medical genetics clinics for adult patients (Wexler, 1979). A very helpful lay group called Committee to Combat Huntington Disease has been active over the past decade in working with families, in seeking the interest of researchers, and in campaigning for humane and useful social services.

HUMAN BEHAVIOR GENETICS: CHROMOSOMAL DISORDERS

The third category of clinical genetics is the set of syndromes associated with specific chromosomal aberrations. The most important clinically are trisomy 21 (Down syndrome), 45 XO (Turner syndrome), 45 XXY (Klinefelter syndrome), and 47 XYY.

Down Syndrome

Trisomy 21 is the most common single cause of mental retardation, accounting for about 10% of severe cases. An English physician named Down noted a century ago that some children with mental retardation had striking physical features, the combination of which constitutes a syndrome: a characteristic facies, large tongue, hypotonia, and certain congenital anomalies. In 1959, shortly after the techniques of spreading and identifying human chromosomes had advanced far enough to count the human set accurately as 46, cells from patients with Down syndrome were shown by Lejeune to have 47 chromosomes, with a triploid set of one of the smallest chromosomes, # 21. The mean IQ for older patients is about 25, with only a rare patient exceeding an IQ of 50. The children seem to do fairly well in the first 2 or 3 years, but then fall progressively behind their peers in developmental milestones. Their personalities are distinctively friendly and cheerful, which is an advantage in their care.

The most important aspect of the clinical approach to prevention of Down's syndrome is the knowledge that the risk of this chromosome aberration rises dramatically with the age of the mother: about 1/2000 at age 20, 1/1000 at age 30, 1/500 at age 35, at least 1/100 by age 40, and 1/40 by age 45. More than 90% of cases are due to trisomy 21, occurring as a result of nondisjunction; the pair of # 21 chromosomes in the mother (or, less frequently, in the father) fails to separate normally, and fertilization with a normal sperm (or of a normal egg) leads to a fertilized egg (zygote) with three # 21 chromosomes. The number of children born with Down syndrome can be decreased simply by social practices that reduce the average age of mothers or that discourage women from having children after age 35. Alternatively, it is now well established that pregnancies can be tested by determining the chromosome karyotype on cells from amniotic fluid sampled at about 14 weeks of pregnancy; couples who can accept the termination of pregnancy for an affected fetus avail themselves of this technological advance. Fortunately, the vast majority of those who utilize amniocentesis to test for Down syndrome in the fetus receive reassuring news. The risk at age 35 for a woman with no previous history and the risk at any younger age for a woman who has already had a child with Down syndrome due to trisomy 21 is only 1%. Thus, 99/100 times the news from amniocentesis and chromosome analysis is that the baby is unaffected with Down syndrome. Of course, the baby may still have any other problems not tested for, just as other babies may.

Although 90% of Down syndrome cases are due to trisomy 21, the remaining cases are due to mosaicism or to translocation. Especially in the translocation cases (46 chromosomes with an extra # 21 hooked onto another chromosome, usually # 14), it is essential to determine whether either parent is a carrier of a balanced translocation (45 chromosomes with one representing two hooked together). If so, the risk for subsequent pregnancies is much higher, depending on the nature of the translocation and the sex of the parent.

Turner Syndrome (45,XO)

Turner syndrome includes short stature (not over 5 ft tall), lack of sexual maturation due to ovarian dysgenesis, and bony and other somatic abnormalities. Not all have the typical 45,XO chromosome complement; others have mosaicism including 45,XO or some structural abnormalitity of the second X chromosome. Turner syndrome demonstrates the importance of having the second X chromosome in normal females, even though the Lyon hypothesis seems to make the second X excess baggage. Gartler, Liskay, and Gant (1973) proved that X inactivation does not occur in the primary oocyte; lack of the second X is related somehow to the ovarian dysgenesis. It is not known whether some processes in the brain also require both X chromosomes to be active in females. Of course, the original description of the inactive X (the Barr body) was made in cat spinal neurons; it is possible that certain cerebral regions might behave differently.

The reason for such interest in the brain in Turner syndrome patients is compelling. Early IQ tests were said to show "mild mental retardation" in these girls and young women. However, it soon became clear that IQ tests and specific tests of space/form perception show a highly localized cognitive defect. In most cases, they are unable to copy simple designs, such as an arrowhead or a hexagon. At the clinical level, gynecologists have volunteered that these patients stand out as having difficulty navigating the usual maze of clinic partitions. On other components of the IQ tests they perform as well as controls do.

The variability in expression of the 45,XO karyotype is extraordinary. Some of these women are recognized only because of inferitility and short stature; Turner syndrome accounts for up to one-third of patients with primary amenorrhea in infertility clinics. Others are diagnosed at birth because of webbed neck, edema of the hands and feet, or anomalies of the blood vessels or kidneys. However, 98% of 45,XO fetuses do not survive to birth! This remarkable finding shows how little we presently understand about the role of the second X chromosome or perhaps more general chromosome imbalance in human fetal development.

Klinefelter Syndrome (47,XXY)

Klinefelter syndrome consists of testicular atrophy and infertility, occasional gynecomastia, tall eunuchoid appearance, and variable behavior, ranging from altogether normal men through individuals with mild to moderate mental deficiency to those with psychopathic and criminal behaviors. Just as 45,XO was detected among women because of lack of the typical chromatin body or Barr body on stained smears of buccal mucosa cells, Klinefelter syndrome or 47,XXY was recognized by presence of the Barr body in cells from males. Screening for the XXY karyotype, therefore, can be carried out with the buccal smear method. In one study of 942 mentally abnormal inmates with a tendency to criminal behavior, 12 (1.3%) were found to have XXY and 7 (.7%) had the XXYY karyotype, compared with .2% and .02%, respectively, in the general population (Jacobs, Brunton, Melville, Brittain and McClement, 1965). It has been suggested that the psychopathology is secondary to mental deficiency or to hypogonadism, rather than an independent result of the chromosomal abnormality. XXYY males are more likely than XXY individuals to have mental deficiency, as are persons with even greater sex chromosome imbalance, such as XXXY.

Rather than sampling a prison or psychiatrically abnormal population, Nielsen, Sørensen, Thielgaard (1969) evaluated hypogonadal male patients with 46XY ($N = 16$) and with 47XXY ($N = 34$) karyotypes at a sterility clinic in Denmark. The XXY patients had significantly more psychiatric symptoms and were differentiated from the XY patients by signs of immaturity, insecurity, boastful and self-assertive behavior, and a record of legal offenses. Differences in testis size, gynecomastia, and IQ were not related to the indices of psychopathology.

With testosterone therapy, secondary sexual development as in a normal male can be stimulated, with good physical and psychological results. However, the hyalinization of the testes cannot be reversed, and fertility is not possible. The overall frequency of XXY births is about one in 450 males, with slightly increased risk with advancing maternal age. When amniocentesis and prenatal diagnosis is performed to check for Down syndrome, the physicians and the parents should be aware that other chromosome aberrations may be found instead. The XXY karyotype is one of the most common fortuitous findings; couples will vary in their responses to information about the risk of mild mental retardation and of possible predisposition to psychopathology associated with this cause of male infertility. Whether the frequency of this disorder or its medical and psychological effects warrants population screening by amniocentesis is an unresolved question that raises many social and ethical problems.

The XYY Karyotype

The story of the XYY karyotype is one of the most curious in human behavior genetics. XYY karyotypes were first reported in association with a variety of gonadal abnormalities. In 1965, an excessive incidence of XYY males was described after screening very tall men in maximum-security prisons in Scotland (Jacobs et al., 1965). Presumably because men are considered more aggressive than women and because an extra Y chromosome seemed to be an intuitively reasonable basis for greater height and greater aggressiveness, stories from Australia and France about accused murderers having XYY karyotypes made front-page news in the United States. A mass murderer of eight Chicago nurses was publicized, wrongly, as a "possible XYY." (Neglected in all those early reports was the equally high risk, as it turns out, for XXY males).

Behavior geneticists took special interest in the XYY syndrome for another reason. A psychosocial evaluation with family data on nine XYY and 18 XY prisoners at Carstairs, Scotland, indicated that XYY criminals could be distinguished from XY counterparts by a lack of broken families, a lack of criminal records among their siblings, a tendency to get into trouble with the law earlier in their teens with crimes against property rather than people, and a greater lack of concern about their criminal behavior (Price and Whatmore, 1967). In other words, such individuals seemed to represent chromosomal accidents that made them "black sheep" of otherwise upstanding families. The analogy to severe mental retardation syndromes was obvious: A severely mentally retarded child in a family of normal parents and siblings is often the result of a particular chromosome or metabolic abnormality, whereas a mildly or moderately retarded child in a family with parents and siblings of similar IQ reflects the interplay of multiple genetic and environmental influences (Penrose, 1963; Roberts, 1952).

It must be emphasized that subsequent studies have failed to confirm the claimed differentiation between XYY and XY criminals. Furthermore, population screening by the laborious preparation of full chromosomal karyotypes has demonstrated that one in 800 males is XYY, many times the frequency of tall criminals in Western society. Thus, the XYY karyotype appears to be associated with a severalfold increased risk of psychopathology and criminality, but the vast majority of XYY males fall within the normal range of behavior. No parent populations are known to have an increased risk of producing XYY children (except for XYY males themselves).

HUMAN BEHAVIOR GENETICS:
GENERAL COMMENTS

The first objective of behavior genetics in medicine is to make an accurate and precise diagnosis. Delineation of specific syndromes involves a combination of clinical acumen, family assessment, and laboratory investigations. At a phenomenological level, there is often dispute between the "lumpers" and the "splitters" over which clinical disorders should be categorized together. As underlying mechanisms are discovered, the precision of diagnosis rises. Examples given in this chapter illustrate well the difference in precision between the schizophrenias and the causes of hyperphenylalaninemia or homocystinuria. Heterogeneity of mechanisms is extremely important to accurate clinical diagnosis, meaningful genetic counseling, and appropriate selection of therapies.

The second objective of behavior genetics in medicine is to establish the role of genetic factors in disorders with significant behavioral components. The traditional methods, described in detail in this book, of family studies, twin studies, segregation analysis, and fitting of data to models for monogenic and polygenic inheritance are all employed. Some examples have been given in the preceding sections. In addition, the use of the adoption or half-sibling method is especially useful in behavioral disorders. This method permits separation of inherited factors from those environmental factors shared in a family, at least those environmental factors acting after the time of separation of the subject from parents or other relatives. The adoption method has been particularly influential in studies of schizophrenia and has been applied to many other major behavioral phenotypes over the past decade, including alcoholism (Goodwin et al, 1973).

The third objective is to devise and assess appropriate therapies. In the monogenic disorders discussed previously, for which the most precise clinical and genetic diagnoses are feasible, some very specific therapies have been devised. By contrast, for common psychiatric disorders such as affective disorders (including manic–depressive illness), schizophrenias, alcoholism, epilepsy, and minimal brain dysfunction, therapies are empirical, often trial-and-error approaches. There is a heavy and increasing emphasis on treatment with pharmacological agents. For that reason, one of the most fruitful areas for clinical and biochemical investigation in human behavior genetics is the very great individual variation in therapeutic and adverse effects of behavior-modifying drugs. These drugs are used so commonly that clinical investigations are ethical and, indeed, sorely needed. Pharmacogenetics provides a practical arena for the study of genetic and environmental (drug) interactions, enhanced by the fact that sites of action for many of the drugs are well known.

As my pharmacology professor in second-year medical school stated; "If a drug is said to be 10% effective, it is important to find out whether 10% of patients had a complete response, or essentially all patients had a negligible response." Differences in therapeutic effectiveness of a drug across a series of patients may be due to: (1) heterogeneity of the underlying causes and mechanisms of a specific clinical phenotype, a problem in diagnosis; (2) differences in rate (or pathway) of metabolism, a problem in dosage; or (3) differences in susceptibility to the action of the drug at the critical tissue sites, a problem in selection of therapeutic agent (Omenn, 1978b, 1981b). Of course, studies of the sensitivity of brain enzymes or receptors to direct effects of various drugs or drug metabolites are limited by the inaccessibility of the human brain for clinical investigation. Therefore, a high premium should be placed on the identification of peripherally assayable enzymes or receptors that are analogous to the brain functions. Considerable investigation in humans and animals is then required to determine whether the peripheral system is under the same genetic control as the analogous system in the brain. Four such systems have been investigated: catechol–O–methyl transferase (COMT) in red blood cells; dopamine–beta–hydroxylase (DBH) in plasma; monoamine oxidase (MAO) in platelets; and the cellular reuptake system for 5–hydroxytryptamine (serotonin) in platelets. Some aspects of genetic variation and much study of variation of these activities in behavioral disorders have been reported (Gershon, Matthysse, Breakefield, and Ciaranello, 1981).

Finally, the fourth objective of behavior genetics in medicine is to help the patient or person at risk and the family to understand the disease, predict the course of the illness for those affected, and cope with the many medical, psychological, and social problems that may arise. Part of this role is the traditional assessment of the "prognosis," the future course for patients with a particular disorder. Part of it is psychological support and insight. Part of this function emphasizes the evaluation and discussion of recurrence risks for family members. Recurrence risk may be based on straightforward segregation analysis in the case of monogenic disorders that are diagnosable at birth or even *in utero*. For dominant autosomal disorders, such as Huntington disease, which typically develop clinical signs only in adulthood, more complicated genetic counseling is required, including use of Bayesian methods to estimate risk. For example, for a person at 50% risk of developing Huntington disease (one parent affected) but already 45 years old (midpoint of age-of-onset curve for showing symptoms), the risk for still developing signs of this disease is less than 50%, but more than 25%; the risk is one-third, as can be shown graphically below:

	carrier of HD gene $(P = \frac{1}{2})$	not carrier of HD gene $(P = \frac{1}{2})$
Symptoms before age 45 $(P = \frac{1}{2}, \text{ if HD})$	////////	
No symptoms before 45 $(P = \frac{1}{2}, \text{ if HD})$		

As marked, only one of the four equally probable cells has been excluded, leaving the risk at 33%.

For X-linked recessive disorders, more elegant segregation analysis methods have been developed and applied, as well as linkage methods, as relatively good markers are known to be distributed along the X chromosome. For certain X-linked diseases, such as the Lesch–Nyhan syndrome, specific biochemical diagnosis is feasible, even during pregnancy, by test of amniotic fluid cells for enzyme activity (HGPRT). For hemophilia A, precise biochemical diagnosis is feasible after birth, but not yet during pregnancy. For Duchenne muscular dystrophy, no good biochemical diagnosis is feasible before development of signs of the disease, because the underlying mechanism is not known. Often individuals come for genetic counseling who have a lifelong experience with the particular disease. For example, adults at risk for Huntington disease may have seen their own parents disintegrate as a result of the disease for 20 years or more. Young women who have spent all the years they can remember helping to care for a brother disabled from hemophilia or Duchenne muscular dystrophy will fear that they carry the gene and might pass it on to their sons. Their dread of the problem they have experienced so deeply may inhibit normal interest in marrying and having children. For such women the availability of much better treatment in the case of hemophilia and the availability of prenatal testing to determine the sex of the baby (to identify males who may carry the gene and develop the disease and distinguish them from females who may carry the gene but will not develop the disease) have opened important new options and changed the prognosis for them and their children.

For chromosomal disorders, precise laboratory diagnoses are feasible and much is being learned about the natural course of people with specific chromosomal abnormalities. For the sex chromosome anomalies (XO, XXY, XYY) there have been, as cited previously, important surprises, as information was gained about the range of expression of the karyotype in larger numbers of people. The recurrence risks depend on complicated segregation patterns for abnormal chromosome complements, plus the influence of certain external factors such as autoimmunity.

For the common behavioral disorders, as noted, genetic factors are clearly influential, but patterns of inheritance are anything but clear. Pooled data and sophisticated analyses fit best a polygenic or multifactorial model. In these cases, when specific heterogeneous causes cannot be identified, the only recourse for genetic counseling about risk of recurrence is to utilize the empirical risk estimates from direct observation in large numbers of seemingly similar families.

All these aspects of patient interests come together in the process of genetic counseling. Psychological aspects of genetic counseling are extremely important, especially for disorders affecting behavior, sense of self-worth, and sense of personal intactness. The closing section of this chapter is devoted to these considerations.

GENETIC COUNSELING: PSYCHOLOGICAL CONTEXT
AND CONTENT

Genetic counseling has been defined recently by an American/Canadian Ad Hoc Committee on Genetic Counseling (1975). Their essential point is that genetic counseling is a communication process which deals with the human problems associated with the occurrence, or the risk of occurrence, of a genetic disorder in a family. This process involves an attempt by appropriately trained persons to help the individual or family (1) comprehend the medical facts, including the diagnosis, the probable course of the disorder, and the available management; (2) appreciate the way heredity contributes to the disorder, and the risk of recurrence in specified relatives; (3) understand the options for dealing with the risk of recurrence; (4) choose the course of action which seems appropriate to them in view of their risk and the family goals and act in accordance with that decision; and (5) make the best possible adjustment to the disorder in an affected family member and/or to the risk of recurrence of that disorder. This influential formulation is notable for its lack of mention of prevention of genetic disease, long an explicit goal of genetic counseling. Now the emphasis is on communication—on counseling as a communication process—"to help the family or individual   comprehend . . . appreciate . . . understand . . . choose . . . act . . . make the best possible adjustment." As Epstein noted (in Kessler 1979), there are multiple explanations for the shift in emphasis from prevention to communication. First, prevention of most genetically determined defects is unattainable—either for theoretical or practical reasons. Second, not all individuals at risk for transmitting a particular genetic disorder desire to prevent its occurrence. Although the prevention of a genetic disorder might seem desirable to others, including the genetic counselor, it is often the case that persons at risk will not share that opinion. This discrepancy is especially likely over conditions that the patient considers "relatively mild" and over conditions that the patients themselves have; they have coped with the disease and would not wish to have prevented themselves from being born, even in those diseases for which no treatment is available. Also, these individuals may be more optimistic than the health professionals about the potential progress from medical science in developing new and effective treatments.

A third reason for the emphasis on communication is that appropriate and adequate counseling is desperately needed even in many cases in which the family or patient decides upon prevention of a genetically determined defect or disease. Tension between marital partners and depression over decisions made, including whether to have abortions or not to have abortions, are common. Finally, the genetic counseling must give special attention to the psychological burdens of the family in dealing with the affected person in the family, not just the risks of future cases. Often there is much guilt about possible responsibility in "causing" or contributing to the occurrence and complications of the genetic disorder. Parent–child relationships are complex in any case; the presence of a serious illness, especially one "transmitted" from the parent, exacerbates those tensions.

An excellent and eclectic book on psychological dimensions of genetic counseling has been published recently by Kessler (1979). Kessler groups major psychological issues of genetic counseling in the areas of sense of health and illness, procreation, pregnancy, abortion, and parenthood. The process of communication between the counselor and the counselee draws special attention not only for its features of communication, decision making, and coping, but also for the professional challenge of the relationship. The traditional model of the physician–patient relationship is inadequate for the complex objectives of genetic counseling. Physicians are generally trained to take command, to help patients, for example, by reducing fractures or removing inflamed organs. By contrast, the objectives in genetic counseling—beyond the essential role of making the correct and precise diagnosis—hinge on promoting autonomy and giving priority to individual rather than overall societal views and needs. Yet, many patients or counselees will turn to the physician and ask, "What would you do, doctor?" They will often ask the same question of other health professionals, as well. A conflict arises between the desired nondirective counseling and the willingness to give a direct and honest response. Respect for the very different circumstances of different people, circumstances only they can fully appreciate, strongly favors the nondirective position.

Fascinating challenges arise in counseling for specific common genetic problems. Because of the power of the method, amniocentesis and prenatal genetic diagnosis is receiving much attention. Obstetricians may refer their patients for amniocentesis with a casual explanation that amniocentesis is really just a routine or simple medical procedure. Such an approach is intended to be reassuring, in light of the overwhelming probability that no abnormality will be found with the usual tests for chromosomes (looking for Down syndrome) and for a circulating protein called alpha-fetoprotein (looking for neural tube defects or open spine). However, the impact on those women and their husbands whose fetus is found to have an abnormality may be so great that careful advance work is necessary in all cases in order to prepare them for this possibility. Couples need to understand prior to the amniocentesis that the results might force them to have to decide whether to continue or terminate the pregnancy. Consenting to the amniocentesis procedure raises the threat that one will have to consider betraying deeply held beliefs against abortion. At the same time, the couple must face their feelings of having a child with severe defects and little prospect of a productive life. It is not easy to decide how deeply all of these conflicting views should be explored in advance of the amniocentesis, as the vast majority of couples will receive reassuring information. Some substantial discussion is clearly required, however, because the couple must wait about 3 weeks for the results of the tests, and they are almost certain to imagine the worst at some time in that period.

Much psychological assessment has been reported about the effects of dealing with Down syndrome (Antley, in Kessler, 1979). Many parents refuse to believe that the child will fail to develop normal intellectual capabilities. Some refuse to acknowledge that the child is retarded at a later age. Some parents warmly embrace the child in the family, even building the family activities and relationships around the child, whereas others despair of the child and deny the child the love and warmth to which he or she will respond. Many parents seek soon after the diagnosis is made to find another opinion in hopes of disproving the diagnosis and averting and denying their experience in having parented an affected child. Much counseling is required to reach reality-based decisions.

PKU and its treatment introduce additional concepts in the psychology of genetic counseling. In striking contrast with Down syndrome, effective treatment can be offered these patients. However, the treatment is extremely demanding of the parents, requiring a rigorous dietary and medical follow-up program. This regimen reinforces the parent's sense of being responsible for the child's "defect." In some cases, the resulting conflict leads to avoidance behavior and a lax attitude toward adherence to the diet and medical visits. The special diet itself makes many parents negative, because of its unpleasing taste. Of course, if the child does not adhere to the diet or does not seem to be developing normally, the parent will have increased guilt and the child's sense of being defective is reinforced.

Probably the most dramatic examples of psychological content in genetic counseling come from Huntington disease. As noted previously, this disorder usually becomes manifest in adults in their most productive years and after they have already had a family of their own. Wexler (1979) reported her conclusions from in-depth interviews of 35 persons at risk for Huntington disease, all between the ages of 20 and 36. Every disease is associated with particular images and fears; for cancer, threat of pain and suffering; for multiple sclerosis, unpredictability; for Huntington disease, the time bomb. Individuals at risk fear the intellectual deterioration, personality changes, socially embarrassing choreiform movements, incontinence, and especially the extreme dependency involved in becoming chronically ill. All the individuals interviewed had known their affected parent and had watched that parent change and decline from a familiar, healthy person to someone somewhat unrecognizable, with uncontrollable behavior, bizarre movements, and slurred speech. For many, these changes took place during the child's formative years, leaving a distorted understanding of the transformations that had claimed their beloved parent. In many cases the nature of the child's early exposure to the disease appeared to be critical in determining their adult adjustment to their own genetic risk. No matter how mature and well adjusted they were to the presence of the illness in the family, the very nature of the symptomatology in this disease struck at the core of their physical and psychological self-esteem.

Single individuals questioned whether anyone could ever love and value them enough to want to share that dreaded 50% risk, a kind of genetic "Russian roulette." Some felt that so great a burden of responsibility should not be inflicted on someone they loved. Thus, even though 50% of the children of affected persons will not themselves get the disease, all are disabled by the risk.

Waiting for evidence of whether the disease will develop or not is described as living in limbo. There is no laboratory or clinical test that can predict who carries the HD gene. If an at-risk person shows normal swings in mood or attitude, others or they themselves may view such behavior as an ominous sign that the disease is developing. Persons who do not understand or do not accept the genetic basis may consider the disease a form of punishment for unpleasant behavior, or a family "curse" (the sins of the fathers visited upon the sons, etc).

I have often asked at-risk children of parents diagnosed as having Huntington disease whether they would want to be tested should a reliable test be developed. Tremendously disparate answers are given, even within a single sibship. Among four sibs aged 17 to 28, the following responses were given: (1) "No way; you won't be seeing me again"; (2) "I'd be willing to help you check out the test, but I don't want to know the results"; (3) "If you would explain it all to me first, how it works, how reliable it is, and all that, of course I'd want the test and I'd want to know the results"; and (4) "I just don't know." Such counseling sessions are wrenching experiences.

SUMMARY

Clinical genetics has blossomed as a field of medicine in the past 15 years as a result of scientific progress and practical applications to the diagnosis, treatment, and counseling of patients and their families. Behavioral and neurological disorders constitute a very substantial fraction of patients' reasons for seeking genetic counseling, about 50% in our medical genetics clinic at the University of Washington Hospital. This chapter has reviewed the behavioral genetics of selected disorders that are multifactorial, monogenic, and chromosomal in origin. The importance of understanding the underlying genetic and biochemical mechanisms has been stressed and illustrated. The methods described in other chapters and the mechanisms investigated in animal systems have had useful applications in medical genetics. In turn, careful evaluation of patients with specific genetic disorders affecting behavior provides important clues to the understanding of brain function and the care of patients with more common psychological and psychiatric problems. Finally, the psychological context and content of genetic counseling is discussed and illustrated.

Geneticists—like poets—have a consuming interest in the uniqueness of the individual. The complex interplay of genetic and environmental influences, of genetic and cultural evolution, has its richest expression and most poignant consequences in human behaviors.

REFERENCES

Ad Hoc Committee on Genetic Counseling. Genetic counseling. *Amer. J. Human Genetics,* 1975, *27,* 240–242.

Bessman, S. P. et al. Diet, genetics, and mental retardation: Interaction between phenylketonuric heterozygous mother and fetus to produce nonspecific diminution of IQ: Evidence in support of the justification hypothesis. *Proc. Natl. Acad. Sci. U.S.* 1978, *75,* 1562–1566.

Blass, J. P., & Gibson G. E. Abnormality of a thiamine-requiring enzyme in patients with Wernicke-Korsakoff syndrome. *New Engl. J. Med.,* 1977, *297:* 1367–1370.

Cartwright, G. C. Current concepts—diagnosis of treatable Wilson's disease. *New Engl. J. Med.,* 1978, *298,* 1347–1350.

Comings, D. E. The ups and downs of Huntington's disease research. *Amer. J. Human Genetics,* 1981, *33,* 314–317.

Ehrhardt, A. A. & Money, J. Progestin-induced hermaphroditism: IQ and psychosexual identity in a study of ten girls. *J. Sex Research,* 1967, *3,* 83–100.

Fialkow, P. J. Thyroid autoimmunity and Down's syndrome. *Ann. N.Y. Acad. Sci.,* 1970, *171,* 500–511.

Francone, C. A. My battle against Wilson's disease. *Amer. J. Nursing,* 1976, *76,* 247–249.

Freeman, J. M., Finkelstein, J. D., & Mudd, S. H. Folate-responsive homocystinuria and "schizophrenia." *New Engl. J. Med.,* 1975, *292,* 491–496.

Gartler, S. M., Liskay, R. M., & Gant, N. Two functional X chromosomes in human fetal oocytes. *Exp. Cell Res.,* 1973, *82,* 464–466.

Gershon, E. S., Matthysse, S., Breakefield, X. O., & Ciaranello, R. D. (Eds.). *Genetic research strategies for psychobiology and psychiatry.* Boxwood Press, Pacific Grove, California, 1981.

Goodwin, D. W., Schulsinger, F., Hermansen, L., Guze, S. B., & Winokur, G. Alcohol problems in adoptees raised apart from alcoholic biological parents. *Arch. Gen. Psychiat.,* 1973, *28,* 238–243.

Hamerton, J. L., Canning, N., Ray, M., & Smith, S. A cytogenetic study of 14,609 newborn infants. I. Incidence of chromosome abnormalities. *Clinical Genetics,* 1975, *8,* 223–243.

Heston, L. L. The genetics of schizophrenia and schizoid disease. *Science,* 1970, *167,* 249–256.

Jacobs, P. A. Human chromosome heteromorphisms (variants). *Prog. Medical Genetics,* 1977, *2* (new series), 251–274.

Jacobs, P. A., Brunton, M., Melville, M. et al. Aggressive behavior, mental subnormality, and the XYY male. *Nature,* 1965, *208,* 1351–1352.

Kessler, S. *Genetic counseling: Psychological dimensions.* Academic Press, New York, 1979.

Kopin, I. J. Neurotransmitters and the Lesch-Nyhan syndrome. *New Engl. J. Med.,* 1981, *305,* 1148–1150.

Lesch, M., & Nyhan, W. L. A familial disorder of uric acid metabolism and central nervous system function. *Amer. J. Med.,* 1964, *36,* 561–568.

Lloyd, K. G., Hornykiewicz, O., Davidson, L., et al. Biochemical evidence of dysfunction of brain neurotransmitters in the Lesch Nyhan syndrome. *New Engl. J. Med.,* 1981, *305,* 1106–1111.

McGuire, L. S., & Omenn, G. S. Congenital adrenal hyperplasia. I. Family studies of IQ. *Behav. Genetics,* 1975, *5,* 165–173.

McGuire, L. S., Ryan, K. O., & Omenn, G. S. Congenital adrenal hyperplasia. II. Cognitive and behavioral studies. *Behav. Genetics*, 1975, *5*, 175–188.

McKusick, V. A. *Heritable disorders of connective tissue, 4th ed, Homocystinuria*. Mosby, St. Louis, 1972.

McKusick, V. A. *Mendelian inheritance in man. Catalogs of autosomal dominant, autosomal recessive, and X-linked disorders*. Johns Hopkins Press, Baltimore, 5th ed., 1978.

Milunsky, A. *The prevention of genetic disease and mental retardation*. W. B. Saunders, Philadelphia, 1975.

Money, J., & Lewis, V. IQ, genetics and accelerated growth: adrenogenital syndrome. *Bull. Johns Hopkins Hosp.* 1966, 118, 365–373.

Nielsen, J., Sørensen, A., Theilgaard, A., et al. A psychiatric-psychological study of 50 severely hypogonadal male patients, including 34 with Klinefelter's syndrome, 47,XXY. *Acta Jutlanda*, 1969, *41*, 1–183.

Nyhan, W. L. The Lesch-Nyhan syndrome. *Develop. Med. Child Neurol.*, 1978, *20*, 376.

Omenn, G. S. Dysmentation from metabolic alterations. *Textbook of endocrinology*. 6th ed, R. H. Williams, Ed., chapter 13. W. B. Saunders, Philadelphia, 1981. (a)

Omenn, G. S. Some practical strategies for clinical research. *Genetic research strategies for psychobiology and psychiatry*. E. S. Gershon et al. (Eds.), Boxwood Press, Pacific Grove, Calif., 1981. (b)

Omenn, G. S. Prenatal diagnosis of genetic disorders. *Science*, 1978, *200*, 952–958.

Omenn, G. S. Psychopharmacogenetics: an overview and new approaches. *Human Genetics*, 1978b, suppl. *1*, 83–90.

Omenn, G. S. Inborn errors of metabolism: clues to understanding human behavioral disorders. *Behav. Genetics*, 1976, *6*, 263–284.

Omenn, G. S., & Motulsky, A. G. Intrauterine diagnosis and genetic counseling: implications for psychiatry in the future. In American Handbook of Psychiatry, vol VI, 3rd ed. D. A. Hamburg & H. K. H. Brodie, Eds. Basic Books, New York, 1975.

Omenn, G. S., & Motulsky, A. G. Biochemical genetics and the evolution of human behavior. In *Genetics, environment and behavior: Implications for educational policy*. L. Ehrman, G. S. Omenn, & E. Caspari, Eds. Academic Press, New York, 1972.

Paine, R. S. The variability in manifestations of untreated patients with phenylketonuria (phenylpyruvic acid). *Pediatrics*, 1957, *20*, 290–301.

Penrose, L. S. *The biology of mental defect*, 3rd ed. Grune & Stratton, New York, 1963.

Price, W. H., & Whatmore, P. B. Behavioral disorders and pattern of crime among XYY males identified at a maximum security hospital. *Brit. Med. J.*, 1967, *1*, 533–536.

Roberts, J. A. F. The genetic of mental deficiency. *Eugenics Rev.*, 1952, *44*, 71–83.

Robinson, N. M., & Robinson, H. B. *The mentally retarded child: A psychological approach*. McGraw-Hill, New York, 1976.

Rosenthal, D. *Genetic theory and abnormal behavior*. McGraw-Hill, New York, 1970.

Slater, E., & Cowie, V. *The Genetics of mental disorders*. Oxford Univ. Press, 1971.

Stanbury, J. B., Wyngaarden, J. B., & Fredrickson, D. S. (Eds.). *The metabolic basis of inherited disease*, 4th ed. McGraw-Hill, New York, 1978.

Vogel, F., & Motulsky, A. G. *Human genetics: problems and approaches*. Springer-Verlag, New York, 1979.

Wexler, N. S. Genetic "Russian roulette": the experience of being "at risk" for Huntington's disease. In S. Kessler (Ed.), Genetic counseling: Psychological dimensions. Academic Press, New York, 1979.

Wright, S. W., & Tarjan, G. Phenylketonuria. A.M.A. J. Dis. Child., 1957, *93*, 405–419.

6
Genetics of Schizophrenia

Ingrid Diederen

Miami University

Schizophrenia is one of the more complex and less understood of the psychological disorders. For many years, studies have explored the genetic aspect of this syndrome in an effort to provide a better understanding of its etiology, symptoms, course, and development. Much of what is involved in schizophrenia still remains unclear, although great strides have been accomplished by research in the area.

In trying to understand the etiology of psychiatric disorders, one cannot avoid studying gene–environment interactions. Just as genes influence the susceptibility of an individual to the development of almost any disorder, environmental factors are also seen as important etiological contributors to disturbance. Consequently, whether or not genetic or environmental variation is involved in the development of schizophrenia is no longer considered to be an important point of contention in this area of research. Instead, most of the current investigations appear to be concerned with acquiring a more thorough understanding of the magnitude and relevance of both environmental and genetic variation and the ways in which these interact to establish the various manifestations of schizophrenia.

Today there seems to be little doubt that a genetic basis for schizophrenia exists. However, the specific schizophrenia-producing genetic components and the method by which they are transmitted remain unclear. Genetic investigations have involved primarily three methodologies: family studies, twin studies, and adoption studies. Despite some difficulties, these methods have obtained strong evidence pointing toward the involvement of genetic factors in the development of schizophrenia. Currently, the amount of interest and speculation in this area of research is high, and on the basis of findings from previous studies, sophisticated models and new, more innovative research techniques are being developed. This chapter is limited to a representative but necessarily brief review of the major conceptual, diagnostic, and methodological difficulties inherent in this area of investigation, the evidence for the existence of genetic factors in the etiology of schizophrenia, and the models for a genetic mode of transmission that have arisen from these lines of research.

CONCEPTUAL ISSUES

One of the more difficult aspects of this field remains the lack of knowledge concerning the nature of schizophrenia. To date, no homogeneous group of schizophrenics has been isolated and no unitary etiology for the disorder has been found (Spring, 1981). The possibility exists that schizophrenia is not a single, unitary disorder, but may consist of a number of different entities. Schizophrenic heterogeneity could be present at several levels. Seemingly similar phenotypic expressions may be due to quite different etiologies. As a result, the reliance upon symptomatology for the differentiation of schizophrenic subtypes may be grossly misleading. To complicate matters further, the complexity of gene operations makes it difficult to delineate the various interacting etiological components of a schizophrenic disorder. The nature of the activating environment and the timing of that activation in addition to the gene could all combine to produce a single phenotype (Cancro, 1979). Therefore, a range of possible phenotypic outcomes exists for every gene as a function of the activating environment and the time of activation. Furthermore, depending on the environmental circumstances, there are a number of different genes that can produce a single phenotype. Consequently, several genetic mechanisms may underlie the expression of a single characteristic. If schizophrenia is in fact heterogeneous, many of the present research strategies may prove inadequate because most rely on the implicit assumption of homogeneity among patients studied. A schizophrenic sample based on an overinclusive diagnostic category would result in variable schizophrenic performance. As a result, common findings among schizophrenics would be difficult to detect and valid hypotheses may be prematurely discarded on the basis that the data fail to meet statistical significance.

The failure to reach a clear consensus on the nature of schizophrenia has resulted in conflicting viewpoints. To date, genetic studies of schizophrenia have relied almost exclusively on the assumption of a model in which schizophrenia consists of a complex disorder amenable to the same type of study and displaying the same pathogenic characteristics as an inheritable disease (Ødegaard, 1977). Other points of view, especially those among the psychodynamic and contextualist schools of thought, hold that the conceptualization of schizophrenia as a disease entity is inadequate and unjustified. For example, Sarbin and Mancuso (1980) claim that schizophrenia is a label and the product of social processes resulting in the imposition of a moral verdict in response to certain undesirable behaviors. Others (Bateson et al., 1956, 1963; Lidz et al., 1965; Wynne et al., 1958; as reviewed in Shapiro, 1981) have proposed that certain forms of familial interaction predispose an individual to schizophrenia, whereas still others (as reviewed by Shapiro, 1981) have preferred to focus on the schizophrenic's intrapsychic world and dynamics.

An interesting model postulated by Meehl (1962) incorporates both genetic and environmental factors to explain the occurrence of schizophrenia. Meehl suggests that all that is inherited in schizophrenia is an integrative neural defect that he labels *schizotaxia*. The imposition of a social learning history on schizotaxic individuals results in a schizotypic personality organization. That is, all schizotaxic individuals obtain a schizotypic personality organization regardless of environmental conditions. Most of these individuals will remain compensated. However, those who encounter certain detrimental influences (e.g., schizophrenogenic mothering) will decompensate into schizophrenics. In light of this model, only a few of those individuals who have inherited the neural defect and are thereby predisposed to schizophrenia, will manifest the disorder.

Despite the wide disparity among theoretical models, most can be classified as elaborate extensions of the nature versus nurture controversy. Ødegaard (1977) summarized the present state of affairs well when he stated: ''As is true of most ideological wars, this one too is somewhat futile and is a typical example of the deplorable situation where two fighting parties speak mutually unintelligible languages and often do not even listen to each other [p. 354].'' It is apparent that this debate cannot be adequately resolved until a broader base of knowledge is developed. Until that time, it may be wise to bear in mind that genetic research does not necessarily negate endeavors supporting contextualist and psychodynamic conceptions. Instead, genetic research addresses both environmental and genetic factors, and is interested in delineating the effects of each separately and in interaction as they pertain to schizophrenia.

DIAGNOSIS

Another research difficulty related to that of schizophrenic heterogeneity pertains to clinical diagnosis. The new DSM-III (American Psychiatric Association, 1980) allows for considerable heterogeneity as to who will be diagnosed schizophrenic; persons need to fulfill only a certain number of criteria out of a standard set of possibilities in order to receive the diagnosis. Furthermore, schizophrenia cannot be diagnosed unless signs of the disorder have been present for at least 6 months. Illnesses of shorter duration, or illnesses that have resulted in the face of a psychosocial stressor and appear to be transient in nature are no longer included. In fact, these disorders are not even considered to be subtypes of schizophrenia.

These changes in diagnostic criteria have left schizophrenia researchers in a peculiar bind since they now must try to define the onset of a schizophrenic disorder. In the past, the onset of schizophrenia was demarcated by the appearance of psychotic symptoms. However, it is well known that the onset of schizophrenia is often quite insidious, and that prodromal features commonly predate psychotic symptoms. Logically, it would be best if we could define onset as the point in time when one could predict with virtual certainty that a schizophrenic psychosis would occur. Unfortunately, we still lack the necessary knowledge to make such a prediction, and the definition of onset remains strictly arbitrary. Consequently, the conceptual problem of trying to determine criteria for delineating signs along a continuum marking transitions from a schizophrenia-prone personality structure to a prodomal syndrome to a full-blown psychosis still exists (Spring, 1981).

Because of these diagnostic difficulties, a standard definition for what constitutes schizophrenia remains to be determined. The absence of a standard definition remains a major hindrance to the progression of this field. The severity of the problem is due to the fact that every research project must of necessity concern itself with identifying those people who have the disorder. Therefore, the homogeneity of the sample studied and the rigor of the investigation carried out on the sample depend on the accuracy of the diagnosis. Unfortunately, there is no independent validation and only moderate reliability for diagnosing a schizophrenic disorder (Cancro, 1979). Even the combination of moderate interrater reliability and independent validation appears to be inadequate. Consequently, there is no truly homogeneous sample in schizophrenia research. Instead, investigators are forced to contend with the heterogeneous groupings that fit the schizophrenic label, the abundance of contradictory data, and the numerous disagreements over research findings. As long as this state of affairs exists, studies are limited to statistical rather than clinical significance.

At present, many diagnostic difficulties result because little is known about the unity or diversity of schizophrenia. In order to gain a clearer understanding of the nature of the disorder, many investigations have been conducted to determine the degree to which the range of phenotypic expression in schizophrenia is paralleled by genetic heterogeneity. These included certain studies that were designed to investigate whether these subtypes were fundamentally alike or different.

One such study by Winokur, Morrison, Clancy, and Crowe (1974) found that hebephrenic probands had more than three times as many schizophrenic relatives as paranoid probands. These authors determined that both children and other secondary cases of schizophrenia in the families of hebephrenics were more likely to be diagnosed hebephrenic as opposed to paranoid schizophrenic. The same type of pattern was established for paranoid schizophrenics. The children and other secondary cases of schizophrenia in the paranoid probands were more likely to manifest paranoid rather than hebephrenic symptoms. These familial data led the authors to conclude that process schizophrenia is not likely to be a unitary illness. Instead, they postulated the existence of two types of schizophrenia in the population studied, one of which may manifest itself mostly as hebephrenia but also occasionally as paranoid schizophrenia. The second type would manifest itself exclusively as paranoid schizophrenia in which there are relatively few schizophrenic relatives. Together, these findings suggest that there may be only one type of hebephrenic schizophrenia, but two types of paranoid schizophrenia, one related genetically to the hebephrenic subtype.

Another study by Kety et al. in 1975 demonstrated that the diagnoses of chronic, latent or borderline, and uncertain schizophrenia tend to cluster in the biological relatives of schizophrenic adoptees (in Kessler, 1980). Similarly, a study by Slater and Cowie (1971) examined the classical pre-Kraepelinian categories and found a significant association between schizophrenic probands and their schizophrenic offspring with respect to type of category. Kringlen in 1968 corroborated these results and also demonstrated that the entire set of Kraepelinian subtypes is represented in the children of every type of schizophrenic proband (in Gottesman & Shields, 1976).

These studies are typical of investigations in this area. Their results seem to indicate that some groups of disorders are genetically related, whereas the relationships among others remain unclear. Despite some contradictory findings, the most common results from these studies provide evidence for a common genetic basis among various forms of schizophrenic disorders. In general, the findings indicate that cases of all subtypes occur in the families of each schizophrenic subtype. They also suggest that only a modest tendency toward subtype resemblance exists within families (Gottesman & Shields, 1976).

In addition to considering the unity or diversity of schizophrenia, one must also consider the relationship between schizophrenia and other psychiatric disturbances when making clinical diagnoses. This is especially important if one wants to formulate diagnostic criteria that are successful at differentiating among various psychoses.

At present little is known about the specificity of the schizophrenic disorder. However, there is some evidence suggesting a genetic relationship between schizophrenia and other psychiatric disorders. In a study by Reed, Hartley, Anderson, Phillips, and Johnson (1973), a high risk of psychoses for the siblings of psychotics was found. When the investigators looked at specific disorders, they discovered that the risk for first-degree relatives of affective probands to develop a similar disorder was less than their risk of becoming schizophrenic. There is also some indication that the risk of schizophrenia for the offspring of paranoid and hebephrenic schizophrenics is lower than the risk for children of manic-depressives (Gottesman & Shields, 1976). Even some of the very earliest studies (Rudin, 1916, in Gottesman & Shields, 1976) discovered that schizophrenic and affectively ill individuals were about equally prevalent in the families of schizophrenics. Today as well, despite efforts to select clinically homogeneous probands, the occurrence of different types of psychoses in the same family is difficult to avoid.

The suggestion that there may be a similar genetic basis to schizophrenia and other forms of psychoses raises the possibility of overlap in the symptomatology of various psychiatric disturbances; and indeed, this has been found to be the case. Another problem exists in the possibility that symptomatology is not directly related to genetic inheritance. As early as 1917, Bleuler speculated that secondary symptoms such as delusions and hallucinations are not inherited (in Rosenthal, 1978). He also suggested that the primary symptoms do not correspond to the transmitted anomaly itself. As a result of these possibilies, there are some major difficulties involved in basing diagnostic criteria on clinical symptoms. However, there are now some indications that future diagnostic categories will be based on biochemical indicators rather than behavioral symptoms in order to decrease the heterogeneity of individuals in specified subgroups of psychopathology (Maugh, 1981).

Using biological markers could be advantageous in several ways. First, it could avoid the genetically heterogeneous diagnostic groupings based on standard methods of symptom classification. It is well known that even relatively well-formulated clinical subtypes of schizophrenia present diverse symptomatology. As a result, it is impossible to differentiate between genotypic and phenotypic variations under the present diagnostic systems. The use of biologically independent variables in formulating diagnoses may promote the detection of biological etiologies and the formation of more homogeneous psychiatric diagnostic categories. Furthermore, individuals who exhibit divergent or less severe clinical symptoms and are thereby excluded from present investigations could be included if they met a biological criterion. This in turn would allow the high-risk population (those individuals with the salient factor) to be identified and studied. The use of a biologically based strategy would also permit the study of nonhospitalized and unmedicated individuals, thereby avoiding the inclusion of these confounds within investigations (Buchsbaum & Haier, 1978).

In addition to developing biologically based criteria, many other efforts are being made to improve the diagnostic system. For example, the United States–United Kingdom Diagnostic Project, the International Pilot Study of Schizophrenia, and the World Health Organization have provided standardized input on patients for formulating diagnoses, experts to address diagnostic issues, and cross-national reference points in order to establish criteria that will improve the reliability and validity of the diagnosis of schizophrenia on an international basis (Gottesman & Shields, 1976).

Additional reasons for optimism stem from the new trends in schizophrenia nosology. These include an increased reliance on affective symptoms as diagnostically useful, as well as a shift toward incorporating the schizo-affective disorders within the affective diagnoses. In addition, the broad use of the schizophrenic label seems to be decreasing as a function of less emphasis being placed

on classic schizophrenic symptoms, and more effo
operationally defined criteria tested for reliability
present it is still difficult to reach clear and def
dealing with schizophrenia because of the many di?
with this disorder, the increasing refinement of di
opment of more sophisticated research techniques
of removing the current obstacles to gathering (
conducting therapeutic–preventive research.

FAMILY STUDIES

Despite the numerous conceptual and diagnostic difficulties inherent in schizophrenia research, much evidence has been gathered to substantiate the existence
of genetic components in the transmission of the disorder. The idea that various
forms of mental illness are heritable has existed for many centuries. However, it
was not until early in this century that researchers were able to provide actual
documentation of this fact. In 1916, Rudin published the first consanguinity
study that found a significant increase in the prevalence rate of schizophrenia
among relatives of index cases in comparison to the rate of the general population
(as reviewed by Cancro, 1979; Gottesman & Shields, 1972, 1976; Usdin, 1975).
More specifically, he discovered an age-corrected risk for schizophrenia of
4.48% and 4.12% for other psychoses in the siblings of schizophrenics whose
parents were unaffected. Many other studies (Cancro, 1979; Defries & Plomin,
1978; Fuller & Thompson, 1978; Gottesman & Shields, 1972, 1976; Kessler,
1980; Rosenthal, 1971, 1977; Usdin, 1975) have confirmed Rudin's basic finding of a higher rate of schizophrenia in the relatives of samples of schizophrenics
than in the samples from the general population. From these studies the rate of
schizophrenia in the general population has been calculated to range from .35 to
2.85%, with a mean around .85% (Kessler, 1980).

The findings from the consanguinity studies also reveal a clear correlation
between the degree of blood relationship in relatives of an individual affected
with schizophrenia and the risk of those relatives for becoming schizophrenic.
More specifically, those relatives who share fewer genes in common with an
affected individual show a lower expectancy rate for the disorder than those who
share more genes. In addition, these studies have demonstrated that there are no
significant differences between prevalence rates for siblings and other first-
degree relatives (Cancro, 1979). Although the difference found in the rate of
schizophrenia between parents (5.5%) and siblings and children of schizophrenic
probands (10.2% and 13.9% respectively) appears to be large and therefore
contradictory to the foregoing statement, this difference has been explained by
the fact that these parents are largely selected for a degree of mental health
through having been married and having children (Gottesman & Shields, 1972).

rrection is made for this factor, the risk in parents of schizophrenics
s close to that of siblings and children. Furthermore, the somewhat
er risk in children than in siblings of schizophrenics is thought to be due to
mpling error (Gottesman & Shields, 1972).

Additional evidence for a genetic factor in the transmission of schizophrenia is found by comparing the different rates of prevalence for the disease among offspring of matings of various psychotic combinations of parents. Estimates have been obtained indicating that the risk for schizophrenia increases as the genetic loading increases. The best estimate of the risk of schizophrenia in the sibling of a schizophrenic is 9.7% when neither parent has the disorder. This risk increases to 17.2% for the siblings of the proband when one parent also has schizophrenia. When both parents are schizophrenic, the risk to their children ranges between 36.6% and 46.3% (Erlenmeyer-Kimling, 1968, see also Rosenthal, Wender, Kety, Schulsinger, Welner, & Østergaard, 1968; Slater, 1968). Although it can be argued that this raise in risk from 17.2% to 46.3% for the siblings of a schizophrenic may be due to the extreme environmental conditions created by the mating of two schizophrenics, a study by Kallman (1938) reported that a comparable disturbed setting created by the mating of a schizophrenic and a psychopath yielded a sibling risk for schizophrenia of only 15% (in Gottesman & Shields, 1972). The basic findings of these studies have been further substantiated by the results of more recent investigations. For example, a study by Reed et al. (1973) reported that the risk to the offspring when neither parent was psychotic increased relative to the number of aunts and uncles who were psychiatrically disturbed.

The difference among estimates of risk is a fact significant for genetic theories. Not only does this difference demonstrate that the risk of mental illness increases as a function of increased genetic loading, but the large amount of variation in the offspring (Bastiansen & Kringlen, 1972; Bleuler, 1972; Elsasser, 1952; all in Gottesman & Shields, 1976) undermines the prediction of environmental theories that virtually all children from dual matings of schizophrenics would be emotionally disturbed.

Another line of evidence for the role of genetic factors in schizophrenia was provided by Manfred Bleuler, who applied the methodology of consanguinity studies to clinical symptomatology and course over time (Usdin, 1975). He discovered that the biological relatives of schizophrenics who also had the disorder tended to have similar ages of onset, presenting clinical pictures, lengths of duration of illness, and outcomes. These findings can be interpreted in either an environmental or genetic fashion. However, a genetic explanation appears to carry more weight because many of the relatives in the study were geographically separated and in some cases did not even know each other. Therefore, the findings of the study indicate that genetic factors play a role in several clinical parameters of schizophrenia.

Although these family studies have contributed much toward the understanding of schizophrenia and have pointed to the role of genetic factors in the development of this disorder, some major difficulties are inherent in their methodology. The earliest family studies did not question the assumption that schizophrenia is heritable and therefore failed to consider the possibility that environmental factors such as family interactions also contribute to the etiology of the disorder. As a result, these studies provided sources of data on the empirical risk of the occurrence of schizophrenia in different kinds of relatives, but did not elucidate the specific etiology involved in creating those risks. In other words, these studies could indicate if a trait ran in families, but could not specify what aspect of the family determined the trait. Obviously, families share more than genetic similarity, they also share a variety of social, psychological, and cultural experiences. Consequently, the results of these studies can just as validly be interpreted as supporting an environmental position.

Recently, some studies taking an environmental approach have investigated family interactions on the premise that certain conditions of family life can predispose an individual to schizophrenia (Liem, 1980). The majority of these investigations have focused on deviant role relationships and disordered communication processes among family members. The findings of these studies generally indicate that deficits in family communication are highly correlated with schizophrenia and high risks for schizophrenia in offspring. In addition, there are some indications that families of schizophrenics contain poorly articulated and rigid role structures in the areas of dominance, autonomy, and sexuality. Despite the success of this type of research, it is unable to establish the occurrence of schizophrenia as a strictly environmental outcome. In fact, the evidence that is presently available tends to suggest that both genetics and family process play a role in the development of schizophrenia. In light of this evidence, the current mode of investigation has changed to incorporate biological, genetic, psychological, and social factors in the context of complex, multifactorial models of causation when approaching family variables in relation to schizophrenia (Liem, 1980).

TWIN STUDIES

It was not until the advent of twin studies that the role of genetic and environmental factors in schizophrenia could be more clearly defined. The power of the twin method resides in the environmental similarities and genetic differences between monozygotic and dizygotic twins. Both types of twins share the same intrauterine environment simultaneously with their co-twin. However, monozygotic twins share the same fertilized ovum initially, whereas dizygotic twins are no more genetically alike than ordinary siblings. As a result, one can assume the presence of a genetic factor in determining a trait if the concordance rate for that particular trait is significantly higher for monozygotic as opposed to dizygotic twins.

The methodology of twin studies also provides certain advantages. For example, through the comparison of intrapair resemblance, it becomes possible to identify subtypes or components of schizophrenia that may be more under genetic or environmental control than others. Also, by noting the variability of abnormal personality traits in the monozygotic co-twins of typical schizophrenics, it becomes feasible to identify schizophrenic "equivalents" or schizotypes. Furthermore, the comparison of monozygotic twins who differ in outcome will permit the formulation of inference about the relationship between life experiences and outcome (Gottesman & Shields, 1972; Wahl, 1976).

It must be kept in mind that the twin study method relies on certain assumptions. These specify that monozygotic twins as such are not specifically predisposed to schizophrenia, and that the environments of monozygotic twins are not systematically more similar than those of dizygotic twins in such a way as to foster the disorder (Gottesman & Shields, 1972). Also, when assessing the outcome of these studies it is important to be aware of the usage of statistical age corrections and their effects on research findings. Age corrections incorporate estimates of a lifetime risk for developing schizophrenia in order to compensate for a current sample while the disorder exhibits a variable age of onset. Although several methods have been developed, none is entirely satisfactory, and all present only varying degrees of approximation to the true concordance rate. Furthermore, it is essential to realize that the absolute size of the concordance rate is not predictive of the relative importance of the genetic contribution to the trait in question. Concordance rates can be expressed in several fashions. The basic difference in methods depends on whether concordance is expressed in terms of the pair or the individual. The casewise or proband method refers to the proportion of cases with an affected partner, whereas the pairwise rate considers the proportion of pairs in which both twins are affected. The individual or proband concordance method yields a higher percentage than a pairwise concordance. Therefore, only those studies that use the same method are comparable (Kessler, 1980).

The most influential early twin studies were conducted by Kallman (1938) and Slater (1953) (both in Cancro, 1979). Even though the two studies contained important differences and suffered from methodological problems, both found a significantly higher concordance rate for schizophrenia among monozygotic twins. By the end of Kallman's study and with some diagnostic revision, but without statistical age correction, 59% of 174 monozygotic pairs and 9% of 517 dizygotic pairs were concordant for what Kallman labeled "definite schizophrenia." In Slater's study, 65% of 37 monozygotic pairs and 14% of 58 same-sexed dizygotic pairs were found to be concordant (as reviewed by Gottesman & Shields, 1972). Both studies used the pairwise method for calculating the concordance rates. Findings from other earlier studies (as reviewed by Gottesman & Shields, 1972, 1976; Kessler, 1980; Rosenthal, 1977) were basically consistent with these results.

More recently, several studies have been conducted that are methodologically powerful, and that provide highly significant results. One of these studies, reported by Tienari (1963), was conducted on an exclusively male sample in

Finland (in Cancro, 1979; Gottesman & Shields, 1976). Initially, this investigator found a zero concordance rate for monozygotic twins. However, by 1976 the pairwise concordance rate for the sample had risen to 16%. Furthermore, with the use of the proband method, the concordance rate for the sample was found to be between 35% and 53% depending upon which borderline cases were included (Cancro, 1979). As a result, even though it took some time, a significant monozygotic–dizygotic difference was demonstrated.

A study by Kringlen (1967) was conducted in Norway and included both male and female twins (in Gottesman & Shields, 1972; Cancro, 1979). His study was especially important for several reasons. His sample was well constructed and contained the largest number of psychotic twins personally investigated among the more recent studies. His investigation was very comprehensive and included case history, pedigree, clinical analysis, rating scale data, and all functional psychoses in addition to schizophrenia. Using the more conservative pairwise method, Kringlen found a significant difference between monozygotic and dizygotic twins with concordance rates ranging from 60% for monozygotic probands rated as severe to 25% for the most benign. His results indicated no difference between the concordance rates of monozygotic twins as a function of gender. He reported that whether the sexes of the dizygotic twins were the same or different, the concordance rates did not differ, but were equal to those found in ordinary siblings. These results counter findings from some earlier studies and undermine the psychodynamic thesis that concordance for schizophrenia will increase without regard to genetic similarity, but relative to the amount of likeness and resulting identity confusion among siblings (Jackson, 1960).

The findings of another important twin study were reported by Fischer, Harvald, and Hauge in 1969 (in Gottesman & Shields, 1972). This study was conducted in Denmark upon a sample that was very representative with respect to twin types and sexes of probands. Only those probands who met strict criteria for chronic or process schizophrenia were included. Even so, schizophreniform, paranoid, and atypical psychoses in co-twins led to data analysis for quantitative concordance for other than process schizophrenia. As a result of this study, Fischer et al. discovered a pairwise concordance rate of 24% for monozygotic twins and 10% for dizygotic twins. In addition, the authors calculated the rates according to the proband method and obtained a 36% concordance rate for monozygotic and 18% concordance rate for dizygotic pairs. When those co-twins who exhibited a form of schizophrenic disorder different from process schizophrenia were included in their calculations, the investigators obtained proband concordance rates of 56% and 26% for monozygotic and dizygotic twins respectively. These figures appear to be good approximations of the true incidence rate, as the twins used in this study were followed long enough to virtually eliminate the need for any statistical age correction.

Fisher (1973) (in Gottesman & Shields, 1976; Cancro, 1979) went on to report that monozygotic twin pairs concordant for schizophrenia were not at an increased risk for the disorder when compared to discordant monozygotic twins.

She did not discover a relationship between severity of disturbance in the proband and the risk of developing psychosis in the co-twin. She did find however, that the offspring of concordant twins were not at an increased risk for schizophrenia when compared to the offspring of discordant twins. In addition, the likelihood of a discordant twin member producing a schizophrenic child was equal to that of the schizophrenic member of the twin pair. These findings point toward a genetic hypothesis and indicate that the genetic transmission of schizophrenia is not always manifested directly.

The Maudsley Hospital-based study of Gottesman and Shields (1972, 1976) conducted in England obtained similar results to those of the studies reported earlier. These investigators also obtained a significantly increased monozygotic concordance rate over that found in dizygotic twins. Similarly, Pollin, Allen, Hoffer, Stabenau, and Hrubec (1969), in a study based on an American sample, found a higher concordance rate for monozygotic as opposed to dizygotic twins. More specifically, the concordance rate for schizophrenia in monozygotic twins was found to be 3.3 times greater than the rate in dizygotic twins. What is interesting about this study is that the results were based on an exclusively male sample consisting of almost 16,000 twin pairs who had successfully passed army induction exams and who had served in the armed forces during the period spanning World War II and the Korean War. The calculations for the concordance rates were based on chart diagnoses obtained from VA records or responses to mailed questionnaires. As a result, the consistency of the findings from this study appear somewhat amazing when one considered the size of the sample, its apparent skewness toward healthy individuals, and the tendency for wartime nonpsychiatric physicians to use the diagnosis of schizophrenia more loosely. This last factor tends to lower the homogeneity of the schizophrenic sample and the accompanying concordance rates for schizophrenia. Nevertheless, a monozygotic to dizygotic twin concordance ratio of 3.3 was obtained, and the authors concluded that the genetic loading for schizophrenia was higher than for any other psychiatric or medical diagnosis considered in their investigation.

At present, all the major twin studies carried out worldwide have demonstrated higher concordance rates for monozygotic as opposed to dizygotic twins (Kessler, 1980). The monozygotic concordance rates from studies conducted before 1965 tended to average around 60% or more, whereas the results from more recent studies are usually lower. These differences have been attributed to variations in methodological and diagnostic procedures. Even so, the consistency of twin studies' findings despite diversity in design and execution provides strong evidence for a hereditary factor in the etiology of schizophrenia. From these studies it is apparent that identical twins per se are not more prone to develop the disease. However, if one twin does develop the disorder, the co-twin is at an increased risk over an ordinary sibling or dizygotic twin. In addition, a frequent finding from twin studies has been that concordance is higher for index cases with severe forms of schizophrenia (Rosenthal, 1970). Furthermore, the findings from these studies contradict the suggestion that the stress inherent in rearing twins plays a role in the development of schizophrenia by demonstrating that both monozygotic and dizygotic twin pairs are not at an increased risk for the disorder. In fact, the research literature in general suggests that life events and stress do not account for the development of a schizophrenic disorder (Rabkin, 1980).

Despite the usefulness of the twin method in demonstrating that genetic factors play a role in the etiology of schizophrenia, it is still important to consider the ways in which environmental factors may have contributed to the findings reported. Jackson (1960) suggested that monozygotic twins may be more prone to develop schizophrenia because of identity confusion and/or weak identity formation. However, this idea has failed to be substantiated, as no excess of monozygotic twins has been discovered in schizophrenic samples (Rosenthal, 1960). The possibility of one twin identifying with the other, in this case disturbed twin, has also been considered as a psychological factor that might lead to higher concordance rates in monozygotic twins. Usually this is referred to as *folie à deux,* or psychosis by association, a clinical entity that would result in clinical resemblances between the psychoses of identical twins. Again, there appears to be a lack of evidence to support this position. There are simply too few cases of shared delusions or other symptoms of intense identification of one schizophrenic twin with the other to support the theory (Gottesman & Shields, 1972).

Another environmental explanation that has been put forward postulates that both members of monozygotic twin pairs have a greater likelihood of being exposed to a schizophrenia-inducing situation than do members of dizygotic twin pairs. This position would entail the assumption that monozygotic twins are treated more alike than dizygotic twins. In fact, Kallman (1946) reported higher concordance rates in twin pairs living together for 5 or more years before the onset of the disorder than in pairs separated during that period (Gottesman & Shields, 1972). Even so, studies conducted on identical twins where one or both members later became schizophrenic (studies reviewed by Gottesman & Shields, 1972, 1976; Jackson, 1960; Kessler, 1980) reveal similar concordance rates for those reared apart when compared to those reared together. In addition, pairs of monozygotic twins reared apart reveal much higher concordance rates for schizophrenia than dizygotic twins reared together (Erlenmeyer-Kimling, 1976).

One criticism of the twin method addresses the assumption that the sharing of the intrauterine environment simultaneously by both members of the monozygotic twin pair markedly reduces the developmental and environmental variations for the members of the pair. Instead, it is often the case that one twin member is put at a disadvantage and receives less of the limited space and nutrients available than the other twin. Consequently, the disadvantaged twin may appear weaker and receive special care and attention from the parents. This parental behavior may in turn foster dependency and submissiveness in the individual.

In a study conducted by Pollin et al. 1966, (in Torrey, 1977) the conclusion was reached that low birth weight was an important contributing factor to the development of the schizophrenia. However, this finding did not appear to be replicated in other studies (as reviewed by Gottesman & Shields, 1977). For example, Kringlen (1967) (in Gottesman & Shields, 1972, 1977) found no general tendency for submissive, mother-dependent twins to become schizophrenic. In fact, his data suggest that birth weight and obstetric complications are not correlated with the later development of schizophrenia. Instead, he concluded that the twin who is predisposed to become schizophrenic, is predisposed in a nonspecific manner. The conclusions in this area remain unclear however. Other more recent analyses (as reviewed by Torrey, 1977) reveal that the schizophrenic twin is often lighter at birth in discordant twin pairs. Even so, the differences in birth weights are usually small, and the schizophrenic twin's weight is usually within the normal range. Despite the continuing controversy in this matter, it is important to remember that even if some conclusive evidence could be obtained with regard to birth weights, it would still throw little light on the etiology of schizophrenic disorders because lowered birth weights can result from numerous factors and may create any number of reactions that may or may not be etiologically linked to schizophrenia.

Taken together, the findings from the twin study method have strengthened the view that genetic factors are significantly involved in creating the higher monozygotic concordance rates reported in the literature. This is not to say that environmental variables are not important in the etiology of schizophrenia. However, the findings do point out the vital contribution twin studies have made for the genetic position in this area of research.

ADOPTION STUDIES

In the adoption strategy there are basically four approaches that can be used to separate genetic factors from other possible etiological factors related to familial behavior and environment (Rosenthal, 1971). In the adoptees' families' method, schizophrenic adoptees are selected as index cases to be compared to a control group. The control group consists of other adoptees who are free of any psychiatric history but who match the index cases with respect to age, sex, age at transfer to adoptive parents, and social class of adoptive parents. The difference in incidence of schizophrenia is then determined for both the biological and adoptive relatives of the index and control groups.

The adoptees' study method consists of finding schizophrenics whose children were placed for adoption at an early age so as to compare the prevalence ratings of these children with either a matched control group or the population in general. A third strategy, the adoptive parents' method, starts with the adoptee schizophrenic index cases and then compares the incidence of pathology in their biological and adoptive parents.

Cross-fostering is perhaps the most sophisticated research strategy among adoption studies. With this research design two different groups of adoptees are identified: those who have a biological parent who is schizophrenic, but whose adoptive parents are free of psychiatric illness, and those whose biological parents have no psychiatric history, but who are adopted by a schizophrenic. The two groups are then compared with respect to incidence of schizophrenia and other psychiatric disturbances.

One study that uses the adoptees' method was conducted by Heston (1966). He found that the offspring of severely schizophrenic mothers who were given up for adoption within the first 3 days of life still grew up to have schizophrenia at the same rate as expected for those reared by their schizophrenic mothers.

In addition, the offspring from schizophrenic mothers showed considerably more abnormalities other than schizophrenia than would be expected for the general population. A later and more elaborate study conducted on a Danish population (Rosenthal, Wender, Kety, Schulsinger, Welner & Østergaard, 1968) confirmed Heston's results.

Another Danish study conducted by Kety, Rosenthal, Wender, & Schulsinger (1968) identified adoptees who became schizophrenic and then through the use of documents determined the status of their biological and adoptive parents, siblings, and half-siblings. The purpose of the study was to discover if the schizophrenia in the adoptees was a function of schizophrenia in the adoptive or biological families. Again, supporting evidence was found for a genetic factor in the transmission of schizophrenia.

It is important to note that Rosenthal et al. (1968) and Kety et al. (1968) used a broad diagnostic concept, the "schizophrenic spectrum" in their investigations. Rosenthal included borderline and schizoid or paranoid diagnoses along with the diagnosis of schizophrenia; and Kety included chronic, acute, or borderline schizophrenia, definite or uncertain, in addition to inadequate personality in the schizophrenic spectrum. Both studies reported that schizophrenic spectrum index cases showed an increase in the rate of schizophrenic spectrum disorders only in their biological relatives. Besides indicating that schizophrenia is heritable, these studies also offer evidence that the genetic potential for schizophrenia may manifest itself in ways other than overt schizophrenia.

Since these findings were reported, Kety has conducted interviews with the relatives of schizophrenics and the control adoptees in the field, in addition to increasing his sample size (Gottesman & Shields, 1976). Similarly, Rosenthal increased his sample size of index adoptees and interviewed the spouses of the schizophrenic parents. In general, the more recent data from these studies confirmed the findings of the earlier reports. More specifically, there were more schizophrenic and schizophrenia-related disorders in the biological relatives of schizophrenic adoptees and in the adopted-away children of schizophrenic parents than in the control groups (Gottesman & Shields, 1976).

A unique study by Wender, Rosenthal, Key, Schulsinger, & Welner (1974), which used the cross-fostering technique, also found a greater prevalence of psychopathology among the adopted-away offspring of schizophrenics, but no such increase among the cross-fostered group. The investigators concluded that genetic factors play a role in the etiology of schizophrenia whereas familial psychopathology does not.

An earlier study conducted by Wender, Rosenthal, and Kety (1968) looked at the parents of adult schizophrenics who had been given up for adoption as infants, and compared them with parents who had reared their own, now adult schizophrenic offspring. These two groups were also compared to the adopting parents of normal young adults. The findings of this research indicated that the biological parents of schizophrenics were considerably more disturbed than the adopting parents of schizophrenics, and that the adopting parents of schizophrenics were exhibiting a slightly greater degree of psychopathology than the adopting parents of normal young adults. Consequently, both the psychosocial and genetic hypotheses were supported by this research.

In a later study based on the original test data, protocols were blindly analyzed, and both the adoptive and biological parents of schizophrenic patients were successfully differentiated from the adoptive parents of normal offspring (Wynne, Singer, & Toohey, 1976, in Lidz, 1976). An earlier series of experiments using projective techniques in blind predictions of psychiatric patients and their families was also completed. In addition to matching the tests of the patients with their families, test data from other members of the family were successfully employed to deduce the diagnosis, form of thinking, and degree of disorganization in the young adult psychiatric patient (Singer & Wynne, 1965). These findings provided substantial support for the environmentalist position and seemed to indicate that certain types of familial communication patterns are possible etiological contributors to schizophrenia. The debate between those supporting a contextualist position and those espousing a genetic model raged on, as so often happens in this area of investigation, without arriving at a firm resolution. Part of the difficulty in interpreting this type of research stems from the inability to avoid genetic and psychological confounds and the failure to differentiate schizophrenogenic behavior from the behavior of schizophrenics. In addition, problems arise when the findings from certain studies may be attributable to several hypotheses. For example, the Wynne et al. (1976) results may just as easily be attributed to psychological changes in the parents resulting from rearing a schizophrenic child, or genetic factors accountable for both the psychopathology and communication patterns observed. That is, even if communication deviances are etiologically significant for schizophrenia, they do not rule out the possible important of genetic factors as influential in creating those deviances (Kety, Rosenthal, Wender, & Schulsinger, 1976). In any case, further supportive evidence for a genetic position was obtained in a later replicative study by Wender, Rosenthal, Rainer, Greenhill, and Sarlin (1977) using the same design as previously but including a systematic sample. The findings from this study revealed that the biological parents of schizophrenics were more disturbed than adoptive parents of either schizophrenics or normals; and that the adoptive parents did not differ on degree of psychopathology.

Despite all the advances made by the adoption research, it is still important to realize that even the most sophisticated adoption studies are limited in their resolving power. As of yet, confounds and other research difficulties cannot be avoided. This is in part due to the many practical considerations that must be dealt with when using this type of research strategy. For instance, only certain children are permitted to be legally adopted, and sometimes these are abnormal before placement. Furthermore, adopting parents as a group usually create a restricted range of rearing environments and share a similar socioeconomic status with the biological parents. In addition, fostered children are not included in the initial sample; and the spouses of schizophrenics whose children are adopted probably include a higher proportion of socially inadequate individuals than those spouses of schizophrenics whose children are not adopted (Gottesman & Shields, 1976). Although these limitations do not erode the principal conclusions obtained through adoption studies, they do limit their generalizability.

In spite of the various difficulties inherent in this research method, the adoption research has made a vital contribution to the investigation of the etiology of schizophrenia. Perhaps most importantly, it has disconfirmed the hypothesis that the high rate of schizophrenia observed in the offspring of schizophrenics is necessarily due to being raised by a schizophrenic parent. As a result of their findings, the adoption studies along with the family and twin studies have all but dispelled any doubt that a genetic factor is operating in the transmission of schizophrenia. This seems pretty remarkable when one considers that this genetic factor has surfaced in spite of all the problems in the formulation of individual diagnoses, the heterogeneous nature of the schizophrenic syndrome, and the interindividual differences that must be present in any population labeled schizophrenic.

GENETIC MODELS

With the knowledge that hereditary factors play an important role in the development of schizophrenia, scientists have begun to search for what exactly is transmitted in the disorder. However, so far the search for an adequate model has not yielded conclusive results. What has happened, is that investigators have proposed several theoretical models and have explored the fit between the theoretical predictions made from these models and the existing data.

The proposed models for the genetic transmission of schizophrenia can be roughly divided into two types: monogenic and polygenic. Monogenic models rest on the assumption that the action of a single or a few major genes leads to a specific metabolic deficit that in turn causes schizophrenia. Polygenic models propose that the action of many genes, each with a small effect, can, given a certain combination, produce the disorder (Fuller & Thompson, 1978; Gottes-man & Shields, 1972, 1976; Keith, Gunderson, Reifman, Buchsbaum, & Mosher, 1976; Kessler, 1980; Rosenthal, 1977).

Since Rudin conducted his study in 1916, it has been apparent that the mode of genetic transmission in schizophrenia does not seem to follow any simple Mendelian pattern. In all genetic studies, the number of cases of schizophrenia found in the relatives of schizophrenic probands has always been less, but never higher than would be expected with a simple Mendelian distribution. Therefore, one can conclude that the incidence of schizophrenia was not due to random error nor could it be explained by a simple monogenic theory. Consequently, investigators have had to propose the existence of modifying factors to accompany monogenic theories in order to account for the lack of consistency with Mendelian ratios (Fuller & Thompson, 1978).

In 1965, Slater proposed a major partially dominant gene model that included the concept of penetrance (Keith et al., 1976). Penetrance refers to the percentage of cases that manifest the trait in question when the necessary genetic components are present. According to Slater's modified dominance theory, virtually all schizophrenics inherit the same abnormal gene in a single or double dose, but only one out of four will manifest the disorder, because the gene expression is only partially penetrant (Keith et al., 1976). Later, modifications of this model by other investigators such as Slater and Cowie (1971), and Kidd and Cavalli-Sforza (1973) produced variations in the posited penetrance of the dominant gene in the heterozygote.

Another model developed by Kallman in 1953 proposes a recessive single gene that can be modified by another variable called the "constitutional resistance to manifestation" (in Keith et al., 1976). This variable was considered to be a polygenetically determined trait. Those individuals who were homozygous with respect to the single gene were thought to be much more likely to develop schizophrenia than those individuals who were heterozygous. However, the degree of constitutional resistance present could affect the tendency to develop schizophrenia even when the individual was homozygous. Through this model, which allows for various combinations of homozygosity and heterozygosity interacting with high or low levels of constitutional resistance, Kallman hoped to account for different degrees of severity in the manifestation of the disorder.

One criticism that has been leveled against these monogenic models is that they cannot stand without some kind of support from such concepts as penetrance or resistance to manifestation. These types of concepts are especially problematic because the specific mechanisms proposed have never been demonstrated, do not further the understanding of the etiology of schizophrenia, and fail to provide a means of disproving a particular model (Keith et al., 1976). Consequently, these types of concepts have often been labeled as ad hoc explanations whose only purpose is to bring order and consistency to a theoretical model.

When one works with polygenic models many of the difficulties inherent in the monogenic models disappear but others arise. Mendelian distributions are no longer required and the expectancy rates for schizophrenia, which were lower than expected for monogenic positions, are not contraindicative of polygenic theory. Furthermore, there is good scientific precedent for this type of theory, as many traits among many different kinds of species have been shown to have complex non-Mendelian, multifactorial genetic determinants (Keith et al., 1976).

One interesting model advanced by Falconer (1965) uses a quantitative genetic approach and makes the assumption that underlying the etiology of the disorder is a variable distributed along a continuum (Rosenthal, 1977). This variable, called liability, includes both innate tendencies and external circumstances that may be predisposing to the development of the disorder. The variation of liability is distributed in the form of a normal, bell-shaped curve. At one end of this curve is a point called the threshold, which demarcates those individuals who are affected by the disorder from those who are not. This point is positioned on the curve according to the prevalence of the disorder in the general population. Once the prevalence rates among different classes of relatives of affected individuals are determined, it becomes possible to use the model to calculate the heritability of the liability of the disorder in question.

In 1967, Gottesman and Shields were the first to apply Falconer's model to schizophrenia (Rosenthal, 1977). They obtained many heritability values using different samples of subjects, and these values ranged between 45 and 106%. Accordingly, the authors acknowledged that the method was subject to errors, and that the underlying assumptions of the model had not all been fully met. Even so, the authors noted that the findings did indicate the heritability of liability to schizophrenia across samples to be substantial. However, they also qualified these findings by pointing out that heritability is not a property of the trait alone, but also of the population sampled and of environmental circumstances.

Another polygenic theory has been proposed by Cancro (1979). His model assumes the existence of a continuum with regard to phenotypic expression in the schizophrenic syndrome, and does not assume that the phenotype is inherently abnormal. Furthermore, he postulates that several phenotypes are involved in schizophrenia. That is, different subgroups of schizophrenia are assumed to be involved in different combinations or patterns of phenotypes. In addition, there are a number of different genes that can produce each phenotype depending on the environmental conditions present. Therefore, two individuals who share the same genetically loaded trait may have obtained that characteristic by different means. Obviously, Cancro's model is quite complex. Not only does he have the possibility of a particular phenotype developing from different pathways, he also allows for the heterogeneity of the schizophrenic disorder by assuming that multiple traits are involved.

Although these polygenic models contain several advantages over the monogenic models, they are also more loosely formulated and consequently more difficult to test. Furthermore, the complexity of the models precludes the application of conventional methods of data analysis. Instead, new statistical methods have to be devised and applied.

Studies measuring the fit of theoretical models to the ever increasing amount of data are continually taking place. For instance, Kidd and Cavalli-Sforza (1973) reviewed the then current research data and tested the goodness of fit of these data to both a monogenic and polygenic model, using a threshold approach. On the basis of their analysis, neither a monogenic nor polygenic model could be discarded, as both seemed to fit the available data equally well. This led the authors to speculate that any intermediate model between the two extremes investigated would also fit the data satisfactorily. In fact, they postulated that any model of genetic heterogeneity would fit, with improvement in fit increasing as more parameters are included in the calculations.

Interestingly, when Mathysse and Kidd (1976) expanded on this approach, they found that neither of the two extreme genetic hypotheses were adequate when trying to account for the observed data. The single major locus or monogenic model provided estimates that were too low, whereas the multifactorial model provided estimates that were too high relative to the observed incidences of schizophrenia among monozygotic twins, and offspring of parents who were both affected. Only a model with two interacting genes could possibly account for the incidences observed in the input data. Even so, both the monogenic and multifactorial models did predict genetic heterogeneity among schizophrenics and provided some interesting perspectives on the disorder.

When the input data were fitted according to the monogenic model, the genetic composition varied along with the frequency of the schizophrenia-producing allele. Every individual has two alleles or genes, one from each parent, at each locus. At low frequencies of the allele, the development of the disorder seems to be largely due to environmental factors. At relatively high frequencies of the allele, the proportion of heterozygotes increases whereas the number of phenocopies decreases. Furthermore, the proportion of homozygotes remains low no matter what the allelic frequency. Consequently, the single major locus model predicts that genes play an extremely important role on those occasions when they are present in double dose, but those cases would be rare even among persons manifesting the disorder.

Under the polygenic model, however, Mathysse and Kidd found that 9.1% of the schizophrenic population has such a high genetic risk that environmental factors play virtually no role in the development of the disorder. That is, this model proposes that in one of every 11 schizophrenics, the disorder is almost totally genetically determined. Obviously the results are still unclear and the conclusions remain tentative. Arguments over genetic hypotheses will probably continue to lack meaning as long as virtually all models can account for the data. Until the research data becomes homogeneous, future investigations will be unable to rely on the use of statistical techniques alone to clarify which genetic model is most applicable to schizophrenia.

Although monogenic and polygenic models represent the genetic extremes of inheritance, there are also models of genetic heterogeneity that rest on the principle that more than one gene at one locus is needed to produce schizophrenia. Although appearing similar to polygenic theory, the two theories are different because polygenic models require that more than one locus is necessary for the development of the disorder. Furthermore, the effects of different schizophrenia-producing genes are not cumulative in genetic heterogeneity theory as they are in polygenic models, and the heterogeneity model does not propose a continuum. Instead, each gene is believed to be separate and distinct in its effects. Consequently, a particular combination of genes is not required to produce the disorder, but rather the appropriate specific genes (Keith et al., 1976).

A heterogeneity model had been proposed by Karlsson (1966, in Rosenthal, 1977). He postulated a two-gene theory in which a mechanism of modified dominance was involved. The theory allowed for the prediction of distinctive personality types from specific genotypes. However, there does not appear to have been any independent validation of this model.

Taking a different approach to trying to uncover the mode of inheritance in schizophrenia, Mitsuda (1967, in Rosenthal, 1977) examined different types of schizophrenics and then classified them according to outcome. Cases for the apparent mode of transmission were classified into three groups: dominant, recessive, and intermediate. His findings suggested that the dominant gene group contained more prognostically favorable cases, whereas the recessive group had more that were unfavorable, and the intermediate group fell somewhere in between. As a result, he decided on the existence of a heterogeneity model for which three modes of genetic transmission were possible.

At present, theories of heterogeneity are very popular. However, if they are correct, they imply that the likelihood of discovering the mode of genetic transmission in schizophrenia is very low. On the other hand, monogenic theories appear more inviting because they allow for the possibility of identifying exactly what is transmitted in schizophrenia, thereby offering hope that therapeutic and preventive techniques will be developed sometime in the future (Fuller & Thompson, 1978). In light of these differences, it may be more worthwhile to direct research toward trying to disprove the single-mode theory, if only to provide some direction as to which model is "true."

It is apparent that no single model for the mode of genetic transmission in schizophrenia has been universally accepted. There are even disagreements among investigators espousing the same type of model. For example, among the monogenic theorists there is no clear consensus regarding the mode of action of a single gene. However, there does seem to be agreement among the research results that a relatively low frequency of homozygotes exists in the population, and that the manifestation rate is very high for these individuals.

At present, it is still not possible to decide on exactly what is inherited in schizophrenia and how it is transmitted. However, the studies conducted to date, along with the improvements for research strategies that they foster, provide the necessary impetus for more sophisticated model building and testing. These in turn, provide reason for optimism that this area of investigation is worthwhile, and that it will lead to the eventual discovery of the mode of transmission in schizophrenia.

METHODOLOGICAL DIFFICULTIES

As we have seen, there are many methods that have been employed in the study of genetics and schizophrenia. Twin studies have been especially valuable for assessing the importance of biological factors in an illness. Of course, their value is still dependent on the validity of the diagnoses of zygosity and disorder. Despite their usefulness, they still present several problems for data collection and analysis. In particular, the usage of different definitions of concordance rates can be particularly problematic. Furthermore, monozygotic twins often differ on biological traits (Kidd & Matthysse, 1978). Other factors such as obstetrical complications and different birth weights may interact with susceptibility to psychiatric disorders such as schizophrenia.

Separation studies, including studies of adopted children, cross-fostering studies, studies of half-siblings, and studies of monozygotic twins reared apart, have proven invaluable by demonstrating unequivocally that genetic factors are involved in the transmission of schizophrenia. Even so, these studies in themselves cannot elucidate what genetic and biological factors are operating and how they contribute to the development of a schizophrenic disorder. Instead, the best contemporary use of separation studies may be to shed more light on the relative importance and specificity of certain environmental factors in relation to schizophrenia. This could possibly be accomplished through the comparison of rearing environments with populations at high and low genetic risk (Kidd & Matthysse, 1978). In this manner, a negative correlation between genetic risk and a specific environmental factor would suggest a causal role for that factor.

Nuclear families have constituted the most popular form of data for genetic studies. This type of data is easily related to the epidemiology of the disease, but often poses difficulties for genetic analyses and interpretations. Part of the difficulty stems from the researchers' failure to distinguish the parent–offspring and sibling–sibling relationships. Instead, risks are calculated only for first-degree relatives, while the finer distinctions are ignored. Failure to note the sex of both proband and relative is also common and may affect morbid risk calculations (Kidd & Matthysse, 1978). Until these factors are taken into account and sufficient specifications by relationship are made, these types of studies will remain limited in utility and ability to discriminate between proposed genetic models and types of heterogeneity.

Another valuable but often neglected resource in the investigation of genetic factors in schizophrenia is the use of multigenerational pedigrees. One major advantage of this approach is that it allows for the potential identification of homogeneous phenotypic subgroups on the basis that any rare disorder found in several members of one family is likely to have the same etiology in all. Furthermore, pedigrees are amenable to the study of genetic linkage and the testing of genetic models (Kidd & Matthysse, 1978).

Sampling procedures also pose major problems in this area of research. For example, Lewine, Watt, and Grubb (1981) point out that the current use of parent–child concordance rates may result in the use of a sample that consists of schizophrenics with an unusually strong genetic and environmental predisposition to the disorder. These authors have also demonstrated how the overrepresentation of females among index parents may introduce systematic sex differences in schizophrenic characteristics as well as biased prevalence rates. This in turn may have negative implications for the interpretations based on this type of research. For example, current risk studies rely on theoretical concepts and assessment procedures based for the most part on adult male populations. At present, it is still unknown how generalizable these conceptualizations are to female schizophrenics. Furthermore, it is difficult to ascertain if findings regarding the offspring of schizophrenic women are comparable to the findings for offspring of schizophrenic men. Although we know that the risk for schizophrenia is the same for both men and women, the evidence indicating that schizophrenia may take different forms in men and women suggests that different etiological factors may be of importance for the sexes. Furthermore, the preponderance of a reactive, atypical, or schizo-affective type of schizophrenia in women suggests that careful diagnoses of index parents and offspring according to schizophrenic subtype are necessary to clarify the nature of schizophrenia, including its possible heterogeneity. In addition, the fact that varying sampling procedures may result in differing concordance rates does not permit easy comparison of results among studies (Rosenthal, 1961, 1962).

The importance of studying families over unrelated individuals seems to be well established. As such, it appears to be the optimal strategy for genetic research on schizophrenia at this time. In addition, it is apparent that many unknowns remain in this area of investigation. Therefore, until such a time as our methods of analyses and diagnostic capabilities have improved sufficiently, one needs to remain aware of the tentativeness of our present conclusions as to the exact nature and transmission of a genetic factor in schizophrenia.

COMMENT

Anytime one delves into the schizophrenia literature, one is struck by the over-whelming amount of information available in this area. Given the wealth of knowledge that has accumulated in this field, it seems almost ironic that schizo-phrenia remains one of the most complex and least understood of all the psycho-logical disorders.

It is obvious that none of the theoretical models proposed to date is suffi-ciently capable of explaining schizophrenia. Each one of them is lacking in certain respects. For example, most research studies dealing with genetic or family theories cannot explain the occurrence of schizophrenia as either a strictly environmental or genetic outcome. Instead, the evidence tends to suggest that both genetics and family process play a role in the development of the disorder. Similarly, although biochemical and neurophysiological theories hold much promise with regard to deriving useful diagnostic categories, their focus appears too narrow to explain the full scope of the schizophrenic syndrome. On the other hand, psychological and psychoanalytic theories appear comprehensive in their provision of meaningful ways for conceptualizing and attending to schizophrenic process. Because no one can perceive the schizophrenic's subjective experience, these theories appear to try to understand its nature and etiology through an intuitive or empathic approach. Unfortunately, these theories are difficult to investigate and therefore it is difficult to obtain enough empirical data to support these positions.

In trying to compensate for these deficits, much of the emphasis in research designs has been placed on analysis and description leading to a wealth of confusing and often contradictory data. This condition appears to be typical of the early stages of development of any complex field of knowledge; accordingly, many would argue that a need for synthesis that would integrate apparently opposing theories now exists. However, a synthesis of incomplete information would not necessarily lead to a better understanding of schizophrenia. By ignor-ing some of the dichotomies and polarities in the data, by narrowing the focus to trying to find the interconnections among the information presently available, some valuable information might be lost. At the same time, one wants to avoid becoming engulfed in spending too much time and effort in the analysis of small details, as is often the case with research designed to support a particular theory. It all boils down to the dilemma of not being able to formulate a proper integra-tive principle or theoretical model until one has at least enough information to delineate the concept one is involved with, and one cannot obtain that informa-tion unless a great deal of varied research encompassing many different areas of investigation is conducted.

In order to reach the goal of a more comprehensive theoretical framework for schizophrenia, it seems important to do several things. First, it appears vital that theory, research, and intervention remain interconnected so that any discovery in one area will influence the other. At present, dangerous rifts among these areas appear to be developing, wherein practitioners apply interventions with little or no theoretical basis and theoreticians form postulates without any specific delineation for integration into treatment or research strategies. Furthermore, workers in the field need to view the different models as being limited in their explanatory power, and not as particularly right or wrong. Similarly, we cannot afford to dismiss any of the material presently available, but instead should try to appraise and evaluate it objectively without any preconceived ideas or bias stemming from our own theoretical models.

REFERENCES

American Psychiatric Association. *Diagnostic and statistical manual of mental disorders* (3rd ed.). Washington, D.C.: APA, 1980.

Buchsbaum, M. S., & Haier, R. J. Biological homogeneity, symptom heterogeneity, and the diagnosis of schizophrenia. *Schizophrenia Bulletin, 1978, 4,* 473–475.

Cancro, R. Genetic evidence for the existence of subgroups of the schizophrenic syndrome. *Schizophrenia Bulletin, 1979, 5,* 454–459.

Defries, J. C., & Plomin, R. Behavioral genetics. In M. R. Rosenzweig & L. W. Porter (Eds.), *Annual review of psychology* (Vol. 29). Palo Alto: Annual Review, 1978, 473–515.

Erlenmeyer-Kimling, L. Studies on the offspring of two schizophrenic parents. In D. Rosenthal & S. S. Kety (Eds.), *The transmission of schizophrenia.* Oxford: Pergamon, 1968.

Erlenmeyer-Kimling, L Schizophrenia: A bag of dilemmas. *Social Biology, 1976, 23,* 123–134.

Fuller, J. L., & Thompson, W. R. *Foundations of behavior genetics.* St. Louis: Mosby, 1978.

Gottesman, I. I., & Shields, J *Schizophrenia and genetics: A twin study vantage point.* New York: Academic Press, 1972.

Gottesman, I. I., & Shields, J. A critical review of recent adoption, twin, and family studies of schizophrenia: Behavioral genetics perspectives. *Schizophrenia Bulletin, 1976, 2,* 360–398.

Gottesman, I. I., & Shields, J. Obstetric complications and twin studies of schizophrenia: Clarifications and affirmations. *Schizophrenia Bulletin, 1977, 3,* 351–354.

Haier, R. J. The diagnosis of schizophrenia: A review of recent developments. *Schizophrenia Bulletin, 1980, 6,* 417–425.

Heston, L. L. Psychiatric disorders in foster home reared children of schizophrenic mothers. *British Journal of Psychiatry, 1966, 112,* 819–1825.

Jackson, D. D. A critique of the literature on the genetics of schizophrenia. In D. D. Jackson (Ed.), *The etiology of schizophrenia.* New York: Basic Books, 1960.

Keith, S. J., Gunderson, J. G., Reifman, A., Buchsbaum, S., & Mosher, L. R. Special report: Schizophrenia 1976. *Schizophrenia Bulletin, 1976, 4,* 510–565.

Kessler, S. The genetics of schizophrenia: A review. *Schizophrenia Bulletin, 1980, 6,* 404–416.

Kety, S. S., Rosenthal, D., Wender, P. H., & Schulsinger, F. The types and prevalence of mental illness in the biological and adoptive families of adopted schizophrenics. In D. Rosenthal & S. S. Kety (Eds.), *The transmission of schizophrenia.* Oxford: Pergamon, 1968.

Kety, S. S., Rosenthal, D., Wender, P. H., & Schulsinger, F. Studies based on a total sample of adopted individuals and their relatives: Why they are necessary, what they demonstrated and failed to demonstrate. *Schizophrenia Bulletin, 1976, 2,* 413–428.

Kidd, K. K., & Cavalli-Sforza, L. L. An analysis of the genetics of schizophrenia. *Social Biology, 1973, 20,* 254–265.

Kidd, K. K., & Matthysse, S. Research designs for the study of gene–environment interactions in psychiatric disorders. *Archives of General Psychiatry*, 1978, *35*, 925–932.

Lewine, R. R., Watt, N. F., & Grubb, T. W. High-risk-for-schizophrenia research: Sampling bias and its implications. *Schizophrenia Bulletin*, 1981, *7*, 273–280.

Lidz, T. Commentary on "A critical review of recent adoption, twin, and family studies of schizophrenia: Behavioral genetics perspectives." *Schizophrenia Bulletin*, 1976, *2*, 402–412.

Liem, H. Family studies of schizophrenia: An update and commentary. *Schizophrenia Bulletin*, 1980, *3*, 429–454.

Matthysse, S. W., & Kidd, K. K. Estimating the genetic contribution to schizophrenia. *American Journal of Psychiatry*, 1976, *133*, 185–191.

Maugh, T. H. Biochemical markers identify mental states. *Science*, 1981, *214*, 39–41.

Meehl, P. E. Schizotaxia, schizotypy, schizophrenia. *American Psychologist*, 1962, *17*, 827–838.

Ødegaard, O. Comments on the genetics issue. *Schizophrenia Bulletin*, 1977, *3*, 345–347.

Pollin, W., Allen, M. G., Hoffer, A., Stabenau, J. R., & Hrubec, Z. Psychopathology in 15,909 pairs of veteran twins. *American Journal of Psychiatry*, 1969, *126*, 597–609.

Rabkin, J. G. Stressful life events and schizophrenia: A review of the research literature. *Psychological Bulletin*, 1980, *87*, 408–425.

Reed, S. C., Hartley, C., Anderson, V. E., Phillips, V. P., & Johnson, N. A. *The psychoses: Family studies.* Philadelphia: Saunders, 1973.

Rosenthal, D. Confusion of identity and the frequency of schizophrenia in twins. *Archives of General Psychiatry*, 1960, *3*, 297–304.

Rosenthal, D. Sex distribution and the severity of illness among samples of schizophrenic twins. *Journal of Psychiatric Research*, 1961, *1*, 26–36.

Rosenthal, D. Problems of sampling and diagnosis in the major twin studies of schizophrenia. *Journal of Psychiatric Research*, 1962, *1*, 116–134.

Rosenthal, D. *Genetic theory and abnormal behavior.* New York: McGraw–Hill, 1970

Rosenthal, D. *Genetics of psychopathology.* New York: McGraw–Hill, 1971.

Rosenthal, D. Searches for the mode of genetic transmission in schizophrenia: Reflections and loose ends. *Schizophrenia Bulletin*, 1977, *3*, 269–276.

Rosenthal, D. Eugene Bleuler's thoughts and views about heredity in schizophrenia. *Schizophrenia Bulletin*, 1978, *4*, 476–477.

Rosenthal, D., Wender, P. H., Kety, S. S., Schulsinger, F., Welner, J., & Østergaard, L. Schizophrenics' offspring reared in adoptive homes. In D. Rosenthal & S. S. Kety (Eds.), *The transmission of schizophrenia.* Oxford: Pergamon, 1968.

Sarbin, T. R., & Mancuso, J. C. *Schizophrenia: Medical diagnosis or moral verdict?* New York: Pergamon Press, 1980.

Shapiro, S. A. *Contemporary theories of schizophrenia.* New York: McGraw–Hill, 1981.

Singer, M. T., & Wynne, L. C. Thought disorder and family relations of schizophrenics. *Archives of General Psychiatry*, 1965, *12*, 201–212.

Slater, E. A review of the earlier evidence on genetic factors in schizophrenia. In D. Rosenthal & S. S. Kety (Eds.), *The transmission of schizophrenia.* Oxford: Pergamon, 1968.

Slater, E., & Cowie, V. A. *The genetics of mental disorders.* New York: Oxford University Press, 1971.

Spring, B. Stress and schizophrenia: Some definitional issues. *Schizophrenia Bulletin*, 1981, *7*, 24–33.

Torrey, E. F. Birth weights, perinatal insults, and HLA types: Return to "original din." *Schizophrenia Bulletin*, 1977, *3*, 347–351.

Usdin, G. *Schizophrenia: Biological and psychological perspectives.* New York: Brunner/Mazel, 1975.

Wahl, O. F. Monozygotic twins discordant for schizophrenia: A review. *Psychological Bulletin*, 1976, *83*, 91–106.

Wender, P. H., Rosenthal, D., & Kety, S. S. A psychiatric assessment of the adoptive parents of schizophrenics. In D. Rosenthal & S. S. Kety (Eds.), *The transmission of schizophrenia*. Oxford: Pergamon, 1968.

Wender, P. H., Rosenthal, D., Kety, S. S., Schulsinger, F., & Welner, J. Crossfostering: A research strategy for clarifying the role of genetic and experiential factors in the etiology of schizophrenia. *Archives of General Psychiatry*, 1974, *30*, 121–128.

Wender, P. H., Rosenthal, D., Rainer, J. D., Greenhill, L., & Sarlin, B. Schizophrenics' adopting parents. *Archives of General Psychiatry*, 1977, *34*, 777–785.

Winokur, G., Morrison, J., Clancy, J., & Crowe, R. Iowa 500: The clinical and genetic distinction of hebephrenic and paranoid schizophrenia. *Journal of Nervous and Mental Disease*, 1974, *159*, 12–19.

7

Genetics and Intelligence[1]

Sandra Scarr
Yale University

Louise Carter-Saltzman
University of Washington

INTRODUCTION

The idea of genetic differences as a source of individual and group differences in intelligence is one of the most controversial in the history of psychology. Some experts advocate the view that no evidence to date should compel anyone to accept the idea that genetic differences have anything to do with individual or group differences in anything we measure as intelligence (Kamin, 1974, 1981; Schwartz & Schwartz, 1974; Taylor, 1980). Other experts claim that the evidence points to a very high degree of genetic determination of differences in intelligence (Eysenck, 1973, 1979, 1980; Jensen, 1973a, 1978a). Most investigators in behavior genetics conclude from the evidence that about half ($\pm$ .1) the current differences among individuals in United States and European white populations in measured intelligence result from genetic differences among them (Loehlin, Lindsey, & Spuhler, 1975; Nichols, 1978; Plomin & DeFries, 1980; Scarr, 1981).

It is curious that so many experts examining the same evidence could reach such different conclusions. To understand the controversy and the differing views, we must examine some of the history of investigations of genetic differences in intelligence and take a new look at the major pieces of evidence. Further, we describe the varied approaches in genetics to the study of human behavior, for they are profoundly different and often confused by behavioral

[1]A slightly different version of this chapter appears in R. S. Sternberg (Ed.), *Handbook of Human Intelligence*, Cambridge: Cambridge University Press. Reprinted by permission of Cambridge University Press.

217

scientists. This chapter reviews the older and the more recent evidence on genetic and environmental differences in general and specific abilities and proposes a future for research.

The chapter must seem to be redundant with the many reviews of genetics and intelligence that have appeared in the last 10 years. The psychological community has been informed by reviews of the literature by Bouchard, 1976; DeFries & Plomin, 1978; Jencks, 1972; Jensen, 1973a; Kamin, 1974; Lindzey, Loehlin, Manosevitz & Thiessen, 1971; Loehlin et al., 1975; McCall, 1975; McClearn, 1970; McGuire & Hirsch, 1977; Nichols, 1978; Scarr-Salapatek, 1975; Thiessen, 1970. The genetics community has been informed by reviews of genetics and intelligence literature, by Anderson, 1974; Childs, Finucci, Preston, & Pulver, 1976; Lewontin, 1975; Morton, 1972.

There seem to us three legitimate reasons to produce yet another review of the literature on genetics and intelligence: (1) readers of this volume do not have access to the reviews, which seems unlikely; (2) startling new data or ideas have appeared since the last reviews—in part true!—or (3) we have a new, beneficial perspective to bring to the review of the literature, a vanity we would like to entertain.

What We Used to Know

Nearly 40 years ago, the distinguished psychologist, R. S. Woodworth, was asked by the Social Science Research Council to review the research on *Heredity and Environment* (Woodworth, 1941) that had recently been obtained from several studies of twins and foster (adopted) children and studies of nursery schools. There was much dispute at the time among the Stanford (Terman, McNemar, & Burks), Minnesota (Leahy & Goodenough), and Iowa (Wellman, Skodak, & Skeels) groups over the interpretation of evidence favoring heredity or environment as the predominant source of intellectual differences in the U.S. white population. Woodworth began his evaluative volume with the following statement, which serves as a suitable introduction for this chapter as well:

> If the individual's hereditary potencies could somehow be annulled he would immediately lose all physiological and mental characteristics and would remain simply a mass of dead matter. If he were somehow deprived of all environment, his hereditary potencies would have no scope for their activity and, once more, he would cease to live. To ask whether heredity or environment is more important to life is like asking whether fuel or oxygen is more necessary for making a fire. But when we ask whether the *differences* obtaining between human individuals or groups are due to their differing heredity or to differences in their present and previous environments, we have a genuine question and one of great social importance. In a broad sense both heredity and environment must be concerned in causing individuals to be different in ability and personality, but it is a real question whether to attach more importance to the one or the other and whether to look to

eugenics or euthenics for aid in maintaining and improving the quality of the population [p. 1].

After a judicious review of the evidence from studies of twins reared together and apart, from studies of foster children, orphanage children, and those with and without nursery school experience, Woodworth wrote a conclusion that stands today as a fine summary of what we know about genetic and environmental differences in intelligence. We reprint it here in full to inform the reader that the research of the past 40 years has only firmed the conclusions and elaborated the theoretical and methodological bases for those judgments. Mostly, we have had to rediscover these facts, because the intervening intellectual history buried them in an avalanche of naive environmentalism (Scarr & Weinberg, 1978). It is true that methods of analysis and the measurement of intelligence have both advanced over those of the 1920s and '30s but, curiously, the conclusions have not changed. According to Woodworth (1941):

> As the preceding survey had found repeatedly, there are serious difficulties in the way of separating the factors of heredity and environment when our interest lies in such traits as human intelligence and personality. There are sampling difficulties, inadequacies in even the best available tests for mental abilities, and much vagueness as regards the proper measures of environment. Also we have no direct indication of the individual's hereditary constitution, of his particular combinations of genes. It is not to be wondered at if the results of elaborately planned investigations leave us unsatisfied and uncertain. A few findings do seem to be well assured.
>
> Heredity and environment differ as between families, and also as between the children of the same parents growing up in the same home. By noticing how much siblings differ in comparison with children from the community at large we can estimate the total effect of intra-family variation in heredity and environment combined, as compared with the inter-family variation. We find the inter-family component in the total variance of the population to be smaller than the intra-family component. From the examination of foster children compared with own children in similar homes we gather that the inter-family differences are due partly to differences in heredity and partly to differences in home environment and about equally to the two factors. That is, own children from different homes differ in part because their families have somewhat distinctive heredity and in part because the home influences are different.
>
> As to the intra-family differences, the fact that there are some even between identical twins reared together proves that such differences are due in part to environment. But the relatively small differences between these twins leave the major part of the intra-family differences still undissected. Since siblings in general differ in heredity, they differ correspondingly in the effective environment, dependent as that is on their own characteristics. The environmental factors that differ as between children in the same home are often too subtle to be easily controlled or measured, and no promising beginning has been made toward estimating their respective shares in the production of individual differences. Differences between

own children in the same home are sometimes the result of prenatal and natal accidents. But for the rest they are due to the combination or interaction of heredity and environment, and that is about all we can say at present.

The most striking feature of these results is the small share that can be attributed to inter-family differences in environment. Not over a fifth, apparently, of the variance of intelligence in the general population can be attributed to differences in homes and neighborhoods acting as environmental factors. The reason is probably to be sought in the large degree of uniformity of environment produced by the schools and other public and semipublic agencies. It is still possible that raising the intellectual level of the environment would raise the general level of intelligence, while not by any means annuling the individual differences due to heredity.

The gains of foster children and of other children in changed and improved environments have been much less striking than might have been expected. About 5 or 10 points in IQ is all that can be claimed for the average gain, with much individual variation above and below this average. Even this amount of gain is not established beyond doubt—nor, to be sure, is it proved that still better environments would fail to register much larger gains. Somewhat larger gains and losses have indeed been indicated in some of the identical twin pairs who received very unequal educational opportunities.

An important result of several foster-child studies is the good showing made by many children whose own parents are rated very low in the socio-economic scale. Instead of saying that these children have made good in spite of poor heredity, we much conclude that their heredity was good or fair in spite of the low status and unsatisfactory behavior of their own parents. Their heredity was obviously good enough to permit them to do what they have actually done. By this test of accomplishment some children of feebleminded parents are proved to have average heredity. But to infer that all or even most children of inferior parents are possessed of average heredity would be going far beyond the present evidence, because of the elimination of especially unpromising children that has always occurred before the samples were made available to the investigators. To assure a gifted young couple that they could do as much for the next generation by adopting any "normal" infant as by having a child of their own would be a scandalous exaggeration of the known facts [pp. 84–86].

This compact summary of knowledge about genetic and environmental differences deserves several readings, for Woodworth makes virtually all the general points that one would want to make today about intellectual differences:

1. Both heredity and environment contribute to differences among and within families.

2. Intrafamily differences are subtle and undissected. Sibling differences depend in part on their own genetic differences and in part on their environments, with which we now know birth order differences are correlated.

3. Interfamily environmental differences among U.S. whites are a very small part of the total individual variation in IQ, probably, as Woodworth says, be-

cause of the uniformity of environments produced by public schooling and other public agencies.

4. It is still possible to raise the general level of intelligence by improving the environment, but that will not by any means annul individual differences due to heredity.

5. The gains made by adopted children reared in improved environments are not dramatic (unless they are minority children, who were not studied at that time), and differences between separated MZ twins were not large unless their educational differences have been large.

6. But children from very disadvantaged backgrounds often proved to be of normal intelligence when given the opportunities of foster homes. The notions of the 30s about fixed heredity were obviously wrong. On the other hand, Woodworth concludes, two intellectually gifted parents could not be advised to adopt any "normal" infant and expect that child to be an intellectual match for a likely offspring of their own. Heredity does play a substantial role in determining individual differences in intelligence.

A Brief Political History of Genetic Differences in IQ

There is a remarkable ratio of cant to data in this field. Many people express opinions (and write polemics in reviews), and relatively few people do empirical research. Since the mid-70s the ratio is improving as more investigators see the possibilities of important research questions and approaches, particularly through the study of adopted children (De Fries & Plomin, 1978; Scarr-Salapatek, 1975).

To understand the controversy over the very study of genetic variance in intelligence, one has to place oneself in a political frame of mind and believe that genetic differences have dire implications. On the one hand, one may decide to suppress the evil consequences of such knowledge and its possible uses. On the other hand, one could consider oneself the appointed guardian of an awful truth: that genetic differences among human beings and among groups of people are so pervasive, so terrifyingly strong that the knowledge is essential to bring before the public for their consideration in social policy issues. Now, the stage is set for a confrontation, a noisy conflict that has persisted from the early 20th century to the present (see Block & Dworkin, 1976, for a good collection of articles).

Terman (1922), Yerkes (1923), and Brigham (1923) engaged in polemic debate with the great literary columnist, Walter Lippmann (see Cronbach, 1975, for a reference list, 1922–23). Lippmann argued for the malleability of intelligence and the role of cultural differences in IQ test scores. Like any sensible liberal, Lippmann recognized that the tests sampled culturally bound knowledge and skills. Terman, Yerkes, and Brigham pigheadedly argued for the immutability of intelligence and cultural fairness of their testing programs. They lost (Block & Dworkin, 1976; Cronbach, 1975). Later, Jensen was engaged by another great columnist, Joseph Alsop, who effectively tackled his views on racial differences in intelligence. As Cronbach said, psychologists who gain the limelight often lose their academic heads.

Within the academic community, Neff (1938) challenged the hereditarian views of the Stanford school and discounted the research of Terman and Burks. The Iowans (Skeels, 1938; Wellman, 1940) advocated the environmental effects of early interventions, whereas McNemar (1940) wrote scathing commentaries on their statistics. On the race issue, Herskovitz (1926) and Klineberg (1963) were especially forceful in their views against the idea of racial differences in intelligence. Their evidence was scanty, but their values were strong. Later "validation" of their hopes and beliefs was offered by the American Anthropological Association, which adopted a resolution declaring all races biologically identical. The last was in response to Jensen and the furor over his 1969 article in the *Harvard Education Review* (see Jensen, 1973b).

More recently, E. O. Wilson (1977) was moved to write:

> Discussion of the inheritance of human intelligence consists of two slippery slopes joined by a razor's edge. One slope descends to antiracism at any intellectual cost, the other to intellectual freedom at any social cost. The shabby misuse of IQ testing in the support of past American racist policies has created understandable anxiety over current research on the inheritance of human intelligence. But the resulting personal attacks on a few scientists with unpopular views has had a chilling effect on the entire field of human behavioral genetics and clouds public discussion of its implications. . . .
>
> The political capture of a discipline in the United States is highly unlikely, but feelings run high. No controversy within the academic world has been more cruelly divisive; none better illustrates the maxim that tragedy is a clash of rights. There appears to be only one solution to the dilemma: a return to an uncompromising ethic of objectivity, based on a careful decoupling of the collection and analysis of data from the discussion of their social and political implications.

A similar view was expressed by a prestigious group of geneticists:

> The application of the techniques of quantitative genetics to the analysis of human behavior is fraught with complications and potential biases, but well-designed research on the genetic and environmental components of human psychological traits may yield valid and socially useful results and should not be discouraged [Statement adopted by the Genetics Society of America, 1975].

Some doubt that well-designed research on the genetic variation in human behavior is possible. Lewontin (1975), for example, sets forth requirements for the design of adoption studies that cannot be met in any population: randomizing genotypes over environments, both of which are representative of the entire range of the population to which inferences are to be made. He concludes that:

"the failure to adhere to clean experimental design renders all work uninterpreta ble. It is simply not true that approximate designs give approximate results." Furthermore, these same critics doubt that the study of genetic variation in human behavior is worth any effort. He adds: "Finally, from a scientific stand-point or from one of valid inferences about social policy, the problem of assaying genetic components of IQ test differences seems utterly trivial and hardly worth the immense effort that would be needed to carry out decent studies[p.403]."

Kamin (1981) echos similar sentiments:

> The great merit of Scarr's plentiful empirical research lies, in my view, in the demonstration that no scientific gain is to be had from further "behavior genetic" research on the heritability of IQ. The same data set from which Scarr concludes that IQ is substantially heritable can also be used—since Scarr is willing to share her raw data—to show that IQ is not at all heritable. The data are not, after all, the product of clearly designed and well controlled experimentation. They are neces-sarily correlational data, collected in difficult and inevitably flawed field settings. The patterns discerned within such data are many, and complex. The interpretation of these complex patterns, I believe, must reflect the investigator's theoretical bias (in Scarr, 1981, p. 468).

Layzer (1972, 1974) and Feldman and Lewontin (1975), for example, deny any legitimate use for the information about the "heritability" of behavioral phenotypes in human populations. They point to the possible existence of geno-type–environment correlations and interactions as rendering any studies of ge-netic and environmental variances indeterminant. They further argue that studies of sources of variation in human populations are misleading and have no serious implications for any scientific or practical purpose. According to Feldman and Lewontin (1975):

> We must distinguish those problems which are by their nature numerical and statistical from those in which numerical manipulation is a mere methodology. Thus, the breeding structure of human populations, the intensities of natural selec-tion, the correlations between mates, the correlations between genotypes and en-vironments, are all by their nature statistical constructs and can be described and studied, in the end, only by statistical techniques. It is the numbers themselves that are the proper objects of study. It is the numbers themselves that we need for understanding and prediction.
>
> Conversely, relations between genotype, environment, and phenotype are at base mechanical questions of enzyme activity, protein synthesis, developmental movements, and paths of nerve conduction. We wish, both for the sake of under-standing and prediction, to draw the blueprints of this machinery and make tables of its operating characteristics with different inputs and in different milieus. For these problems, statistical descriptions, especially one-dimensional descriptions like heritability, can only be poor and, worse, misleading substitutes for pictures of the machinery. . . . At present, no statistical methodology exists that will enable us

ge of phenotypic possibilities that are inherent in any genotype,
nique of statistical estimation provide a convincing argument for a
ism more complicated than one or two Mendelian loci with low and
rance. Certainly the simple estimate of heritability, either in the
w sense, but most especially in the broad sense, is nearly equivalent
to ... ation at all for any serious problem of human genetics [pp. 1167–
1168].

What Are the Questions?

One's evaluation of the answers that can be supplied by statistical and mechanistic models surely depends on the question that one is trying to answer. For
questions about the current intellectual state of human populations, the distribution of intelligence, and the likely success of improving intellectual phenotypes
through intervention with *known* environmental manipulations, one would want
a statistical model of contemporary sources of variance in the population.
Knowledge of evolutionary history, selection pressures, or enzyme activity at a
few loci will not address such questions. Nor will appealing to the unpredictable
effects of yet-to-be-devised interventions help to deal with the problems of the
here and now.

The analogy to another population statistic, birthrate, is a good one. Suppose
that one proposed to study the birthrate of a developing country in order to devise
a family planning program. The logic of Feldman and Lewontin's argument (see
also Medawar, 1977) against studying sources of variance in populations would
oppose such a study, because: (1) statistical studies will not inform us about the
mechanisms and physiology of reproduction; (2) the correlations between reproduction and social structure render the birthrate a meaningless statistic; and
(3) knowledge of birthrate will not permit us to advise the Joneses or the Smiths
about their reproductive plans. We submit that each of these objections is based
on questions not addressed by studies of population variability or reproductive
rate; different questions require different studies.

It seems that some scientists fear that knowledge of the current sources of
intellectual differences in a population will foreclose attempts to search for ways
to improve the intellectual status or distribution of resources of the population. If
current differences in intelligence are attributable half to genetic differences,
about 10% to differences among family environments, and the rest to differences
among individuals within families, are we led to abandon a commitment to
improve children's lives? We fail to see the connection.

As Anderson (1974) pointed out:

Genetics as a discipline cuts across the four levels of biological study—molecular,
cellular, organismal, and populational. At each level testable hypotheses can be

stated and the results at one level may lead to questions whi'
at other levels.

Genetic studies treat variability as the primary foc'
merely as noise to be eliminated or disregarded. This variau.
both within families and between families. If only one pair of genes ıs .
incidence among relatives will follow simple ratios, but with multigenic inhe..
there is a more complex set of expectations [pp. 20–21].

Consider the following questions that one might ask about the role of genetic differences in intelligence:

1. How do genes affect intelligence? (Which pathways of gene–protein–enzyme activity to physiology and brain function cause differences in intelligence?

2. What are the sources of individual (and group) differences in intelligence? (What are the sources of *variation* in a population at the present time?)

3. Why and in what ways does human intelligence differ from that of other primates? (What is the evolutionary history of primate species?)

4. Why is there a distribution of individual differences in intelligence within a population, and perhaps among populations? (What is the evolutionary history and structure of human populations?)

Developmental Versus Individual Differences

Confusion about the different nature of each of these questions has led many critics of the study of genetics and intelligence to commit amazing feats of illogic. Most common is the confusion of Questions 1 and 2 in the psychological literature. It is asserted that one cannot study the sources of individual differences in a population because both genes and environments are required for individual development. Well, yes, we all assume that development requires both genes and environments that act together to program human development. But, as Woodworth said, that is not an answer to Question 2, about sources of variation. Many species-typical developments depend on the same or functionally equivalent genotypes and environments for development within bounds that are normal for the species. Individual differences depend on functional differences in genotypes and/or environments that cause noticeable variations in phenotypes (Scarr, 1979).

Populations can be described only in statistical terms, because they are *distributions* of individuals. Although one can reify some frequent "type" as representative of a population, it is quickly apparent that more is lost than gained by typological thinking (Hirsch, 1967). As Dobzhansky (1950) and his students (McDonald & Ayala, 1974) have demonstrated, a naturally occurring population ₒost often genetically diverse, and the more environmental diversity contained ₁thin its range of habitats, the more genotypically and phenotypically diverse the individuals with a population will be. An evolutionary history of selection pressure does not generally lead to the elimination of all variability, for a variety of reasons (see Kidd & Cavalli-Sforza, 1973). Thus, an understanding of the role of genetic differences in human intelligence necessitates questions about the *distribution* of intelligence in the population. There are many possible questions about the evolution of intelligence, genetic drift, local adaptations, adaptation tc various niches within a population, developmental changes in the distribution of intelligence, and so forth, but one question surely is, "Why do individuals within a population vary in intelligence at the present time?"

Uses and Abuses of the Term *Intelligence*

It has become fashionable to adopt a nihilistic position regarding the nature and definition of intelligence; it is often said that no one knows what intelligence is. In fact, *everyone* knows what intelligence is—a "fuzzy" construct! We create confusions by using the term *intelligence* at several levels that are not necessarily related to one another. There are four levels to which the term can be applied or misapplied. In our view, "intelligence" should be reserved for the individual level of cognitive functioning.

At a cultural level, the prescribed and habitual solutions to problems make better or worse use of the available resources in relation to the demands of size and density of the population. The *adaptation* of a cultural group can be evaluated by the degree to which there is a balance between the needs of the population and the available resources that the group knows how to use (Marvin Harris, 1975). "Intelligent" cultures are better described by their adaptation.

At the level of social organization, one can speak of the structures and functions of the organization as working more or less effectively to deal with the problems presented by the society. Effective social organizations have structured social roles and allocate those roles appropriately to individuals. Thus, intelligence at the societal level is better described as the *organization and allocation* of social roles.

In small groups, there are also social roles to be allocated, but the structure of such groups is often informal and shifting. Incumbents change roles over short periods of time. One can evaluate the effectiveness of such groups by criteria of problem solving, use of time, satisfactions of members, and so forth, as is frequently done in the literature on street gangs, boy scout troups, and other informal small groups. At the group level, intelligent groups solve everyday problems (which movie to see, when to rumble) in ways that satisfy most members and make better than worse use of time and resources.

Also in small groups, one can look at the adequacy of individuals' role performance. Intelligent behavior in small groups requires that one take into account the social and material resources in the setting. On the basketball court, playing intelligently does not mean that one takes every available shot oneself. Rather, one is most effective if one's shot-taking decisions take into account the positions and shooting prowess of one's teammates. Similarly, intelligent behavior in many other group situations requires that one withhold some uses of one's individual cognitive skills, so that the group may function most effectively. This means that small groups are not likely to be good settings in which to sample individual intelligence.

We should reserve the term *intelligence* for those individual attributes that center around reasoning skills, knowledge of one's culture, and ability to arrive at innovative solutions to problems. This is a quick and dirty list of attributes for the term, but the main point is that "intelligence" should be reserved to describe cross-situational attributes in individuals that they carry with them into diverse situations.

Intelligence and Social Competence

Social competence is most clearly defined as an individual's success in filling social roles. The most socially competent people are those who can fill many social roles well; less competent people are those who have few options in social roles and who fill them badly. This definition confounds breadth of role options with performance in roles, but some weighted combination is necessary to capture what is meant by social competence. Some people fill a few roles well but are quite limited in their options. Others have qualifications to fill many roles but do none of them well.

People who are intelligent by the individual definition are likely to have greater social competence, because breadth of role options is related to intelligence, as is goodness of role performance. Intelligence is, however, only one component in social competence, perhaps a major one but not unique. There is probably a middling correlation between the two. People who are considered socially competent need not have high intelligence. Depending on the roles they choose or have thrust upon them, their performance may be quite adequate with average or low average intellectual levels.

Looking unintelligent in social roles depends on a *mismatch* of the role to the person; for example, putting a professor psychology in to coach a basketball team or putting most basketball coaches into the professor's job. Looking intelligent in social roles is not what we mean by "intelligence," a term reserved for those mental processes mentioned previously.

People at any but the more retarded intelligence levels should be able to find social roles in which they fit and behave competently. People of lesser intelligence will not have the breadth of options that more intelligent people have, nor are they as likely to fill the roles they do have as competently, on the average, as more intelligent people; they can, however, be socially competent at some level.

Political Problems with the Term *Intelligence*

As we all recognize, the term, intelligence, is surrounded with a halo of valuation that threatens to include all manner of virtues. Many avoid the term in writing and public speaking, because of the unusual amount of surplus meaning. This is a case of bad money driving out good, because there is a perfectly good domain of behaviors to be described as "intelligence." We use terms, such as cognitive skills, intellectual skills, abstract reasoning, and so forth, because others may take offence at the possible exclusion in measures of intelligence of virtues they value and at which their friends and relatives seem to be accomplished.

Not all good things about people are "intelligence," nor is the relationship of measured intelligence to the allocation of resources as strong as it is to educational level, reading achievement, and mathematics skills. The allocation of social and economic resources seems to be tied more closely to educational landmarks and motivation to work hard than to intelligence per se.

Acknowledging the public problems with the use of the term, intelligence, we propose that it be available to close friends and colleagues to apply to those individual, mental processes in the domain described earlier.

In sum, the field of intelligence is not served well by blurred distinctions among levels of analysis nor among meanings of the term, intelligence. We should find other words to apply to cultures, social organizations, small groups, and individual behaviors in social contexts. Although the term has an aura of virtue, its application to other contexts or behaviors will not improve anyone's functioning.

Intelligence at a Population Level

A serious view of intelligence should include a consideration of the population level. Imagine a complex environment inhabited by people whose problems of adaptation are diverse: some simple, some complicated, some requiring general skills, some requiring specialized skills. The population that is best adapted to a typical human environment includes people with diverse skills. The benefits of their diverse talents are shared through their social interdependence. This is a far more realistic picture of human evolution and the role of intelligence than the Social Darwinian notion of a best adapted phenotype.

In a real human society, there is no one best adapted phenotype. We need people who are good at many different things. Intelligence as a population concept refers to the distribution of skills in a population of humans who work and live together in interdependent, socially organized ways.

The application of a population concept to intelligence in modern industrial societies is easy. If we had to do our own home building the result would indeed be primitive, because we do not have skills that are comparable to those of a specialized architect, builder, carpenter, or plumber. Neither do they know how to deliver human services or investigate educational and psychological problems, for which they depend on people like us. Not only are these differently developed skills, but they require different kinds of intelligence. A well-adapted society, then, is one that fosters diversity of intelligence.

Even in less industrial societies, there is division of labor that makes the diversity of intelligence an adaptive advantage. In the most primitive form, there is at least a division of labor by sex. It seems to us not accidental that women all over the world are socialized to be nurturant, whereas men are socialized to be more independent. Although these particular forms of the division of labor are no longer fashionable in postindustrial societies such as the United States (and we agree they should not be, as child rearing and participation in the economy need no longer be gender related), it is likely that throughout human evolution, until this century, there was an important division of labor by gender, each sex requiring a somewhat different set of intellectual and social skills.

In contemporary societies that are preindustrial, the division of labor is not only along gender lines. People who are good storytellers often have suitable roles, as do those who are particularly good builders, farmers, fishermen (rarely women), embroiderers, weavers, and so forth. It is interesting that one modern government program that was readily accepted by the Quechua women on the eastern slopes of the Andes in Ecuador was the sponsorship of especially good weavers to teach the young weavers in other villages. Women weavers who attained status in their own villages for their skill at devising and weaving intricate patterns were traditionally the teachers of young women in other villages, so that the government program was compatible with a long-established recognition of this form of intelligence. Husbands and children of particularly good weavers suffer their absences with pride in their accomplishments (personal knowledge from field work, S. Scarr).

A complex society requires variation in intelligence for optimal adaptation. Given that a distribution of skills is required, the next question concerns levels of skills. One would generally think that individuals are better adapted if they have the highest possible levels of intelligence. Certainly, it is maladaptive to be seriously retarded and to be unable to take care of oneself. Above that level, at an intellectual level of mild retardation, it is not so clear that a population is better adapted without some individuals of low intelligence. It is easier to defend the idea that a population needs some people of high intelligence, who define and solve problems for the society as a whole. People of more average abilities carry out the major work of the society. There may still be roles in every society for the mildly retarded, who complain less about tedium, who are willing to do jobs that few others want and to get satisfaction from them. We are speculating in saying that the continued presence of large numbers of mildly retarded persons in every society suggest that their fitness, as witnessed by their above-average reproductive rate, means that they have productive or at least adapted roles in most societies.

Despite one's conclusion about the contribution of the mildly retarded to the overall adaptation of a population, we believe that populations require a range of intellectual levels and specific skills to be optimally adapted. There is no one best adapted phenotype, in the range of average to very superior levels of intelligence or with respect to specific profiles of skill.

MOLECULAR AND POPULATION GENETICS

Mechanisms and Variance

When the layperson thinks of the term *genetic,* he or she is likely to think of the material segments of chromosomes that produce some mysterious products that govern our growth and development. The lay model is based on a linear, causal stream of determinism that flows from molecular to molar levels of analysis. This molecular model has been developed through the study of major aberrations in the genetic code that lead inexorably to noticeably different, abnormal phenotypes. The scientific success of molecular genetics is a sociological phenomenon, worthy of study in its own right. Our purpose here is to describe the phenomena in intelligence to which molecular models do and do not (yet?) apply and to propose the efficacy of the lesser known evolutionary population genetic models for complex phenotypes, such as intelligence.

There are more than 150 known gene defects that are associated with mental retardation (Anderson, 1972). There are many ways for development to go awry. At the minimum, we know that this many genes must perform properly in the chorus for the performance of normal intelligence to go on stage. Genes that fail to code for enzymes required for normal metabolism (PKU, galactosemia, Lesch–Nyhan syndrome, etc.) are inborn errors of metabolism, many of which have mental retardation as a feature. Other single-gene disorders code for incorrect enzymes, fail to code for proper proteins, or produce indigestible products somewhere in the gene action pathway. The literature in genetics is replete with examples of rare and specific gene defects that lead to mental retardation. But what does this tell us about the development of normal intelligence?

We propose that models of gene action gone awry tell us very little about the role of genetic differences in intelligence per se. Perhaps, they serve to confirm the views that genetic variants are present in the human organism, that normal intellectual development is a complex feat of genetic programming, and that genetic variation "counts" in explaining normal variation in intelligence. If one begins with an evolutionary view of intelligence, however, all those points are obvious from the start.

Apart from single gene models, one cannot fail to note the extensive literature on chromosomal abnormalities. Yes, having too much or too little chromatin will affect intelligence, always for the worse. Is it surprising that having the species-typical amount of genetic material should make the development of intelligence in the normal range more likely or that having an abnormal amount of chromatin makes mental retardation more likely? Having too many or too few sex chromosomes is not as disastrous to intelligence as too many or too few other chromosomes (the last is evidently lethal as there are no living examples). Intellectual development is adversely affected by the wrong amount of genetic material, which evidently messes up the program for brain development.

Gene Action and Intellectual Development

If gene action pathways in human development were known, writing this chapter would be a simple reporting rather than speculative construction. In fact, only bits and pieces of the genetics of developmental processes are known. The basic DNA–RNA protein synthesis code is well established. Knowledge of fetal development at a morphological level is fairly complete. But how does morphological development over the fetal period, and indeed the life-span, relate to protein synthesis at a cellular level? What causes some cells to differentiate and develop into the cortex and others into hemoglobin? And how do gene action and morphological development relate to intellectual development from birth to senescence? How do cells, which all originate from the same fertilized ovum and carry the same genetic information, come to program development into different organs and systems and in different behavioral states of development?

It has been hypothesized (Jacob & Monod, 1961) that several different kinds of genes exist: structural genes to specify the proteins to be synthesized; operator genes to turn protein synthesis on and off in adjacent structural genes; and regulator genes to repress or activate the operator and structural genes in a larger system (Jacob & Monod, 1961; Lerner, 1968; Martin & Ames, 1964). The instructions that a cell receives must be under regulatory control that differentiates the activity of that cell at several points in development.

Genes and chromosome segments are "turned on" at some but not other points in development. Enlargements of a chromosome section (called puffs) have been observed to coincide with RNA synthesis in the cell. Puffs occur on different portions of the chromosomes at different times in different cells, indicating the existence of regulatory mechanisms in development.

Regulatory genes are probably the ones responsible for species and individual differentiation through control of the expression of structural genes. Most of the structural genes, which are directly concerned with enzyme formation, are common to a wide array of species and function in approximately the same way. They provide the fundamental identity of life systems. The diversity of individuals and species is due in large part to the regulatory genes that modify the expression of basic biochemical processes. According to Thiessen (1972):

> In other words, the greatest proportion of phenotypic variance, at least in mammalian species, is probably due to regulatory rather than structural genes—genes that activate, deactivate, or otherwise alter the expression of a finite number of structural genes [p. 124].

Several cellular regulatory mechanisms have been suggested (Lerner, 1968). First, the cytoplasms of different cells contain different amounts of material and may contain different materials. As cell division proceeds, daughter cells receive unequal amounts of cytoplasm, which may relate to their progressive differentiation. Second, the position of the developing cells may influence their course. Outer cells may have different potentialities for development than those surrounded by other cells.

Third, the cell nuclei become increasingly differentiated in the developmental process. Progressively older nuclei have a more limited range of available functions; they become more specialized in the cell activities they can direct. Specialization of nuclei is related to the differentiation of organs and functions in different portions of the developing organism.

The regulation of developmental processes over the life-span is accomplished through the gene-encoded production of hundreds of thousands of enzymes and hormones. During embryogenesis there are precise correlations between changes in enzyme concentrations and development. As Hsia (1968) has concluded:

> For example, *cholinesterase* activity shows particularly close relationships with neural development. As early as the closure of the neural tube, high cholinesterase activity has been found in association with morphogenesis of the neuraxis. . . . Nachmansohn has shown that cholinesterase is synthesized in the developing nervous system of the chick embryo exactly at the time that synapses and nerve endings appear [pp. 96–97].

Any behavior represented phenotypically by the organism *must* have a genetic and organismic substrates. It does not appear without CNS regulation, and CNS regulation does not occur without brain myelenization, synaptic transmission, and previous experience encoded chemically in the brain.

Enzymatic differentiation is specific to the stage of development, the specific organ, specific regions within organs, and the type of enzyme. Development proceeds on a gene-regulated path by way of enzymatic activity. Generalizations are very risky from one point in time to another and from one organ part to another.

There are several enzyme systems that are active in the embryo but that disappear with the cessation of growth. Other enzymes that are absent or present in low activity in the embryo greatly increase in activity at the time an organ becomes functionally mature. These enzymes then remain active throughout life to regulate functional organ activity. A third class of enzymes is activated only with maturation and remains active the rest of adult life (Hsia, 1968, pp. 96–107).

Interference with regulatory mechanisms at a cellular or organ system level can result in a variety of phenotypic abnormalities. The result of interference is often related to the time it occurs during development. For example, male rabbit fetuses castrated on the 19th day of gestation resemble a female at birth. Castration on any day up to the 24th results in a gradation of femininity, but if castration is performed on the 25th day or later, there is no effect on the development of male genitalia. Figure 7.1 is a schematic presentation of the biochemical development of the embryo and the influence of environment at all levels of development.

Hormonal activity is critically important to the stimulation of protein synthesis, and to the differentiation of male embryos from the basic female form. Minute quantities of fetal testosterone at critical periods in development affect genital differentiation as well as CNS differences that seem to last a lifetime

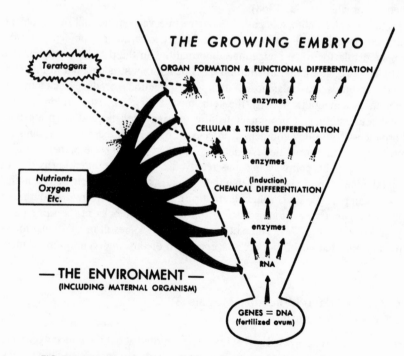

FIG. 7.1. Model of the biochemical development of the growing embryo and the influence of environment at all levels of development. (From Hsia 1968, after Wilson.)

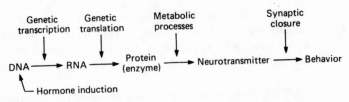

FIG. 7.2. Model of hormone–gene flow from cellular to behavioral levels. (From Thiessen, 1972).

(Levin, 1967). The variety of hormone that stimulates protein synthesis includes growth hormones as well as sex hormones, cortisone, insulin, and thyroxine (Thiessen, 1972, p. 95). A model of hormone–gene flow is presented by Thiessen, as shown in Fig. 7.2.

There are many known ways in which normal development can be disrupted at a biochemical level. Defects in the biochemical pathways between gene action and normal cell metabolism number in the hundreds. In the glucose to glycogen pathway alone, there are seven independent genetic errors that result in different genetic anomalies (Hsia, 1968).

Environmental pathogens can, of course, intervene in normal development. Radiation, infectious diseases, drugs, and other specific environmental factors are responsible for some congenital abnormalities in the developing fetus.

The effect of ionizing radiation on CNS development is detailed in Fig. 7.3. Rubella, mumps, toxoplasmosis, and viral infections produce characteristic anomalies when contracted by the fetus in the first trimester of pregnancy. Mental retardation is a prominent feature of many genetic and environmental disturbances in the developmental process. Single genes currently account for more than 159 abnormalities (Anderson, 1972) of mental development.

Another genetic pathway that has received considerable attention is that of phenylalanine. Although many behavioral scientists recognize that a block in this pathway can produce PKU (phenylketonuria), most are not aware that four other identifiable genetic syndromes result from additional blocks in the same pathway, as shown in Fig. 7.4. Three of the genetic blocks result in additional forms of mental retardation, but all are susceptible to dietary intervention in infancy.

The Void between Molecular Models and Normal Variation

Given the extensive knowledge that has been acquired about the molecular nature of gene action and the many forms of mental retardation associated with single-gene and chromosome defects, what implications are there for the understanding of normal intellectual variation? On this subject, those behavior geneticists and geneticists who prefer mechanistic models become understandably vague. They can offer only the hope that future knowledge will provide full accounts of the mechanisms underlying normal intelligence and hence variations in the normal range. One hope is for major genes, a few of which might be found to account for a major portion of the variability in intelligence. Even though hundreds of gene loci may be involved in the program for normal intelligence, perhaps only two or three control half or more of the phenotypic variation.

The normal distribution of quantitative traits need not depend on small contributions of many genes; in fact, three or four loci without complete dominance will generate a nicely Gaussian curve of phenotypes, as shown in Fig. 7.5.

The search for major genes is an active area of research; with the development of linkage studies and pedigrees, it is hoped that specific gene loci can be located that make the largest contributions to intellectual variation. The next section examines critically that line of research. We express strongly negative opinions about the mechanistic model; readers may well disagree.

Arguments for Population Genetic Models

Recently, Andrew Pakstis, a graduate student with Scarr at the University of Minnesota, discovered several important linkages between genetic blood group

Age (Days) Mouse	Man	Embryo (mm.)	Nervous System	Other
0–9	0–25		No damage	
9	25½	2.4	Anencephaly (extreme defect of forebrain)	Severe head defects
10	28½	4.2	Forebrain, brain stem, or cord defects	Skull, jaw, skeletal, visceral defects, anophthalmia
11	33½	7.0	Hydrocephalus, narrow aqueduct, encephalocele, cord, and brain stem defects	Retinal, skull, skeletal defects
12	36½	9.0	Decreasing encephalocele microcephaly, porencephaly	Retinal, skull, skeletal defects
13	38	12.0	Microcephaly, bizarre defects of cortex, hippocarpus, callosum, basal ganglia, decreasing toward term	Decreasing skeletal defects

FIG. 7.3. Timetable of radiation malformations in mice and man. (From Hsia, 1968, after Hicks.)

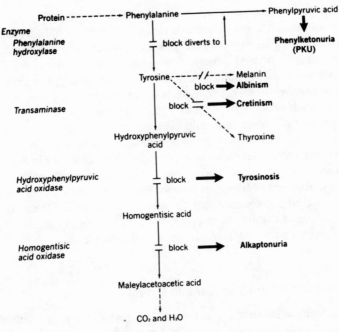

FIG. 7.4. Genetic blocks in the metabolism of phenylalanine. (From Lerner, 1968.)

markers and skin color reflectance. The blood group markers are independently segregating, single genes, located on three of the 60 or so chromosome segments that make up the new combination of genetic material a child receives from his or her parents. Evidently, three of the four to six genes for skin color are located on the same chromosome segments as three of the blood-group markers. We know this to be the case, because siblings who have the same blood-group genes on these segments also have more similar skin color than brothers and sisters who have different blood-group markers. Our statistical procedures show that these results would occur by chance only once in several million trials.

What are we to make of these results? It seems probable that three of the genes that code for metabolic processes leading to the production of melanin in skin cells have been found. Most geneticists, and indeed most other scientists, would mentally model the gene action from the DNA to the melanin in a linear, causal chain, with each step in the sequence permitting the occurrence of the subsequent steps, given the presence of appropriate enzymes at each step. In fact we already have a well-detailed metabolic pathway for one—perhaps, the major—pathway for the production of melanin, as shown in Fig. 7.4. More importantly, the idea that specific genes can be identified does not strike us as outlandish.

In addition, our analysis has identified two major genes for abstract reasoning. Oh, you say, that's another matter. Genes for behavior—how outrageously reductionist! The reason for our hasty rejection of a linear causal chain for the production of abstract reasoning abilities is, of course, that many environmental events have important effects on the development of abstract reasoning abilities from the earliest years to adolescence, when we measured them.

The rejection of the idea of ''genes *for* behavior'' is also based on authorities in the field, who explicitly denounce any such notion. The accepted account for the relation between gene action and behavioral development was summarized in these words by Delbert Thiessen (1972):

> The lengthy, often tortuous, path from DNA specificity to metabolic synchrony explains why behavior must be considered a pleiotropic reflection of physiological processes. Gene influence in behavior is always indirect. Hence the regulatory process of a behavior can be assigned to structural and physiological consequences of gene action and developmental canalization. The blueprint for behavior may be a heritable characteristic of DNA, but its ultimate architecture is a problem for biochemistry and physiology. Explaining gene–behavior relations entails knowing

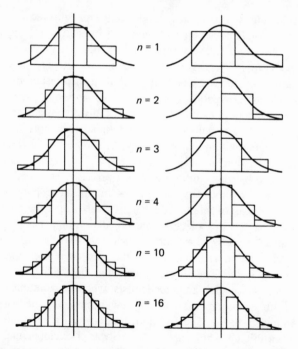

FIG. 7.5. Gradual approximation to normal curve of phenotypes controlled by *n* pairs of genes. Left: no dominance. Right: complete dominance for all genes. (After Lush, 1945.)

every aspect of the developmental pattern: its inception, its relation to the environment, its biochemical individuality, and its adaptiveness [p. 87].

The advice of the experts is to apply a Mendelian, linear, causal model to the gene action that produces melanin (and single–gene, mental defects) and to reject that same model for normal abstract reasoning. Note, however, that Thiessen retains a commitment to explaining causally the genetic architecture of behavior, even if the causal sequence is fraught with developmental uncertainties.

Our argument is that in neither the case of skin color nor of abstract reasoning is a mechanistic model appropriate to account for the phenotypes we observe. Mechanistic models are hopelessly reductionist and will not "explain" human behavior or any other complex phenotype. Furthermore, they are misleading in their emphasis on efficient or proximal causes.

The Schizophrenia of Genetics

In genetics, as in other sciences, the natural-science model is the dominant force. Molecular models of gene-action pathways is to genetics as mechanics is to physics. The gene-action pathway illustrated in Fig. 7.4 is an example of the mechanistic model in genetics. Parallel to mechanistic models, however are the statistical models of evolutionary biology, with their manifestations in ethology, behavior genetics and population genetics. Darwinian models are probabilistic and, in that sense, indeterminant on the level of individual organisms. The level of analysis is a breeding population, whose members are more likely to mate, reproduce, live, or die within that population than others. The essential evolutionary concepts, variation and selection, refer to differences among individuals, and selection is said by most authorities to act primarily at an individual level. Individuals are merely temporary receptacles for the gene pool; the level of interest is that aggregate of interbreeding individuals called the population.

The levels of analysis in molecular and population branches of genetics are entirely different. The methods are totally different. And the emphases in their explanatory models are at variance, to say the least. It is not surprising that a sociological look at genetics as a field reveals that molecular geneticists talk to and publish with cell biologists and biochemists, whereas evolutionary biologists communicate with demographers and the few behavioral scientists who will listen. The Animal Behavior Society, and the Society for the Study of Social Biology represent this nexus of population analysts. Developmental geneticists talk to pediatricians and physiologists but rarely to those concerned with human behavioral development.

Mather (1971) described in uncritical terms the contrast between Mendelian interest in deterministic genetic models and biometrical (or population) genetic concerns with probabilistic models of genetic variation in populations:

> The Mendelian approach depends on the successful recognition of clearly distinguishable phenotypic classes from which the relevant genetical constitution can be inferred. It is at its most powerful when there is a one-to-one correspondence of phenotype and genotype, though some ambiguity of the relationship, as when complete dominance results in the heterozygote and one homozygote having the same phenotype, is acceptable [p. 351]. . . . The biometrical approach is from a different direction starting with the character rather than the individual determinant. It makes no requirement that the determinants be traceable individually in either transmission or action. It seeks to measure *variation* in a character and then, by comparing individuals and families of varying relationship, to *partition* the differences observed into fractions ascribable to the various genetical (or for that matter nongenetical phenomena [p. 352].

Mather does not warn us fully of the uncertainties in the pathways from genes to the character of interest, even when there is "one-to-one correspondence of phenotype and genotype." Take, for example, the well-known case of mental retardation resulting, it is said, from the absence of the enzyme phenylalanine hydroxylase in the presence of the protein phenylalanine, ingested in many foods. This is the same pathway illustrated in Fig. 7.4. Is there a one-to-one correspondence between genotype and phenotype for the recessive homozygote? Of course not; the distribution of intellectual attainment among children with PKU, treated or untreated, is considerable. Although untreated children are severely retarded, *on the average,* individuals vary from unmeasurably low IQ levels to dull normal, a range of 60 to 80 points! Yet, all have the same DNA sequence that fails to code for the enzyme to metabolize phenylalanine. Somewhere, at the cellular level, at the more general physiological levels, and at the level of organism–environmental transactions there are uncertainties in the development of the phenotype. The "expression" of PKU varies among individuals in a probability distribution.

Let us turn now to the effects of diet on the "expression" of PKU. As popularly understood, the virtual elimination of phenylalanine from the diet of infants and young children without the enzyme phenylalanine hydroxylase prevents mental retardation. Does this mean that all treated children develop the same intellectual level? Of course not. There is, on the average, a beneficial effect of reduced phenylalahine on the intellectual development of PKU children, but individual IQs vary from moderately retarded to bright average, again a range of more than 60 IQ points. We are not referring here to the ontogenetic effects of diet administered at earlier and later periods of infancy—there surely is a more beneficial effect of earlier than later diet regimen—but to the variability among children who all received the diet from the first months of life.

Let us now examine the notion of dominance and recessivity, as Mather describes it. The heterozygotes, in the case of PKU, are the carrier parents, whose intellectual distribution is normal, varying from retarded to superior. Their average IQ level is like that of the dominant homozygote, the normal population. In other words, there is what Mather calls complete dominance in PKU. (But see Bessman, Williamson, & Koch, 1978 for a study of carriers whose IQ levels are 10 points lower than noncarrier siblings' IQs.)

Unfortunately, that satisfyingly simple, mechanistic view applies only at the behavioral level of analysis. If one examines the level of phenylalanine hydroxylase activity in the heterozygote, it is *midway* between the two homozygotes. When loaded with phenylalanine, the heterozygote metabolizes the protein at a rate about half that of the normal homozygotes. This convenient difference between normals and carriers is the basis for the detection and genetic counseling of carriers who want to be parents. But what does this fact do to the notion of "complete dominance"? The idea of dominance and recessivity depends entirely on the level of analysis of the phenotype, so that it is impossible to speak of a one-to-one correspondence between genotype and phenotype. In addition, of course, there is individual variability among heterozygote carriers in their enzyme activity, from nearly normal to nearly defective levels. And so it is with all known genetic defects: The variability among those identically afflicted at the single locus is considerable.

There are many other examples of the points illustrated by PKU. Down's syndrome children, with an extra chromosome 21, are far more alert, function at much higher levels of personal care, and communicate far better with others, when they are reared by affectionate caretakers, rather than impersonal custodians. Black Africans have lower rates of hypertension under the dietary and stress conditions of village than of urban life. And, to return to the example of skin color, sunlight, hormones associated with age and gender, and probably diet, all affect the degree to which melanin is produced in the skin cells. Women, on the average, are paler than men, and people darken with puberty. Clearly some genotypes produce more melanin than others under the same exposure to sunlight. Vary the exposure to sunlight, however, and you will discover that some genotypes who were *paler* phenotypes than others under *lower* exposure to sunlight are now *darker* than others under higher levels of sunlight. They have a greater *responsiveness* to the sunlight and produce melanin at rates that vary more directly with exposure. Others (of us) produce little melanin regardless of exposure levels. This homely example points to a more profound objection to mechanistic models of gene action.

To wit, the mechanistic notion of one-to-one correspondence between genotype and phenotype is largely a myth, fashioned from models for the blood and serum proteins, where the relationship between genotype and phenotype seems *not* to be vulnerable to interventions from other genes or from environments at any level. As is often the case with perfect models, they have very limited applicability. For intellectual phenotypes in the normal range the relationship between genotype and phenotype can be explained only partially by mechanistic models of gene action.

Fudge Factors

Molecular genetics, like all sciences, has "fudge factors" to account for phenotypic variability. "Penetrance," "expression," and "buffering" are concepts invoked to explain (?) why the same gene, coding for the same disorder, fails to produce identical phenotypes. All these concepts attempt to account for the *probabilistic* relationship between genotype and phenotype. Such concepts are needed because the mechanistic model invoked to link gene action to phenotype is a reductionist one. By reducing the level of analysis, both methodological and explanatory, to gene action, such models assert a form of determinism that excludes so many possible, intervening events that they are largely indeterminant when applied to phenotypic variability in the real world.

The disadvantages of mechanistic models in psychology and behavior genetics may far outweigh their explanatory assets. Mechanistic models are not now, and never will be, in our view, preferable to probabilistic models to account for the genetic determination of abstract reasoning, skin color, or mental retardation associated with PKU. Such models "work" only under conditions where all other possible effects on phenotypic development are held constant, as they never are in the real world. To account for the variation that exists in vivo, in situ, requires probabilistic models with greatly lessened generality. *Local conditions*—the particular combinations of events that affect development—and the *time frame* of effects—the ontogenetic time during which the phenotype develops and is susceptible to change—both limit the applicability of mechanistic models to explain the relationship of genotype to phenotype.

Linear causal models that begin with the gene and proceed directly to the phenotype—whether physical or behavioral—explain *some* of the variance among people's observable traits. We do not deny the results of Pakstis' linkage study. Chromosomes have material reality, and on several of those tangible chromosome segments are located DNA sequences that affect the development of skin color and abstract reasoning (McKusick & Ruddle, 1977). But mechanistic models do not emphasize the most important facts about the relationship of genotype to phenotype: *that developmental environments shape phenotypic development in such ways as to render that relationship entirely probabilistic and not determinant, within even our wildest dreams of specifying all of the events that shape (intellectual) development.* Even if one knows the genotype, as in simple Mendelian disorders such as PKU, the genetic background of individuals and their particular developmental histories will so alter the expression of the single gene as to make population thinking preferable even here. In the case of normal intellectual variation, there is little hope that we ever will know the genotype in a mechanistic sense. Even if the gene action pathways for the hundreds of loci were known, and the systematic interactions among the loci known, the relationship between genotype and phenotype would still be rendered indeterminant for individuals by idiosyncratic genetic and environmental events.

Contrary to the claims of Feldman and Lewontin (1975), Monod (1971), and many others, the most fruitful questions about the nature of genetic and environmental differences in intelligence are asked and answered at a population level. The question, "To what extent are existing intellectual differences among individuals due to genetic and to what extent to current environmental differences in a specified population?", is scientifically important and has many possible implications for the design of environmental programs to enhance people's lives (see DeFries, Vandenberg, & McClearn, 1976; Nichols, 1978; Scarr & Weinberg, 1978; and Willerman, 1979, for a variety of opinions on this matter). Questions about individual variability in normal intelligence at a mechanistic level of determinism have not been fruitful and do not promise to become so.

Genetic Fixity

A great danger in the application of mechanistic, linear models to the genetic study of behavior is that they lead to erroneous ideas about the *fixity* of genetic effects. They permit malleability in the development of the phenotype only incidentally, as residual, unexplained error.

As Paul Weiss (1969) so eloquently said about genetic determination:

> The term "genetically determined" means three different things to three different groups of people: (1) the broad-gauged student of genetics, who is thoroughly familiar with the underlying facts and uses the term simply as a shorthand label to designate unequivocal relations between certain genes and certain "characters" of an organism; (2) scientists in various other branches who are not familiar with the actual content of the term and accept it literally in its verbal symbolism; and (3) the public at large, to whom the term frequently imparts a fatalistic outlook on life, frustrating in its hopelessness, of an inexorably pre-set existence and fixed course towards a pre-ordained destiny [pp. 33–34].

By contrast, probabilistic models of development and of population dynamics may include mechanistic parts, where applicable, in the larger theory of change. But in most evolutionary accounts, there is little emphasis on efficient and proximal causes. Rather, evolutionary changes are interpreted in terms of reproductive fitness and adaptation, final causes. In Aristotelian logic, final causes are reasons or goals for a particular adaptation. Mechanistic models emphasize what Aristotle called efficient causes, which are the proximal, immediate antecedents for an event. This difference in emphasis between mechanistic and probabilistic models has been the source of a great deal of the anguish about sociobiology. The idea of human behavioral adaptations having evolved, genetic bases that limit variability and bias learning has been interpreted within the mechanistic model of efficient causes to mean that human behavior is on a "fixed course toward a pre-ordained destiny," to quote Paul Weiss.

As Li (1971) has pointed out: "Environmentalists sometimes misunderstand the very implications of population genetics, thinking that heredity would imply 'like class begets like class.' Probably the opposite is true. Only very strong social and environmental forces can perpetuate an artificial class; heredity does not[p. 172]."

Li has presented a simple but comprehensive polygenic model for intelligence that explains parent–child regression, variability within families, and the other phenomena observed for phenotypic IQ. The most important simgle consequence of the genetic model is that for any given class of parents, their offspring will be scattered in various classes; conversely, for any given class of offspring, their parents will have come from various classes. This effect is shown in Fig. 7.6.

Parents at the high and low *extremes* of the distribution contribute offspring primarily to the upper or the lower *halves* of the distributions, whereas parents in the middle of the distribution contribute children to all classes in the distribution. On the average, the children will have less extreme scores than their parents, but the total distribution of phenotypic IQ will remain relatively constant from one generation to another (unless selective forces or radical environmental changes intervene).

To the redistribution of offspring from parental to offspring classes in each generation, Li adds the Markov property of populations: "The properties of an

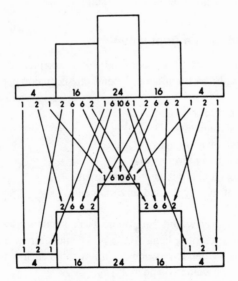

FIG. 7.6.   The distributions of offspring and parents in five phenotypic classes in a random mating population. (From Li, 1971.)

individual depend upon the state (in this case, genotype) in which he finds himself and not upon the state from which he is derived. A state is a state; it has no memory [p. 173]."

Thus, the distant descendants of Jean Bernoulli are distributed into the various classes of mathematical ability in exactly the same way as the distant descendants of one whose mathematical ability was subnormal. Family members who are as much as six to eight generations apart are practically u..related even though they retain the same family name.

Whether present-day family groups and social classes are genetically artificial groups is debatable (Eckland, 1979; Herrnstein, 1971; Scarr & Weinberg, 1978), because one assumption of Li's model is random mating, which is violated by an IQ correlation of about .40 between parents. Even under conditions of high assortative mating, however, there is considerable regression of offspring scores toward the population mean and the majority of IQ variation is found among the offspring of the same parents. Fear not! The population may be relatively stable but individual determinism is not yet upon us.

ANALYSES OF VARIATION IN HUMAN INTELLIGENCE

At a population level of analysis, the goal is to estimate the magnitude of genetic and environmental variances in a given interbreeding group. Both genetic and environmental variances may be analyzed into several components, although not without some difficult and questionable assumptions.

Individual Variation in a Population

The relative contributions of genetic and environmental differences to phenotypic diversity within a population depend on six major parameters:

1. Range of genotypes.
2. Range of environments.
3. Favorableness of genotypes.
4. Favorableness of environments.
5. Covariance of genotypes and environments.
6. Interactions of genotypes and environments.

The range of genotypes and environments can independently and together affect the total variance of a behavioral, polygenic trait in a population. The mean favorableness of genotypes and environments can independently and together affect the mean values of phenotypes.

Two separate problems are involved in understanding the effects of mean favorableness and ranges of genotypes in a population: first, gene frequencies, and second, the distribution of genes among the genotypes. Gene frequencies are affected by two principal processes: differential reproduction or *natural selection* and *sampling errors*. Genotype frequencies are affected by *assortative mating*. Two populations (or two generations of the same population) may have equal gene frequencies but different genotype frequencies if assortative mating for a behavioral trait is greater in one population than the other.

Genotypic Range and Favorableness

Natural Selection. Changing environmental conditions, such as the introduction of more complex technology, may affect the rate of reproduction in different segments of the IQ distribution in a generation. We know, for example, that severely mentally retarded persons in the contemporary white populations of Europe and the United States do not reproduce as frequently as those who can hold jobs and maintain independent adult lives (Bajema, 1968; Higgins, Reed, & Reed, 1962). Being severely retarded renders one less likely to be chosen as a mate and less likely to produce progeny for the next generation.

If one segment of the phenotypic IQ range has been strongly and consistently selected against, as severely mentally retarded persons are in contemporary industrial populations, then the range and favorableness of the total gene distribution will be slowly changed. If, in another population, high phenotypic IQ were disadvantageous for mate selection and reproduction, then the genic distribution would be reduced at that end. It does not seem likely that high phenotypic IQ has ever been strongly selected against within a population.

Sampling. Gene frequencies can also be affected by genetic drift, random sampling error. Not every allele at every gene locus is equally sampled in every generation through reproduction. Rare genes especially may disappear through random failure to be passed on to the next generation, and the frequencies of other alleles may be randomly increased or decreased from generation to generation.

A special case of restriction in genic range is nonrandom sampling from a larger gene pool in the formation of a smaller breeding group. If, for example, an above-median sample from the IQ group migrated to a distant locale and bred primarily among themselves, the gene frequencies within the migrant group might vary considerably from those of the nonmigrant group, all other things being equal.

Assortative Mating. The distribution of genes in genotypic classes within a population can vary because of assortative mating. To the extent that "likes" marry "likes," genetic variability is decreased within families and increased among families. Assortative mating for IQ also increases the standard deviation of IQ scores within the total (white) population by increasing the frequency extremely high and extremely low genotypes for phenotypic IQ. On a random mating basis, the probability of producing extreme genotypes is greatly reduced because extreme parental genotypes are unlikely to find each other by chance. The sheer frequency of middle-range genotypes makes an average mate the most likely random choice of an extreme genotype for both high and low IQ.

Because children's IQ values are distributed around the mean parental value (with some regression toward the population mean), the offspring of such matings will tend to be closer to the population mean than offspring of extreme parental combinations. The phenotypic distribution under conditions of random mating will tend to have a leptokurtic shape with a large modal class and low total variance.

Environmental Range and Favorableness

The range of environments within a population can also affect phenotypic variability. Uniform environments can restrict phenotypic diversity by eliminating a major source of variation. Because environments can be observed and manipulated, there are many studies on infrahuman populations to demonstrate the restriction of variability through uniform environments (Manosevitz, Lindzey, & Thiessen, 1969).

Far more important, however, for the present discussion is the favorableness dimension of the environment. Environments that do not support the development of a trait can greatly alter the mean value of the trait. If environments in the unfavorable range are common to all or most members of a population, then the phenotypic variance of the population can be slightly reduced, whereas the mean can be drastically lowered.

The most likely effects of very suppressive environments are that they lower the mean of the population, decrease phenotypic variability, and consequently reduce the correlation between genotype and phenotype (Henderson, 1970; Scarr-Salapatek, 1971b). A contrast can be made between uniform environments that support the development of a particular behavior and suppressive environments that may also be uniform but not supportive of optimal development (Nichols, 1970). Uniform environments of good quality may reduce total variability and raise the mean of the population.

The ranges of genotypes and environments and the favorableness of the environment control a large portion of the total phenotypic variance in IQ. The two additional factors, covariance and interaction, are probably less important (Jinks & Fulker, 1970), at least within the white North American and European populations.

Covariation

Covariation between genotypes and environments is expressed as a correlation between certain genotypic characteristics and certain environmental features that affect phenotypic outcome (e.g., the covariance between the IQs of children of bright parents, which is likely to be higher than average, and the educationally advantaged environment offered by those same parents to their bright children). Retarded parents, on the other hand, may have less bright children under any environmental circumstances, but also supply those children with educationally deprived environments. Covariation between genotype and environment may depend also on the genotype and the kind of response it evokes from the environment. If bright children receive continual reward for their educationally superior performance, while duller children receive fewer rewards, environmental rewards can be said to covary with IQ. Finally, children build their own niches, selecting those aspects of the environment that are most compatible with their phenotypes, including genotypes (Plomin, De Fries, & Loehlin, 1977).

Interaction

Covariance is sometimes confused with interaction but they are quite different terms. When psychologists speak of genetic–environment interaction, they are usually referring to the reciprocal relationship that exists between an organism and its surround. The organism brings to the situation a set of characteristics that affect the environment, which in turn affects the further development of the organism and vice versa. This is not what quantitative geneticists mean by interaction. The psychologists' term would better be *transaction* between organism and environment, because the statistical term, genetic–environment *interaction* refers to the *differential* effects of various organism–environment transactions on development.

Behavioral geneticists, whose experimental work is primarily with mouse strains and *Drosophila,* often find genotype–environmental interactions of considerable importance. The differential response of two or more genotypes to two or more environments is interaction. In studies of animal learning, where genotypes and environmental conditions can be manipulated, so-called maze-dull rats who were bred for poor performance in Tryon's mazes, were shown to perform as well as so-called maze-bright rates when given enriched environments (Cooper & Zubec, 1958) and when given distributed rather than massed practice (McGaugh, Jennings, & Thompson, 1962).

Studies of genotype–environment interaction in human populations are quite limited. Biometrical methods that include an analysis for interaction have failed to show any substantial variance attributable to nonlinear effects on human intelligence (Erlenmeyer-Kimling, 1972; Jinks & Fulker, 1970). This is not to say that genotype–environment interaction may not account for some portion of the variance in IQ scores in other populations or in other segments of white populations (e.g., disadvantaged).

Heritability

Heritability is a summary statement of the proportion of the total phenotypic variance that is due to additive genetic variance (narrow heritability) or to total genetic variance (broad heritability). Heritability (h^2) is a *population statistic,* not a property of a trait (Fuller & Thompson, 1960). Estimates of h^2 vary from population to population as genetic and environmental variances vary as proportions of the total variance. (For the calculation of various kinds of heritability estimates, see Falconer, 1960.)

The six parameters of individual variation within a population (ranges and favorableness of genotypes and environments, covariation and interactions) discussed here, are the major contributors to the total phenotypic variance in any population. The proportions of genetic variance and environmental variance may well vary from one population to another depending on the ranges and favorableness of the two sets of variables, their covariances and interactions. The variance terms and heritability statistics are frequently used in twin and family studies to estimate the relative importance of genetic and environmental differences to account for phenotypic IQ differences.

Methods of Analysis for Twin and Family Studies

The simplest, and most defensible method is to calculate a statistical measure of resemblance between pairs of persons in two sets who differ in their degrees of genetic and/or environmental relatedness, and to stop there. No one objects to the calculation of regressions coefficients for child IQ on parental IQ or the use of intraclass correlations for twins and siblings. These are basic measures of the degree to which pairs of individuals in various relationships resemble each other, compared to randomly paired people. One may even test for the significance of the difference between coefficients.

Comparisons of coefficients of resemblance between relatives of different degrees or persons of the same genetic relatedness reared in environments with greater or lesser similarity are based on some assumptions that legitimate the comparison. In the case of twin comparisons of MZ and DZ pairs, the major assumption is that the environments of MZs do not create additional similarities over those of DZs. This assumption has been tested in several ways and found to be satisfactory, as is reviewed in a later section. In the case of comparisons of adopted and biologically related children, a similar objection can be raised to the comparability of environmental variances that are similar for the two kinds of relatives. Again this assumption has been tested and found satisfactory (Scarr, Scarf, & Weinberg, 1980).

From the comparison of coefficients of resemblance for persons of different degrees of relatedness, one can calculate a "heritability" coefficient. The assumptions at this stage are many and difficult to defend. Is there parental assortative mating, which decreases the genetic variance within families for DZ twins and siblings and increases their genetic resemblance? What can one assume about the role of gene–environment correlation and interaction (nonlinear type) in phenotypic resemblance? What about broad "heritability" versus narrow "heritability"; the former can be calculated, under certain assumptions from a variety of family data, the latter from parent–child data, again under certain assumptions.

Rather than detail here the many assumptions and objections to "heritability" analyses, we refer the reader to sources listed under the next section, on biometrical methods.

Biometrical Analysis

The next, large step in complex analysis of kinship data are models of several or many degrees of relationship simultaneously. We are not experts on biometrical methods. For mathematically sophisticated treatments of this extensive literature, the reader is referred to articles by the Birmingham group (Eaves, 1975, 1976; Eaves, Last, Martin, & Jinks, 1977; Jinks & Fulker, 1970; Martin, 1975; Martin, Eaves, Dearsley & Davies, 1978), the Hawaii group (Morton & Rao, 1977), the North Carolina group (Elson & Stewart, 1971; Haseman & Elston, 1970), and the Stanford group (Cavalli-Sforza & Feldman, 1973, 1977). One group of economists (Taubman, Behrman, & Wales, 1978) are also involved in similar analyses of twin data. Telling criticisms of the biometrical enterprise have been made by Goldberger (1975, 1978).

Jencks (1972) was the first to our knowledge to apply Sewell Wright's path models to the complement of family and twin data sets from U.S. studies. Since that time, there has been considerable interest in expanding the application of path models to the study of population variation. Although Jencks misspecified a parameter in his model (Loehlin et al., 1975), the method is increasingly used. An example of a path model of parent–child resemblance is given in Fig. 7.7 (from De Fries, Kuse, & Vandenberg, 1979).

The additive genotypes for two phenotypes of the parent (Ax and Ay) leads to those of the children (Axo and Ayo). In this model, the parent phenotypes do not directly affect the child's phenotypes, nor do the environments of the parents directly affect the child's environments or phenotypes. Do you buy these assumptions? If not, you may draw a different model. And so it goes. The authors suggest that this model is applicable to studies where the parents do not provide the rearing environment.

The objections that Goldberger raises with the biometrical approach are that the assumptions that must be made are largely indefensible and that the statistical manipulations of the data exaggerate the already large statistical errors in twin and family correlations. By relaxing any of the assumptions, or making others, Goldberger has shown that quite different results can be obtained. We do not care to enter this fray.

In simple lay language, biometrical models attempt to estimate genetic and environmental parameters from sets of family and twin data, primarily intelligence tests, by making assumptions about the nature of the genetic and environmental effects (such as, is there gene–environment correlation that acts from

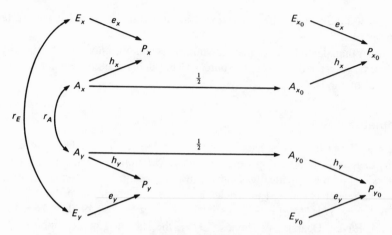

FIG. 7.7.   Path diagram of bivariate resemblance between parental phenotypes (P_x and P_y) and offspring phenotypes (P_{x_o} and P_{y_o}) (From DeFries & Vandenberg, 1979.)

parental phenotype to child phenotype or from parental genotype to child phenotype?). With simultaneous equations and matrix algebra, they solve for the best fits to the available data. Unfortunately, the data are not, as a group, robust enough to withstand extensive manipulations without yielding solutions that can be seriously challenged by doubters. Nor are the models defensible without recourse to the data.

Lewontin (1974) tackled Morton (1974) and his colleagues (Rao, Morton, & Yee, 1974), and by implication many other biometricians, for an alleged confusion of the analysis of variance and the analysis of causes. The latter, Lewontin says is what we "really" want to know about genetic effects on human development, because only by a knowledge of the mechanisms underlying genetic effects can we arrive at "correct schemes for environmental modification and intervention [p. 409]." Analysis of the variation in a population cannot do this, because "its results are a unique function of the present distribution of environment and genotype [p. 409]." Only in the case where there is perfect or nearly perfect additivity between genotypic and environmental effects, he says, can the analysis of variance estimate functional relationships between genotypes and environments. Lewontin doubts that linear combinations of gene and environmental effect are the rule and presents many examples of interactions, both hypothetical and from rates of survival for *Drosophila* larvae.

One must ask how likely it is that genotype–environment interactions account for a major portion of the variance in human intelligence. Erlenmeyer-Kimling (1972) reviewed the scanty literature on possible interaction effects of genes and environments on human intelligence and found no evidence. Neither have there been good methods for testing for interaction effects. The issue is whether or not it is plausible that, with the exception of extreme deprivation or extreme over-stimulation (which could be defined in sensory and affective terms), some children develop better phenotypes under impoverished than under enriched environments, whereas others profit more from sensory and affectively rich than from poor environments?

In our view, those who propose that genotype–environment interactions are major determinants of intellectual variation in populations are more interested in putting roadblocks in the way of studies of normal human variation than in clarifying the scientific issues. It is not accidental that they call upon a mechanistic model for "real" understanding of human variation and deny the legitimacy of statistical models, which are feasible approaches to understanding individual differences that exist in a population at the present time. In fact, Lewontin confuses the individual level of analysis, at which mechanistic studies are appropriate, with the population level of analysis, for which analysis of variation is the appropriate model. For social policy planning of intervention programs, the latter may in fact be more important in deciding whether and how to distribute *known* resources. We agree that knowledge of the mechanisms by which genetic effects are translated into phenotypic behaviors would be handy in designing intervention programs, but treatment of the mentally ill and the mentally retarded has proceeded with reasonably good effects without benefit of mechanistic knowledge in most cases. Life goes on even without complete information.

Many have argued that knowledge of sources of variation in a population is irrelevent to social policy (Jensen, 1973b; Scarr-Salapatek, 1971a). The empirical, trial-and-error approach would be necessary in any case to test the effects of an intervention. Let us say here that we do not agree and will deal with this issue in the concluding section of the chapter.

TWIN AND FAMILY STUDIES

In this section we present the heart of the research on genetic differences in normal intelligence. Necessarily, the many studies will be condensed into fewer paragraphs, beginning with studies of familial mental retardation, followed by a review of twin and family studies of normal intelligence. Because of the recent surge in studies of adoptive families, particular attention will be given to these remarkable new data.

Mental Retardation: A Population View

Assuming that 2% of the general population is retarded, Reed and Anderson (1973) estimated that 17% of the retarded children have a retarded parent. If no retarded persons were to reproduce, the frequency of retardation in the next generation would drop to 1.7%. If the retarded were to reproduce at the same rate as the general population, retardation would rise to 2.2%, and 26% of the retarded children would have a retarded parent.

The increased risk of retardation in families is graphically shown in Fig. 7.8. The risk of familial mental retardation rises from 2% for all children without retarded parents or siblings to more than 70% when both parents are retarded and there is at least one retarded sibling. These data are drawn from the Reed and Reed (1965) study of some 80,000 individuals, sampled in the State of Minnesota as the descendants (mostly distant) of institutionalized retardates from the early years of this century.

It has not been possible to differentiate a major gene model from a polygenic one in the determination of familial mental retardation. The possibility of making such a discrimination depends on the rarity of the trait, and mental retardation is too common to make the choice among models possible (Anderson, 1974, p. 33). Pauls (1972) studied nearly 6000 individuals, neither of whose parents was retarded, and calculated a heritability estimate for retardation of .62, very close to contemporary estimates for intelligence in general. There was no evidence from the analysis of the family data that major genes accounted for a major part of the retardation.

Quantitative Genetic Theory

As Anderson indicated in his review (1974) several features of the literature on mental retardation are relevant to a quantitative genetic theory: First, the frequency of retardation is higher among males; second, retarded females are more likely to reproduce than retarded males; third, the risk of retardation is higher among the offspring of female than of male retardates. In summary, the risk for retardation is highest among the sons of affected mothers and lowest among the daughters of affected fathers. As Anderson indicates, these results are congruent with a hypothesis that familial mental retardation results from multigenes with a sex-modified threshold, which predicts that, when a trait is more common among

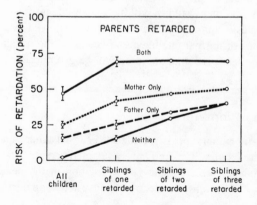

FIG. 7.8.   Risk of mental retardation among offspring, by presence or absence of retardation in parents and siblings. The vertical bars for the first two points in each line indicate ± 1 standard error. The standard errors for the remaining points are larger, but are difficult to estimate. (From Anderson, 1974.)

males, the least risk will be for the female relatives of retarded males. Because more severely affected females than males reproduce, the offspring of retarded mothers have a higher rate of retardation than the offspring of (less severely) retarded males. The females have, on the average, greater "genetic loading" for retardation, so that part of the maternal effect for retardation may be genetic. Another part may be the rearing environment, which is more related to the mother's mental status than the father's.

Familial Versus Clinical Retardation

Although there are overlapping or indistinct boundaries between categories, most workers in the field of retardation hold that there are two broad categories of retarded individuals: the lower grade mental defectives with major chromosome, genetic, or traumatic disabilities, and the higher grade, "familial" retardates, without histories of specific etiology or known defect—just much lower than average IQ and social competencies. Roberts (1952) found that the siblings of children with severe mental retardation are more likely to have normal IQ scores than the siblings of familial mental retardates. If the causes of severe defect were rare chromosomal single-gene and environmental events, then the sibs would be unlikely to be affected; if the higher grade mental retardates are the lower end of the normal curve of polygenic inheritance for intelligence, then their sibs would be likely to have lower IQs as well.

Kamin (1974, pp. 136–141) questioned the methods of the Roberts study and ridiculed the results. To reexamine Roberts' hypotheses with new data, Johnson, Ahern, and Johnson (1976) looked at the sibling and parents of 289 retarded probands, reported by Reed and Reed (1965). Forty-seven of the probands had no IQ score or no sibling; data on the 242 remaining probands and their siblings are presented in Table 7.1.

It is clear from the table that higher grade retardates have more retarded siblings than lower grade retardates ($X^2 = 26.22$, p. .001). Of the siblings of retardates with IQs less than 40, 21.5% are retarded. Of the siblings of retardates with IQ scores about 40, 31.0% are retarded. Also in keeping with Roberts' results, the siblings of the higher grade retardates were more likely to be of higher grade retardation themselves. If severely retarded probands had affected sibs, the sibs were likely to be severely retarded themselves.

The authors suggested that the groups, clinical and familial retardates, were not as clearly divisible as is sometimes suggested, for the severely retarded had some higher grade retarded sibs. But the parents of the retardates with IQ scores below 40 were far less likely to be described occupationally as indigent or unskilled than the parents of higher grade retardates, thus supporting the notion of familial retardation. The percentage of parents who were themselves retarded also differed between the groups: of the severely retarded probands, 12.7% had both parents defective and 28.8% had one parent defective; of the higher grade

TABLE 7.1

Status of Full Siblings of Reed and Reed Probands by IQ Level of Probands

| IQ level of probands | Number of probands[a] | Sibs | | | Total number of sibs | Percent retarded of retarded, normal, and unknown sibs | Number of probands with one or more retarded sibs | Percent of probands with retarded sibs | Mean IQ of retarded sibs of probands[d] |
		Dead[b]	Retarded	Normal or of unknown ability[c]					
0-19	47	81	34	169	284	16.70	17	36.17	28.21 ($N = 19$)
20-29	32	57	41	104	202	28.28	14	43.75	34.21 ($N = 19$)
30-39	31	33	35	127	195	21.60	17	54.84	35.78 ($N = 9$)
40-49	54	65	77	195	337	28.31	35	64.85	47.56 ($N = 18$)
50-59	37	40	42	119	201	26.09	21	59.46	58.12 ($N = 17$)
60-79[e]	41	56	68	106	230	39.08	33	80.49	60.50 ($N = 24$)
Total	242	332	297	820	1449	26.59	137	56.61	45.32 ($N = 106$)

[a]Only probands with IQ scores and full siblings are included in this table.

[b]Includes recorded miscarriages, stillbirths, infant and neonatal deaths.

[c]Internal evidence (Reed and Reed, 1965, Table 27, p. 39) indicates that persons of unknown ability nearly always are of normal ability, despite the fact that too little is known about them to state that they are of normal ability.

[d]Only sibs with known IQ scores are included.

[e]Only one proband was above IQ 69.

retardates, had two parents defective and 33.8% had one parent defective. Interestingly, the defective parents of the severely retarded probands were equally divided between mothers and fathers, whereas the mothers of the higher grade retardates were more likely to be retarded than the fathers, in keeping with the polygenic theory developed by Anderson (1974), which posits a greater genetic loading for retarded women than men.

Twin Studies of Intelligence

Robert Nichols (1978) compiled 211 studies of intelligence and abilities that compare the resemblance of identical (MZ) and fraternal (DZ) twins. His results for 1100 to 4500 pairs of MZs and a like number of DZs are given in Table 7.2 (his Table 1).

For general intelligence, the mean correlation for MZ twins, taken from a variety of tests, is .82; for DZ twins, .59. Although the MZ correlation exceeds the DZ coefficient by .22, it is clear that being genetically related and reared as twins in the same family are potent determinants of individual differences in measured intelligence. From the comparison of MZ and DZ correlations from the 30 studies, one can calculate an estimate of the broad heritability of IQ in the white U.S. population of adolescents (who made up most of the studies' subjects) as somewhere between .3 and .7, with a most likely value in the .5 range given a correction for assortative mating of the parents and without correcting for

TABLE 7.2
Mean Intraclass Correlations from Twin Studies of
Various Traits

Trait	Number of studies	Mean Intraclass Cor.		Difference $r_{MZ} - r_{DZ}$	
		r_{MZ}	r_{DZ}	Mean	Stand. Dev.
Ability					
General Intelligence	30	.82	.59	.22	.10
Verbal Comprehension	27	.78	.59	.19	.14
Number and Mathematics	27	.78	.59	.19	.12
Spacial Visualization	31	.65	.41	.23	.16
Memory	16	.52	.36	.16	.16
Reasoning	16	.74	.50	.24	.17
Clerical Speed and Acc.	15	.70	.47	.22	.15
Verbal Fluency	12	.67	.52	.15	.14
Divergent Thinking	10	.61	.50	.11	.15
Language Achievement	28	.81	.58	.23	.11
Social Studies Achievement	7	.85	.61	.24	.10
Natural Science Ach.	14	.79	.64	.15	.13
All abilities	211	.74	.54	.21	.14

the reliability of the tests. Nichols corrected for the unreliability of measurement and calculated a most likely estimate of genetic differences as accounting for about .60 to .70 of the IQ variation.

As anyone fond of distributions and sampling theory would have smilingly predicted, the studies summarized by Nichols (1978) form a distribution of results. The population of studies of general intelligence and other abilities have means and variances of their own. In Fig. 7.9 Nichols indicates the results of each study with a circle and, by an arrow, the weighted mean correlation coefficient for MZ and DZ twins for each measure. As anyone who is knowledgeable about statistical distributions would expect, there are some outlying values and more studies clustered nearer to the middle of the distribution. In these cases, however, there are some unusually low correlation values, outlying by themselves, which raises suspicions about measurement, test administration, sampling restriction, and the like.

Critics of the twin study method have seized upon the outliers in distributions of studies to raise questions about the overall pattern of results. Kamin (1974), in particular, has made much of equivocal findings from some twin and adoption

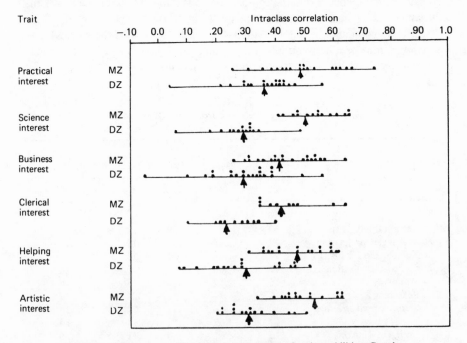

FIG. 7.9.   Intraclass correlations from twin studies of various abilities. Correlations obtained in each study for MZ (identical) and DZ (fraternal) twins are indicated by dots; the mean correlation, weighted by the number of cases, is indicated by an arrow below the horizontal line representing the range of correlations for each trait. (From Nichols, 1978.)

studies (Kamin, 1981). McAskie and Clarke (1976) use the variance of parent–child study results for the same purpose. It seems to us that a more sophisticated, statistical look at the twin results is persuasive evidence for the greater average similarity of MZ than DZ co-twins for measures of general and specific abilities. The greater similarity of MZ twins is usually interpreted as due to their greater genetic similarity, but that conclusion is based on a critically important assumption: that the environments of MZ twins do not bias their behavioral similarities by their more similar treatment. The objection to the twin study method is that MZ twins are made more similar by the greater similarity of their treatment by others, based on their striking physical similarity.

Environmental Similarity of MZ and DZ Twins

Identical twins, it is said by critics of twin studies, are treated more similarly by their parents and others than fraternal twins; therefore, the usually greater behavioral similarity of MZ twins is due not to their greater genetic relatedness but to the more similar environmental response to their identical appearance (Kamin, 1974).

Three approaches have been taken to testing the effects of greater environmental similarity of MZ than of DZ twins. First, Scarr (1968) and Scarr and Carter-Saltzman (1979) compared the actual intellectual, personality, and physical similarities of twins who were correctly and incorrectly classified as MZ and DZ by themselves, their parents, and others. There appeared to be little bias from the belief in zygocity, as the incorrectly classified MZs and DZs were as similar on most measurements as the correctly classified pairs. On intellectual measures, belief in zygocity had no significant effect on actual similarity. On personality and physical measures, both actual zygocity and belief in zygocity were related to measured similarities. It is hard to imagine that twins grow taller or shorter, have greater or lesser skeletal maturity because someone believes them to be identical or fraternal twins. Thus, we concluded that for personality and physical measures, actual similarity is a basis for the judgment of zygocity, not likely the reverse.

Lytton (1977) has taken a second approach to the issue of environmental similarity between identical and fraternal twins. With extensive observations of the parental response and initiation of interactions with very young twins, Lytton showed that the parents of MZs treat their children more similarly than do the parents of DZs, because the identicals give the parents more similar stimuli to respond to. He observed no difference in parental treatment of MZs and DZs that would create additional similarities or differences to bias comparisons between the types of twins.

A third approach to the study of the role of more similar environment for identical than fraternal twins was developed by Plomin, Willerman, and Loehlin (1976). Their reasoning was as follows: If identical twins are more similar *because* they are treated more similarly, then those identicals who experience more similar environments should be more similar than those identicals who experience less similar environments. Sharing the same room, friends, class-rooms, and receiving similar parental treatment, for example, should increase the behavioral similarity of some MZs over that of others. In short, those environ-mental similarities that differ *between* identical and fraternal twins ought to affect the degree of similarity *among* identical twins as well. (Fraternal twins were not used, because some pairs are genetically more similar than others, a confounding factor in any analysis of greater and lesser environmental similarity. To some extent, greater genetic similarity may lead some DZ pairs to select and receive more similar environments than others). The result of Plomin et al.'s analysis of the effect of greater similarities on the environmental factors that differentiate betweer. MZ and DZ twins was clear: Greater environmental similarity does not inflate actual similarities among identical twins on intellectual or personality dimensions. Rather, it seems likely that more similar genotypes develop greater behavioral similarity and select more similar environments more often than less similar (e.g., DZ) genotypes. The greater environmental similarity of MZ than DZ twins is, therefore, primarily a result and not a cause of behavioral similarity.

Results Without Sir Cyril Burt

Following on the summary by Nichols (1978) it seems appropriate to mention the famous review of the literature on genetic differences in intelligence by Erlen-meyer-Kimling and Jarvik (1963). They presented a chart of 52 studies, clearly excluding some of the twin studies included by Nichols with low correlations, and showing the median correlations for the various degrees of kinship and rearing closeness. Unfortunately, their chart included data from the "studies" of Sir Cyril Burt, now considered questionable at least, fraudulent at most (Hearn-shaw, 1979).

The older, selected studies of genetic differences in measured intelligence supported a conclusion that about 70% of the variance among individuals in the white populations of the United States and Great Britain were due to genetic variation. The omission of Burt's data does little to change that picture, as Rimland and Munsinger (1977) showed in their annotated graph (here Fig. 7.10) from Erlenmeyer-Kimling and Jarvik (1963).

The regularity of these data and the magnitude of the differences among kinship groups, with some variance due to rearing together, supported a strong genetic hypothesis. More recent data, as will be shown, support a more moderate position on the heritability of general intelligence.

Rowe and Plomin (1978) compared Burt's data to those surveyed by Jencks (1972), based entirely on U.S. samples. Their data, given in Table 7.3, again support more moderate estimates of heritability than Burt's data.

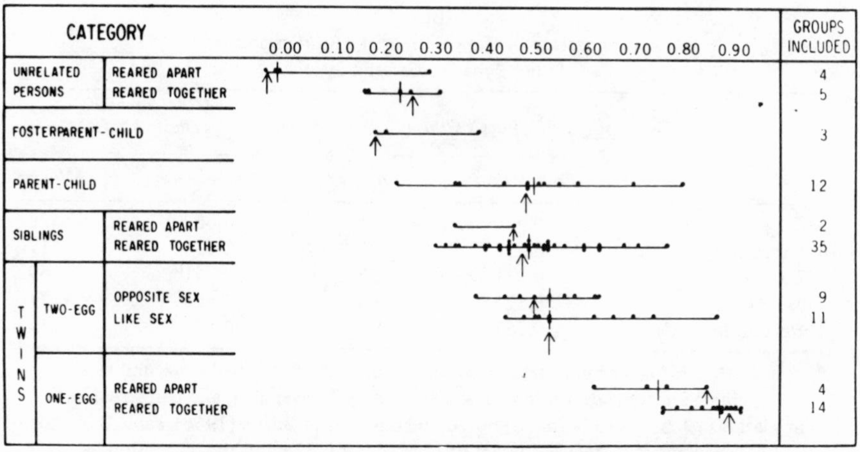

FIG. 7.10.   IQ correlations of unrelated persons, first degree relatives, and twins reared together and apart. Vertical line indicated median value and arrow indicated values reported by Burt (From Rimland & Munsinger, 1977).

Twin Family Studies

A particularly interesting design for the estimation of genetic and environmental effects on intelligence is the study of families of identical twins, who are themselves adults with offspring. Suppose that an identical twin pair of males marries two unrelated women and each pair has two children. The children of one are equally related genetically to the co-twin, with whom the children do not live, and not related at all to the spouse of the co-twin. Thus, these families provide examples of parent–child pairs together and in different households (albeit correlated environments). Furthermore the children of the MZ twins are more than ordinary first cousins; they are half-siblings, having genetically the same father and a different mother. The design also provided for the study of ordinary full sibs, spouse correlations, and maternal versus paternal effects, depending on the gender of the MZ twin pair. Nance and Corey (1976) have proposed models for the analysis of such data, and Rose (1979) has reported initial data from 65 such family constellations for the Wechsler Block Design. The data for Block Design and fingerprint ridge count are given in Table 7.4.

One notable feature of the results is that the regression of genetic offspring on genetic parent (whether actual parent or twin uncle/aunt) is only slightly lower when they do not reside in the same household, whereas the regression of offspring on the genetically unrelated spouse of the twin is zero. The twin, sibling, and half-sibling data are all consistent with the parent–child results and heritability estimates in the .4 to .6 range. A comparison of the Block Design

TABLE 7.3
A Comparison of Data from Burt's Studies and from
Other Behavior Genetic Studies

	Burt's studies[a]		Other behavior genetic studies	
	Correlation	N/pairs	Correlation	N/pairs
MZ together	0.930	95	0.857[b]	526
DZ together	0.537	127	0.534[c]	517
MZ apart	0.841	53	0.741[d]	69
Sibs together	0.525	264	0.545[e]	1671
Unrelated together	0.267	136	0.376[f]	259

[a]From Burt (1966). The correlations for the group tests, individual tests, and final assessments were converted to z scores, weighted by N, averaged, and converted back to correlations. Jensen (1974, p. 18) noted one misprint in Burt's (1966) Table 2 in which the final assessment correlation was 0.453 instead of 0.534 and the correction was noted in our calculations.

[b]From Jencks (1972). Table A-6, column headed "Identical Twins." Correlations from studies on IQ (excluding Burt's data) were converted to Z scores and averaged within each study reported; then the weighted average for all studies was computed and the resultant z score was converted back to a correlation. The data included in this calculation were Holzinger, Otis IQ and Binet IQ; Newman, Freeman, and Holzinger, Stanford-Binet IQ and Otis IQ; Schoenfeldt. Project Talent IQ Composite; Blewett, Thurstone PMA; Eysenck and Prell, Wechsler-Bellevue; and Herman and Hogben, Otis Advanced Group Test.

[c]From Jencks (1972), Table A-6, column headed "Fraternal Twins." Same studies as in b above.

[d]From Jencks (1972), Table A-12. Included were Juel-Nielsen, Raven's Progressive Matrices, Wechsler-Bellevue Verbal, and Wechsler Bellevue Performance: Newman, Freeman, and Holzinger, Stanford-Binet IQ and Otis IQ; and Shields, Non-verbal Intelligence.

[e]From Jencks (1972), Table A-7. Included were Conrad and Jones; Hart; Madsen: McNemar; Outhit; and Hildreth.

[f]From Jencks (1972), Table A-8 (panel 1). Included from panels 1 and 2 were Stanford-Binet data from Freeman, Holzinger, and Mitchell; Leahy; and Skodak; and from panel 1, Burks's Stanford-Binet data.

and fingerprint ridge count data show a very similar pattern of family resemblance, with the physical measure being more heritable (an estimated range of .68 to .84). The study of twin families seems to be a very appealing one and a design that will yield more defensible estimates of genetic variance than the study of twins alone.

Another study that incorporates both twins and siblings is the Louisville Twin Study (Wilson, 1977). The twins and one of their siblings were tested around the age of 8 years with the WISC. The data are shown in Table 7.5 (combined results of Wilson's Tables 2 & 3). The correlations for siblings are very close to those of DZ twins, with whom they are genetically related as sibs and who are genetically

TABLE 7.4

Regression and Correlation Analyses of Block
Design Test Scores and Finger Print Ridge Counts

	Block Design		Ridge Count	
	Coefficient	N	Coefficient	N
Regressions				
Son/daughter on father/mother	0.28 ± 0.04	572	0.42 ± 0.05	564
Nep/niece on twin uncle/aunt	0.23 ± 0.06	318	0.37 ± 0.05	310
Nep/niece on spouse uncle/aunt	-0.01 ± 0.06	241	-0.06 ± 0.07	247
Offspring on midparent	0.54 ± 0.07	254	0.82 ± 0.07	254
Correlations				
Monozygotic twins	0.68 ± 0.06	65	0.96 ± 0.03	60
Full siblings	0.24 ± 0.08	297	0.36 ± 0.08	296
Half-siblings	0.10 ± 0.12	318	0.17 ± 0.12	310
Father-mother	0.06 ± 0.10	102	0.05 ± 0.10	98

related to each other as ordinary sibs. (Twins were designated *A* and *B* according
to the alphabetical order of their legal first names.) Identical (monozygotic) twins
are far more similar in WISC scores than any of the sibling groups. Wilson
concludes, and we agree:

> The concordance for full-scale IQ and verbal IQ showed no significant change from
> dizygotic twins to sibling pairs to twin–sibling sets, so the unique experiences of
> being born and raised as twins did not promote significantly greater similarity in
> IQ. Nor did the differential experiences of twin versus singleton lead to greater
> disparities in school-age IQ among the twin–sibling sets. From this perspective, the

TABLE 7.5

Within-Pair Correlations on the Wechsler Scale for
Different Sets of Siblings and Twins

	Full Scale IQ	Verbal IQ	Performance IQ	Difference between Verbal and Performance IQ	No. of Sets
Sibling-sibling pairs	.46	.42	.37	.21	56
Sibling and MZ twin A	.41	.46	.24	.24	65
Sibling and MZ twin B	.46	.48	.29	.26	65
Sibling and DZ twin A	.47	.35	.47	.17	53
Sibling and DZ twin B	.28	.39	.09	.24	53
Dizygotic twins (8 years)	.45	.41	.41	.27	71
Monozygotic twins (8 years)	.82	.79	.67	.49	86

In those families where two siblings were available, the sibling closest to age 8 was used.

concordance among age-matched zygotes from the same family was negligibly related to the experiences of being a twin or a singleton. The more potent determinants appeared to be the proportion of shared genes plus the common family environment [p. 214].

Not only are the full-scale, verbal and performance scores per se heritable in a range of .5 to .8, but the *pattern* of verbal and performance scores also shows evidence of moderate heritability. The fourth column in the table gives the correlations of the difference between the two subscores for the related pairs. The first-degree relatives' patterns are correlated about .24, whereas the identical twins' patterns are correlated .49.

MZ Twins Reared "Apart"

We, like De Fries, Vandenberg, and McClearn (1976), believe that adoption studies offer the best evidence for genetic differences in intelligence. Although most laypeople find the study of identical twins reared apart most compelling, there are reasons of nonrandom selection and nonrandom assignment to environments that render the study of MZs apart less useful than research on adopted children. If there were a study of identical twins reared in uncorrelated environments, genetic differences would be controlled, whereas both within-family and between-family environmental variables are free to vary. This would be an ideal study of genetic differences. Unfortunately for science, there are simply too few pairs of MZs reared apart, too peculiarly sampled, to make these subjects useful to social science. The most notable studies of MZ twins reared in separate households for at least most of their growing years (but not entirely separated and not reared in uncorrelated environments) are those of Newman, Freeman, and Holzinger (1937), and Shields (1962). The average IQ correlation of MZs reared in correlated but not the same households is .76, considerably higher than DZ twins reared in the same household. Thus these studies provide some evidence for the importance of genetic differences in intelligence.

Adoptive Family Studies

Adopted children, on the other hand, provide almost as useful data as the rare identical twins reared apart, and they are far more available. Adopted children are not genetically descended from the family of rearing, so that environmental differences between families are not confounded with genetic differences in the children, if the adopted children are randomly placed by adoption agencies. Theoretically, regressions of adopted child outcomes or adoptive family characteristics will provide genetically unbiased estimates of true environmental effects in the population. Unfortunately, adoptive families are selected by agencies for being above average in many virtues, including socioeconomic status. Thus, they are always an unrepresentative sample of the population to which one would like to generalize. Although it is possible that the adoptive family coefficients on background are good estimates of the population values, it is difficult to know without modeling the way in which the families were selected. An easier corrective for the possible bias of selected adoptive families is to have a comparison sample of biologically related children in the same adoptive families or a sample of biological families that are similarly selected.

A complete adoption study design would include comparable information on the intelligence of the natural parents of the adopted children. No study to data has reported IQ data on both the natural mothers and fathers of adopted-away children, but the prospective adoption study being conducted by Robert Plomin and his colleagues at the University of Colorado will provide information on the IQs of both natural parents, as well as the adoptive parents and the adopted children. The studies of Skodak and Skeels (1949) and Horn, Loehlin, and Willerman (1979) reported IQ test scores on the natural mothers of the adopted children. Other studies have been forced to use educational level of the natural parents as indices of intelligence, because there were no IQ tests given to most of the natural parents.

Bias in the Adoption Study Methods?

Comparisons of adopted and biologically related relatives assume that the greater behavioral similarity usually found among biological relatives is due to their greater similarity. Critics of behavior genetic methods assert, to the contrary, that important biases creep into comparisons of genetically related and unrelated families or members of families through parental and child expectations of greater similarity among biological than in adoptive relatives. If biological parents see themselves in their offspring and expect them to develop greater similarity to the parents, then the children may develop more similarly in many ways. Adoptive parents, knowing that there is no genetic link between them and their children, may expect less similarity and thus not pressure their children to become like the parents. The greater expectation of similarity among biological than among adoptive relatives could well bias the comparisons of the genetically related and unrelated families, confounding genetic relatedness with environmental pressures toward similarity that run in the same direction. The greater behavioral similarity of biological relatives might be due as much or more to parental and child expectations as to differences in genetic relatedness.

To test the hypothesis that knowledge of biological or adoptive status influences actual similarity, we (Scarr, Scarf, & Weinberg, 1980) correlated absolute differences in objective test scores with ratings of similarity by adolescents and their parents in adoptive and biological families. Although biological family members see themselves as more similar than adoptive family members, there are also important generational and gender differences in perceived similarity that cut across family type. There is moderate agreement among family members on the degree of perceived similarity, but there is no correlation between perceived and actual similarity in intelligence or temperament. However, family members are more accurate about shared social attitudes.

Knowledge of adoptive or biological relatedness is related to the degree of perceived similarity, but perceptions of similarity are not related to objective similarities and thus do not constitute a bias in comparisons of measured differences in intelligence or temperament in adoptive and biological families.

Three Famous Adoption Studies

As we indicated by the extensive quotations from Woodworth at the beginning of the chapter, the adoption studies of the 1920s and '30s were crucial to the conceptualization of his conclusions about the role of genetic and environmental differences in intelligence in the white, nonethnic population of the United States at that time. Despite their detractors (Kamin, 1974), they are remarkable studies, which the contemporary psychologist is unlikely to appreciate without more intimate contact with them.

Barbara Burks (1928, 1938), Alice M. Leahy (1935), and Marie Skodak and Harold Skeels (Skeels, 1938; Skodak, 1938; Skodak & Skeels, 1949) contributed unique and valuable information to the nature–nurture debate. Their studies supported the following points: (1) the above-average intellectual level of adopted children reared in advantaged homes, and hence the malleability of IQ scores; and (2) the lesser resemblance of adopted than natural children to their adoptive parents, and hence the role of genetic resemblance in intellectual resemblance. It is worthwhile to look more closely at these three studies, particularly in light of Kamin's (1974) criticisms of them.

The Pioneer, Barbara Burks

Burks (1928) set out in her dissertation for Stanford University to answer the question, "To what extent are ordinary differences in mental level due to nature and to what extent are they due to nurture?" Even in 1928, she said; "Few scientific problems have been the subject of so much speculation and controversy. . . . This is probably attributable to two facts: the practical and theoretical significance of the problem itself, and the extreme difficulty of gathering data which cannot be applied with more or less plausibility to the support of either the nature or nurture hypothesis [p. 219]."

In her review of the literature of the time, Burks refers to Galton's *Hereditary Genius,* to other studies of familial genius and feeblemindedness, to the consistent decrease in correlations among relatives for "psychical traits" as genetic relatedness decreases; the consistency in individual test scores over time; and to the marked differences in average intelligence among racial and social class groups. All these phenomena, she said, might conceivably be due either to hereditary or environmental differences or to both. What was (is) needed, Burks proposed, was one additional experimental step for the whole set of data "to become invested with definite meaning"—the isolation of the effects of heredity and environment.

To this end, she conducted her study of 214 adoptive (called foster) families and 105 control (biologically related) families in California, from San Francisco to San Diego. The foster families were selected through adoption agencies' records, to which Burks applied seven criteria of age at adoption, ethnic background, intactness of the family, and accessibility. Control families were matched for intactness, the child's age, sex, and preschool experience, and the parents' ethnicity, locality, type of neighborhood, and father's occupational field. Burks and her two assistants traveled the state to visit each family and to spend 4 to 8 hours assessing the mental level of the child and both parents, the cultural and material level of the home, and parental ratings of the children's character and temperament. In her report of the study, Burks was meticulous in telling the reader of the problems of sample recruitment, assessment procedures, and the like. Her data analyses were brilliant and far advanced over most behavioral scientists of the time. Her use of multiple correlations and path analysis, then being developed by the geneticist Sewell Wright at Wisconsin, should amaze contemporary readers, for there were no electronic calculators or computers available to her; all the computations were done by hand. More important, her conclusions were appropriate and balanced, despite the gross misrepresentations of her work in secondary sources:

> Reference should be made to the educational opportunities of the children examined, which were good, . . . If the children had varied considerably in educational opportunity, . . . and if, in addition, home environment and educational opportunity had been correlated, it would have been quite difficult to separate the effects of the two upon the mental variability or our children. . . .
>
> Thus, the study is based upon children homogeneous as to race and educational opportunity; sufficiently homogeneous in health and physique to avoid confusion; and about as variable in hereditary endowment and in home environment (including kindred social mores) as white children of ordinary communities.
>
> The study does not purport to demonstrate what proportions of the *total* mental development of an individual are due to heredity and to environment. Biologists have frequently pointed out the futility of attempting such a demonstration, since *any development whatever would be impossible without the contributions of both nature and nurture.* But if we direct our attention to the contributions of ordinary differences in heredity and ordinary differences in environment to *mental differences* (i.e., I.Q. variance), it is possible to draw some significant conclusions. The causes which affect human differences, rather than the causes which conditon the absolute developmental level of the human species have, after all, the more vital bearing upon social and educational problems.
>
> Given a group of school children such as our subjects (which surely are representative of the largest single element in the American juvenile population), it will

later be seen that the date gathered in this investigation lead to the conclusion that *about 17 percent of the variability of intelligence is due to differences in home environment.* It will further appear that the best estimate the data afford of the extreme degree to which the most favorable home environment may enhance the I.Q., or the least favorable environment depress it, is about 20 I.Q. points. This amount is larger, no doubt, than some of the firmest believers in heredity would have anticipated, but smaller than the effects often attributed to nurture by holders of an extreme environmentalist's view. To the writer, these results constitute an important vindication of the potency of home environment. But even more significant appear to be the implications of these basic results, e.g., that *not far from 70 percent of ordinary white school children have intelligence that deviates less than 6 I.Q. points up or down from what they would have if all children were raised in a standard (average) home environment;* that, while home environment in rare, extreme cases may account for as much as 20 points of increment above the expected, or congenital, level, heredity (in conjunction with environment) may account in some instances for increments above the level of the generality which are five times as large (100 points) [p. 222–223].

The major data that led Burks to these conclusions are shown in her Tables XI and XXXI, reprinted here as Tables 7.6 and 7.7. In the first table, the distribution of Stanford–Binet IQ scores of the foster and control children are given. The mean of the adopted children's scores was 8 points below that of the control children, but the adopted children's scores were half a standard deviation above the population mean of California schoolchildren on whom this version of the

TABLE 7.6
Intelligence Distribution of Children in I.Q.

	Foster	Control		Foster	Control
175–179	–	–	105–109	32	15
170–174	–	–	100–104	32	13
165–169	–	–	95– 99	27	5
160–164	1	–	90– 94	16	5
155–159	1	1	85– 89	7	–
150–154	1	2	80– 84	2	–
145–149	1	1	75– 79	3	2
140–144	–	2	70– 74	2	–
135–139	3	7	65– 69	1	–
130–134	7	3	60– 64	–	–
125–129	8	10	55– 59	–	–
120–124	16	13	50– 54	1	–
115–119	24	18	45– 49	–	–
110–114	28	8	40– 44	1	–
Mean	107.4	115.4			
S. D.	15.09	15.13			
N.	214	105			

Stanford–Binet was standardized. Burks noted that, based on the facts about the natural parents of the adopted children, their expected mental level was not more than 2 to 3 points above 100. "But the average IQ level actually found in this group was 107. Can this discrepancy be accounted for through superior environmental advantage? Probably it can be [p. 304]." She estimated that the total complex of environmental variables in the adoptive homes was between one-half and one standard deviation above the average. Because the multiple correlation of home environment with foster children's IQ scores was .42 (corrected for attenuation), a positive increment of one-half to one SD in home environment would predict a rise in child IQ of 3 to 6 points, or very close to what was actually found.

The second table shows the correlations between the major parental and home variables and child's IQ. Even a glance will reveal that the IQ scores of biological offspring bear closer resemblance to their parents' intelligence and to the home environment provided by the parents than do the scores of adopted children to their adoptive parents. From these correlations, Burks calculated the multiple correlations of home environment and child IQ.

In the same volume in which Barbara Burks published her research, Freeman, Holzinger, and Mitchell (1928) described their very large study of foster children adopted at any time between early infancy and late adolescence, at an average of 4½ years. Freeman et al., in contrast to Burks, found that foster children bore a remarkable intellectual resemblance to their adoptive parents. Although the authors concluded that home environments accounted for the high degree of resemblance for unrelated parents and children, the specter of the highly selective placement of older children haunted the study. Although it is difficult to predict the intellectual level of an infant, it is not difficult to estimate the IQ of a 5-year-old when he or she has been in the care of an agency for an average of 11 months before placement in the adoptive home.

Alice M. Leahy's Study

Alice M. Leahy (1935), the author of the second important study of adopted children, summarized her predecessors' work as follows:

> In contrast to the low coefficients of correlation between test intelligence of child and foster home found by Burks, Freeman secured coefficients that ranged from .32 to .52 when certain subclassifications were used, and .48 for his entire population. From this he concluded that environment is capable of exercising an influence on mental ability commensurate with that established for true parent and child in which both heredity and environment are operative [p. 248]."

Leahy concluded that selective placement in the Freeman et al. study raised serious and unanswerable questions about the results and planned her own study to avoid the problems of selective placement. She selected families whose

TABLE 1.1

Child's I.Q. Correlated With Environmental and Hereditary Factors*

Factor	Type of r	Foster			Control		
		r	P.E.	N	r	P.E.	N
Father's M.A.	P.M.	.07	.05	178	.45	.05	100
Mother's M.A.	P.M.	.19	.05	204	.46	.05	105
Mid-parent M.A.	P.M.	.20	.05	174	.52	.05	100
Father's vocabulary	P.M.	.13	.05	181	.47	.05	101
Mother's vocabulary	P.M.	.23	.04	202	.43	.05	104
Whittier index	P.M.	.21	.04	206	.42	.05	104
Whittier index (using 5-yr.-olds only)	P.M.	.29	.08	63	–	–	–
Culture index	P.M.	.25	.05	186	.44	.05	101
Culture index (using 5-yr.-olds only)	P.M.	.23	.08	60	–	–	–
Grade reached by father	P.M.	.01	.05	173	.27	.06	102
Grade reached by mother	P.M.	.17	.05	194	.27	.06	103
Parental supervision rating 3 or 4 vs. 5 or 6	B.	.12	.05	206	.40	.09	104
Income	P.M.,K.	.23	.05	181	.24	.06	99
No. of books in home library	P.M.,K.	.16	.05	194	.34	.06	100
Owning or renting home	B.	.25	.07	149	.32	.10	100
No. of books in child's library	P.M.,K.	.32	.04	191	.32	.06	101
Private tutoring (in music, dancing, etc.)	B.						
Boys		.06	.10	77	.43	.11	46
Girls		.31	.08	108	.52	.09	56
Five-year-girls only		.50	.12	31	–	–	–
Home instruction by members of household (hrs. weekly)	P.M.						
Ages 2 and 3		.34	.04	181	-.05	.07	101
Ages 4 and 5 (children over 5)		.15	.06	129	-.03	.08	71
Ages 6 and 7 (children over 7)		.03	.07	88	.24	.09	46
Ages 2 and 3 (5-yr.-olds only)		.18	.09	51	–	–	–
Ages 4 and 5 (5-yr.-olds only)		.13	.09	52	–	–	–
Father's rating of child's intelligence	P.M.	.49	.04	164	.32	.06	98
Mother's rating of child's intelligence	P.M.	.39	.04	181	.52	.05	101

*The following abbreviations are used in this table: M.A. for mental age. P.M. for product-moment correlation. B. for biserial correlation. K. for Professor Kelley's auxiliary score method.

adopted children were placed before 6 months of age and control (biologically related) families with identical environmental distribution. She, like Burks, limited her sample to white Northern European children, legally adopted by married couples who were still together at the time of the study. She matched the control group on the child's sex and age, parents' school attainments, father's occupation, and residence in towns where educational opportunities would be advantageous.

Like Burks, Leahy combed her state of Minnesota to find and interview the 193 adoptive and 193 control families. Three interviews and test sessions were held with each family, for a total of 5 to 7 hours. Both parents and children were given mental tests and interviews to assess the psychological status of the child and the qualities of the home environment. Also, like Burks', this work was Leahy's PhD dissertation, an effort that shames most contemporary theses.

By limiting her sample to very early adoptions and by sampling in the state of Minnesota, Leahy excluded all children who were for any reason suspected of possible mental abnormality, either because of parental problems or because of early suspicions about the child (pp. 273–274). The average Stanford–Binet IQ scores of her adopted and control children were nearly identical, as shown in her Table 3, reproduced here as Table 7.8.

TABLE 7.8
Distribution of IQ of Adopted and Control Children

Stanford-Binet IQ		Adopted children	Control children
160-164		0	2
155-159		1	1
150-154		0	0
145-149		1	2
140-144		2	0
135-139		1	6
130-134		6	7
125-129		11	19
120-124		26	13
115-119		21	19
110-114		31	23
105-109		33	21
100-104		23	26
95- 99		21	24
90- 94		6	17
85- 89		10	5
80- 84		0	7
75- 79		1	2
	M	110.5	109.7
	SD	12.5	15.4
	N	194	194

The IQ scores of both groups of children, reared in very similar, above-average families, were ⅔ of a standard deviation above average. It seems that both groups were environmentally advantaged and genetically select, because the children of feebleminded persons, the insane, and others suspected of immoral (lower-class) behaviors would not have been permitted early adoption under the state laws of the 1920s.

Leahy analyzed her data with correlations for parent–child resemblance in the adoptive and biologically related families. Her Table 11, reproduced here as Table 7.9, gives the results of her study in a form similar to that of Burks' Table XXXI. The results of this study support Burks' conclusions about the predominant role of heredity for individual differences in measured intelligence. The greater resemblance of biologically related children to their parents led Leahy to conclude that:

1. Variation in IQ is accounted for by variation in home environment to the extent of not more that 4%; 96% of the variation is accounted for by other factors.

2. Measurable environment does not shift the IQ by more than 3 to 5 points above or below the value it would have had under normal environmental conditions.

3. The nature or hereditary component in intelligence causes greater variation than does environment. When nature and nurture are operative, shifts in IQ as

TABLE 7.9
Child's IQ Correlated with Other Factors
(r corrected for unequal range in child's IQ)

Correlated Factor	Adopted children			Control children		
	r	P.E.	N	r	P.E.	N
Father's Otis score	.19	.06	178	.51	.04	175
Mother's Otis score	.24	.06	186	.51	.04	191
Mid-parent Otis score	.21	.06	177	.60	.03	173
Father's S.B. vocabulary	.26	.06	177	.47	.04	168
Mother's S.B. vocabulary	.24	.06	185	.49	.04	190
Mid-parent S.B. vocabulary	.29	.06	174	.56	.03	164
Environmental status score	.23	.06	194	.53	.03	194
Cultural index of home	.26	.06	194	.51	.04	194
Child training index	.22	.06	194	.52	.04	194
Economic index	.15	.06	194	.37	.04	194
Sociality index	.13	.06	194	.42	.04	194
Father's education	.19	.06	193	.48	.04	193
Mother's education	.25	.06	192	.50	.04	194
Mid-parent education	.24	.06	193	.54	.03	194
Father's occupational status	.14	.06	194	.45	.04	194

great as 20 IQ points are observed with shifts in the cultural level of the home and neighborhood.

Any additional feature of Leahy's analyses, which were in general less sophisticated than Burks' earlier work, was the comparison of IQ scores of the adopted and control children by the occupational status of their fathers in the family of rearing. Her remarkable results led Barbara Burks (1938) to reanalyze her own data in similar fashion. Burks' tabulations of her own and Leahy's data are given in Table 7.10.

Leahy found that the average IQ scores of adopted children reared in professional families exceeded those of adopted children reared in slightly skilled and unskilled families by 5 IQ points. Children born and reared in families of the same social class groups differed on the average by 16.5 IQ points. Burks found the same 5-IQ-point difference for adopted children and a 12-point difference for genetic offspring of similar families. These data laid the groundwork for contemporary studies of the effects of family environments (Scarr & Weinberg, 1978; Scarr & Yee, 1980).

Skodak and Skeels' Natural Mothers and Adopted Children

Perhaps, the most widely cited of the early adoption studies in the contemporary literature is that of Skodak and Skeels (1949), a follow-up study on 100 children adopted away from mothers whose IQ they had tested at the time of the child's delivery. Like Burks, they found a substantial gain in the children's IQ scores over what would have been expected had the children been reared by their mothers. Jensen (1973c) has estimated that the children would have averaged IQ 96 under natural-parent rearing, whereas they actually scored IQ 107 at the average age of 13 years, when tested on the same version of the Stanford–Binet on which their mothers had averaged IQ 86. On the 1937 version of the Binet they scored an average of IQ 117. Altogether their intellectual development in the adoptive families was prodigious and spoke well for the adoptive home environments. Table 7.11 (their Table 4) gives the results.

The relative benefit of the home environments was, however, very much tempered by the genetic endowment of the child. Not only did the children's IQ scores correlate more highly with the biological mothers' than with the educational levels of the adoptive families, but there was a definite association between the IQ level of the child and the IQ level of the natural mother, regardless of the advantages of the home environment. Skodak and Skeels' Table 15 (here Table 7.12) shows the results of comparisons between children of mothers of "inferior and of above average intelligence." By 4 years of age (Test 2), the children of natural mothers with IQ scores above average were performing on IQ tests well above the children of mothers with low scores. Although the samples are small, the data are compelling.

TABLE 7.10

Means and Dispersions of Intelligence Scores by Occupational Group

| | Stanford Study occupation of father (or foster father) | | | | Minnesota Study occupation of father (or foster father) | | | | |
	Profess. I	higher bus., semi-prof. II	Lower bus. III	Skilled labor IV	Profess. I	Bus. Mgr. II	skilled trades & clerical III	Semi-skilled IV	Slt. skilled & day labor V
Foster children									
Mean (IQ)	109.1	108.6	108.0	104.6	112.6	111.6	110.6	109.4	107.8
S.D. (IQ)	17.2	14.5	14.3	16.7	11.8	10.9	14.2	11.8	13.6
No.	32.0	47.0	41.0	43.0	43.0	38.0	44.0	45.0	24.0
Control children									
Mean (IQ)	118.7	118.5	115.5	106.1	118.6	117.6	106.9	101.1	102.1
S.D. (IQ)	15.4	12.2	18.6	12.4	12.6	15.6	14.3	12.5	11.0
No.	18.0	33.0	27.0	18.0	40.0	42.0	43.0	46.0	23.0
Foster parents*									
Mean	221.8	207.3	201.2	184.7	59.6	59.6	49.6	39.7	38.4
S.D.	22.6	30.8	29.7	30.3	8.0	6.7	11.9	12.3	11.2
No.	24.0	40.0	34.0	34.0	–	–	–	–	–
Control parents*									
Mean	221.6	221.8	192.0	176.2	64.6	57.1	51.8	44.0	38.3
S.D.	24.4	30.4	33.4	31.6	5.4	10.0	11.5	11.5	9.0
No.	18.0	32.0	27.0	18.0	–	–	–	–	–

*In the case of the Stanford study, data are for Stanford-Binet mental age in months of foster fathers and control fathers. In the case of the Minnesota study, data are for mid-foster parent and mid-parent point score on the Otis Test of Mental Ability.

TABLE 7.11
IQ Scores of Adopted Children at Ages from
Two Years to Thirteen Years

Test	Age	Mean IQ	SD	Range	Median
I	2 yrs. 2 mo.	117	13.6	80-154	118
II	4 yrs. 3 mo.	112	13.8	85-149	111
III	7 yrs. 0 mo.	115	13.2	80-149	114
IV (1916)	13 yrs. 6 mo.	107	14.4	65-144	107
IV (1937)	13 yrs. 6 mo.	117	15.5	70-154	117

From Skodak and Skeels (1949) (Their Table 4)

TABLE 7.12
Comparisons between Children of Mothers of Inferior
and of Above Average Intelligence

Case No.	True mother's IQ	True mother's educ.	Foster mid-par. educ.	Foster father occup.	Test I	Test II	Child's IQ Test III	Test IV	Test IV'37
				Group A					
8B	64	8	16	I	126	125	114	96	106
10B	64	11	8	III	125	109	96	87	100
18B	65	8	9	VI	114	102	112	122	118
53G	63	8	13	III	127	121	119	101	111
54G	67	9	12	III	116	113	113	91	102
58G	54	8	13	III	117	114	119	98	113
60G	66	8	10	V	105	109	90	105	115
67G	65	6	12	IV	110	111	114	95	103
70G	63	1	10	II	110	113	107	101	118
76G	67	7	15	I	109	92	87	74	84
82G	53	3	12	IV	81	87	80	66	74
Mean	63	7	12	3.2	113	109	105	96	104
Median	64	8	12	III	114	111	96	96	106
				Group B					
17B	128	12	12	III	120	128	148	127	145
22B	109	13	11	III	102	107	113	108	130
57G	109	13	16	III	99	126	139	132	130
61G	109	13	15	II	112	113	125	128	135
71G	113	12	19	II	128	112	114	114	122
72G	110	12	8	VI	116	92	105	103	104
73G	105	8	9	IV	125	111	129	110	131
87G	109	13	11	III	128	145	125	119	133
Mean	111	12	12.5	3.3	116	117	125	118	129
Median	109	12.5	11.5	III	117	112.5	125	117	130

Using the Skodak and Skeels data and scores from her own University of California Guidance Study, Majorie Honzik (1957) published two of the most reprinted graphs in the history of psychology. Undeterred by predecessors, we reprint them once again for their dramatic illustration of the effects of genetic resemblance on intellectual resemblance. Figures 7.11 and 7.12 (Honzik's Figures 2 & 4) show that with increasing age, biological offspring come to resemble their parents. Adoptive parents' educational levels (here a proxy for intelligence) are hardly related to the intelligence of the children they rear, unless there is a genetic relationship. These data are entirely consonant with our later studies of adoptive and biologically related families with young children and with adolescents. To paraphrase the punch line; "As a twig is bent, so grows the tree, so long as the twig is genetically related to the bender."

Surely, the other moral from this study concerns the high average IQ scores of the children. If we assume that the natural mothers were an intellectually average sample of the general population, as was later found (Pearson & Amacher, 1956) for unmarried mothers in Minnesota for 1948–52, then the adopted children's IQ scores are not surprisingly high. But the IQ scores of the mothers are unwarrantedly low, because the tests were given soon after the delivery of the child and relinquishment. Who would not be depressed? In addition, one might venture the guess that the largely rural Iowa mothers were culturally disadvantaged on the Stanford–Binet; their genetic value did not shine through on the test as much as it would have, had they been reared in circumstances more congruent with the samples of knowledge on the test.

Kamin (1974) poses objections to each of the three studies described here. Matching adoptive and control families on socioeconomic variables left other

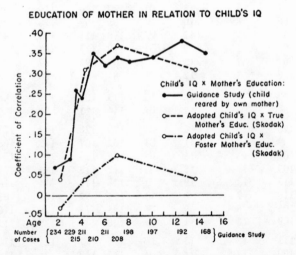

FIG. 7.11.   Education of mother in relation to child's IQ. (From Honzik, 1957).

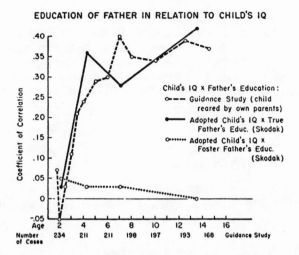

FIG. 7.12. Education of father in relation to child's IQ. (From Honzik, 1957).

variables unmatched. Adoptive parents were several years older on the average than biological parents; income levels for older fathers were higher, even when occupational levels were matched; adopted children had fewer siblings than children in biological families. Kamin implies that these sources of differences between biological and adoptive families contribute to, or indeed account for, the large differences in resemblance between the adoptive and biological families. More recent research will speak to many of his proposed "explanations," but even within these older studies one can ask how reasonable it is to raise questions about the effective environments of adopted and biologically related families. Is there any evidence that parent–child resemblance depends on the number of siblings, when the family sizes varied from 1 to 4 or 5? And isn't it just as likely that adopted children in smaller families would come to resemble their parents more, because of closer association with the adults in the household? Aren't older parents supposed to spend more time with their children? Hence again, one would predict greater resemblance between the adopted children and their older, less frivolous parents. After all, adoptive parents chose to have the child, which certainly cannot be said for many biological parents!

Alas, it is useless to argue with ad hoc, environmentalist arguments that employ adventitious variables to discredit well-designed research. Peter Urbach (1974) summed up the plight of naive environmentalism in the face of the evidence for *some* genetic variance in intelligence:

The hereditarian programme has anticipated many novel facts. . . . When the environmentalist programme has attempted to account for the novel facts produced by the hereditarian programme, it has been unable to do so except in an ad hoc fashion [pp. 134–135]. In this part of my paper I have considered some of the predictions made by the environmentalist programme, especially in regard to social class and racial differences in average IQ. I have shown that almost none of these predictions has been confirmed and that when predictions have failed, environmentalists have rescued their theories in an ad hoc fashion. This patching-up process has left the environmentalist programme as little more than a collection of untestable theories which provide a "passe partout which explains everything because it explains nothing" [p. 253].

Recent Twin and Family Studies

In this section we review several studies of twins and siblings from the past 4 years (1976–80). After a brief overview of the findings, we take closer looks at the Hawaii Family Study, the Texas Adoption Project, and the Minnesota Adoption Studies.

Plomin (1980) compiled recent studies of twins and families and reported family correlations for seven groups, as shown in Table 7.13. Unfortunately, the "intelligence" measures are diverse and the age groups of relatives vary, so that it is difficult to interpret the results; for example, the higher DZ same-sex twin correlation (.62 for late adolescents who took the NMSQ examination) compared with the nontwin sibling coefficient (.31 for early adolescents on the first principle component of the Hawaii Family Study battery of 15 cognitive tests). Plomin's comparisons would imply that, unlike Wilson's (1977) results, the twin experience per se increases similarity above that of ordinary siblings, but the samples, measures, and social class distributions are different, so that comparisons are difficult to interpret.

It is impressive that so many pairs of relatives have been studied in the past 4 years. In fact, the three family studies reviewed here in more detail have contributed *all* nontwin data, with the exception of 94 pairs of adoptive mothers and children (Fisch, Bilek, Deinard, & Chang, 1976).

The Hawaii Family Study of Cognition

In the Hawaii Family Study, a battery of 15 tests of specific cognitive abilities was administered to members of 1816 intact families with 6581 members living on the island of Oahu. Families consisted of both biological parents and one or more children 13 years or older. The families were recruited through church groups, community groups, and various other means.

The scores for the cognitive battery were age corrected, and the resemblances of various biological family members were calculated. The study is reported most fully by DeFries, Johnson, Kuse, McClearn, Polovina, Vandenberg, and Wilson (1979). That report presents spouse correlations, single parent–single child correlations, regressions of midchild on midparent, and sibling correlations for the two largest ethnic groups in Hawaii—Americans of European ancestry (AEA) and of Japanese ancestry (AJA).

The goal of the study was to ascertain the degree of "familiality" for a battery of specific cognitive abilities. There was no variation in the degree of familial relatedness or the degree of environmental similarity.

TABLE 7.13
Correlation Coefficients for New "Intelligence" Data

	Correlation	N/pairs
Genetically identical		
Same individual tested twice	.87	456
Identical twins reared together	.86	1,300
Genetically related (first degree)		
Fraternal twins reared together	.62	864
Nontwin siblings reared together	.31	455
Parent-child living together	.35	2,715
Parent-child separated by adoption	.29	342
Genetically unrelated		
Unrelated children reared together	.25	553
Adoptive parent-adopted child	.15	1,578

[a]The correlation of .87 was obtained over an average period of 15 months for scores of 456 individuals on the first principal component derived from the 15 tests of specific cognitive abilities used in the Hawaii Family Study of Cognition (Kuse, 1977).
[b]From Loehlin and Nichols (1976, Table 4-10).
[c]From DeFries et al. (1979, Table IX). Average weighted correlation for 216 brother-sister pairs, and 125 sister-sister pairs in 830 families of European ancestry for the first principal component score derived from scores on the 15 tests of specific cognitive abilities in the Hawaii Family Study of Cognition.
[d]From DeFries et al. (1979, Table VIII). Average weighted correlation for 672 father-son pairs, 692 mother-daughter pairs, 666 mother-son pairs, and 685 father-daughter pairs in 830 families of European ancestry for the first principal component score derived from scores on the 15 tests of specific cognitive abilities in the Hawaii Family Study of Cognition.
[e]From the Texas Adoption Study (DeFries & Plomin, 1978, Table 9)
[f]Average weighted correlation for three recent adoption studies: Texas Adoption Study, .28 for 282 pairs of unrelated children reared in the same family (DeFries & Plomin, 1978, Table 9); a transracial adoption study, .33 for 187 pairs (Scarr & Weinberg, 1977, Table 6); and an adoption study of adolescents, -.03 for 84 pairs (Scarr & Weinberg, 1978).
[g]Average weighted correlation for four recent adoption studies: Texas Adoption Study, .18 for 541 adoptive mother-adopted child pairs, .12 for 454 adoptive father-adopted child pairs (DeFries & Plomin, 1978, Table 9); a transracial adoption study, .23 for 109 adoptive mother-adopted child pairs, .15 for 111 adoptive father-adopted child pairs (Scarr & Weinberg, 1977, Table 4); an adoption study of adolescents, .09 for 184 adoptive mother-adopted child pairs, .16 for 175 adoptive father-adopted child pairs (Scarr & Weinberg, 1978); and a study by Fisch and coworkers, .08 for 94 adoptive mother-adopted child pairs (1976).

TABLE 7.14

Single-Parent/Single-Child Correlations for Cognitive Test and
Principal Component Scores in Total Hawaiian AEA and AJA Samples[a]

Tests and composites	AEA				AJA			
	Father-son	Mother-daughter	Mother-son	Father-daughter	Father-son	Mother-daughter	Mother-son	Father-daughter
Tests								
Vocabulary	0.31	0.30	0.35	0.32	0.38	0.36	0.40	0.31
Visual Memory (immediate)	0.03	0.11	0.08	0.10	0.01	0.07	0.08	0.17
Things	0.19	0.27	0.22	0.19	0.18	0.19	0.15	0.23
Mental Rotations	0.20	0.30	0.13	0.20	0.20	0.11	0.17	0.24
Subtraction and Multiplication	0.29	0.24	0.21	0.14	0.19	0.22	0.12	0.09
Elithorn Mazes ("lines and dots")	0.11	0.17	0.16	0.07	0.13	0.09	0.15	0.19
Word Beginnings and Endings	0.21	0.25	0.19	0.27	0.25	0.22	0.25	0.17
Card Rotations	0.26	0.34	0.21	0.23	0.24	0.17	0.10	0.11
Visual Memory (delayed)	0.14	0.10	0.16	0.22	0.13	0.10	0.07	0.07
Pedigrees	0.27	0.33	0.28	0.29	0.18	0.38	0.27	0.27
Hidden Patterns	0.24	0.32	0.24	0.27	0.09	0.19	0.13	0.22
Paper Form Board	0.28	0.33	0.29	0.35	0.21	0.29	0.20	0.27
Number Comparisons	0.25	0.24	0.21	0.17	0.25	0.07	0.20	0.24
Social Perception	0.10	0.17	0.11	0.14	0.10	0.22	0.10	0.16
Progressive Matrices	0.23	0.25	0.32	0.25	0.09	0.25	0.24	0.20
Composites								
Spatial	0.33	0.38	0.29	0.31	0.26	0.22	0.20	0.32
Verbal	0.24	0.26	0.29	0.32	0.38	0.36	0.33	0.31
Perceptual Speed and Accuracy	0.30	0.29	0.22	0.17	0.25	0.08	0.13	0.15
Visual Memory	0.11	0.12	0.15	0.18	0.06	0.10	0.11	0.10
First principal component (unrotated)	0.30	0.40	0.35	0.35	0.25	0.34	0.31	0.27
Number of pairs	672	692	666	685	241	248	244	237

[a] AEA and AJA are Americans of European and Japanese ancestry, respectively. Correlations greater than 0.09 are significantly ($p \leq 0.05$) different from zero for AEA subjects, whereas those greater than 0.14 are significant ($p \leq 0.05$) for AJA subjects.

Spouse Correlations

The resemblance between husbands and wives in this study was lower than has been reported in many other studies (Jensen, 1978). Jensen reported a mean spouse correlation of .45 for 43 studies of intelligence. The age-adjusted spouse correlations for the Hawaii study's first principle component (most similar to IQ) were .23 and .15 for the AEA and AJA samples, respectively. The spouse correlations for the individual tests in the battery ranged from − .08 (Subtraction and Multiplication) to .29 (Pedigrees, a verbal reasoning task). The authors believe that their spouse correlations are lower than the reported literature, because: (1) the age corrections lowered spouse correlations because spouses are highly correlated for age; (2) spouses may not be as highly correlated for specific abilities as for g or verbal intelligence, although their general factor correlates .73 with the WAIS full-scale IQ; (3) the families who participated may have been drawn from a restricted range of the intellectual population distribution. As volunteers for social science studies are always self-selected samples, it is very likely that the volunteers for the Hawaii study also represented a biased sample.

Parent–Child Correlations

Table 7.14 gives the correlations of single parents and single children for the test battery, the composite factors, and the first principle component (the closest measure to IQ). As one can see, the correlations for vocabulary and the first principle component are fairly high (.3 to .4), whereas the family resemblance for many of the other tests are low. Because the samples are very large, the standard errors around the correlations are quite low. It seems from this study and from the Minnesota Adoption Study (Carter–Saltzman, 1978) of adolescents' and their parents' specific cognitive abilities that biological family correlations for specific abilities are dramatically lower than for measures of g.

The regression of offspring (either single-child or midchild) on midparent value is a direct estimate of h^2, i.e.,

$$b_{\mathrm{op}} = \frac{\frac{1}{2}V_A (1 + t)}{\frac{1}{2}V_P (1 + t)} = \frac{V_A}{V_P} = h^2$$

The regression of midchild on midparent value was used as the principal index of parent–child resemblance. The authors point out, however, that between-family environmental influences may be an important source of variance for tests of mental ability. For this reason the regression coefficients presented here should be regarded only as measures of familiality and not as direct estimates of heritability.

The corrected coefficients in Table 7.15 may be compared across tests for evidence of differential familiality. The Spearman rank correlation between the

TABLE 7.15

Regressions of Midchild on Midparent for Cognitive Test and
Principal Component Scores in Total Hawaiian AEA and AJA Samples[a]

	Uncorrected		Corrected[b]	
Tests and composites	AEA	AJA	AEA	AJA
Tests				
Vocabulary	0.64	0.55	0.67	0.57
Visual Memory (immediate)	0.15	0.12	0.26	0.21
Things	0.41	0.35	0.55	0.47
Mental Rotations	0.43	0.40	0.49	0.45
Subtraction and Multiplication	0.38	0.34	0.40	0.35
Elithorn Mazes ("lines and dots")	0.24	0.23	0.27	0.26
Word Beginnings and Endings	0.39	0.42	0.55	0.59
Card Rotations[c]	0.46	0.30	0.52	0.34
Visual Memory (delayed)	0.31	0.18	0.50	0.29
Pedigrees	0.52	0.45	0.72	0.63
Hidden Patterns[c]	0.45	0.27	0.49	0.29
Paper Form Board	0.51	0.46	0.61	0.55
Number Comparisons	0.38	0.29	0.46	0.36
Social Perception	0.26	0.18	0.38	0.26
Progressive Matrices[c]	0.52	0.24	0.60	0.28
Composites				
Spacial[c]	0.60	0.42	0.64	0.45
Verbal	0.54	0.48	0.61	0.55
Perceptual Speed and Accuracy	0.41	0.34	0.46	0.38
Visual Memory	0.31	0.18	0.43	0.25
First principal component[c] (unrotated)	0.60	0.42	0.62	0.43
Number of families	830	305	830	305

[a]AEA and AJA are Americans of European and Japanese ancestry, respectively. Standard errors for AEA and AJA coefficients range from 0.04 to 0.05 and from 0.06 to 0.08.
[b]Corrected for test reliability.
[c] AEA and AJA regression coefficients significantly ($p \leq 0.05$) different.

AEA and AJA corrected coefficients across the 15 individual tests of the total
AEA and AJA samples is .76. With regard to the four varimax rotated principal
component scores, the regressions are moderately high for verbal and spatial
measures and somewhat lower for visual memory and perceptual speed in both
ethnic groups. In contrast to this finding of differential familiality, Loehlin and
Nichols (1976) found no evidence of differential heritability among various
mental ability tests when they analyzed extensive twin data. Such results suggest
that it may be between-family environmental variance that differs in relative
importance among various measures of mental ability.

As noted previously, the unrotated first principal component score may be
used as a measure of general intelligence. Parent–offspring regressions (uncorrected) for this measure are .60 and .42 for the total AEA and AJA samples,

respectively. These values are similar to the estimate of .46 obtained by Williams (1975) for WAIS/WISC full-scale IQ. Because the regression of off-spring on midparent value provides an "upper-bound" estimate of heritability, assuming positive environmental covariance, these results indicate that the heritability of general intelligence in these populations is less than .6.

Sibling Resemblance

If between-family environmental influences contributed equally to parent–child and sibling resemblance, the difference between single-parent/single-child and sibling correlations could be used to obtain an estimate of dominance variance. However, because siblings share a rearing environment, whereas parents and children do not, it seems logical to assume that between-family environmental influences may contribute more to sibling than to parent–child resemblance.

Sibling correlations for the total AEA and AJA samples are presented in Table 7.16. Because the AJA sibling correlations are based on relatively small sample sizes, they are less reliable. The AEA correlations are somewhat more reliable and correspond reasonably well with the single-parent/single-child correlations. Median AEA sibling correlations are .25 (brother–sister), .26 (brother–brother), and .16 (sister–sister); corresponding values for single-parent/single-child correlations are .23 (father–son), .25 (mother–daughter), .21 (mother–son, and .22 (father–daughter). This comparison suggests that between-family environmental influences on specific cognitive abilities are not more important for sibling resemblance and that dominance variance for such characters is not relatively large.

The Hawaii Family Study has also reported on the nearly identical factor structures of cognitive abilities in the two major ethnic groups, AEA and AJA (DeFries, Vandenberg, McClearn, Kuse, Wilson, Ashton, & Johnson, 1974) and across the ages of the children and parents sampled in their study.

The authors have also tried to estimate genetic and environmental parameters from their data on biological families, but we find the assumptions tenuous and the results questionable. For these results, see DeFries, Johnson, et al. (1979).

The Hawaii Family Study is the largest study of its kind: the specific cognitive abilities of hundreds of families in two large ethnic groups in a setting where the environments of the participants are not notably disadvantaged. The lack of relatives who vary in genetic or environmental relatedness limits the inferences that can be made from the study, however.

Texas Adoption Project

The Texas Adoption Project was begun in 1973 soon after the discovery of an adoption agency that had routinely administered over 1000 IQ and personality tests to the unwed mothers in their care. The agency administered these tests in

TABLE 7.16
Sibling Intraclass Correlations (± SE) for Cognitive Test and Principal Component Scores in Total Hawaiian AEA and AJA Samples[a]

Tests and composites	AEA			AJA		
	Brother-sister	Brother-brother	Sister-sister	Brother-sister	Brother-brother	Sister-sister
Tests						
Vocabulary	0.33 ± 0.06	0.32 ± 0.09	0.38 ± 0.09	0.51 ± 0.11	0.17 ± 0.15	0.39 ± 0.16
Visual Memory (immediate)	0.21 ± 0.07	0.04 ± 0.09	0.02 ± 0.08	0.06 ± 0.12	-0.08 ± 0.12	-0.08 ± 0.13
Things	0.17 ± 0.07	0.28 ± 0.10	0.14 ± 0.08	0.43 ± 0.11	0.05 ± 0.14	0.26 ± 0.17
Mental Rotations	0.25 ± 0.07	0.35 ± 0.09	0.16 ± 0.09	0.18 ± 0.12	0.16 ± 0.15	0.24 ± 0.17
Subtraction and Multiplication	0.29 ± 0.07	0.38 ± 0.09	0.26 ± 0.09	0.38 ± 0.12	0.14 ± 0.15	0.35 ± 0.17
Elithorn Mazes ("lines and dots")	0.11 ± 0.07	0.06 ± 0.09	0.07 ± 0.08	0.34 ± 0.12	0.07 ± 0.14	0.00 ± 0.14
Word Beginnings and Endings	0.27 ± 0.07	0.33 ± 0.09	0.25 ± 0.09	0.30 ± 0.12	0.11 ± 0.15	0.41 ± 0.16
Card Rotations	0.33 ± 0.06	0.17 ± 0.09	0.25 ± 0.09	0.27 ± 0.12	0.33 ± 0.15	0.26 ± 0.17
Visual Memory (delayed)	0.14 ± 0.07	0.19 ± 0.09	0.13 ± 0.08	0.21 ± 0.12	0.01 ± 0.13	-0.05 ± 0.13
Pedigrees	0.22 ± 0.07	0.28 ± 0.10	0.16 ± 0.09	0.28 ± 0.12	0.02 ± 0.14	0.36 ± 0.17
Hidden Patterns	0.30 ± 0.07	0.28 ± 0.10	0.18 ± 0.09	0.43 ± 0.11	-0.05 ± 0.12	0.04 ± 0.15
Paper Form Board	0.35 ± 0.06	0.26 ± 0.10	0.20 ± 0.09	0.28 ± 0.12	0.08 ± 0.14	0.19 ± 0.17
Number Comparisons	0.27 ± 0.07	0.19 ± 0.09	0.23 ± 0.09	0.38 ± 0.12	0.08 ± 0.14	0.14 ± 0.16
Social Perception	0.12 ± 0.07	0.09 ± 0.09	0.14 ± 0.09	0.31 ± 0.12	0.27 ± 0.15	0.19 ± 0.17
Progressive Matrices	0.20 ± 0.07	0.19 ± 0.09	0.16 ± 0.09	0.33 ± 0.12	0.12 ± 0.15	0.41 ± 0.16
Composites						
Spatial	0.36 ± 0.06	0.29 ± 0.10	0.25 ± 0.09	0.39 ± 0.12	0.17 ± 0.15	0.29 ± 0.17
Verbal	0.27 ± 0.07	0.30 ± 0.10	0.26 ± 0.09	0.55 ± 0.10	0.21 ± 0.15	0.44 ± 0.16
Perceptual Speed and Accuracy	0.33 ± 0.06	0.29 ± 0.10	0.28 ± 0.09	0.34 ± 0.12	0.19 ± 0.15	0.32 ± 0.17
Visual Memory	0.22 ± 0.07	0.19 ± 0.09	0.10 ± 0.08	0.16 ± 0.12	0.01 ± 0.13	-0.10 ± 0.12
First principal component (unrotated)	0.31 ± 0.06	0.36 ± 0.09	0.25 ± 0.09	0.53 ± 0.11	0.01 ± 0.13	0.44 ± 0.16
Number of families	216	114	125	66	44	37

[a] AEA and AJA are Americans of European and Japanese ancestry, respectively.

order to provide their clients with occupational and educational counseling. The test data were also used to provide adoptive parents with some general information concerning the background of their adoptive children.

An unwed mother from the adoption agency was included in the sample of biological mothers only if an IQ score was available for her in the files and she had been tested between 1963 and 1971. Of the 1381 eligible unwed mothers with IQ test scores, 364 were included in the final sample. The 300 adoptive families were administered IQ tests by 22 different licensed psychologists across the State. The average age of the adopted children was about 8 years, with a range of 2 to 20 years. The adoptive families were well above average socioeconomically and intellectually. In this study the biological mothers of the adopted children also came from advantaged families, because the private home for unwed mothers from which the sample was drawn asked that the families of the unwed mothers contribute "significant amounts of money to offset the costs of caring for their daughters [Horn et al., 1979, p. 182]." The average IQ of the natural mothers of the adopted children in this sample was 108.7 on the Beta. The average IQ scores of the adoptive parents on the same test was 113.8 and on the WAIS 113.9. The IQ scores of the adopted and biological offspring of the adoptive families are given in Table 7.17.

Unlike the Burks (1928) and the Minnesota Adoption Studies described later, but consonant with Leahy's (1935) results, the Texas Adoption Project found no average difference between the IQ scores of the adopted and biological children of the same families.

Parent–Child Correlations

Table 7.18 gives the correlations for adoptive and biological parents with their children. Although the biological relatives tend to have higher correlations with their offspring than the adoptive relatives, all of the correlations are quite low. The magnitude of the IQ correlation of the biological mother with her adopted-away child is as high as the correlations of the adoptive parents with their own

TABLE 7.17
IQs of Children, by Test[a]

	Adopted children			Biological children		
	Mean	SD	N	Mean	SD	N
All children						
WAIS	111.0	8.69	5	112.9	8.60	22
WISC	111.9	11.39	405	111.2	11.55	123
S-B	109.2	13.22	59	113.8	11.18	19

[a]WAIS given to children aged 16 or older. WISC to children aged 5–15. Stanford-Binet to children aged 3 and 4—with one or two exceptions at borderline ages. S–B IQs adjusted (see text).

TABLE 7.18
Correlation of Parent's Beta IQ with Child's IQ Tests[a]

	Child test			
	Wechsler performance IQ		Wechsler or Binet total IQ	
Correlational pairing	r	N	r	N
Adoptive father and				
biological child	0.29	144	0.28	163
adopted child	0.12	405	0.14	462
Adoptive mother and				
biological child	0.21	143	0.20	162
adopted child	0.15	401	0.17	459
Unwed mother and				
her child	0.28	297	0.31	345
other adopted child in same family	0.15	202	0.19	233
biological child in same family	0.06	143	0.08	161

[a]N's refer to the number of pairings (= the number of children)—the same parent may enter more than one pairing. In the case of twins, the second twin was excluded from the unwed mother-other child comparisons.

offspring. There is evidence for considerable selective placement in these families, as the IQ score of the biological mother of one adopted child in a family is correlated .19 with the IQ score of another (unrelated) adopted child being reared in the same family and .08 with the biological child of the family that is rearing her adopted-away child.

Sibling Correlations

The overall IQ correlation of biological offspring of the adoptive parents reared together was .35. Genetically unrelated children reared together had IQ scores that correlated .26, a statistically lower coefficient, but quite high! Table 7.19 gives the sibling data.

The parent–child data from this study suggest moderate heritability for IQ. The means of the groups of unwed mothers, adoptive parents, and biological and adopted children of the adoptive families do not agree. The authors speculate about the reasons for the higher than expected IQ levels of the adopted children, and suggest several reasons for the lack of mean differences between the adopted and biological children of the adoptive families.

A path model is presented for the parent–child data. As the authors point out, their initial path model may not be the most appropriate analysis of the data, but it implies a narrow heritability (additive genetic variance only) of .45 to .53; because there is very little estimated dominance variance, the same estimates are appropriate for broad h^2, as well.

TABLE 7.19
IQ Correlations Among Biological and Adoptive Siblings[a]

Correlational pairing	Verbal IQ		Performance IQ		Wechsler or Binet IQ	
	r	df/df w b	r	df/df w b	r	df/df w b
Among biological children	0.14	40/35	0.33	40/35	0.35	46/39
Among adopted children	0.19	132/121	0.05	132/121	0.22	167/150
Between biological and adopted	0.21	159/97	0.24	159/97	0.29	197/116
All unrelated children	0.21	266/195	0.18	266/195	0.26	330/235

[a]r = intraclass or interclass correlations (see text). df$_w$ = degrees of freedom within families = $\Sigma (n_i - 1)$, where n_i is the number of children in family i entering into the correlations. df$_b$ = degrees of freedom between families = number of families entering into the correlation − 1. For twins, only the first member of the pair was included.

Environmental and Genetic Effects on Adoptees' IQ Scores

Willerman (1979) divided the Texas adoptees into two extreme groups by natural mothers' IQ scores. For the low IQ group he selected natural mothers with IQ scores below 95, and for the high IQ group mothers with IQ scores of 120 or above. Table 7.20 shows the adoptive midparent IQ, the mean IQ scores of the offspring of the high and low IQ natural mothers and the percentage of the adopted children with IQ scores below 95 or equal to or above 120.

Although there is some evidence of selective placement for the adopted children, because the adoptive parent groups differ by 4 points in IQ, the 13-point IQ difference between the offspring of high and low IQ mothers shows that individuals of different genetic backgrounds differ in their responsiveness to the generally good environments of adoptive homes. Although the offspring of the low IQ mothers score above the population mean, they are not as bright as the offspring of high IQ mothers in similar environments.

TABLE 7.20
IQs of Adoptees as a Function of Biological Mother's IQ

Biological mother (Beta)	Adoptive midparent (Beta)	Adoptee (WISC/Binet)	Adoptees ≥ 120 IQ	Adoptees ≤ 95 IQ
Low IQ (N = 27; M = 89.4)	110.8	102.6	0%	15%
High IQ (N = 34; M = 121.6	114.8	118.3	44%	0%

Note. WISC = Wechsler Intelligence Scale for Children; Binet = Stanford-Binet Intelligence Scale.

The Minnesota Adoption Studies: Transracial Adoption

Two adoption studies were launched in 1974 for two quite different purposes.

The *transracial adoption study* was carried out from 1974 to 1976 in Minnesota to test the hypothesis that black and interracial children reared by white families (in the culture of the tests and of the schools) would perform on IQ tests as well as other adopted children (Scarr & Weinberg, 1976).

In the transracial families were 143 biological children, 111 children adopted in the first year of life (called the Early Adoptees) and 65 children adopted after 12 months of age—up to 10 years at the time of adoption. Most of the later adoptees were in fact placed with adoptive families before 4 years of age, but they were not the usually studied adopted children who have spent all their lives past the first few months with one adoptive family. As we described in an earlier paper (Scarr & Weinberg, 1976), the later adoptees have checkered preadoptive histories.

The 101 participating families included 176 adopted children, of whom 130 were socially classified as black (29 with two black natural parents and 101 with one black natural parent and one natural parent of other or unknown racial background), and 25 as white. The remaining 21 included Asian, North American Indian, and Latin American Indian children. All the adopted children were unrelated to the adoptive parents. Adopted children reared in the same home were unrelated, with the exception of four sibling pairs and one triad adopted by the same families, who were excluded from the analyses of family similarity.

IQ Levels of Family Members

Both the parents and the natural children of the families were found to score in the bright average to superior range on age-appropriate IQ tests. The black and interracial adopted children were also found to score above the average of the white population, regardless of when they were adopted. The black children adopted in the first 12 months of life scored on the average at IQ 110 (Scarr & Weinberg, 1976). This remarkable result was interpreted to mean that adopted children were scoring at least 20 point above comparable children being reared in the black community. We interpreted the dramatic change in the IQ scores, and school performance of the black and interracial children to mean that: (1) genetic racial differences do not account for a major portion of the IQ or academic test performance difference between racial groups; and (2) black and interracial children reared in the culture of the tests and the schools perform as well as white adopted children in similar families (Burks, 1928; Horn et al., 1979; Leahy, 1935; Scarr & Weinberg, 1978). The adopted children scored 6 points below the natural children of the same families, however, as Burks (1928) and our second adoption study also found.

Parent–Child Correlations

Table 7.21 shows the correlations of the parents and children in the transracial adoption study. The adoptive families had adopted at least one black child, but there were also other adopted children and many biological offspring of these same parents. The children ranged in age from 4 to about 18. Because of the age range, children from 4 to 7 years were given the Stanford–Binet, children from 8 to 16 the WISC, and older children and all parents the WAIS. The adopted children averaged age 7, and the natural children about 10.

Table 7.21 shows the parent–child IQ correlations for all the adopted children in the transracial adoptive families, regardless of when they were adopted. The total sample of adopted children is just as similar to their adopted parents as the early adopted group! The midparent–child correlation for all adoptees is .29, and for the early adoptees, .30. Mothers and all adopted children are equally similar, and fathers more similar than they are to the early adopted children.

Table 7.21 also shows the correlations between all adopted children's IQ scores and their natural parents' educational levels. Because we did not have IQ assessments of the natural parents, education is used here as a proxy. Despite this limitation, the correlations of natural parents' education with their adopted-away offspring's IQ scores are as high as the IQ correlations of biological parent–child pairs and exceed those of the adopted parent–child IQ scores. The midnatural parent–child correlation of .43 is significantly greater than the midadopted parent–child of .29.

Because the adoptive parents are quite bright, their scores had considerably restricted variance. In Table 7.21 the correlations between parents and their natural and adopted children are not corrected for restriction of range in the

TABLE 7.21
Comparisons of Biological and Unrelated Parent-Child IQ
Correlations in 101 Transracial Adoptive Families

	N (pairs)	r
Parents-Biological Children		
Adoptive mother-own child	141	.34
**Natural mother-adopted child	135	.33
Adoptive father-own child	142	.39
**Natural father-adopted child	46	.43
Parents-Unrelated Children		
Adoptive mother-adopted child	174	.21 (.23)*
**Natural mother-own child of adoptive family	217	.15
Adoptive father-adopted child	170	.27 (.15)*
**Natural father-own child of adoptive family	86	.19

*Early Adopted Only (N = 111)
**Educational level, not IQ scores

parents' IQ scores. When corrected, the correlations of biological offspring with their parents rise to .49 and .54, and the midparent–child correlation is .66. Adopted child–adoptive midparent IQ resemblance rises to .37 (Scarr & Weinberg, 1977). When the IQ scores of the parents are corrected for restriction of range, the magnitude of the resemblance between biological parents and children reared together exceeds that of the natural parents' educational level and the IQ scores of the adopted-away offspring, but the latter are still higher than the correlations of corrected IQ score correlations for the adoptive parents and adopted children.

The correlations between natural parents of adopted children and the biological children of the same families is an estimate of the effects of selective placement. If agencies match educational and social class characteristics of the natural mothers with similar adoptive parents, then the resemblance between adoptive parents and children is enhanced by the genetic, intellectual resemblance of natural and adoptive parents. Selective placement also enhances the correlation between natural parents and their adopted-away offspring, because the adoptive parents carry out the genotype–environment correlation that would have characterized the natural parent–child pairs, had the children been retained by their natural parents. Thus, neither the adoptive parent–child correlations nor the natural parent–adopted child correlations deserve to be as high as they are. In another paper (Scarr & Weinberg, 1977), we adopted the solution proposed by Willerman 1977, to subtract half the selective placement coefficient of .17 from both the adoptive parent–child correlation and the natural parent–adopted child correlation. There are other corrections that could be justified by the data set, but we will leave the "ultimate" solution(s) to biometricians. Our simple figuring of these data yields heritabilities of .4 to .7.

Sibling Correlations

In Table 7.22, the sibling correlations reveal a strikingly different picture. Young siblings are quite similar to each other, whether genetically related or not! The IQ correlations of the adopted sibs, genetically unrelated to each other, are as high as those of the biological sibs reared together. Children reared in the same family environments and who are still under the major influence of their parents score at similar levels on IQ tests. The IQ correlations of the adopted sibs result in small part from their correlations in background, such as their natural mothers' educational levels (.16) and age at placement in the adoptive home (.37), which is in turn related to the present intellectual functioning of the children—the earlier the placement the higher the IQ score. Age of placement is itself correlated with many other background characteristics of the child and is a complex variable (Scarr & Weinberg, 1976). It seems that some families accepted older adoptees and others didn't, and that the families differed on the average in the rearing environments they provided. But note that the correlation among

TABLE 7.22
Sibling Correlations
Natural and All Adopted Children of Adoptive Families

Natural Sibs	N (Pairs)	r
All IQ scores	107	.42
Stanford-Binet	10	.50
WISC + WAIS	63	.54
Natural Sib-Adopted Sib		
All IQ scores	230	.25
Stanford-Binet	57	.23
WISC + WAIS	63	.20
Natural Sib-Early Adopted Sib (All IQ scores)	34	.30
All Adopted Sibs		
All IQ scores	140	.44
Stanford-Binet	36	.31
WISC + WAIS	50	.64
Early Adopted Sibs (All IQ scores)	53	.39

the early adopted siblings is fully .39! Even among the families who had early adoptees, differences in family environments and selective placement account for an unexpectedly large resemblance between unrelated children.

The major point is that the heritabilities calculated from young sibling data are drastically different from those calculated from the parent–child data. As Christopher Jencks pointed out in his earlier book (1972), the correlations of unrelated young siblings reared together do not fit any biometrical model, because they are too high. This study only makes the picture worse.

The Adolescent Adoption Study

This study was conceived to assess the cumulative impact of differences in family environments on children's development at the end of the child-rearing period (Scarr & Weinberg, 1978; Scarr & Yee, 1980). All the adoptees were placed in their families in the first year of life, the median being 2 months of age. At the time of the study they were 16 to 22 years of age. We administered the short form of the WAIS to both parents and to two adolescents in most of the 115 adoptive families. A comparison group of 120 biological families had children of the same ages. Both samples of families were of similar socioeconomic status, from working to upper middle class, and of similar IQ levels, except that the adopted children scored about 6 points lower than the biological children of similar parents. Table 7.23 gives these results.

TABLE 7.23

Means, Standard Deviations, and Correlations of Adoptive and Biological Family Characteristics

Biological Children (N = 237)

	1	2	3	4	5	6	7	8	9	10	11	Mean	S.D.
1 Child's IQ												112.82	10.36
2 Father's Education	.26											15.63	2.83
3 Mother's Education	.24	.51										14.68	2.24
4 Father's Occupation	.10	.61	.36									62.47	24.73
5 Family Income	.22	.44	.39	.47								24,987.34	8,770.43
6 Birth Rank	-.19	.01	.02	.01	.00							1.62	0.63
7 Family Size	-.21	-.36	-.36	-.30	-.25	.08						3.85	1.48
8 Father's IQ	.39	.56	.43	.37	.38	-.00	-.30					118.02	11.66
9 Mother's IQ	.39	.24	.46	.13	.19	.03	-.10	.20				113.41	10.46
10 Natural Mother's Age	-.10	-.04	.03	.12	-.02	-.11	-.04	-.10	.03				
11 Natural Mother's Education	.21	.33	.24	.29	.43	.09	.14	.20	.10	.07			
12 Natural Mother's Occupation	.12	-.00	.13	.11	.06	-.06	.11	.11	.15	.28	.33		

Adopted Children (N = 150)

	1	2	3	4	5	6	7	8	9	10	11
Mean	106.19	14.90	13.95	60.30	25935.00	1.43	2.87	116.53	112.43	22.46	11.97
S.D.	8.95	3.03	2.06	24.14	10196.78	0.57	1.20	11.36	10.18	5.80	1.66

$r \geq .16$, $p < .05$

Parent–Child and Sibling Correlations

Table 7.24 gives the parent–child and sibling correlations for the WAIS IQ and the four subtests on which it is based. The parent–child IQ correlations in the biological families are what we were led to expect from our earlier study and others—around .4 when uncorrected for the restriction of range in the parent's scores. The adoptive parent–child correlations, however, are lower than those of the younger adopted children and their parents. And the IQ correlation of adopted children reared together is zero! Unlike the younger siblings (who, after all, are also of different races), these white adolescents reared together from infancy do not resemble their genetically unrelated siblings at all.

The IQ "heritabilities" from the adolescent study vary from .38 to .61, much like the parent–child data in the study of younger adoptees, but very unlike those data on younger sibs.

Our interpretation of these results (Scarr & Weinberg, 1978) is that older adolescents are largely liberated from their families' influences and have made choices and pursued courses that are in keeping with their own talents and interests. Thus, the unrelated sibs have grown less and less alike. This hypothesis cannot be tested fully without longitudinal data on adopted siblings; to date all the other adoption studies sampled much younger children, at the average age of 7 or 8. We can think of no other explanation for the markedly low correlations between the adopted sibs at the end of the child-rearing period, in contrast to the several studies of younger adopted sibs, who are embarrassingly similar.

Effects of Family Background on IQ, Aptitude, and Achievement Scores

For contrast with the material that is forthcoming, let us look first at the effects of family environments on young adoptees' IQ scores differences. Table 7.25 shows two regression equations, one for the biological children and one for the early adopted children of the transracial adoptive families. The predictive variables are more substantially related to the IQ scores of the biological children, with the R^2 of .30, compared to an R^2 of .156 for the young adoptees. The major difference in the two equations is the predictive value of the parents' IQ scores for the biological children's IQ scores. The IQ scores are correlated, of course, with parental demographic characteristics, whose coefficients are pulled in a negative direction when they coexist in the equation.

Now, let us look at similar data for the adolescent adoptees and their biological, comparison families. The adolescents' IQ, school aptitude, and achievement test scores were regressed on family demographic characteristics, sibling order, and parental IQ. The adopted adolescents' scores were regressed on those variables plus the natural mother's age, education, and occupational status. The goal of these analyses was to estimate how much the indexed differences in family environments contribute to individual differences in IQ and school test

TABLE 7.24

Correlations Among Family Members in Adoptive and Biologically-Related Families (Pearson Coefficients on Standardized Scores by Family Member and Family Type) for Intelligence Test Scales

Child Score	Reliability (*)	Biological (120 families)				Adoptive (104 families)			
		MO	FA	CH	MP	MO	FA	CH	MP
Total WAIS IQ	(.97)	.41	.40	.35	.52	.09	.16	-.03	.14
Subtests									
Arithmetic	(.79)	.24	.30	.24	.36	-.03	.07	-.03	-.01
Vocabulary	(.94)	.33	.39	.22	.43	.23	.24	.11	.26
Block Design	(.86)	.29	.32	.25	.40	.13	.02	.09	.14
Picture Arrangement	(.66)	.19	.06	.16	.11	-.01	-.04	.04	-.03

___ = *biological* > *adoptive correlation, p* < .05

Sample Sizes: Pairs of Family Members

	Biological				Adoptive			
	MO	FA	CH	MP	MO	FA	CH	MP
Children	270	270	168	268	184	175	84	168

Assortative Mating

	Biological FA-MO	Adoptive FA-MO
WAIS IQ	.24	.31
Arithmetic	.19	-.04
Vocabulary	.32	.42
Block Design	.19	.15
Picture Arrangement	.12	.22
Sample Size	120	103

MO = mother-child; FA = father-child; CH = child-child; MP = midparent-child
*reliability reported in the WAIS manual for late adolescents

TABLE 7.25
Regressions of Child IQ on Family Demographic Characteristics,
and Parental IQ in Transracial Adoptive Families
with their Own Children

	Biological Children (143)		Early Adopted Children (111)	
	B	beta	B	beta
Mother's IQ	.474	.32	.141	.13
Father's IQ	.513	.40	-.028	-.02
Father's Education	.682	.14	.389	.09
Mother's Education	-.943	-.15	1.501	.25
Father's Occupation	-.174	-.23	.008	*
Family Income	.445	.06	-.371	-.06
Total R^2	.301		.156	
Shrunken R^2	.269		.116	

*$F < .01$, variable did not enter the equation.

scores. The contribution of genetic differences to test score differences is grossly underestimated by this procedure, because the only parental scores available are WAIS IQ for the biological parents. There are no comparable data on the natural parents of the adopted children nor are there school test scores on any of the parents. Nonetheless, it is interesting to examine the pattern of R^2's obtained from the regression of the IQ, aptitude, and achievement scores on social and genetic background. Table 7.26 gives a summary of the regression analyses. (Detailed versions of the regressions are given in Scarr & Yee, 1980).

Let us concentrate on the adoptive families first. Because the parents in this case provide only the social environment, it is possible to estimate the effects of differences in these environments, which range socioeconomically from working to upper middle class. The R^2 values, shrunken for each equation, give the estimated percentages of variance in test scores accounted for by socioeconomic differences *between* families—that is, those social environmental features that siblings share—and by environmental differences between siblings *within* the same families, which are indexed here by sibling order (in biological families this would be called birth order).

Between-Family Effects

The most striking result is that differences in adoptive families' income, parental education, fathers' occupations, and parents' IQ scores account for none of the variance in their adolescents' IQ scores. In fact, the uncorrected R^2 for the regression of adopted adolescents' IQ scores on their adoptive parents' characteristics is only .02, which shrinks to $-.01$ with correction. This means that differences among families' social class and intellectual *environments* have virtually no effect on IQ differences among their children at the end of the child-

TABLE 7.26
R^2 Estimates of the Effects of Social Environmental and
Genetic Differences on IQ, Aptitude, and Achievement
Test Scores (Stepwise Regressions)

				Shrunken R^2's				
		WAIS	*Aptitude*			*Achievement*		
		IQ	*Verbal*	*Num.*	*Total*	*Read*	*Math*	*Total*
Adopted Adolescents N =		*150*	*147*	*128*	*128*	*140*	*128*	*128*
Step								
	Social Environmental Indices							
1.	Between Families[1]	-.01	.05	.03	.04	.09	.08	.10
2.	Within Families[2]	.02	.02	.00	.01	.01	.05	.03
	Total Environment	.02	.07	.03	.05	.10	.13	.13
3.	*Genetic Indices*[3]	.06	.08	.02	.05	.07	.07	.09
	Total R^2	.08	.15	.05	.10	.17	.20	.22
Biological Adolescents N =		*237*	*231*	*158*	*158*	*195*	*187*	*187*
Social Environmental Indices and Genetic Indices								
1.	Between Families[1]	.26	.19	.13	.18	.14	.14	.18
2.	Within Families[2]	.03	.04	.04	.07	.01	.02	.02
	Total R^2	.29	.23	.17	.25	.15	.16	.20

Notes   1 = parental education, father's occupation, family income, parental WAIS
            IQ's
         2 = sibling order
         3 = natural mothers' education, occupation, and age (to correct for young
            mothers)

rearing period. By comparison, the same variables accounted for 11.6% of the IQ variance among the younger adopted children.

The same analysis for the biologically related adolescents is given at the bottom of Table 7.26. In contrast to .01, their corrected R^2 is .26 for the same measures of between-family differences in social class and parental IQ. This value is identical to the shrunken R^2 for the younger sample of biological children in the transracial adoptive families. In the case of biological children, of course, these differences between families are due to both environmental and genetic differences, the latter being of overwhelming importance in explaining the IQ differences both among younger children and adolescents in these families.

As we move from IQ to school test scores, there are three important trends to notice: (1) the effect of differences in social environments *between* families increases as the tests sample more recently taught material; (2) natural mothers' genetic contribution to test score differences is similar and moderate across the various tests; and (3) the contribution of biological parents' IQ scores to their offsprings' test score differences is far less for school aptitude and achievement tests than for IQ tests.

The first point is that the major difference is explained variance between IQ and school achievement test scores is that social class differences—that is, differences *among* families—account for the majority of the explained variability in achievement scores and virtually none of the IQ differences. It is the social environment differences *among* the adoptive families, indexed by parental demographic characteristics, that contribute most to school achievement differences among the adopted adolescents. In one sense, then, school achievement tests are more biased against working-class environments than are IQ tests!

Natural Mothers' Effects

To test the second point, the effects of genetic differences among the adopted adolescents, the index of genetic differences is admittedly very weak. We have information on only one of the natural parents, and that information is limited to educational and occupational level at the time of the child's birth, and age, which was entered into the regression equations to correct for any underestimation of younger mothers' educational and occupational levels. Regardless of the limitations of those variables, one can see from Table 7.26 that natural mothers' characteristics are substantially related to their offspring's intellectual achievements, even though any variance due to selective placement has been removed by entering social environmental variables into the equations first.

Biological Parents' IQ Effects

On the third point, the predictive power of biological parents' IQ scores, the detailed tables of regression analyses (available in Scarr & Yee, 1980) show that parental IQs decline from 15% of the variance in adolescents' IQ scores (holding everything else in the equation constant) to less than 2% of the variance in aptitude and achievement test scores (again holding constant education, income, and other variables). Parental IQ is by far the best predictor of IQ differences among biologically related children, but parental education and family income are as good predictors of school aptitude score differences and better predictors of school achievement scores. This does not mean that the genetic differences are less important for aptitude and achievement scores, as we can note from both the natural mothers' data and from sibling correlations of test scores to be reported. But it does mean that parental IQ differences are more closely related to their offspring's differences in IQ than in school achievements. If we had obtained reading and mathematics achievement scores for the parents, however, it may well be that the between-family genetic differences would remain relatively constant across the kinds of tests, whereas the impact of social environments would rise, giving a higher total between-family R^2 for achievement than for IQ test scores. From the adopted family results, it is clear that environmental differences among families are a trivial source of IQ differences and a substantial source of differences in school test scores.

Sibling Correlations

Another method for checking on the effects of family environment on test scores is to calculate the correlations between pairs of siblings who are genetically unrelated but who have been reared together from early infancy, as are our adopted children. Their sibling correlations are given in Table 7.27, with the corresponding biological sibling correlations for comparison.

As one can see, the effects of being reared in the same household, neighborhood, and schools are negligible unless one is genetically related to one's brother or sister. The correlations of the biological siblings are modest but statistically different from zero.

With the most simpleminded version of the heritability coefficient and an assumption that parental assortative mating is the same for aptitude and achievement as for IQ, we multiply the difference between the biological and adopted silbings' correlations by 1.6. The heritability estimates vary from .22 to .61, with a median of .37. Although these values are not .8, as some would claim, neither are they zero. There seems to be no consistent difference in heritability by the kind of test.

The negligible differences in heritability for IQ, aptitude, and achievement scores in this study of late adolescents is congruent with Lloyd Humphreys' findings of equal heritabilities for all cognitive measures in the Project Talent data (Humphreys, 1980) and the Texas Adoption Study result of equal sibling resemblances of IQ and school achievement measures in a sample of younger children (Willerman, Horn, & Loehlin, 1977). In other words, there seems to be

TABLE 7.27
Sibling Correlations of IQ, Aptitude, and Achievement Test
Scores of Adopted and Biologically-Related Adolescents

	Biological		Adopted		$h^2 = 1.6(r_{bio}\text{-}r_{adopt})$
	N(pairs)	r	N(pairs)	r	
WAIS Verbal	168	.23	84	.07	.26
Performance	168	.21	84	.07	.22
IQ	168	.35	84	-.03	.61
Aptitude, Verbal	141	.29	68	.13	.26
Numerical	61	.32	49	.07	.40
Total	61	.32	49	.09	.37
Achievement, Reading	106	.27	73	.11	.26
Math	104	.35	58	-.11	.53
Total	104	.33	58	-.03	.58

no greater sibling resemblance for one or another kind of intellectual achievement, when they are all *g* loaded. Humphreys and we agree, however, that some specific skills may have different heritabilities.

It seems to us that the effects of family environments vary with the age of the child and the material sampled on the test. Younger children seem to be far more influenced by differences among families. Children reared in working-class families are more disadvantaged in comparison to upper-middle-class children when the tests sample specifically and recently taught material, that is, by school achievement tests rather than IQ tests. And, finally, we hope you will agree that the evidence from these studies argues for a heritability of intellectual measures in the .4 to .7 range, and not .8.

Other Adoption Studies

One of the most interesting and novel adoption studies was done in France (Schiff, Duyme, Dumaret, Stewart, Tomkiewicz, & Feingold, 1978), of working-class families who had relinquished one child for adoption but retained others to rear themselves. The authors found a small sample of such adopted children of working-class families who had been adopted into upper-middle-class families. The average IQ of the adoptees was 110.6, whereas their nonadopted siblings averaged only 94.7. Similarly, in school the adoptees seldom failed grades (13%), compared to the nonadopted sibs, 55% of whom had failed. The authors concluded that there are no important genetic differences between social groups for factors relevant to school failure. This result from France is in direct contradiction to the social class differences found in the Texas and Minnesota adoption studies. Willerman (1979) suggested that the range of environments in the U.S. studies may be too narrow to provide a test of such social class environmental effects. One is reminded that population statistics are entirely sensitive to the ranges and variances of the genetic and environmental sources of variation, which may well differ from population to population.

From the Netherlands, Claeys (1973) reported on the Primary Mental Abilities and field independence of 84 adopted children. The average level of performance of the adopted children was about 10 points above the levels that are average for the social class from which they came, and equal to or better than the level of biological offspring of the families of the same social level as the adoptive families. Like Schiff et al. in France, Claeys found no evidence for genetic differences between social class groups. One wonders if social mobility by ability has been so restrained in Europe that there is no association between intelligence and social class, as there seem to be in U.S. studies.

Fisch et al. (1976) used another version of an adoption-study design, comparing 94 children adopted by nonrelatives with 50 children kept by their biological mothers, who eventually married men who, as stepfathers, adopted the children. Each adopted child in both groups was matched on gestational age, birth weight, sex, and socioeconomic status of the mother at the time of the birth. Although the authors collected data on physical growth and health status as well, we include here the information on IQ and school achievement. The adopted children did not have higher Stanford–Binet IQ scores at 4 years or WISC IQ's at 7 than their controls matched on their natural mothers' socioeconomic status at the birth; in addition, their IQ scores were not as high as those of the biological offspring families of the same socioeconomic status as their adoptive families. The adopted children did perform better than controls of the SES of their biological mothers on two WISC subtests and on measures of reading and spelling, indicating the effects of adoptive homes on school achievement.

Racial Differences in Intelligence

> In view of all the most relevant evidence which I have examined, the most tenable hypothesis, in my judgment, is that genetic, as well as environmental, differences are involved in the average disparity between American Negroes and whites in intelligence and educability, as here defined. All the major facts would seem to be comprehended quite well by the hypothesis that something between one-half and three-fourths of the average IQ difference between American Negroes and whites is attributable to genetic factors, and the remainder to environmental factors and their interaction with the genetic differences [Jensen, 1973, p. 363].

The evidence to which Jensen refers is: (1) the unbiased nature of cognitive tests (see also Jensen, 1980); (2) the heritability of individual differences within each racial group studied; (3) the inability of environmental factors that account for individual variation within racial groups to account for racial differences between groups; and (4) the poor performance of U.S. black groups compared to Indians and Mexican–Americans, whose social conditions are even worse than those of blacks. Jensen admits that none of these arguments addresses directly the issue of genetics racial differences between blacks and whites in the U.S.

In 1975 Loehlin et al. published their SSRC-sponsored review of the literature on racial differences in intelligence. Their equivocal conclusion that genetic differences may or may not be involved in intellectual differences between the races led many social scientists to accept their view as the most respectable scientific stance of the day. Since that time, three investigations on the possible genetic origins of racial differences in performance on school and IQ tests have rejected the hypothesis of genetic differences as the major source of intellectual differences between the races. First, a study of transracial adoption (Scarr & Weinberg, 1976) showed that black and interracial children reared by socioeconomically advantaged white families score very well on standard IQ tests and on school achievement tests. Being reared in the culture of the test and the school resulted in intellectual achievement levels for black children that were comparable to adopted white children in similar families. Therefore, it is highly unlikely that genetic differences between the races could account for the major portion of the usually observed differences in the performance levels of the two groups.

A second study on the relation of black ancestry to intellectual skills within the black population (Scarr, Pakstis, Katz, & Barker, 1977) showed that having more or less African ancestry was not related to how well one scored on cognitive tests. In other words, holding constant social identity and cultural background, socially classified blacks with more African ancestry scored as highly on the tests as blacks with less African ancestry. A strong genetic difference hypothesis cannot account for this result.

Briefly, blood groups were used to estimate the proportion of each person's African and European ancestry. This is roughly possible because the parent populations differ in the average frequencies of many alleles at many loci and differ substantially at a few loci. Therefore, if a person had a particular allele, we were able to assign a probability that he or she inherited that gene from one of the two populations. Although there is undoubtedly a large error term in these estimates, they had several satisfactory characteristics, such as appropriately large sibling correlations and correlations with skin color. What is most important here is that the estimates of ancestry did not correlate with any measures of intellectual performance in the black sample. Thus, we concluded that degree of white ancestry had little or no effect on individual levels of performance within the black group. We must look to other explanations.

The third study was of black and white twins in Philadelphia (Scarr & Barker, 1981). Briefly, the black 10- to 16-year-olds scored one-half to one standard deviation below the whites on every cognitive measure. The social class differences between the races were not sufficiently large, as Jensen has reminded us, to account solely for the magnitude of this performance differences between the racial groups. The major hypothesis was that black children have less overall familiarity with the information and the skills being sampled by the tests and the schools. By using twins in this study we were able to examine three implications of cultural differences compared to a genetic differences hypothesis. The major predictions of the cultural differences hypothesis are:

1. Black children will score relatively worse on these tests that are more culturally loaded than on more "culture-fair" tests when the instructions for all tasks are equally understood.

2. The cultural differences of the blacks constitute a "suppressive environment" with respect to the development of the intellectual skills sampled by typical tests, and therefore black children will show less genetic variability in their scores and more environmental variability (Scarr-Salapatek, 1971a).

3. Differences among black children will be more dependent on differences among their family environments in the extent to which they aid children in the development of test-relevant skills, and therefore (a) the twin correlations will be higher for black twins; and (b) there will be less difference between MZ and DZ coefficients in the black than in the white groups.

Three major predictions of a genetic differences hypothesis are:

1. Black children will score relatively worse on those tests that are loaded more highly on a g factor than on more verbal, culturally loaded tests.
2. The proportions of genetic and environmental variability will be the same in both racial groups.
3. Family environments will be no more important in black than in white racial groups in determining individual variation.

We believe that the pattern of results supports a general cultural difference hypothesis far better than a genetic differences view. The major intellectual results are:

1. Black children have lower scores on all the cognitive tests, but they score relatively worse on the more culturally loaded on the conceptual tests.
2. The cognitive differences among the black children are less well explained by genetic individual differences, by age, and by social class differences than are those of the white children.
3. The similarity of the black co-twins, particularly the DZs, suggests that being reared in different families determines more of the cognitive differences among black than white children, but that those between-family differences are not those usually measured by SES variables in the white community.

Therefore, we conclude that the results of this study support the view that black children are being reared in circumstances that give them only marginal acquaintance with the skills and the knowledge being sampled by the tests we administered. Some families in the black community encourage the development of these skills and knowledge, whereas others do not. In general, black children do not have the same access to these skills and knowledge as white children, which explains the lower performance of black children as a group. The hypothesis that most of the differences among the cognitive scores of black and white children are due to genetic differences between the races cannot, in our view, account for this pattern of results. Therefore, in three studies, the hypothesis of genetic differences between the races fails to account for the IQ performance differences.

SEPARATE COGNITIVE ABILITIES

Discussions about the behavior genetics of separate cognitive abilities often focus on data from multivariate studies indicating that some abilities that correlate highly with one another are more heritable than other clusters of abilities. There is considerable dispute about the strength of such a claim and about its implications for a theory of intelligence. Many behavior geneticists continue to doubt that different kinds of intellectual functioning are differentially heritable (Loehlin & Nichols, 1976; Nichols, 1978). Despite the general finding of moderate to high correlations among tests of specific cognitive abilities, the existence of sharp discontinuities of intellectual performance have kept alive the hope that there may be discrete and uncorrelated components of information processing that contribute to what we loosely construe as intelligence. The existence of "idiots savants," individuals with subnormal psychometric IQ scores but who excel in specific areas such as music, art, or numerical abilities, has been cited as testimony to the separate abilities approach to intelligence (Anatasi & Levee, 1959; Minton & Schneider, 1980). Additional support for this position has come from evidence of differential rates of decline with age for verbal and spatial abilities (Botwinick, 1977; Wechsler, 1950).

There would be no good rationale for asking questions about genetic and environmental mechanisms for separate cognitive abilities if we believed that all cognitive abilities developed in the same way and responded identically to differences in genetic constitution, biological factors, and environmental circumstances. We know, however, that this is not the case. There is good evidence for discontinuities in cognitive development, not only from the factor analytic studies that are commonly cited, but from studies of cognitive dysfunction. We know, for example, that women with Turner's syndrome, a chromosomal anomaly involving the X chromosome, although normal or above normal in verbal function, often show deficits in spatial thinking. If this genetic anomaly had deleterious effects on intellectual functioning in general, we would expect to see an overall decrement in cognitive skills. Similarly, if the physical consequences of Turner's syndrome (short stature, webbed neck, absence of secondary sexual characteristics, infertility, and others) led to treatment by parents, siblings, and teachers that would depress cognitive performance, it would be expected that the performance depression would not be restricted to spatial thinking.

Overall, attempts to specify and elaborate the routes from genes to behaviors have not met with great success when general level of intellectual functioning has been the object of study. It is possible, however, that the mechanistic approach will have greater utility with respect to specific cognitive abilities. There are certain clinical syndromes with known or suspected genetic etiologies (Turner's syndrome, autism, dyslexia) in which only certain kinds of cognitive functions are disrupted while others are quite intact. There are other conditions with known genetic antecedents in which the precise nature of the cognitive impairment has not been fully explored (PKU, Down's syndrome, Klinefelter's syndrome, and Huntington's chorea, for example). Although the prospect of carrying out the kinds of detailed biochemical analyses on probands and their families may seem arduous and painstaking, it is only by doing such studies that we can hope to understand the biological substrates of abnormal cognitive development and develop effective treatments for affected individuals.

The bulk of recent behavior–genetic research on IQ and general intellectual ability has focused on normal development rather than on clinical populations. The same has not been true in studies of specific cognitive abilities. Researchers seem here to fall into two camps: those who take a population approach and administer cognitive tests to large numbers of individuals who differ in the extent to which they share common genes and common rearing environments; and those who sample from populations with a specific deficit, and either study those individuals intensively or study those individuals *and* their relatives, in an attempt to explore possible biological and sociocultural mechanisms underlying the condition. With the first approach, the goal is to determine what proportion of total variance can be attributed to genetic and nongenetic factors. Although the second approach may also attempt to explain some variance in the population, the major focus seems to be one of developing explanatory, mechanistic models that might lead to effective interventions, at whatever level.

Multivariate Studies of Specific Cognitive Abilities

As mentioned earlier in the chapter, there is no convincing evidence that any one kind of cognitive ability is substantially more heritable than any other. Although some studies have reported greater familiality for verbal abilities, others have found no significant differences.

The most comprehensive and largest twin studies to data have come up with mixed results, and have led us to conclude that the twin data cannot be used to support the notion that there are genetically mediated differences in the relative degree of familial resemblance across different cognitive abilities (Bruun, Markkanen, & Partanen, 1966; Loehlin & Nichols, 1976; Schoenfeldt, 1968). Although the heritability estimates for specific and general cognitive abilities from twin studies do not differ significantly (.42 for specific abilities; .48 for general IQ), an examination of the magnitudes of the MZ and DZ correlations suggests that within-family environmental influences may affect specific cognitive abilities more than general cognitive functioning (Plomin et al., 1980).

Family studies have also come up with inconsistent findings. Williams (1975) reported that family resemblance was greater for Verbal than Performance subscales on Wechsler tests in a study of 55 10-year-old boys and their parents. Verbal test correlations were not consistently higher than correlations for other tests, however, in Loehlin, Sharan, & Jacoby's (1978) study of cognitive performance in 192 Israeli families with two children aged 13 or older.

Data from the largest family study to date, the Hawaii Family Study, discussed in a previous section, have been combined with data from a study of 209 Korean families (Park, Johnson, DeFries, McClearn, Mi, Rashad, Vandenberg, & Wilson, 1978) in Fig. 7.13, in at attempt to present comparable information (the same measures were used in both studies) on three different ethnic groups. After age corrections had been made, midchild scores were regressed on mid-

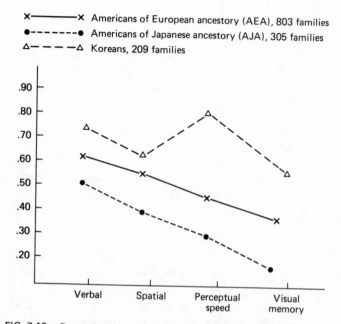

x————x Americans of European ancestory (AEA), 803 families
●-------● Americans of Japanese ancestory (AJA), 305 families
△— — — —△ Koreans, 209 families

FIG. 7.13.   Regressions of midchild on midparent for four factors scores in three ethnic groups. (From DeFries et al., 1970; Park et al., 1978).

parent scores for individual cognitive tests and for factor scores. Both the Americans of European ancestry and the Americans of Japanese ancestry had somewhat higher familial similarity on Verbal and Spatial factors than on Perceptual Speed and Visual Memory (Fig. 7.14). This pattern is not evident for the Korean sample, however, in which, although the value for Visual Memory is the lowest of the four factors, all factors reflect substantial familiality. Park et al. (1978) have suggested that the discrepancy in results might be partially explained by two factors. First, the Korean families were tested in nuclear family groups rather than in large groups consisting of several families, as in the Hawaii study. This would tend to increase between-family differences. Second, assortative mating coefficients for the Korean sample were higher than for the Hawaiian sample. This may be the result of the cultural practice of matchmaking in Korea, and could result in greater genetic variance in that population.

Few adoption studies have looked at specific cognitive abilities, but those that have are consistent with the twin and family studies in that they report significantly higher correlations for biological relatives for adoptive relatives, but do not present convincing evidence that one kind of cognitive process is more heritable than another (Carter-Saltzman, 1978; Claeys, 1973; Scarr & Weinberg, 1978). A reexamination of Table 7.24 will make it clear that although biological relatives are more similar than adoptive relatives on WAIS subtests, the patterns

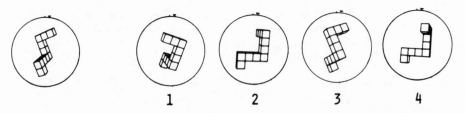

FIG. 7.14.   Sample item from Vandenberg and Kuse's (1978) Mental Rotations Test. (From Vandenberg & Kuse, 1978.)

of correlations (especially when assortative mating coefficients are taken into account) do not allow us to make any strong statements about differential heritability of specific abilities.

Spatial Thinking

It is probably fair to say that spatial abilities have received more attention from behavior geneticists than any other kind of cognitive processes, and there are several reasons why this is so. First, there are substantial individual and group differences in some kinds of spatial thinking. In particular, strong and consistent sex differences in "spatial visualization" have been replicated many times and have led to the formulation and testing of specific genetic models to account for the distributional and family correlational patterns obtained. Second, psychologists have been somewhat more precise in their definitions of spatial ability than in definitions of other cognitive abilities, so the phenotype under consideration can be clearly differentiated from other cognitive phenotypes. Factor analytic studies have greatly facilitated this differentiation, and have even enabled us to distinguish among various kinds of spatial thinking (*not* all of which show the aforementioned sex differences). Finally, some clinical subgroups have been found to show striking patterns of deficits and advantages in spatial thinking. In Turner's syndrome, a known genetic disorder, there is a definite disturbance of spatial thinking. Knowledge that Turner's syndrome women differ from other women by the absence of sex chromatin in their cells (they are missing the second X chromosome) has led investigators to focus attention on possible X-chromosome involvement in spatial thinking. Autistic children, on the other hand, often are exceptionally good at spatial tasks, but do poorly on measures requiring verbal skills. Autism appears to be a disorder with multiple causes, but a genetic etiology seems probable for some children.

Definition of Spatial Thinking

Factor analytic studies since the 1930s have provided strong support for the existence of two separate kinds of spatial abilities, visualization and orientation (McGee, 1979). In a recent review of human spatial abilities, McGee (1979) defines *visualization* as involving: "the ability to mentally manipulate, rotate, twist, or invert a pictorially presented stimulus object. The underlying ability seems to involve a process of recognition, retention, and recall of a configuration in which there is movement among the internal parts of the configuration . . . or the recognition, retention, and recall of an object manipulated in three-dimensional space . . . or which involves the folding or unfolding of flat patterns [p. 893]." *Orientation* is said to involve: "the comprehension of the arrangement of elements within a visual stimulus pattern and the aptitude to remain unconfused by the changing orientation in which spatial configuration may be presented [McGee, 1979, p. 893]." Whether or not either of these abilities bears any relationship to the ability to navigate through three-dimensional space in everyday life has not been investigated.

There are still several unanswered questions about the specificity of the two spatial factors. Many of the studies from which claims of such distinct abilities were drawn involved subjects limited in age range (usually age 10 to 20), and few studies included female subjects or analyzed the results separately by sex. Several investigators have reported positive and significant correlations between the visualization and orientation factors (Borich & Bauman, 1972; Goldberg & Meredith, 1975; Karlins, Schuerkoff, & Kaplan, 1969; Yen, 1975).

Sex Differences

Differences between the sexes, favoring males, have been found across a wide range of spatial tasks by many different investigators. The strongest and most consistent differences have been discerned on tasks that have a visualization component, and this difference is rarely detected before puberty. Although some early sex differences in spatial abilities have been reported (see Vandenberg & Kuse, 1979, for review) the most discriminating tests that tap visualization have proved too difficult for both young boys and girls. An example of such a task is illustrated in Fig. 7.14. A subject is presented with a two-dimensional representation of a three-dimensional figure and is then asked to choose, from an array of similar representations, those that are rotations of the original prototype. The rotations can be either in the picture plane, perpendicular to the picture plane, or conceivably diagonal to the picture plane. This particular measure is Vandenberg & Kuse's (1978) pencil and paper adaptation of Shepard and Metzler's (1971) task that was designed to address questions about information-processing time as a function of degrees of rotation. The most striking sex differences emerge in adolescence, when males move well ahead of their female peers, most of whom never manage to catch up (Harris, 1977, 1979; Maccoby & Jacklin, 1974). It should be noted, of course, that not all females lack spatial visualization ability, although there are more males than females in whom this talent is expressed.

Questions about the nature of the mechanisms underlying this sex difference still remain unresolved, despite considerable vigilance on the part of several investigators. Hypotheses about the effects of sociocultural, physiological, endocrinological, and genetic mechanisms have been tested. We review the genetic models that have been proposed and assess the state of the field in the light of family studies that have been conducted to date. It is sometimes forgotten that behavior genetic methodologies allow us to test environmental models as well as genetic models. Some of the data obtained from family studies have indeed been used to explore the power and specificity of environmental effects on the development of spatial thinking.

X-Linkage Model of Inheritance

Results from twin studies and studies of biological and adoptive families have indicated that spatial abilities are as heritable as other kinds of cognitive abilities (Claeys, 1973; DeFries, Ashton, Johnson, Kuse, McClearn, Mi, Rashad, Vandenberg, & Wilson, 1976, 1970; McGee, 1979; Osborne & Gregor, 1968; Vandenberg, 1962, 1967; Williams, 1975).

Due both to the difficulties inherent in accurate assessment of the phenotype—there is, for instance, no absolute cut-off level on a particular test above which one can reliably diagnose the presence of spatial visualizing ability—and in correct designation of genotype, speculations about the genetic mechanisms influencing sex differences have been somewhat vague. Results from a number of early studies lent support to a hypothesis involving X linkage (Bock & Kolakowski, 1973: Corah, 1965; Hartlage, 1970; Guttman, 1974; Stafford, 1961). This hypothesis was most clearly articulated by Bock and Kolakowski (1973), who proposed that spatial visualization ability, although probably influenced by multiple autosomal alleles, was "enhanced" by a recessive gene located on the X chromosome. The frequency of this gene was calculated to be about .50 in North American white populations. As expected with a gene frequency of .50, about 50% of all males and about 25% of all females in these populations would express the trait, as males express whatever is in their single X chromosome, whether it is dominant or recessive, whereas females would be required to have two of the recessive X's in order for the trait to be expressed (Bock & Kolakowski, 1973).

A large number of studies have found support for the X-linkage hypothesis when considering the distribution of scores. The general result has been that only about one-quarter of females scored above the median score for males on a variety of spatial tests (Bock & Kolakowski, 1973; Bouchard & McGee, 1977; Loehlin et al., 1978; O'Connor, 1943; Yen, 1975).

The patterns of family correlations from early studies also supported the X-linkage hypothesis. The expected pattern of correlations is as follows: Equally high mother–son and father–daughter correlations would be expected because a son receives his only X chromosome from his mother, and a father passes on his only X chromosome to his daughter; the correlation between fathers and sons should be very close to zero because sons receive no X chromosomes from their fathers; and the correlation between mothers and daughters should be intermediate, as neither mothers nor daughters necessarily express what is coded on a single X chromosome. Therefore, the order of correlations for an X-linked trait should be as follows: Mother–Son = Father–Daughter > Mother–Daughter > Father–Son. For sibling correlations it is expected that sisters would be most similar, as they share the paternally derived X, and have a 50% chance of sharing an X chromosome from the mother; brothers would have a 50% chance of having the same maternal X chromosome; and brother–sister pairs would be the least similar, as their 50% chance of sharing the maternally derived X is offset by the certainty that the sister will receive an X from the father, whereas the brother will not. The expected order of sibling correlations, then, is: Sister–Sister > Brother–Brother > Sister–Brother.

Results from the early studies supported an X-linked hypothesis for spatial visualization. Since 1973, however, several studies have tested the X-linkage hypothesis, usually with quite large samples of families, and often with the same measures across studies. For more detailed treatments of these investigations, the reader is referred to recent reviews by McGee (1979) and Vandenberg and Kuse (1979). The overwhelming conclusion that has been drawn from the more recent studies is that the X-linkage hypothesis cannot be supported on the basis of the rank order of intrafamilial correlations. Comparability across investigations has been difficult in the past, in part because each researcher used a different test to measure spatial ability. This has not been true in recent behavior-genetic research, and the correlations from family studies that used the same three tests— Mental Rotations, Card Rotations, and Hidden Patterns—are presented in Tables 7.28 and 7.29. Mental Rotations is a test of spatial visualization; Card Rotations measures orientation ability, and Hidden Patterns measures the ability to pick out a simple line figure from a more complex drawing.

Although the recent family correlation results clearly cannot be used to support an X-linkage hypothesis, nongenetic factors could be invoked to explain the poorness of fit. It is probable that only linkage analysis will be widely accepted as a critical test of the hypothesis. The X-linkage model would seem quite viable if it could be demonstrated that relatives who were identical for some marker known to be located on the X chromosome were also more similar in spatial skills than relatives who differed on the critical marker. Only one such experiment has been done to date, and the results gave tentative support to the involvement of the X chromosome in field dependence, but found no evidence for X linkage of visualization abilities. Goodenough, Gandini, Olkin, Pizzamiglio, Thayer, & Witkin (1977) administered a battery of seven spatial ability tests to sons in 67 Italian families with at least three sons. All the sons were tested also for red–green color blindness and typed on the XG(a) blood group, both phenotypes known to be located at different parts of the X chromosome. For each marker phenotype the spatial test correlations of sons who were identical for the marker were compared to correlations of sons who differed from one another. No evidence of linkage was found in the analyses of color blindness, but sons who

TABLE 7.28
Parent-Offspring Correlations of Spatial Scores

	FA-SON	MO-DAU	MO-SON	FA-DAU
Mental Rotations Test[a] (Spatial Visualization)				
DeFries et al. (1979): AEA Hawaii	.20 (672)	.30 (692)	.13 (666)	.20 (685)
Spuhler (1976): Colorado	.25 (81)	.04 (81)	.10 (81)	.32 (81)
Carter-Saltzman (1977): Minnesota	.04 (94)	.23 (119)	.04 (94)	.17 (119)
McGee (1978): Minnesota	.23 (185)	.16 (196)	.20 (204)	.17 (172)
Guttman & Shoham (1979): Israel	.32 (153)	.21 (182)	.12 (168)	.08 (157)
DeFries et al. (1979): AJA Hawaii	.20 (241)	.11 (248)	.17 (244)	.24 (237)
Park et al. (1978): Korean	.22 (99-103)	.46 (113-121)	.26 (100-105)	.41 (107-117)
Card Rotations Test[b] (Spatial Orientation)				
Spuhler (1976): Colorado	.25 (81)	.03 (81)	.16 (81)	.15 (81)
DeFries et al. (1979): AEA Hawaii	.26 (672)	.34 (692)	.21 (666)	.23 (685)
DeFries et al. (1979): AJA Hawaii	.24 (241)	.17 (248)	.10 (244)	.11 (237)
Park et al. (1978): Korean	.12 (99-103)	.61 (113-121)	.36 (100-105)	.54 (107-117)
Loehlin et al. (1978): Israel	.27 (183)	.40 (201)	.27 (183)	.32 (201)
Hidden Patterns Test[c] (Disembedding Figures)				
DeFries et al. (1979): AEA Hawaii	.24 (672)	.32 (692)	.24 (666)	.27 (685)
DeFries et al. (1979): AJA Hawaii	.09 (241)	.19 (248)	.13 (244)	.22 (237)
Park et al. (1978): Korean	.51 (99-103)	.65 (113-121)	.58 (100-105)	.56 (107-117)
Loehlin, Sharan, & Jacoby (1978): Israel	.40 (183)	.22 (201)	.44 (183)	.38 (201)
Guttman & Shoham (1979): Israel	.24 (153)	.18 (182)	.17 (168)	.19 (157)

[a]Vandenberg (1975), and Vandenberg & Kuse (1978)
[b]French, Ekstrom, & Price (1963), and Ekstrom, French, & Harman (1976)
[c]Thurstone & Thurstone (1941)

TABLE 7.29
Sibling Correlations of Spatial Scores

	SIS-SIS	BRO-BRO	SIS-BRO
Mental Rotations Test[a] (Spatial Visualization)			
Yen (1975): California	.41 (103)	.32 (84)	.27 (191)
Bouchard & McGee (1977): Minnesota	.21 (112)	.50 (132)	.33 (249)
DeFries et al. (1979): AEA Hawaii	.16 (125)	.35 (114)	.25 (216)
DeFries et al. (1979): AJA Hawaii	.24 (37)	.16 (44)	.18 (66)
Card Rotations Test[b] (Spatial Orientation)			
DeFries et al. (1979): AEA, Hawaii	.25 (125)	.17 (114)	.33 (216)
DeFries et al. (1979): AJA, Hawaii	.26 (37)	.33 (44)	.27 (66)
Loehlin et al. (1978): Israel	.52 (51)	.44 (42)	.24 (99)
Hidden Patterns Test[c] (Disembedding Figures)			
Loehlin, Sharan & Jacoby (1978): Israel	.55 (51)	.76 (42)	.39 (99)
DeFries et al. (1979): AEA, Hawaii	.18 (125)	.28 (114)	.30 (216)
DeFries et al. (1979): AJA, Hawaii	.04 (37)	-.05 (44)	.43 (66)

[a]Vandenberg (1975) and Vandenberg & Kuse (1978)
[b]French, Ekstrom & Price (1963) and Ekstrom, French & Harman (1976)
[c]Thurstone & Thurstone (1941)

shared the same XG(a) typing were more similar on the two tests of field dependence (Rod-and-Frame Test and Embedded Figures Test) than sons who differed in blood type. It should be noted that the one test included to measure the visualization factor, Stafford's (1961) Identical Blocks Test, provided no support whatever for the X-linkage model. The absence of positive results, however, does not disprove the hypothesis, inasmuch as a gene influencing spatial thinking could be located on the X chromosome, but be quite far from both markers tested. Only positive results can be interpreted with certainty. Clearly this kind of study needs to be done with new and larger samples. It could prove most informative.

Turner's Syndrome and Spatial Performance

Studies of the cognitive performance patterns of women with Turner's syndrome are of particular interest for several reasons. First, Turner's syndrome has a known genetic antecedent involving the X chromosome. In the classic Turner's syndrome, there is a total absence of sex chromatin; that is, the entire second X chromosome is missing, leaving the patient with a total of only 45 chromosomes (45,X). Some Turner's women are mosaics, with some of their cells showing the classic 45,X patterns, and others having the normal chromosome complement. Still others show variations of the short arm of the second X chromosome: deletions, translocations, or complete absence of the arm (Serra, Pizzamiglio, Boari, & Spera, 1978; Stern, 1973). Because of the X-linkage hypothesis about the transmission of spatial abilities, there has been considerable interest in study-ing possible X-chromosome involvement in spatial thinking. If spatial visualiza-tion were X linked one would expect the spatial performance of Turner's syn-drome women to resemble that of normal males, as any information coded on the X chromosome should be expressed. In contrast to that expectation, Turner's women perform exceptionally poorly, on the average, on tests of spatial think-ing. In addition, this deficit seems to be quite specific to spatial thinking; perfor-mance on tests of verbal ability is at or above normal levels.

Garron (1977) reported on a sample of 67 Turner's women, aged 6 to 31 years, and 67 control subjects who were matched for age, race, education, residence, social class, marital status, and ethnic and religious background. All subjects were administered the version of the Wechsler Intelligence Scale that was appropriate to their ages (WAIS or WISC), and Turner's subjects were classified according to karyotype and presences or absence of a variety of physi-cal stigmata. Garron's general conclusions from the study were as follows (p. 125):

1. There is no increased incidence of either severe or moderate general men-tal retardation, as such is usually understood, among persons with Turner's syndrome.

2. The distribution of intelligence, and the presence of specific cognitive deficits, is similar in all groups of persons with Turner's syndrome, regardless of karyotype and/or somatic stigmata.

3. The cognitive deficits are equally characteristic of children and adults with Turner's syndrome, although the expression of these deficits may be influenced by particular stages of intellectual development.

4. The nature of these deficits may be understood better by an emphasis on the cognitive processes involved, rather than simply by an emphasis on the stimulus attributes.

Although direct comparisons of WAIS Verbal IQ, Performance IQ, and Full Scale IQ revealed significant differences between adult proband and control groups (Table 7.30), comparisons of cognitive factors showed significant dif-ferences in favor of controls for Numerical Ability and Perceptual Organization factors, but not for the Verbal Comprehension factor. Analyses of WISC factors yielded the same results.

When one examines the patterns of Verbal and Performance IQ scores for Garron's older (age 16 and up) and younger subjects, the most striking finding is the large discrepancy between Verbal and Performance scales in the adult pro-bands. It is at puberty that large differences between males and females begin to appear on spatial tests (Maccoby & Jacklin, 1974), and female enrollment in mathematics courses begins to decline (Fennema & Sherman, 1977). Indeed, in

TABLE 7.30
Performance Profiles of Turner's Females and Controls
[Garron (1977)]

	VIQ	PIQ	VIQ-PIQ
WISC			
Probands	95.6	89.1	6.5
Controls	99.4	105.5	-6.1
WAIS			
Probands	104.1	88.4	15.7
Controls	109.5	104.0	5.5

Garron's sample the control subjects showed a reversal of pattern from the younger to the older group, although the size of the Verbal–Performance gap remained constant.

In a study of 13 Turner's probands aged 12 to 22 years, and 13 controls matched on Full Scale IQ, age, race, education, parents' marital status, and SES, Silbert, Wolff, and Lilienthal (1977) also found that Wechsler Performance IQs were significantly depressed relative to Verbal IQs in the Turner's sample. In addition to age-appropriate Wechsler IQ tests, subjects were given six tests of spatial perception and organization, three tests of sensory-motor sequencing, three automatization measures, and three auditory tests. Across all the spatial tests administered, Silbert et al. (1977) reported deficits in Turner's patients only for "tasks requiring the integration of spatial elements into synthetic wholes or the remembering of total spatial configurations [p. 19]." They did not find differences on tests that required that spatial arrays be analyzed and broken down into component parts. There were no differences between groups on the automatization measures, but the Turner's patients again did more poorly than controls on all three auditory measures (Seashore Rhythm Test, Seashore Tonal Memory Test, and Auditor Figure–Ground Test).

The authors concluded that Turner's patients may have a selective deficit of right hemisphere functioning that affects synthetic spatial thinking and configurational auditory processing. The idea is intriguing and surely bears further investigation. It seems the next step would be to assess Turner's subjects with some of the standard measures of differential hemispheric processing: verbal and nonverbal dichotic listening tests; tachistoscopic presentation of verbal and nonverbal information to the right and left visual fields; EEG measurements of cortical activity while carrying out different cognitive operations.

As Serra et al. (1978) suggested, it is possible that the expression of spatial cognitive abilities may be dependent on a normal gonadal hormonal environment, and that in turn may depend on sex heterochromatin in the prenatal period.

Hormonal Models for Differences in Spatial Ability

Models linking spatial performance to hormonal events—prenatally, at puberty, and in adulthood—have made a respectable showing over the past 15 years, and most are still viable today. These models have tried to account for cognitive differences as a function of maturation rate (Waber, 1976, 1977), physical androgyny (Klaiber, Broverman, & Kobayashi, 1967; Mackenberg, Broverman, Vogel, & Klaiber, 1974; Petersen, 1976), and hormonal fluctuations during the menstrual cycle (Englander-Golden, Willis, & Deinstbier, 1976; Klaiber, Broverman, Vogel, & Kobayashi, 1974). Although research in this area is of great importance and will surely enhance our eventual understanding of the biological mechanisms underlying cognitive performance patterns, no studies to data have employed behavior-genetic methods. For recent reviews and critiques of these models the reader is referred to Dan (1979), McGee (1979), Petersen (1979), and Waber (1979).

For the present, we discuss briefly a condition of known genetic origin that results in abnormal hormonal concentrations and synthesize some of the hypotheses about the relationships among sex chromosomes, hormones, and cognitive performance.

Adrenogenital Syndrome (AGS)

The Adrenogenital Syndrome, caused by an autosomal recessive gene, is a metabolic dysfunction involving an enzyme insufficiency that causes the adrenal gland to secrete excessive amounts of adrenal androgens throughout the lifetime. Since 1950, cortisone treatment has been available to patients with this disorder, and the treatment leads to a reduction of androgen secretions. Therefore, most current cases have suffered only prenatal exposure to abnormally high androgen levels (Reinisch, Gandelman, & Spiegel, 1979). The existence of both genotypic males and females who have been exposed to elevated androgen levels at a specific phase of development allows us to examine the specific effects of prenatal androgen on cognitive performance patterns.

Overall, in studies with appropriate controls, no significant differences in specific cognitive abilities have been found between AGS subjects and controls (Baker & Ehrhardt, 1974). Females with AGS do not provide evidence that prenatal exposure to excessive androgen specifically enhances spatial ability. Baker and Ehrhardt (1974) used parents and siblings of AGS patients as controls and found no significant differences between any groups on Verbal or Performance IQ. They did, however, find that the AGS subjects, their parents, and their siblings, all performed above the population norms for IQ. This finding raises the intriguing possibility that both unaffected heterozygotes and homozygotes for AGS share some genetically mediated characteristic that is related to high IQ.

Sex Chromosomes, Hormones, and Cognitive Performance

It has been suggested that the Y chromosome retards maturation rate, permitting fuller expression of the genome in males than females (Ounsted & Tayler, 1972). This would presumably be reflected in more complete penetrance in males (already proposed in some of the genetic models reviewed previously), as well as in greater male variance for a wide array of phenotypes. Wilson and Vandenberg (1978) have reported that in the Hawaii study, variance was greater for males than females on 11 of 15 cognitive tests (the difference was significant for Mental Rotations, Subtraction and Multiplication, and Word Beginnings and Endings). Existing data sets can be examined for sex differences in variances, and the question can partially be addressed in that manner. In addition, it should be possible to look at the rates of physical maturation and cognitive development in individuals with Y-chromosome anomalies.

McGee (1979) has proposed that there may be an X-linked gene that controls the timing of androgen release at puberty and has related his suggestion to the work of Petersen (1976) and Broverman, Klaiber, Kobayashi, and Vogel (1968), who found that high spatial performance was associated with late maturation and low androgenization (determined by physical characteristics) in males, and with highly androgenized body types in females. This notion is consistent with Waber's (1976, 1977, 1979) reports that for both males and females, late maturers performed at higher levels than early maturers on tests of spatial ability. These formulations would be strengthened considerably by data indicating a strong positive relationship between the timing of puberty and ratings of physical androgeny. One would also want to look at familial patterns in the timing of the onset of puberty to see if there were any evidence of X-linked inheritance. To our knowledge such studies have not yet been undertaken.

Studies of the relationship between cognitive performance profiles and hormonal status have not been numerous, but we do not know of a single one in which affected subjects showed higher spatial than verbal performance. Dawson (1967) has reported higher Verbal than Performance scores on Wechsler tests for four different groups of subjects with sex hormone anomalies: genetic males with testicular feminization (Masica, Money, & Ehrhardt, 1969); males with Klinefelter's syndrome; females with Turner's syndrome; and West African males with gynecomastia (breast enlargement) due to kwashiorkor (Dawson, 1967a, 1967b). As already discussed, even AGS subjects had higher Verbal than Performance scores in those instances where differences were found. It may be that an optimal balance of steroid hormones at one or more periods during development is a prerequisite for the development of high spatial ability. Levine (1969) has suggested that early hormones are crucial during some critical periods of neurological differentiation. He proposed that: "the function of gonadal hormones in infancy is to organize the central nervous system with regard to neuroendocrine control of behavior [p. 15]'' and suggested that the differential responsiveness of males and females to externally administered hormones might be dependent on hormonal events occurring in the prenatal and neonatal periods (Levine, 1969).

Visual Perception and Cognition

The existence of individual differences in visual perception is well established (Berry, 1971; Pick & Pick, 1970; Segall, Campbell, & Herskovits, 1966). Strangely enough, no one seems to have tried to relate those differences—for example, in visual acuity, lens pigmentation, illusion susceptibility, persistence of a visual image, brightness judgment, and many other basic aspects of visual processing—to performance on cognitive tests of visuospatial performance. Instead, the very limited literature on individual differences in vision has concentrated on linking such variation to differences in age (see Pick & Pick, 1970, for review), culture (Deregowski, 1973; Segal et al., 1966) or sex (McGuinness, 1976; McGuinness & Lewis, 1976). Yes, there are individual differences in visual perception, but we don't know how such differences might relate to differences in more cognitive processes.

A few studies have been done to investigate the possibility that visual perceptual mechanisms might be affected by genetic variability. Fuller and Thompson (1960, 1978) have reviewed a set of older twin studies (published between 1939 and 1953) and concluded that, amid a host of methodological flaws and outdated techniques, the studies suggest that genetic factors do influence the determination of some aspects of visual perception. To our knowledge almost all the behavior-genetic research published on this topic since 1953 has concerned susceptibility to visual illusions.

Genetic factors have been found to contribute to susceptibility to both primary illusions (those that show a decrease in errors with age) and secondary illusions (those that show an increase in errors with age), and to an illusion that shows a quite complex relationship to age changes. Twin data supporting this general conclusion have been reported for the Mueller–Lyer illusion (Smith, 1949) and the double trapezium illusion (Matheny, 1973), both primary illusions, and for the Ponzo illusion (Matheny, 1971), a secondary illusion.

In a family study including 203 mother–father–offspring triads and 303 sibling pairs, Coren & Porac (1979) found evidence of significant family resemblance for the Muller–Lyer illusion and for the underestimated segment of the Ebbinghaus illusion. In their very informative review of the possible cognitive and sensory processes underlying illusion susceptibility the authors concluded that several heritable optical and neural mechanisms might be responsible for both Muller–Lyer illusion and the underestimated part of the Ebbinghaus, but that no such mechanisms could be invoked to explain Ebbinghaus overestimation. The reader is referred to Coren and Porac (1979) for a brief review of the studies on heritable variation of visual sensory and neural processes. In an earlier paper (Coren & Porac, 1978) the same authors reported that the magnitude of the Ebbinghaus illusion increases with age for the underestimation portion and decreases with age for the overestimation portion. It is not unreasonable to suspect that age differences are also responsive to genetic factors.

One of the most interesting investigations into the relationship of illusion susceptibility to cognitive processes was a co-twin control study done by Matheny (1972a) as a follow-up to his 1971 paper on the Ponzo illusion. In the earlier paper Matheny (1971) found that MZ twins had significantly smaller intrapair differences on the Ponzo than did DZ twins. He later selected 34 pairs of 9- to 11-year-old MZ twins who were discordant on the magnitude of their susceptibility to the Ponzo illusion and examined their differences on WISC subtests in order to test Pollack's (1969) hypothesis that high illusion susceptibility was related to high performance on tests of numerical sequencing and analogical reasoning. The WISC subtests of interest were digit span and similarities. Pollack's hypothesis was confirmed for the female twins (21 pairs) but not for the males (13 pairs). It is a nice demonstration of the relationship between visual perception and cognition, holding genetic factors constant.

Patterns of visual exploration were investigated in a study of 70 MZ and 50 DZ twin pairs between the ages of 5 and 11 years (Matheny, 1972b). The children were given cards with multiple pictures of common objects arranged in various patterns, and they were asked to name each picture. Ratings of twin similarity were based on: (1) whether the twins started at or near the same place in the visual array; and (2) whether they named the pictures according to similar or different spatial patterns. For both older and younger children the MZ twins were significantly more similar than DZs in their visual exploratory strategies. Matheny interpreted the greater similarity of the MZ pairs as a reflection of similarities in genetically mediated cognitive strategies related to memory and intelligence.

Although there has been relatively little work by behavior geneticists on visual perception, the data that exist suggest that both genetic and nongenetic factors are important in the development of perceptual processes, and that such processes are related to cognitive functioning. The field is wide open for a more programmatic approach to the problem.

Sex Differences

Because of the robust sex difference in spatial visualization, we were led to ask about sex differences in visual perception. Are there any? If so, do we know anything about how such differences develop? It will perhaps be surprising to some students of perception and cognition that there is considerable evidence for the existence of sex differences in visual perception (see McGuinness, 1976a, 1976b, for reviews). A few of the differences are: male superiority in both dynamic and static visual acuity (Burg, 1966; Roberts, 1964); longer persistence of visual sensation (as measured by the Ganzfeld and the afterimage) in males than females (McGuinness & Lewis, 1976); female superiority in some perceptual learning and visual discrimination tasks (Laughlin & McGlynn, 1967; Pishkin, Wolfgang, & Rasmussen, 1967; Stevenson, Hale, Klein, & Miller, 1968); greater amount of lens pigmentation for males (Girgus, Coren, & Porac, 1977).

Investigations of sex differences in susceptibility to visual illusions have indicated that although there are no sex differences for two-dimensional, static illusions (Fraisse & Vautrey, 1956; Porac et al., 1979; Pressey & Wilson, 1978), sex differences have been found for Necker cube reversals (Immergluck & Mearini, 1969), an illusion that involves reorganization of a three-dimensional array.

The finding of sex differences only for a three-dimensional illusion is not at all inconsistent with what has been reported for the spatial cognitive tasks. Although sex differences are sometimes found for two-dimensional spatial tests, those findings are less robust and less consistent across studies. The only family studies on visual illusions to date have been limited to two-dimensional illusions.

This brief review of the literature on processes of visual perception and cognition has made one thing abundantly clear: there is much work yet to be done. We know that there are individual differences in some aspects of visual perception, but we know little about the relationship between visual perception and spatial cognition. We also know that genetic differences contribute to differences in perceptual processes, but again, we do not have the data necessary for determining if the same genetic differences affect both perceptual and cognitive processes.

Conclusions: Spatial Thinking

There are clearly large individual differences in spatial thinking. Because so much research effort has been spent on this area, we have chosen to devote considerable attention to it in our chapter. The results of numerous behavior-genetic studies have led us to conlcude that genetic factors do contribute to the observed differences, although it is difficult to come up with a precise estimate of the magnitude of that contribution.

Of all cognitive abilities that have been psychometrically assessed, none has shown greater sex differences, at least in North American Caucasian populations, than spatial visualization. To date, the mechanisms mediating those differences have not been specifically delineated, although several genetic models have been proposed and tested. Some of these models have involved differential penetrance in males and females and have most recently been subjected to complex statistical analyses.

Perhaps the most intriguing (and most widely tested) hypothesis has been the one invoking X linkage, and that question is as yet unresolved. The most recent family correlational data do not favor the hypothesis that there is a recessive gene on the X chromosome that enhances spatial visualization, but distributional data and the results of recent linkage study still do not allow us to discard the idea completely. Although the pattern of cognitive abilities of women with Turner's syndrome seem in direct opposition to the X-linkage hypothesis, it is possible that early (possibly prenatal) hormonal events may set the state for the subsequent unfolding of a pattern of spatial thinking. Turner's syndrome women may differ both from men and from other women with a normal chromosome complement in such hormonal events.

Because there may be some relationship between hormonal status and spatial thinking, it is also possible that the timing of androgen release at puberty and/or the rate of sexual maturation may be related to level of spatial ability. The extent to which genetic factors are important to such a relationship has not yet been addressed with behavior genetic methods. It is only through a longitudinal behavior genetic study that one would be able to demonstrate a clear relationship between hormonal events, spatial performance (at a later point in time), and genetic constitution.

Studies of Turner's syndrome women have allowed us to isolate and test a variety of factors that might relate to spatial thinking. We should do more cognitive testing of subjects with genetic anomalies, particularly those that influence hormonal status. These may provide us with our best opportunities to find out, via experiments of nature, how genotypes are translated into behavioral phenotypes.

The importance of nongenetic factors in the development of spatial thought should not be slighted and can be addressed through classic behavior-genetic studies, cross-cultural studies, and training studies. A combination of these techniques may even lead to new insights about the selection pressures to which subpopulations of our species were subjected in their evolutionary histories.

THE IMPLICATIONS OF INTELLECTUAL DIVERSITY AND GENETIC DIFFERENCES FOR IMPROVING INTELLIGENCE

One implication of evolutionary theories of intelligence, with their emphasis on adaptation and diversity, is that intelligence will not be randomly distributed in the population. Related persons are more likely to share similar talents, be they high g, musical talent, or spatial skills. Two corollaries of this proposition are: (1) more intelligent children tend to be reared by more intelligent parents; and (2) the effects of attempts to improve intelligence are very difficult to predict, because apparent environmental differences in intelligence are most often confounded with genetic differences in intelligence.

As the justifiably maligned Professor Sir Cyril Burt noted a half century ago, more highly intelligent children come from the working than from the middle class; although there are proportionally more highly intelligent children from middle-class families, the smaller numbers of families whom one would consider middle class means that in absolute numbers working-class parents contribute more talented children to the next generation than do middle-class parents. Nonetheless, high intelligence in children is associated with high intelligence in parents, by genetic and perhaps environmental transmission.

The correlation of parent and child intelligence means that more intelligent children also tend to have more intellectually advantaged rearing environments. In some psychological circles this may be called the "double-whammy" effect. In behavior-genetic circles, it is called a genotype–environment correlation of the passive kind. This means that, when one observes that highly intelligent children have parents who read to them, discuss topics of intellectual interest, and listen to their children's opinions, these parental behaviors are not an arbitrary set of parenting skills applied willy–nilly to any child. Rather they are a set of parenting skills used by intelligent parents toward their usually more intelligent children. One is tempted to hypothesize that, if only less intelligent children had parents who read to them, discussed, and listened to the same extent as the parents of more intelligent children, the less intelligent would be like the more intelligent children. The adoptive studies discussed earlier suggest that the effects of parental child-rearing practices per se may be substantial in the early years and wane with the children's advancing independence of familial influences. Early intervention studies do *not* suggest that the magnitude of benefit from trying to teach parents how to stimulate their children intellectually approaches the level of intelligence in children attained by the natural correlation of bright parents rearing their genetically bright children.

In fact, there may be a trade-off between teaching the child apart from the (relatively dull) parents, when the teachers are bright, university-trained people, and training the less bright parent to teach his or her child, when the effects of the teaching will be realized after hours and after the period in which the intervention has been applied—and even be applied by the parent to other children who did not participate in the intervention. There is a trade-off between the quality and intensity of *nonparental* programs and the quantity and extensiveness of *parental* programs.

What one observes about the natural correlation between parental child rearing and child IQ is dependent on genotypic correlations of parent with child, and it is also dependent on the child's genotype for intelligence. Bright children evoke more intelligent environments from others. They also select more stimulation of an intellectual sort from their environments. Even if we control, in an intervention program, the provisions of stimulating environments, we cannot control who gets what from them. In a micro sense, we cannot control the extra stimulation that a bright child gets from that environment by asking smart questions, following complex arguments and presentations, and evoking intellectual interactions from those around him or her.

Thus, it is hard to predict what effects deliberate attempts to improve the intellectual development of children will have. Some beginnings to model the role of environmental interventions on intellectual development follow. First, let us consider the family-based intervention, of teaching parents how to teach their children better intellectual skills.

Quantitative Prediction

Suppose that we find, as we have in research in Bermuda, that some rough-and-ready measures of mothers' teaching skills are strongly related to children's conceptual development, degree of cooperation with adults in learning tasks, and social competencies. Does this mean that teaching other mothers to be better teachers will enhance their children's intelligence, cooperation, and social competencies? I suppose the answer is *yes* but the question of interest is *how much?*

Measured in standard deviation units, the prediction of improvement of a child's scores is a direct function of the unstandardized coefficient from the regression of child scores on mothers' teaching skills. That is, if, as we found, the regression of children's cognitive and social competencies on their mothers' teaching skills is about .67, then every standard deviation improvement in mothers' teaching skills ought to pay off in ⅔ of a standard deviation in children's skills. Let's be concrete with a familiar scale, that of IQ. The regression equation reported predicts that a 1-SD improvement in mothers' teaching skills yields 10 IQ points in the children, a practically important payoff!

Many studies of family situation in the developmental literature stop right there, with the implication hanging and untested. For example, the Caldwell HOME Scale (Caldwell, 1978) is based on such research, which takes naturally occurring correlations of the sort reported and implies that huge gains are to be found in the improvement of mother's interaction or teaching of the child whose family scores are below average on such scales.

Something must be wrong with the predicted payoff of 10 IQ points for every SD improvement in mothers' teaching skills. No study of intervention has shown such striking effects. What's wrong, of course, is that mother's teaching skills are related to many other facts: her home, her whole environment, and most likely her genetic background. Her genotype is correlated with the environment she provides for herself and the child, whose genotype is correlated with both her genotype and her child's rearing environment in the several ways I have described. For example, mother's IQ is correlated with her 24- to 30-month-old's intellectual or social characteristics; although she may be more irritable than before the child was born, she has probably not lost many skills and certainly has not learned any from him or her. The child's IQ, on the other hand, is more likely to be affected directly by the mother's IQ, both genetically and environmentally. Let's also assume that mother's IQ is not directly affected by her skills in teaching a 2-year-old, but rather that her teaching skills are another manifestation of her developed intellect. These reasonable assumptions give us the following model in Fig. 7.15.

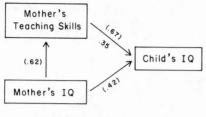

( ) = zero order correlation coefficients

FIG. 7.15. Correlations among child's IQ, mother's teaching skills, and mother's IQ in parentheses; the partial regression coefficient of child IQ on mother's teaching skills with mother's IQ partialled out.

Given the correlation between a mother's teaching skills and her own IQ, and the full impact of the mother's transmission of her IQ to the child by genetic means only (the most conservative assumptions from an intervention point of view), it is still the case that mothers' teaching skills have a partial coefficient of .35 for children's IQ scores; in fact, this figure is obtained after removing the variance in child IQ and teaching skills due to the ethnicity–social class of the family. Thus, an improvement of mothers' teaching skills by 1 SD may well improve children's IQ scores by .35 SD or 5 to 6 IQ points, on the average.

Now let's consider another model where a mother's teaching skills are to some extent a function of the child's responsiveness and cooperation in learning. We know from our mothers' ratings of their children's personalities and from our own ratings of the children's personalities that young children who are seen by their mothers and by us as cooperative, responsive, and not overly active actually receive better teaching in our experimental situation than children who are seen by their mothers and by us as less cooperative, responsive, and overly active. The child plays a role in how well he or she is taught.

Now we have a mediating variable, child personality, between the mother's teaching skills and the child's IQ. Actually, the child's personality affects not just how the mother teaches him or her but how able we are to assess his or her IQ. David Wechsler's view of functioning intelligence as part of general personality and not just cognitive skills is valuable here, because in the real world one has to be able to use effectively what one has cognitively as a necessary but not sufficient set of knowledge and skills for effective intelligence. One's theoretical view of this matter is important for the intervention model one creates to predict the payoff. Consider the following in Fig. 7.16.

The payoff from improving mothers' teaching skills is complicated by the degree to which the program addresses and "improves" the child's cooperation, attention–responsiveness, and lowers his or her activity in learning situations. If the intervention has no effect on children's personality, then the payoff from raising mothers' teaching skills will be considerably less than the .35 SD cited before. (We have not done this calculation.) If, on the hopeful side, mothers who teach better have more attentive, cooperative learners, who will further reinforce

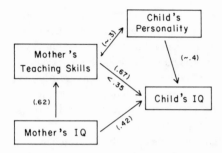

FIG. 7.16. Correlations and regres-
sions of mother and child variables. ( )= zero order correlation coefficients

their mothers' teaching skills, and so forth, the effect of child personality on payoff from the intervention may be minimal or even salutory.

Now let's consider three other possibilities: (1) that mother's improvements in teaching skills are a function (positive or negative) of her initial intellectual level; (2) that the child's gain in IQ is a function of his or her initial intellectual level (positive or negative correlation); and (3) both maternal and child interactions of intellectual level with gain from the program occur. These modifications of the prediction models can be fit on the average results predicted by the former models, but they may be important for policy reasons. If, as rarely happens, those who need the intervention most benefit most from it, one could rejoice especially at the following result, in Fig. 7.17.

Although the best slope might be much the same in the posttest as in the pretest, the R^2 of the child's IQ regressed on maternal teaching skills would be greatly reduced by the restriction in variance caused by the negatively correlated gains of both mother and child with their initial scores.

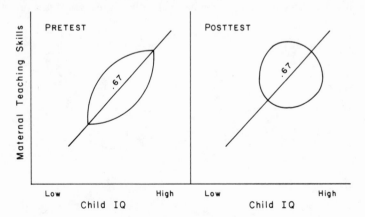

FIG. 7.17. Hypothetical distributions of scores on pre- and posttests if mothers with the lowest initial scores gained the most from intervention.

How about the more usual and opposite result? Although everyone benefits from the intervention to some extent, those that "got" get more from the program, *a la* Sesame Street? Here is that result in Fig. 7.18.

Those who had initially high scores gained a lot from the program; those who had low initial scores gained only a little. The only way to keep the bright from getting more from absolutely everything is to lock them in closets while you teach the rest. Otherwise, they will use their time to learn more and more efficiently than others.

On the other hand, Richard Snow (in press) has shown that certain forms of teaching actually are better geared to slow learners and impede the learning efficiency of faster learners subjected to the treatment. Making instructional programs highly *explicit* and redundant enhances the learning of those who do not ordinarily make connections and generalizations and impedes the learning of those who do. If an intervention program is particularly geared to the learning of mothers who do not teach well because they do not know how to make teaching sufficiently explicit for young children and because they do not spontaneously analyze the tasks that they want to teach, then it may be possible to bore brighter mothers into not gaining and enhance the learning of less bright mothers. Given the social goals of most intervention programs, we ought to gear them to people who need them most, taking into account the functional relationships between the teacher's and the child's characteristics and the functional level of the teacher's intelligence.

Implications

The major implications of this loose thinking about the effects of interventions on the amelioration of intelligence is that: (1) we must take into account the genetic nature of differences in intelligence; and (2) we must recognize that genotype–

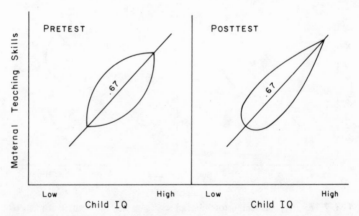

FIG. 7.18. Hypothetical distribution of scores on pre- and posttests if mothers with the highest initial scores gained the most from intervention.

environment correlations are of major importance in understanding the efficacy of interventions. People are different and they possess information from their environments differently. Predictions and expectations from attempts to ameliorate intelligence must take realistic account of these facts.

As Jencks (in press) and Axelrod and Scarr (in press) have noted, the correct statistic for predicting the effects of environmental interventions is not heritability, but according to the latter authors, the reaction range.

What one wants to know about the probable effects of changes in the social environment is the probable improvement in the intelligence of the persons receiving the treatment. Even if the heritability of intelligence were very high, changing the environments of those persons who are forecast to have low intelligence would have indeterminant effects, as long as genotypes of that sort were not regularly exposed to such environments in the populations studied. It is the case that poor genotypes for intelligence are rarely exposed to very stimulating environments and exceptionally good genotypes are rarely exposed to really poor environments. The correlation between genotypes of parents and children guarantee that few gross mismatches will occur. Of course, there are bright parents with retarded children and retarded parents with bright children. Most parents' and children's intellectual levels fall in the middle of the range. But it is rare that parents with midparent IQ scores of, say, 140 have a child who scores below IQ 110. Similarly, it is rare for parents whose average IQ score is 70 to have a child who scores above IQ 110.

Needed: Models of How Genes and Environments Work Together

Certainly, one of the major emotional objections to ideas about genetic differences in behavior is that the term "genetic" implies immutability. Interventions, except dietary and biochemical, seem less likely to improve the lot of the less fortunate, if the cause of their misfortune is genetic.

It is true, of course, that if the sole reason for differences, let us say, between the college entrance rate of lower-class and middle-class youngsters is the economic disadvantage of the former, then providing free tuition and scholarships will equalize their rates of college going. For many economically disadvantaged youth, low interest loans, free tuition, and scholarships do in fact provide the environmental remedy to their problem.

In most cases, however, genetic differences cause and are correlated with environmental differences among people. On the average, lower-class youngsters are less qualified by several criteria to gain entrance into higher education; for example their reading and writing skills are often not as well developed as those of middle-class children, and their work habits in academic settings are less pleasing to school authorities. Although much of the difference in skills and motivation between middle- and lower-class youngsters may arise from differences in their rearing environments, the research on class differences reported earlier in this chapter suggests that genetic differences are implicated as well. In addition, there is a correlation between genetic differences and differences in the rearing histories of middle- and lower-class youth. Although there are many academically able children in lower SES homes, the proportion is smaller than in the middle class, and the environments provided by lower SES families are less conducive, on the average, to the development of academic skills.

Jencks (in press) has provided an interesting scheme for understanding the ways in which genetic and environmental differences can be correlated and in which genetic differences can cause environmental differences.

Jencks argues that, instead of equating implicitly genetic differences with physical causes and environmental with social causes, we should recognize that the two dimensions are independent of one another. Table 7.31 presents a four-fold classification of phenomena affecting IQ test performance.

Cell I includes traditional "genetic" causes of variation, most often defects. Genetic differences are invoked to explain the exogenous cause of the disorder, and physical factors are the endogenous forces to explain the genetic effects. Jencks points out that with earlier screening, the PKU disorder can have minimal effects with diet therapy, so that children born in places where screening is routine will be minimally impaired, whereas others born in places where screening is not done will be mentally retarded. Therefore, the effects of the genetic difference between PKU and normal children interacts with socially determined variations in medical care. Similar examples could be drawn from other genetic disorders with early diagnosis and effective treatment, such as galactosemia. There are also many genetic disorders for which no effective treatment has yet been devised that illustrate this category.

Cell II includes lead poisoning, a condition that has little or no important genetic variation, as far as we know. Exposure to lead poisoning is a physical environmental source of variation. Malnutrition and living near the Love Canal also fit into this category. The physical environment can be detrimental to one's

TABLE 7.31
Sources of Variation in Test Performance

Proximate Endogenous Cause	Exogenous Cause	
	Genetic	Nongenetic
Physical	(I) PKU	(II) Lead Poisoning
Social	(III) Sexism	(IV) Language spoken at home

health and intellectual development. Although there may well be individual genetic differences in susceptibility to damage from lead and other toxic chemicals and from protein–calorie malnutrition, the major source of variation here is the physical environment. Note well, however, that all the cited examples are socially correlated. As Jencks said, virtually all physical determinants of test score differences are themselves socially determined. Social factors can assert their effects physically.

Cell III includes a test score variation that is caused endogenously by social environmental differences but whose exogenous cause is genetic. Sex and race are two prime examples of genetic differences whose relation to IQ test scores may be due more to differences in socialization than to genetic differences in intelligence between groups.

In Cell IV are sources of variation that are causally independent of both genetic differences and physical variations. Jencks' example is the language an individual learns to speak at home. Note that one's native language is sure to be correlated with many social and physical factors, such that Arabic speakers tend to be darker skinned and poorer than most Swedish speakers. Other examples might include the political party of one's choice, one's religion, and preferences for certain foods.

The point of Jencks' presentation is that what we commonly label as "genetic" is spuriously limited to physical modes of effect, thereby disguising the nongenetic physical causes of intellectual variation, and what we label as "environmental" is confused with social causation, which may well have a genetic basis, as in the examples of sex and race.

What we need to know is *how* genes and environments combine to affect intellectual development. It is important to know whether or not a social change will ameliorate a condition, whether its exogenous cause is genetic or not. It is important to know if a genetic difference among people has its effect through physical or social routes. Pellagra and scurvy are fundamentally different from genetically caused vitamin-insufficient metabolisms. Sex differences in spatial abilities are probably different from sex differences in assertativeness, on the social side, and sex differences in shoulder width, on the physical side. We do not know the extent to which socialization affects spatial skills, but we have a fairly good idea that it affects assertativeness. Hormones at puberty have differential effects on the two sexes' shoulder development and may have some effect on spatial abilities. It is important to know.

We (Axelrod & Scarr, in press) agree with Jencks that the extent of heritability is not the information one needs for effective interventions. Rather we need to know *how* (by which routes) genetic differences have their effects. Knowing the degree to which a characteristic in a population differs among individuals for genetic reasons, however, is a starting point—helpful but not sufficient information to plan interventions.

BIBLIOGRAPHY

Alexander, D., & Money, J. Turner's syndrome and Gerstmann's syndrome: Neuropsychologic comparisons. *Neuropsychology,* 1966, *4,* 265–273.

Anastasi, A., & Levee, R. F. Intellectual defect and musical talent. *American Journal of Mental Deficiency,* 1959, 64, 695–703.

Anderson, V. E. Discussion. In L. Ehrman, G. S. Omenn, & E. Caspari (Eds.), *Genetics, environment, and behavior.* New York: Academic Press, 1972.

Anderson, V. E. Genetics and intelligence. In Joseph Wortis (Ed.), *Mental retardation (and developmental disabilities): An annual review* (Vol. 6). New York: Brunner/Mazel, Inc., 1974.

Axelrod, R., & Scarr, S. Human intelligence and public policy. *Scientific American,* in press.

Baker, S. W., & Ehrhardt, A. A. Prenatal androgen, intelligence and cognitive sex differences. In R. C. Friedman, R. N. Richart, & R. L. Vande Wiele (Eds.), *Sex differences in behavior.* New York: Wiley, 1974.

Bajema, C. J. Relation of fertility of occupational status, IQ, educational attainments, and size of family origin: A follow-up study of male Kalamazoo public school population. *Eugenics Quarterly,* 1968, *15,* 198–200.

Bekker, F. F., & Van Gemund, J. J. Mental retardation and cognitive deficits in XO Turner's syndrome. *Maandschr. Kindergeneesk,* 1968, *36,* 148–156.

Bessman, S. P., Williamson, M. L., & Koch, R. Diet, genetics, and mental retardation interaction between phenylketonuric heterozygous mother and fetus to produce nonspecific diminution in IQ: Evidence in support of the justification hypothesis. *Proceedings of the National Academy of Sciences* (USA), 1978, *75,* 1562–1566.

Bishop, P. M. F., Lessof, M. H., & Polani, P. E. Turner's syndrome and allied conditions. In C. R. Austin (Ed.), *Sex differentiation and development.* Cambridge: Cambridge University Press, 1960.

Block, N. J., & Dworkin, G. (Eds.). *The IQ controversy: Critical readings.* New York: Pantheon Books, 1976.

Bock, R. D., & Kolakowski, D. Further evidence of sex-linked major-gene influence on human spatial ability. *American Journal of Human Genetics,* 1973, *25,* 1–14.

Borich, G. D., & Bauman, P. M. Convergent and discriminant validation of the French and Gilford–Zimmerman spatial orientation and spatial visualization factors. *Educational and Psychological Measurement,* 1972, *32,* 1029–1033.

Botwinick, J. Intellectual abilities. In J. C. Birren & K. W. Schaie (Eds), *Handbook of the psychology of aging.* New York: Van Nostrand Reinhold, 1977.

Bouchard, T. J. Genetic factors in intelligence. In A. R. Kaplan (Ed.), *Human behavior genetics.* Springfield, Ill.: Thomas, 1976.

Bouchard, T. J., Jr., & McGee, M. G. Sex differences in human spatial ability: Not an X-linked recessive gene effect. *Social Biology,* 1977, *24,* 332–335.

Brigham, C. C. *A study of American intelligence.* Princeton, N.J.: Princeton University Press, 1923.

Brigham, C. C. Intelligence tests of immigrant groups. *Psychological Review,* 1930, *37,* 158–165.

Broverman, D. M., Klaiber, E. L., Kobayashi, Y., & Vogel, W. Roles of activation in inhibition in sex differences in cognitive abilities. *Psychological Review,* 1968, *75,* 23–50.

Burg, A. Visual acuity as measured by dynamic and static tests: A comparative evaluation. *Journal of Applied Psychology,* 1966, *50,* 460–466.

Burks, B. S. The relative influence of nature and nurture upon mental development: A comparative study of foster parent–foster child resemblance and true parent–true child resemblance. *27th Yearbook of the National Society for the Study of Education,* 1928, 27 (1), 219–316.

Burks, B. On the relative contributions of nature and nurture to average group differences in intelligence. *Proceedings of the National Academy of Sciences,* 1938, *24,* 276–282.

Caldwell, B. M. *Home observation for measurement of the environment.* University of Arkansas, 1978.

Carter-Saltzman, L. Patterns of cognitive abilities in relation to handedness and sex. In M. Wittig & A. Petersen (Eds.), *Determinants of sex-related differences in cognitive functioning.* New York: Academic Press, 1978.

Cavalli-Sforza, L. L., & Feldman, M. Cultural versus biological inheritance: Phenotypic transmission from parents to children (a theory of the effect of parental phenotypes on children's phenotypes). *American Journal of Human Genetics,* 1973, *25,* 618–637.

Cavalli-Sforza, L. L., & Feldman, M. *The evolution of continuous variation III: Joint transmission of genotype, phenotype, and environment.* Unpublished paper. Departments of Genetics and Biological Sciences, Stanford University, 1977.

Childs, B., Finucci, J. M., Preston, M. S., & Pulver, A. E. Human behavior genetics. In H. Harris & K. Hirschhorn, (Eds.), *Advances in Human Genetics* (Vol 7). New York: Plenum Publishing Corporation, 1976.

Claeys, W. Primary abilities and field-dependence of adopted children. *Behavior Genetics,* 1973, *3*(4), 323–338.

Cohen, L. B., & Salapatek, P. *Infant perception: From sensation to cognition.* Vol. II of *Perception of space, speech and sound.* New York: Academic Press, 1975.

Cooper, R., & Zubek, J. Effects of enriched and restricted early environments on the learning ability of bright and dull rats. *Canadian Journal of Psychology,* 1958, *12,* 159–164.

Corah, N. L. Differentiation of children and their parents. *Jounral of Personality,* 1965, *33,* 300–308.

Coren, S., & Porac, C. A new analysis of life-span age trends in visual illusion. *Developmental Psychology,* 1978, *14*(2), 193–194.

Coren, S., & Porac, C. Heritability in visual–geometric illusions: A family study, *Perception,* 1979, *8,*

Cronbach, L. J. Five decades of public controversy over mental testing. *American Psychologist,* 1975, *30*(1), 1–14.

Dan, A. J. The menstrual cycle and sex-related differences in cognitive variability. In M. A. Wittig & A. C. Petersen (Eds.), *Sex-related differences in cognitive functioning.* New York: Academic Press, 1979.

Dawson, J. L. M. Cultural and psychological influences upon spatial–perceptual processes in West Africa: Part I. *International Journal of Psychology.* 1967, 2, 115–128. (a)

Dawson, J. L. M. Cultural and psychological influences upon spatial–perceptual processes in West Africa: Part II. *International Journal of Psychology,* 1967, *2,* 171–185. (b)

DeFries, J. C., Ashton, G. C., Johnson, R. C., Kuse, A. R., McClearn, G. E., Mi, M. P., Rashad, M. N., Vandenberg, S. G., & Wilson, J. R. Parent–offspring resemblance for specific cognitive abilities in two ethnic groups. *Nature,* 1976, *261,* 131–133.

DeFries, J., Johnson, R. C., Kuse, A. R., McClearn, G. E., Polvina, J., Vandenberg, S. G., & Wilson, J. R. Familial resemblance for specific cognitive abilities. *Behavior Genetics,* 1979, *9,* 23–43.

DeFries, J. C., Kuse, A. R., Vandenberg, S. G. Genetic correlations, environmental correlations, and behavior. In J. R. Royce (Ed.), *Theoretical Advances in Behavior Genetics.* Alphen aan den Rijn, The Netherlands: Sythoff and Woordhoff, 1979.

De Fries, J. C., & Plomis, R. Behavioral genetics. *Annual Review of Psychology,* 1978, *29,* 473–515.

DeFries, J. C., Vandenberg, S. G., & McClearn, G. E. Genetics of specific cognitive abilities. *Annual Review of Genetics,* 1976, *10,* 179–207.

DeFries, J. C., Vandenberg, S. G., McClearn, G. E., Kuse, A. R., Wilson, J. R., Ashton, G. C., & Johnson, R. Near identity of cognitive structure in two ethnic groups. *Science,* 1974, *183,* 338–339.

Deregowski, J. B. Illusion and culture. In R. L. Gregory & G. M. Gombrich (Eds.), *Illusion in Nature and Art.* New York: Charles Scribner, 1973.

Dobzhansky, T. The genetic nature of differences among man. In S. Persons (Ed.), *Evolutionary Thought in America.* New Haven: Yale University Press, 1950.

Eaves, L. J. Testing models for variation in intelligence. *Heredity,* 1975, *34,* 132–136.

Eaves, L. J. The effects of cultural transmission on continuous variation. *Heredity,* 1976, *37,* 51–57.

Eaves, L. J., Last, D., Martin, N. G., & Jinks, J. L. A progressive approach to non-additivity and genotype-environmental covariance in the analysis of human differences. *British Journal of Mathematical and Statistical Psychology,* 1977, *30,* 1–42.

Eckland, B. K. Genetic variance in the SES–IQ correlation. *Sociology of Education,* 1979, *52,* 191–196.

Elston, R. C., & Stewart, J. A general model for the genetic analysis of pedigree data. *Human Heredity,* 1971, *21,* 523–542.

Englander-Golden, P., Willis, K. A., & Deinstbier, R. A. *Intellectual performance as a function of repression and menstrual cycle.* Paper presented to American Psychological Association Meetings, September, 1976.

Erlenmeyer-Kimling, L. Gene–environment interactions and the variability of behavior. In L. Ehrman, G. S. Omenn, & E. Caspari (Eds.), *Genetics, environment, and behavior.* New York: Academic Press, 1972.

Erlenmeyer-Kimling, L., & Jarvik, L. F. Genetics and intelligence: A review. *Science,* 1963, *142*(3589), 1477–1479.

Eysenck, H. J. *The inequality of man.* London: Temple Smith, 1973.

Eysenck, H. J. *The structure and measurement of intelligence.* New York: Springer–Verlag, 1979.

Eysenck, H. J. The nature of intelligence. In M. Friedman, J. P. Das, & N. O'Connor (Eds.), *Intelligence and learning.* New York: Plenum, 1980, in press.

Falconer, D. S. *Introduction to quantitative genetics.* New York: Ronald Press, 1960.

Feldman, M. C., & Lewontin, R. The heritability ''hang-up.'' *Science,* 1975, *190,* 1163–1168.

Fennema, E., & Sherman, J. Sex-related differences in mathematics achievement, spatial visualization, and affective factors. *American Educational Research Journal,* 1977, *14,* 51–71.

Ferguson-Smith, M. A. Karyotype–phenotype correlations in gonadal dysgenesis and their bearing on the pathogenesis of malformations. *Journal of Medical Genetics,* 1965, *2,* 142–155.

Fisch, R. O., Bilek, M. K., Deinard, A. S., & Chang, P. N. Growth, behavioral and psychologic measurements of adopted children: The influences of genetic and socioeconomic factors in a prospective study. *Behavioral Pediatrics,* 1976, *89,* 494–500.

Fraisse, P., & Vautrey, P. The influence of age, sex, and specialized training on the vertical–horizontal illusion. *Quarterly Journal of Experimental Psychology,* 1956, *8,* 114–120.

Freeman, F. N., Holzinger, K. J., & Mitchell, B. C. The influence of environment on the intelligence, school achievement and conduct of foster children. *The 27th Yearbook for the National Society for the Study of Education,* 1928, *27*(1), 103–217.

Fuller, J. L., & Thompson, W. R. *Behavior genetics,* New York: John Wiley & Sons, Inc., 1960; Second Edition, 1978.

Garron, D. C. Intelligence among persons with Turner's syndrome. *Behavior Genetics,* 1977, *7,* 105–127.

Garron, D. C., & Vander Stoep, L. Personality and intelligence in Turner's syndrome. *Archives of General Psychiatry,* 1969, *21,* 339–346.

Girgus, J. S., Coren, S., & Porac, C. Independence of in vivo human lens pigmentation from U.V. light exposure. *Vision Research,* 1977, 17, 749–750.

Goldberg, M. B., & Meridith, W. A longitudinal study of spatial ability. *Behavior Genetics,* 1975, 5, 127–135.

Goldberg, M. B., Scully, A. L., Solomon, I. L., & Stenibach, H. L. Gonadal dysgenesis in phenotypic female subjects. *American Journal of Medicine*, 1968, *45*, 529–543.

Goldberger, A. S. Statistical inference in the great IQ debate. *Institute for Research on Poverty Discussion Papers*, 1975, 301–375.

Goldberger, A. S. *Models and methods in the IQ debate: Part I* (revised) (Paper 7801). Madison: University of Wisconsin, Social Systems Research Institute Workshop, February, 1977.

Goldberger, A. S. Pitfalls in the resolution of IQ inheritance. In N. E. Morton & C. S. Chung (Eds), *Genetic epidemiology*. New York: Academic Press, 1978.

Goodenough, D. R., Gandini, E., Okin, I., Pizzamiglio, L., Thayer, D., & Witkin, H. A. A study of X-chromosome linkage with field dependence and spatial visualization. *Behavior Genetics*, 1977, *7*, 373–387.

Guttman, R. Genetic analysis of analytical spatial ability: Raven's progressive matrices. *Behavior Genetics*, 1974, *4*, 273–284.

Hammerton, J. L. *Human cytogenetics* (Vol. 2). New York: Academic Press, 1971.

Harris, Marvin. *Cows, pigs, wars, & witches: The riddles of culture*. New York: Vintage Books, 1975.

Hartlage, L. C. Sex-linked inheritance of spatial ability. *Perceptual and Motor Skills*, 1970, *31*, 610.

Hasman, J. K., & Elston, R. C. The estimation of genetic variance from twin data. *Behavior Genetics*, 1970, *1*(1), 11–19.

Henderson, N.   Genetic influences on the behavior of mice as can be obscured by laboratory rearing. *Journal of Comparative and Physiological Psychology*, 1970, *3*, 505–511.

Hernshaw, L. Structuralism and intelligence. *International Review of Applied Psychology*, 1975, *24*, 85–92.

Hearnshaw, L. *Cyril Burt, psychologist*. New York: Cornell University Press, 1979.

Herrnstein, R. IQ. *The Atlantic*, September, 1971, 43–64.

Herskowitz, M. J. On the relation between Negro–white mixture and standing in intelligence tests. *Pedagogical Seminary*, 1926, *33*, 30–42.

Higgins, J. V., Reed, E. W., & Reed, S. C. Intelligence and family size: A paradox resolved. *Eugenics Quarterly*, 1962, *9*, 84–90.

Hirsch, J. *Behavior genetic analysis*. New York: McGraw–Hill, 1967.

Hirsch, J. Behavior–genetic analysis and its biosocial consequences. In R. Cancro, (Ed.), *Intelligence: Genetic and environmental influences*. New York: Grune & Stratton, 1971.

Honzik, M. P. Developmental studies of parent–child resemblance in intelligence. *Child Development*, 1957, *28*, 215–228.

Horn, J. M., Loehlin, J. C., & Willerman, L. Intellectual resemblance among adoptive and biological relatives: The Texas Adoption Project. *Behavior Genetics*, 1979, *9*(3), 177–207.

Hsia, D. Y. Y. *Human developmental genetics*. Chicago: Yearbook Medical Publishers, 1968.

Humphreys, L. The primary mental ability. In M. Friedman, J. P. Das, & N. O'Connor (Eds.), *Intelligence and learning*. New York: Plenum, 1980.

Immergluck, L., & Mearini, M. C. Age and sex difference in response to embedded figures and reversible figures. *Journal of Experimental Child Psychology*, 1969, *8*, 210–221.

Jacob, F., & Monod, J. Genetic regulatory mechanisms in the synthesis of proteins. *Journal of Molecular Biology*, 1961, *3*, 318–356.

Jencks, C. *Inequality: A reassessment of the effect of family and schooling in America*. New York: Basic Books, 1972.

Jencks, C. Heredity, environment, and public policy. *American Sociological Review*, 1981.

Jensen, A. R. How much can we boost IQ and scholastic achievement? Harvard Educational Review, 1969, *39*, 1–123.

Jensen, A. R. *Educability and group differences*. New York: Harper & Row, 1973. (a)

Jensen, A. R. *Genetics and education.* New York: Harper & Row, 1973. (b)

Jensen, A. R. Let's understand Skodak and Skeels, finally. *Educational Psychologist,* 1973, *10,* 30–35. (c)

Jensen, A. R. The current status of the IQ controversy. *Australian Psychologist,* 1978, *13,* 7–27. (a)

Jensen, A. R. Genetic and behavioral effects of nonrandom mating. In R. T. Osborn, C. E. Noble, & N. Weyl (Eds.), *Human variation: Biopsychology of age, race, and sex.* New York: Academic Press, 1978. (b)

Jensen, A. R. *Bias in mental testing.* New York: Basic Books, 1980.

Jinks, J. L., & Fulker, D. W. Comparison of the biometrical, genetical, MAVA, and classical approaches to the analysis of human behavior. *Psychological Bulletin,* 1970, *73,* 311–349.

Johnson, C. A., Ahern, F. M., & Johnson, R. C. Level of functioning of siblings and parents of probands of varying degrees of retardation. *Behavior Genetics,* 1976, *6*(4), 473–477.

Kamin, L. J. *The science and politics of IQ.* Potomac, Md.: Lawrence Erlbaum Associates, 1974.

Kamin, L. J. Commentary in S. Scarr, *IQ: Race, social class, and individual differences: New studies of old issues.* Hillsdale, N.J.: Lawrence Erlbaum Associates, 1981.

Karlins, M., Schuerkoff, C., & Kaplan, M. Some factors related to architectural creativity in graduating architecture students. *Journal of General Psychology,* 1969, *81,* 203–215.

Kidd, K. K., & Cavalli-Sforza, L. L. An analysis of the genetics of schizophrenia. *Social Biology,* 1973, *20,* 254.

Klaiber, E. L., Broverman, D. M., & Kobayashi, Y. The automatization of cognitive style, androgens, and monoamine oxidase (MAO). *Psycholopharmacologia,* 1967, *11,* 320–336.

Klaiber, E. L., Broverman, D. M., Vogel, & Kobayashi, Y. Rhythms in plasma MAO activity, EEG, and behavior during the menstrual cycle. In M. Ferin, F. Halberg, R. M. Richart, & R. L. VandeWiele (Eds.), *Biorhythms and human reproduction.* New York: Wiley, 1974.

Klineberg, O. Negro–white differences in intelligence test performance: A new look at an old problem. *American Psychologist,* 1963, *18,* 198–203.

Kuse, A. R. *Familial resemblance for cognitive abilities estimated from two test batteries in Hawaii.* Unpublished doctoral dissertation, University of Colorado, 1977.

Laughlin, P. R., & McGlynn, R. P. Cooperative versus competitive concept attainment as a function of sex and stimulus display. *Journal of Personality and Social Psychology,* 1967, *7,* 398–402.

Layzer, D. Science or superstition (a physical scientist looks at the IQ controversy). *Cognition,* 1972, *7,* 265–269.

Layzer, D. Heritability analyses of IQ scores: Science or numerology. *Science,* 1974, *183,* 1259–1266.

Leahy, A. M. Nature–nurture and intelligence. *Genetic Psychological Monographs,* 1935, *17,* 237–308.

Leao, J. C., Vorhess, M. L., Schlegel, R. J., & Gardner, L. I. XX/XO mosaicism in nine preadolescent girls: Short stature as presenting complaint. *Pediatrics,* 1966, *38,* 972–981.

Lerner, I. M. *Heredity, evolution, and society.* San Francisco: Freeman, 1968.

Levine, S. Sex differences in the brain. In J. L. McGaugh, N. M. Weinberger, & R. E. Whalen (Eds.), *Psychobiology.* San Francisco: Freeman, 1967.

Lewontin, R. C. The analysis of causes and the analysis of variance. *The American Journal of Human Genetics,* 1974, *26,* 400–411. (a)

Lewontin, R. C. *The Genetic Basis of Evolutionary Change.* New York: Columbia University Press, 1974. (b)

Lewontin, R. C. Genetic aspects of intelligence. *Annual Review of Genetics,* 1975, 387–405.

Li, C. C. A tale of two thermos bottles: Properties of a genetic model of human intelligence. In R. Cancro (Ed.), *Intelligence: Genetic and environmental influences.* New York: Grune & Stratton, 1971.

Lindsten, J. *The nature and origin of X-chromosome aberrations in Turner's Syndrome.* Stockholm: Almqvist and Wiksell, 1963.

Lindzey, G., Loehlin, J., Monosevitz, M., & Thiessen, D. Behavioral genetics. *Annual Review of Psychology*, 1971, *22*, 39–94.

Loehlin, J. C., Lindzey, G., & Spuhler, J. N. *Raw differences in intelligence.* San Francisco: Freeman, 1975.

Loehlin, J. C., & Nichols, R. *Heredity, environment, and personality: A study of 850 sets of twins.* Austin, Tex.: University of Texas Press, 1976.

Loehlin, J. C., Sharan, S., & Jacoby, R. In pursuit of the "spatial gene": A family study. *Behavior Genetics*, 1978, *8*, 27–41.

Lush, J. L. *Animal breeding plans.* Ames, Iowa: Iowa State College Press, 1945.

Lytton, H. Do parents create, or respond to, differences in twins? *Developmental Psychology*, 1977, *13*(5), 456–459.

Maccoby, E. E., & Jacklin, C. N. *The psychology of sex differences.* Stanford, Calif.: Stanford University Press, 1974.

Mackenberg, E. J., Broverman, D. M., Vogel, W., & Klaiber, E. L. Morning-to-afternoon changes in cognitive performances and in the electroencephalogram. *Journal of Educational Psychology*, 1974, *66*, 238–246.

Manosevitz, M., Lindzey, G., & Thiessen, D. (Eds.). *Behavioral genetics.* New York: Appleton–Century–Crofts, 1969.

Martin, N. G. The inheritance of scholastic abilities in sample of twins. II: Genetic analysis of examination results. *Annuals of Human Genetics*, 1975, *39*, 219–229.

Martin, N. G., Eaves, L. J., Kearsey, M. J., & Davies, P. The power of the classical twin study. *Heredity*, 1978, *40*(1), 97–116.

Martin, R. G., & Ames, B. N. Biochemical aspects of genetics. *Annual Review of Biochemistry*, 1964, *33*, 235–256.

Matheny, A. P. Genetic determinants of the Ponzo illusion. *Psychonomic Science*, 1971, *24*, 155–156.

Matheny, A. P. Cognitive factors associated with the Ponzo illusion: A study using the Coturn method. *Psychonomic Science*, 1972, *29*, 91–93. (a)

Matheny, A. P. Perceptual exploration in twins. *Journal of Experimental Child Psychology*, 1972, *14*, 108–116. (b)

Matheny, A. P. Hereditary components of the response to the double trapezium illusion. *Perceptual and Motor Skills*, 1973, *36*, 511–513.

Mather, K. On biometrical genetics. *Heredity*, 1971, *26*, 349–364.

McAskie, M., & Clark, A. M. Parent–offspring resemblances in intelligence: Theories and evidence. *British Journal of Psychology*, 1976, *67*(2), 243–273.

McCall, R. B. *Intelligence and heredity.* Homewood: Il.: Learning Systems Co., 1975.

McClearn, G. E. Behavioral genetics. *Annual Review of Genetics*, 1970, *4*, 437–468.

McClearn, G. E., & DeFries, J. C. *Introduction to behavioral genetics.* San Francisco: Freeman, 1973.

McDonald, J. F., & Ayala, F. J. Genetic response to environmental heterogeneity. *Nature*, 1974, *250*(5467), 572–574.

McGaugh, J. L., Jennings, R. D., & Thompson, C. W. Effect of distribution of practice on the maze learning of descendents of Tryon maze bright and maze dull strains. *Psychological Reports*, 1962, *10*, 147–150.

McGee, M. G. Human spatial abilities: Psychometric studies and environmental, genetic, hormonal, and neurological influences. *Psychological Bulletin*, 1979, *86*(5), 889–918.

McGuinness, D. Away from a unisex psychology: Individual differences in visual sensory and perceptual processes. *Perception*, 1976, *5*, 279–294. (a)

McGuinness, D. Sex differences in the organization of perception. In B. Lloyd & J. Archer (Eds.), *Exploration of sex differences.* London: Academic Press, 1976. (b)

McGuinness, D., & Lewis, I. Sex differences in visual persistence: Experiments on the Ganzfeld and afterimages. *Perception*, 1976, *5*, 295–301.

McGuire, T. R., & Hirsch, J. *General intelligence (g) and heritability (H²h²)*. New York: Plenum Press, 1977.

McKusick, V. A., & Ruddle, F. H. The status of the gene map of the human chromosomes. *Science*, 1977, *196*, 390–406.

McNemar, Q. E. A critical examination of the University of Iowa studies of environmental influences upon the IQ. *Psychological Bulletin*, 1940, *37*, 63–92.

Medawar, P. B. *Unnatural science*. New York Review, 1977.

Minton, H. L., & Schneider, F. W. *Differential psychology*. Monterey, Calif.: Brooks/Cole, 1980.

Mittwoch, U. *Genetics of sex differentiation*. New York: Academic Press, 1973.

Money, J. Two cytogenetic syndromes: Psychologic comparisons and specific-factor quotients. *Journal of Psychiatric Research*, 1964, *2*, 223–231.

Money, J., & Alexander, D. Turner's syndrome: Further demonstration of the presences of specific cognitional deficiences. *Journal of Medical Genetics*, 1966, *3*, 47–48.

Monod, J. [*Chance and necessity: An essay on the natural philosophy of modern biology.*] (Austryn Wainhouse, Trans.). New York: Knopf, 1971.

Morton, N. E. Human behavioral genetics. In L. Ehrman, G. S. Omenn, & E. Caspari (Eds.), *Genetics, environment, and behavior*. New York: Academic Press, 1972.

Morton, N. E. Analysis of family resemblance. I. Introduction. *American Journal of Human Genetics*, 1974, *26*, 318–330.

Morton, N. E., & Rao, D. C. *Genetic epidemiology of IQ and socio–familial mental defect*. PGL paper. Population Genetics Laboratory, University of Hawaii, 1977.

Nance, W. E., & Corey, L. A. Genetic models for the analysis of data from families of identical twins. *Genetics*, 1976, *83*, 811–826.

Neff, W. S. Socioeconomic status and intelligence: A critical survey. *Psychological Bulletin*, 1938, *35*, 727–757.

Newman, H. G., Freeman, F. N., & Holzinger, K. J. *Twins: A study of heredity and environment*. Chicago: University of Chicago Press, 1937.

Nichols, P. L. *The effects of heredity and environment on intelligence test performance in 4- and 7-year-old white and Negro sibling pairs*. Unpublished doctoral dissertation, University of Minnesota, 1970.

Nichols, R. Twin studies of ability, personality, and interests. *Homo*, 1978, *29*, 158–173.

Nichols, R. C. Policy implications of the IQ controversy. In L. S. Shulman (Ed.), *Review of research in education*. Itasca, Ill.: Peacock, 1979.

O'Connor, J. *Structural visualization*. Boston: Human Engineering Laboratory, 1943.

Ohno, S. *Sex chromosomes and sex-linked genes*. Berlin: Springer–Verlag, 1967.

Osborne, R. T., & Gregor, A. J. Racial differences in heritability estimates for tests of spatial abilities. *Perceptual and Motor Skills*, 1968, *27*, 735–739.

Ounsted, C., & Taylor, D. (Eds.). *Gender differences: Their ontogeny and significance*. London: Churchill Livingstone, 1972.

Pauls, D. *A genetic analysis of mental retardation and high intelligence*. Unpublished doctoral dissertation, University of Minnesota, 1972.

Pearson, J. S., & Amacher, P. L. Intelligence test results and observations of personality disorder among 3594 unwed mothers in Minnesota. *Journal of Clinical Psychology*, 1956, *12*, 16–21.

Petersen, A. C. Physical androgyny and cognitive functioning in adolescence. *Developmental Psychology*, 1976, *12*, 524–533.

Petersen, A. C. Hormones and cognitive functioning in normal development. In M. A. Wittig & A. C. Petersen (Eds.), *Sex-related differences in cognitive functioning*. New York: Academic Press, 1979.

Pick, H. L., Jr., & Pick, A. D. Sensory and perceptual development. In P. H. Mussen (Ed.), *Charmichael's Manual of Child Psychology* (3rd ed.). New York: Wiley, 1970.

Pishkin, V., Wolfgang, A., & Rasmussen, E. Age, sex, amount and type of memory information in concept learning. *Journal of Experimental Psychology,* 1967, *73,* 121–124.

Plomin, R., & DeFries, J. C. Genetics and intelligence: Recent data. *Intelligence,* 1980, *4,* 15–24.

Plomin, R., DeFries, J. C., & Loehlin, J. C. Genotype–environment interaction and correlation in the analysis of human behavior. *Psychological Bulletin,* 1977, *84*(2), 309–322.

Plomin, R., DeFries, J. C., & McClearn, G. E. *Behavioral genetics: A primer.* San Francisco: Freeman, 1980.

Plomin, R., Willerman, L., & Loehlin, J. C. Resemblance in appearance and the equal environments assumption in twin studies of personality. *Behavior Genetics,* 1976, *6,* 43–52.

Polani, P. E. Chromosomal factors in certain types of educational sub-normality. In P. W. Bowman & H. V. Mautner (Eds.), *Mental retardation: Proceedings of the First International Congress.* New York: Grune & Stratton, 1960.

Pollack, R. H. Some implications of ontogenetic changes in perception. In W. J. Flavell & D. Ellcind (Eds.), *Studies in cognitive development: Essays in honor of Jean Piaget.* New York: Oxford University Press, 1969.

Porac, C., Coren, S., Girgus, J. S., & Verde, M. Visual–geometric illusions: Unisex phenomena. *Perception,* 1979, *8,* 401–412.

Pressey, A. W., & Wilson, A. E. Another look at age changes in geometric illusion. *Bulletin of the Psychonomic Society,* 1978, *12,* 333–336.

Rao, D. C., Morton, N. E., & Yee, S. Analysis of family resemblance II. A linear model for familial correlation. *American Journal of Human Genetics,* 1974, *26,* 331–359.

Reed, E. W. Genetic anomalies in development. In Frances, D. Horowitz, E. M. Hetherington, Sandra Scarr-Salapatek, & G. M. Siegal (Eds.), *Review of Child Development Research,* 1975.

Reed, S. C., & Anderson, V. E. Effects of changing sexuality on the gene pool. In F. F. La Cruz & G. D. LaVeck (Eds.), *Human sexuality and the mentally retarded.* New York: Brunner/Mazel, 1973.

Reed, E. W., & Reed, S. C. *Mental retardation: A family study.* Philadelphia: W. B. Saunders, 1965.

Reinisch, J. M., Gandelman, R., & Spiegel, F. S. Prenatal influences on cognitive abilities: Data from experimental animals and human endocrine syndromes. In M. A. Wittig & A. C. Peterson (Eds.), *Sex-related differences in cognitive functioning.* New York: Academic Press, 1979.

Rimland, B., & Munsinger, H. Burt's IQ data. *Science,* 1977, *195,* 248.

Roberts, J. *Binocular visual acuity of adults.* Washington, D.C.: Department of Health, Education and Welfare, 1964.

Robert, J. A. F. The genetics of mental deficiency. *Eugenics Review,* 1952, *44,* 71–83.

Rose, R. J. Genetic variance in non-verbal intelligence: Data from the kinship of identical twins. *Science,* 1979, *205,* 1153–1155.

Rowe, D. C., & Plomin, R. The Burt controversy: A comparison of Burt's data on IQ with data from other studies. *Behavior Genetics,* 1978, *8,* 81–84.

Scarr, S. Environmental bias in twin studies. *Eugenics Quarterly,* 1968, *15*(1), 34–40.

Scarr-Salapatek, S. Race, social class, and IQ. *Science,* 1971, *174,* 1285–1295. (a)

Scarr-Salapatek, S. Unknowns in the IQ equation. *Science,* 1971, *174,* 1223–1228. (b)

Scarr-Salapatek, S. Genetics and intelligence. In F. D. Horowitz (Ed.), *Review of child development research* (Vol. 4), Chicago: University of Chicago Press, 1975.

Scarr, S. *Comments on psychology, behavior genetics, and social policy from an anti-reductionist.* Paper presented at the Second Houston Symposium on Psychology as a Science: Emerging Issues, University of Houston, May, 1979.

Scarr, S. Genetic differences in "g" and real life. In M. Friedman, J. P. Das, & N. O'Connor (Eds.), *Intelligence and learning.* New York: Plenum, 1980.

Scarr, S., & Barker, W. The effects of family background: A study of cognitive differences among black and white twins. In S. Scarr (Ed.), *IQ: Social class and individual differences.* Hillsdale, N.J.: Lawrence Erlbaum Associates, in press, 1981.

Scarr, S., & Carter-Saltzman, L. Twin method: Defense of a critical assumption. *Behavior Genetics,* 1979, *9,* 527–542.

Scarr, S., Pakstis, A. J., Katz, S. H., & Barker, W. B. The absence of a relationship between degree of white ancestry and intellectual skills within a black population. *Human Genetics,* 1977, *39,* 69–86.

Scarr, S., Scarf, E., & Weinberg, R. A. Perceived and actual similarities in biological and adoptive families: Does perceived similarity bias genetic inferences? *Behavior Genetics,* 1980.

Scarr, S., & Weinberg, R. A. Intellectual similarities within families of both adopted and biological children. *Intelligence,* 1977, *1* (2),

Scarr, S., & Weinberg, R. A. The influence of "family background" on intellectual attainment. *American Sociological Review,* 1978, *43,* 674–692.

Scarr, S., & Yee, D. Heritability and educational policy: Genetic and environmental effects on IQ, aptitude, and achievement. *Educational Psychologist,* 1980, in press.

Schoenfeldt, L. F. The hereditary components of the Project TALENT two-day test battery. *Measurement and Evaluation in Guidance,* 1968, *1,* 130–140.

Schiff, M., Duyme, M., Dumaret, A., Stewart, J., Tomkiewicz, S. & Feingold, J. Intellectual status of working-class children adopted early into upper-middle-class families. *Science,* 1978, *200,* 1503–1504.

Schwartz, M., & Schwartz, J. Evidence against a genetical component to performance on IQ tests. *Nature,* 1974, *248*(5443), 84–85.

Segall, M. H., Campbell, D. T., & Herskovits, M. J. *The influence of culture on visual perception.* Indianapolis: Bobbs–Merrill, 1966.

Serra, A., Pizzamiglio, L., Boari, A., & Spera, S. A comparative study of cognitive traits in human sex chromosome aneuploids and sterile and fertile euploids. *Behavior Genetics,* 1978, *8,* 143–154.

Shaffer, J. W. A specific cognitive deficit observed in gonadal aplasia (Turner's Syndrome). *Journal of Clinical Psychology,* 1962, *18,* 403–406.

Shepard, R. N., & Metzler, J. Mental rotation of three-dimensional objects. *Science,* 1971, *171,* 701–703.

Shields, J. *Monozygotic twins brought up apart and brought up together.* London: Oxford University Press, 1962.

Silbert, A., Wolff, P. H., & Lilienthal, J. Spatial and temporal processing in patients with Turner's syndrome. *Behavior Genetics,* 1977, *7,* 11–21.

Skeels, H. M. Mental development in children in foster homes. *Journal of Consulting Psychology,* 1938, *2,* 33–43.

Skodak, M. Children in foster homes. *University of Iowa Child Welfare,* 1938, *15*(4), 191.

Skodak, M., & Skeels, H. M. A final follow-up study of one hundred adopted children. *Journal of Genetic Psychology,* 1949, *75,* 85–125.

Smith, G. *Psychological studies in twin differences.* Lund, Sweden: Gleerup, 1949.

Snow, R. Aptitude–treatment interaction in education. In R. J. Sternberg (Ed.), Handbook of intelligence. Cambridge: Cambridge University Press, in press.

Stafford, R. E. Sex differences in spatial visualization as evidence of sex-linked inheritance. *Perceptual and Motor Skills,* 1961, *13,* 428.

Stempfel, R. S., Jr. Abnormalities of sexual development. In L. E. Gradner (Ed.), *Endocrine and genetic diseases of childhood.* Philadelphia: Saunders, 1969.

Stern, C. *Principles of human genetics (2nd ed.).* San Francisco: Freeman, 1973.

Stevenson, H. W., Hale, G. A., Klein, R. E., & Miller, L. K. Interrelations and correlates in children's learning and problem solving. *Monographs of the Society for Research in Child Development,* 1968, *33.*

Taubman, P., Behrman, J., & Wales, T. The roles of genetics and environment in the distribution of earnings. In Z. Griliches, W. Krelle, H. J. Krupp, & O. Eyne (Eds.), *Income distribution and economic inequality.* New York: Halsted Press, 1978.

Taylor, H. J. *The IQ game: A methodological inquiry into the heredity–environment controversy.* New Brunswick, N.J.: Rutgers University Press, 1980.

Terman, L. M. The great conspiracy. *New Republic,* 1922, *33,* 116–120.

Thiessen, D. D. Philosophy and method in behavior genetics. In A. R. Gilgen, (Ed.), *Scientific psychology: Some perspectives.* New York: Academic Press, 1970.

Thiessen, D. D. *Gene organization and behavior.* New York: Random House, 1972.

Urbach, P. Progress and degeneration in the "IQ debate" I and II. *British Journal of Philosophical Science,* 1974, *25,* 99–135; 235–259.

Vandenberg, S. G., & Kuse, A. R. Mental rotations: A group test of three-dimensional spatial visualization. *Perceptual and Motor Skills,* 1978, *47,* 599–604.

Vandenberg, S. G., & Kuse, A. R. Spatial ability: A critical review of the sex-linked major-gene hypothesis. In M. A. Wittig & A. C. Petersen (Eds.), *Sex-related differences in cognitive functioning.* New York: Academic Press, 1979.

Waber, D. P. Sex differences in cognition: A function of maturation rate? *Science,* 1976, *192,* 572–574.

Waber, D. P. Sex differences in mental abilities, hemispheric lateralization and rate of physical growth at adolescence. *Developmental Psychology,* 1977, *13,* 29–38.

Waber, D. P. Cognitive abilities and sex-related variations in the maturation of cerebral cortical functions. In M. A. Wittig & A. C. Petersen (Eds.), *Sex-related differences in cognitive functioning.* New York: Academic Press, 1979.

Wechsler, D. *The measurement of adult intelligence* (3rd ed.). Baltimore: Williams & Williams, 1950.

Weiss, P. The living system: Determinism stratified. In A. Koestler & J. R. Smythies (Eds.), *Beyond reductionism.* Boston: Beacon Press, 1969.

Wellman, B. L. Iowa studies on the effects of schooling. *Thirty-ninth Yearbook of the National Society for the Study of Education,* 1940, *2,* 377–399.

Willerman, L. Personal communication, 1977.

Willerman, L. Effects of families on intellectual development. *American Psychologist,* 1979, *34,* 923–929.

Willerman, L., Horn, J. M., & Loehlin, J. C. The aptitude–achievement test distinction: A study of unrelated children reared together. *Behavior Genetics,* 1977, *7,* 465–470.

Williams, T. Family resemblance in abilities: The Wechsler scales. *Behavior Genetics,* 1975, *5*(4), 405–409.

Wilson, E. O. Unpublished manuscript, 1977.

Wilson, J. R., & Vandenberg, S. G. Sex differences in cognition: Evidence from the Hawaii family study. In T. E. McGill, D. A. Dewsbury, & B. D. Sachs (Eds.), *Sex and behavior: Status and prospectus.* New York: Plenum, 1978.

Wilson, R. S. Twins and siblings: Concordance for school-age mental development. *Child Development,* 1977, *48,* 211–216.

Witkin, H. A. Individual differences in ease of perception of embedded figures. *Journal of Personality,* 1950, *19,* 1–15.

Woodworth, R. S. *Heredity and environment: A critical survey of recently published material on twins and foster children.* A report prepared for the committee on Social Adjustment. New York: Social Science Research Council, 1941.

Yen, W. M. Sex-linked major-gene influences on selected types of spatial performance. *Behavior Genetics,* 1975, *5,* 281–298.

Yerkes, R. M. Testing the human mind. *Atlantic Monthly,* 1923, *131,* 358–370.

8 Ethology and Behavior Genetics[1]

John L. Fuller

State University of New York at Binghamton

The application of genetics to behavior precedes by millennia the recognition of genetics as an experimental science. Vergil advised Roman herdsmen to breed their well-dispositioned cattle. Plato favored the education of talented men and women in the same institutions to increase the probability of marriage between the best specimens of both sexes. Much later, Darwin recognized behavior as well as structure as a factor in natural selection. The field we now call behavior genetics includes practitioners with backgrounds in genetics, animal behavior, the psychology of individual differences, the neurosciences, pharmacology, and biochemistry. Research necessarily requires competence in disciplines other than genetics in the narrow sense. Defining and measuring behavioral phenotypes is extremely important. Ethologists have made contributions to behavior genetics, but fewer than might be expected considering the scientific problems that are common to the two fields. My purpose is to review some of these past findings and to look toward the future.

Three major themes are prominent in behavior genetics. The first is concerned with the importance of genetic variation within populations as a cause of individual differences in behavior, and with the mode of transmission of these differences. These are often known respectively as heritability and genetic architecture. Both deal with processes within specified populations and are not necessarily generalizable to a species as a whole. A trait may show high heritability in a heterogeneous colony of mice and low heritability in a small closed population;

---

[1]This chapter is adapted from a lecture presented to the XVIth International Ethological Conference in Vancouver, B.C., August, 1979.

337

a trait may appear dominant in one cross, recessive in another (see as an example, Fuller, 1964b).

A second area of major interest is centered on biological contributions to the behavioral development of individuals and their interactions with environmental variables. Genes and genotypes can be regarded as treatments that modify the course of behavioral development. So also are physical factors ranging from nutrition to parental care. Sometimes mutant genes become figurative scalpels that excise specific biochemical processes and enable an investigator to deduce developmental patterns (Benzer, 1973). Genes of this type are rare in natural populations, but are carefully preserved in fruit flies, mice, and a few other species. Numerous comparisons among inbred strains of such species demonstrate that genotypes modify behavioral development even though the species-typical patterns are retained (Fuller & Thompson, 1978). The interplay of genotypes and the physical and social environment during development is a major interest of behavior geneticists.

The third area, evolutionary behavior genetics, is somewhat less developed, but is now attracting more interest. Its devotees are concerned with the molding of the gene pool of a species through behavioral adaptation to environmental opportunities and hazards. Obviously, such a process can occur only if the behavior of interest is heritable and the population is genetically heterogeneous. Behavioral adaptation is not distinct from structural adaptation; the two are simply different aspects of a single process. The case for the interdependent evolution of brain and behavior has been persuasively presented by Jerison (1973). However, the more subtle forms of behavioral variation may depend on minute structural features of the central nervous system that are difficult to observe and quantify. The behavior-geneticist's approach to evolution differs from that of most sociobiologists. The latter group work in the framework of population genetics, and their usual genetic model involves a single locus with two alleles. Williams (1981) clearly proclaims sociobiology's disinterest in the pathways between genes and behavior with these words: "Sociobiology . . . does not address questions of behavioral ontogeny . . . [it] is of little help in establishing links among genes, development, learning and socioculture [p. 257]." In contrast a behavior geneticist (Henderson, 1978) places emphasis on polygenic systems and uses the genetic architecture underlying behavioral variation at different stages of development to deduce the nature of selective forces at different stages of development. Ethologists should be interested in both approaches.

GENES AND ADAPTATION

A species can adapt to a variable environment in one of three ways, two of which are genetic. It can retain a large reservoir of genetic variation, thus increasing the odds that under any stressful condition some members of the group are pre-adapted, and survive. In fact, most species in nature are genetically hetero-geneous (Mayr, 1970; Selander, 1970), and the maintenance of the variability is assumed to be due to natural selection. Genetic uniformity is restricted to small, isolated, inbred populations. Another strategy is for a species to move in the direction of genetically programming a complex nervous system that modifies itself adaptively during development by responding to environmental stimuli. The two modes of adaptation are not mutually exclusive and, in fact, often coexist in different proportions. We must be careful, however, not to infer genetic variability from behavioral variability, or vice versa. As an example, Ebert and Hyde (1976) selected for high and low aggression from a phe-notypically intermediate wild population of mice. The range of phenotypes in the selected lines greatly exceeded that in the foundation population. Underneath the apparent phenotypic similarity of the wild mice was the genetic potential for extreme variation upward and downward. On the other hand, those of us who work with inbred strains of mice, each composed of genetically identical indi-viduals, are often frustrated by the amount of behavioral variability within a single strain. Unfortunately for many experimenters, the behavioral phenotype is not rigidly predetermined by the genotype.

These two strategies of adaptation appear to be consequences of different situations (Plotkin & Odling-Smee, 1979). A species living in a stable, predict-able environment will prosper best with a fixed behavioral repertoire that is precoded in its genotype as a plan for a specialized, unmodifiable nervous system. A species living in an unpredictable environment will do better to have its genotype program the development of a plastic nervous system with the capacity for modification by appropriate stimuli. If part of the environment is predictable, a probably universal situation, selection will favor programming fixed responses for the predictable but not for the unpredictable components of its surroundings. A complete separation of these two modes of adaptation is impossible. Thus, modifiability of behavior by experience along with genetic variation in its efficiency has been demonstrated in insects (McGuire & Hirsch, 1977). And there appear to be biological constraints on learning even in such large-brained creatures as humans (Seligman & Hager, 1972).

How Consistent is Species-Typical Behavior?

This discussion of the relations between behavior genetics and ethology begins with two questions. How uniform is species-typical behavior? How much of its variation is attributable to genetic differences? In general, members of a species share a common ethogram as well as a common structural plan. Even though taxonomists depend primarily on structure as a basis for classification, field observers know that behavior is often a reliable means of distinguishing between physically similar species. The most striking examples are sibling species, particularly common in insects, fish, and amphibians although a few examples are known in birds and mammals (Mayr, 1970). These species, however, are not our primary concern here. There is ample evidence of behavioral polytypy within good species as defined by the free interbreeding of their members. Such polytypy is likely to be associated with social organization or food gathering, and explanations are frequently given in terms of ecological pressures and adaptive response.

Three major hypotheses could explain intraspecific behavioral heterogeneity: (1) variant behavior is transmitted from adults to young by learning; (2) behavioral variation is an expression of genetic heterogeneity within the species; and (3) behavioral variation arises because individuals develop differently depending on the environment they encounter.

Hypothesis 1 is in a broad sense cultural, as it implies the handing down of traditional modes of behaving from generation to generation. Cultural transmission is nongenetic except to the extent that it requires a nervous system with a capacity for learning to model the behavior of others. In humans, cultures are a major source of behavior variation, although their prevalence within our species does not logically exclude a genic contribution to group differences. A well-known example of a group characteristic in animals is variation in food habits among troops of Japanese macaques (Kawamura, 1962). In one group the habit of washing food was apparently begun by one young female and was adopted quickly by her companions. It was not found in other troops living under similar circumstances. Cultural transmission of this type may be more important in nonhumans than is generally recognized, but because it has not been widely investigated from the genetic point of view I shall not consider it further.

Hypotheses 2 and 3 emphasize individualistic responses to environmental challenges rather than modeling behavior on the actions of others. Culture is involved only as a part of the environment in which development takes place. The difference between the hypotheses lies in the role of genetic variation. Hypothesis 2 places a premium on genetic heterogeneity as a means of ensuring that some individuals will be able to survive and propagate in environments that deviate markedly from the species optimum. This is clearly the classical concept of natural selection, and it is supported in a general way by the universality of genetic variability in natural populations.

Hypothesis 3 does not deny the role of natural selection in the past, but for the present it emphasizes the lability of phenotypic development and adaptive responses to particular environmental conditions. For adherents of this view, behavioral variation as we see it within a species is more a matter of differences in individual life histories than in individual genotypes. Actually, the three hypotheses are not mutually exclusive. All are supported by evidence, but their relative importance varies according to circumstances. In the following section I consider some typical examples of behavioral polytypy and try to deduce their probable origins.

Behavioral Polytypy: Observational Studies

An extensive literature deals with variation in the behavior of domesticated and laboratory animals, house mice, rats, dogs, fowl, and fruit flies. In these species it is often possible to separate experimentally the effects of genetic and environmental factors. Genetic variation has been demonstrated for activity level, emotional reactivity, eating and drinking, learning ability, and social behavior. In some domestic species, selection for behavior has produced breeds with special aptitude for certain tasks. These breed differences persist even when animals are reared as nearly as possible in the same manner.

As an example, Fuller (1955) trained five breeds of dogs to walk on a lead from their outdoor living quarters to a laboratory and found striking differences in their reactions. Shetland sheep dogs crowded against their guides, a pattern that strongly resembled the herding of sheep by their working relatives. Basenjis, with a short history of domestication, struggled more violently and bit their leash more frequently than any other breed. Beagles, who have been selected to track small game and to give voice as they follow a trail, were noteworthy for their mournful vocalization. None of these behaviors was restricted to a single breed, and none was completely absent from any breed. Their frequencies and intensities differed enough, however, to define significantly different breed profiles. Crosses between Basenjis and cocker spaniels, breeds with widely differing profiles, yielded hybrids that were intermediate to their parents on all characteristics except vocalization, which was higher than that of either parent (Scott & Fuller, 1967).

The occurence of stable, intraspecific behavioral variation among populations is ethological polytypy. It is not restricted to domesticated animals. Polytypy has been reported within single taxa whose range extends over a wide territory that covers different environmental conditions. Examples include baboons (*Papio anubis*), vermets (*Cercopithecus aethiops*), and langurs (*Presbytis entellus*), whose social organization varies greatly over their ranges and appears to be correlated with the nature of the habitat (Rowell, 1969). Hendrichs (1978) has recorded variations in the social organization of three species of antelope, diddik (*Madoga kirki*), reedbuck (*Redunca redunca*), and waterbuck (*Kobus ellipsiprymnus*), that are associated with population density, that is, in turn, presumably regulated by the supply of food and other resources. Stacey and Bock (1978) report that acorn woodpeckers (*Melanerpes formicovorus*) that ordinarily live in permanent groups and defend year-round territories deviated from this pattern in one location. Here, they mated in pairs and migrated in early winter. Old field mice (*Peromyscus polionotus*) from different geographical areas were found to have identical patterns of sexual behavior, but these varied in quantitative aspects such as latency, frequency, and duration of the components that make up the total mating act (Dewsbury & Lovecky, 1974). There is no evidence that these polytypies are attributable to genetic heterogeneity. In fact, many of the cited authors generally ascribe these variations in social organization and behavior to environmental pressures. The assumption, not always explicitly stated, is that given the same circumstances any member of the species could adopt either of the alternate patterns. However, there have been few tests to disprove a genetic basis, and it would be difficult to make one with many of these species. It is possible in species that can be maintained in large numbers in captivity and bred systematically.

Behavioral Polytypy: Comparative Studies

Important contributions to the genetic analysis of ethological polytypy have been made by King and his students. The genus *Peromyscus* includes many taxa that have been classified at the subgeneric, species group, species, and subspecies level in ascending order of closeness of relationship. King, Price, and Webber (1968) compared animals from nine taxa on five motor behaviors: swimming, running, climbing, digging, and gnawing. They were searching for correlations between behavioral similarities and closeness of relationship as judged by taxonomists. Secondarily, they also looked for correlations between behavior and habitat. For example, burrowing species might be expected to excel in digging, woodland species in climbing. In all tests the only motivation to perform any of the actions was to escape from an unfamiliar place. Differences among taxa were found on every task, but the expectation that the degree of similarity would correlate with the degree of relationship was not fulfilled. Surprisingly, the behavior–habitat correlations were no better.

Although the results of this study were disappointing, King (1977) rethought the problem and carried out other experiments. Instead of comparing taxa on specific motor acts in an artificial setting he designed a set of tests to measure more general behavioral characteristics: speed of adaptation, reflex responses, abstractive and strategic behavior. Again the subjects were *Peromyscus* taxa of varied degrees of relationship and similarity of habitat. The results of this study lead to a revised view of behavioral polytypy as related to genetics. All taxa of *Peromyscus*, and many other "mice," are roughly similar in form and are capable of "rodentlike behavior" such as gnawing, grooming, and nest building. The major differences among them lie in their modes of organization of activities in new situations as they search for food, mates, or shelter. Some taxa become specialized through fixation of genes favoring survival under typical natural conditions; others retain their juvenile plasticity into adulthood and adapt as best they can to situations as they arise. King finds evidence that both Hypothesis 2 (fixed genetic programming) and Hypothesis 3 (flexible neural programming) apply to *Peromyscus*. He sees in mice and men a need to compromise between the forces that reinforce rigidity and those that facilitate flexibility in development.

In a somewhat similar experiment with wild *Mus musculus*, captured in three ecologically distinct areas and reared similarly in a laboratory, Plomin and Manosevitz (1974) found differences in open-field activity, defecation, climbing, running-wheel activity, and nest building. They attributed the differences to genetic factors, although controls for maternal effects and cross-breeding experiments were not included in their study. A repetition with more subjects and additional controls would be useful.

In muroid rodents, male copulatory behavior has been studied from a genetic point of view. Dewsbury (1979b) reported basically similar factor patterns in laboratory rats, the descendants of a wild population of house mice, and deer mice. In spite of the factorial similarity each of these species has a distinctive style of courtship. In more closely related taxa, such as subspecies, male copulatory behavior is much more uniform and for practical purposes qualitatively identical. However, significant quantitative differences in the latencies and durations of phases of mating were found in 11 of 14 measures. Subjects were three subspecies of *Peromyscus maniculatus: P.m. bairdii, P.m. blandus,* and *P.m. gambeli* (Dewsbury, 1979a). Cross-fostering of deer mice on house mice produced minor quantitative effects on latencies, but, according to Dewsbury: "basic motor patterns in cross-fostered mice were identical to those of normally reared animals [p. 158]."

A particularly interesting instance of behavioral polytypy occurs in the squirrel monkey (*Saimiri sciurus*) (Ploog, Hupfer, Jurgens, & Newman, 1975). Two populations of this species differ markedly in the phonic characteristics of their extensive vocabulary of calls. Externally the populations are distinguishable by facial markings. The Roman arch type found in Peru, Costa Rica, and Panama, and the Gothic arch type resident in Columbia and Guyana also differ in their karyotypes. That the dialects differ for genetic rather than cultural reasons was demonstrated by rearing infants with devocalized mothers and finding that they conversed in the dialect appropriate to their chromosomes and emitted these calls in their appropriate context. A personal communication from Newman adds that new research has shown that isolation peeps (IPs, presumably a sign of distress) of these two forms differ so much that analysis of sonograms permits classification of individuals with 99% accuracy. The recorded IPs of eight F_1 hybrids were matched with their phenotypic appearance. Five hybrids with Roman arches emitted typical Roman IPs. The IPs of the three Gothic-type hybrids were variable, one strongly Gothic, one strongly Roman, and one mixed. These observations are insufficient to hypothesize a mode of inheritance, but apparently vocalization patterns are associated to a degree with physical characteristics.

Behavioral Polytypy: Breeding Studies

The studies reviewed in the two previous sections did not make use of the traditional genetic procedures of crossbreeding and selection. Most knowledge of the formal genetics of social behavior comes from experiments with selected breeds or inbred lines of house mice, Norway rats, dogs, domestic fowl, and

Drosophila. Can results obtained with these long-domesticated animals lead to principles that can be extended to their wild relatives?

Lorenz (1965) asserted that results from animal strains that have been bred for a long time in captivity are not applicable to free-living animals: "because captivity changes all hitherto effective selective factors in so profound a way that serious changes must be expected in the genome of the stock after only a few generations [p. 99]." Recent research indicates that Lorenz was too sweeping in this statement. Domestic albino rats transferred from their protective laboratory environment to outdoor pens survived severe northern winters, maintained a stable population, constructed and lived in burrows indistinguishable from those of wild rats, and in general behaved like their wild cousins (Boice, 1977). Given their freedom, the albinos were able to cope with stresses that their recent ancestors had never encountered. Of course, in a completely feral state their impaired vision and conspicuous coloration would make them more vulnerable to predators. The important point of Boice's study is that there was no evidence of a loss of species-typical adaptive behavior. Disuse of skills does not by itself lead to their deletion from the behavioral repertoire.

In a similar vein, Miller (1977) made extensive observations on Peking ducks that have been separated from their wild mallard (*Anas Platyrhyncos*) ancestors for centuries. Pekings differ from mallards in important characteristics: They are too heavy to fly, and they lack the mallard's striking sexual dimorphism in plumage. Yet, when observed in a natural setting during three breeding seasons, they performed all the species-typical displays that have been reported for mallards. Miller (1977) concluded that his data: "offer no justification for the view that domestication has a 'degenerative' effect on social courtship behavior patterns, but are rather supportive of the view that the customary nonoccurence of these displays is a function of the inhibitory or unfavorable environmental context in which the domestic birds usually find themselves [p. 229]."

The observations of Boice and Miller do not negate the existence of behavioral differences between domesticated animals and their wild progenitors (see, for example, Barnett, 1963; Price & Loomis, 1973), but they support the idea that many such differences are the results of selection for a characteristic desired by their breeders: docility in laboratory rats, and flightlessness in birds that are intended for the dinner of their custodians.

Some changes accompanying domestication probably are the effects of the loss of dominant alleles through inbreeding. Lynch (1977) found that inbred wild house mice, compared with random-bred animals from the same foundation stock, had smaller litters, bore lighter offspring, and constructed poorer nests. Random-bred wild mice reared in a laboratory were not larger, more fertile, or better nest builders than standard inbred strains. In fact, their nests were smaller than those of the heterogeneous HS stock produced by first intercrossing eight inbred strains and then continuing with a random breeding system. Lynch explains the hypertrophy of a behavioral trait related to fitness in a stock synthesized from inbred strains that built only normal-sized nests by postulating that some deleterious recessive alleles present in wild mice were eliminated during inbreeding. She believes that the inbred lines that survive the inbreeding process are good material for analyzing the genetic architecture of a species.

Her view is contrary to that of Smith (1978) who found that the heritability of shuttle-avoidance learning was much lower in lines of mice selected from a wild stock than in inbred strains and heterogeneous stocks. He doubts that the genetic architecture of inbred strains and hybrids can be used to deduce evolutionary history. I would question the relevance of shuttle-box avoidance to biological fitness. There is considerable evidence that good performance in the apparatus depends less on learning ability than on the prepotent response to alarm signal, freezing or running (Katzev & Mills, 1974). Domestic rats have been selected for docility, and this may indeed disqualify them for inferring selection pressures in nature that are related to escape from shock.

Some of the changes in behavior that accompany domestication seem to be the effects of inbreeding rather than of conscious or unconscious selection. Connor (1975) compared wild house mice reared in a simulated natural environment with laboratory-reared wild mice and with three standard inbred strains. Wild mice, irrespective of their habitats, were more aggressive toward intruders, vocalized more when handled, and were more difficult to recapture when they were freed in a room. In a second experiment, Connor followed behavioral changes over 10 generations in random-bred wild mice maintained in a natural or a laboratory habitat. A third group from the same original population were inbred by brother–sister matings. No behavioral differences were found between the laboratory and the seminaturally reared random-bred stocks. Maternal effects were looked for in a cross-fostering study, but none were found. In contrast, the inbred wild lines were less aggressive and easier to recapture.

The evidence is convincing that we should not ascribe intraspecific variation in behavior exclusively to genetic or to environmental factors without evidence from experiments designed to separate their effects. Price (1969), on the basis of his studies of the modification of activity by mode of rearing in wild and semi-domesticated *Peromyscus maniculatus bairdii* came to the following conclusions. The genetic requirements for a behavioral character should not change during breeding in captivity unless: (1) it is selected for deliberately; or (2) it is selected for unwittingly because it improves fitness under the conditions of captivity. These conclusions are encouraging. By a judicious mixture of field observations with controlled breeding and rearing in captivity, we someday may be able to determine the extent to which natural variability in species-typical behavior is a function of genetic differences. An example is the work of Moss and Watson (1980) described in Chapter 12, Sociobiology. Genetic heterogeneity is an absolute necessity for evolution; we know that it exists, but we do not know how important it is with respect to differences in behavioral fitness.

GENOTYPE–ENVIRONMENT INTERACTION AND
BEHAVIORAL DEVELOPMENT

Neither genotype nor environment is sufficient to determine a behavioral phenotype. Without a genotype there is no organism. For a zygote or parthenogenetic ovum to develop into an adult there must be an appropriate physical and biological environment. Ethologists are especially interested in the biosocial factors that shape the development of behavior. Indeed, the developmental process is often described as a genotype–environment (GE) interaction. I prefer to designate the general phenomenon of joint involvement as GE coaction, and to confine GE interaction to situations in which a specific environmental change affects organisms differently depending on their genotypes. GE interactions are detected by exposing two or more genotypes to two or more conditions (e.g., stimulated versus nonstimulated) and comparing the behavioral effects. Broadhurst and Jinks (1966) distinguished two ways in which genes might affect development. (1) They could influence its *stability*, particularly during early stages, and effect processes that could potentially influence an individual's behavior for life. A tendency for phenotypic development to proceed on a stable, species-specific course regardless of ambient conditions is called *developmental homeostasis*. (2) At any stage of life, genes and their productions may influence the capacity of an organism to adjust to *changes* in its environment such as removal from a familiar habitat. Broadhurst and Jinks' distinction between stability and change is somewhat arbitrary, but it is still useful. In general, stability refers to processes during early development that have generalized, permanent effects; change involves behavioral adaptations to a specific situation, with less carry-over to other conditions.

GE interaction takes several forms. The most extreme is *disordinal* interaction where the effects of an environmental change are opposite in direction depending on the genotype of the subjects. The possibility of such interactions has been used to discredit the importance of genetic influences on behavior (for example, Feldman & Lewontin, 1975), but I seriously doubt that disordinal interaction is common. In fact, apparent examples may be artifacts of an incomplete experimental design. Figure 8.1 demonstrates that in a 2-genotype–2-treatment experiment, one could come to completely different conclusions regarding interaction depending on one's choice of treatments.

In this figure it is assumed that an organism must receive an appropriate exposure to some stimuli in order to develop optimally. The low end of the environmental scale corresponds to extreme stimulus deprivation; the upper end to severe stress. Strains A and B differ in their optimal stimulus requirements. A requires less than B, but it is also more easily damaged by high levels. Suppose now that we conduct four experiments with these strains choosing only two values of stimulus intensity for each. The chosen values are indicated by x's on the abscissa. Experiment 1 yields a clear interaction; strain A is helped by

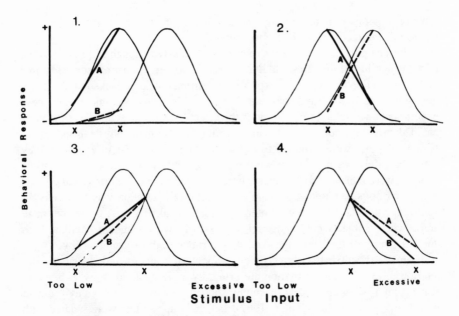

FIG. 8.1.   A variable-threshold model for genotype–environment interactions. The choice of experimental parameters can lead to very different conclusions regarding the existence and nature of GE interaction. Two strains, *A* and *B*, are compared under two levels of stimulation in each of four experiments. Both strains require some stimulus input to respond optimally; both are impaired by excessive stimulation. The two normal curves in each sector of the figure represent the potential response of the two strains over a wide range of stimulus intensity. For each strain optimal response occurs with an intermediate intensity. *A* requires less stimulation than *B* for good performance, but is also more vulnerable to strong stimulation. The straight solid lines show the observed responses when the two stimulation levels designated by *x*'s on the abscissa are chosen for the experiment. In Experiment 1, *A* responds optimally; *B* very little, if at all. In Experiment 2, *A* is impaired by increased stimulation; *B* is improved. GE interaction is found in both experiments, but its nature is very different. In Experiment 3, both strains respond positively to increased stimulation. In Experiment 4, both are impaired by it. In neither 3 or 4 is there much evidence of interaction. Conclusion: Deducing the nature of GE interactions from a simple experimental design is a risky business.

stimulation, strain *B* scarcely at all. Experiment 2 produces a marked disordinal interaction. Strain *A* is impaired as stimulation increases; strain *B* is improved. In Experiment 3, both strains profit from additional stimulation, and there is no interaction. Finally, in Experiment 4 both strains are impaired as stimulation increases. Thus the nature of any interaction found may be a matter of the choice of experimental procedures rather than evidence for fundamentally different ways of responding to a stimulus change. In the figure the basic response of strains *A* and *B* is identical except that their quantitative requirements differ. The remedy, obviously, is to employ a wide range of treatments in any experiment where GE interactions are likely to occur.

Unfortunately, there are few experiments extensive enough to provide a test of the generality of the threshold hypothesis. However, I illustrate some features of GE interaction by a series of experiments carried out at the Jackson Laboratory during the late 1960s (summarized in Fuller, 1967). The basic objective of these experiments was to determine in dogs the effects of varying schedules of experiential deprivation upon activity and reactions to common physical and social stimuli. Puppies were removed from their mothers at 3 weeks of age when they could be safely weaned. They were placed separately (in a few studies, some subjects were pair reared) in cages where they had no physical or visual contact with other dogs or humans. Generally the period of isolation was 12 weeks; in some experiments isolation was interrupted periodically by brief placement in an arena where subjects could interact successively with a human handler, with another puppy of the same age, and with a rubber ball or swinging toy.

From the results of a series of experiments it was concluded that the deleterious effects of early isolation are primarily attributable to an overwhelming fear reaction when an adolescent puppy suddenly faces a complex environment that is totally unfamiliar (Fuller, 1967). Isolated puppies that were sedated with drugs at the time of first emergence from isolation performed almost like normally reared individuals, demonstrating that their motor and perceptual development had been little impaired by living in the small chambers. In most mammals reactive capacities develop gradually from infancy to maturity and the environment that must be dealt with at different ages is matched with the infant's or adolescent's perceptual and motor capacities. Rather small interruptions in isolation were sufficient to counteract the effects of experiential deprivation. Puppies who had 15 minutes per week in our test arena developed social and manipulative skills almost as well as pet-reared animals with several hundredfold more experience (Fuller, 1964a).

A number of our experiments were directed toward defining breed differences in the intensity and duration of the effects of experiential deprivation with beagles and wirehaired fox terriers as the subjects. Beagles have a history of selection for olfactory tracking of small game animals. Litters housed together in group pens seldom fought and dominance hierarchies within litters were weak. In a familiar laboratory setting they gave the impression of relaxation, but they were somewhat shy in unfamiliar surroundings. Housed in the same way, the terriers engaged in fierce battles and formed strong dominance hierarchies within litters. When handled in tests their muscle tension was high, and they were more likely to attack than to retreat from a threat (Scott & Fuller, 1967).

Thus, it was not surprising that the effects of isolation were more striking in beagles than in terriers. In one experiment, beagles, transferred to the test arena in a small cage, often failed to explore the new environment; terriers rapidly overcame an initial hesitation and reacted appropriately to objects, people, and other puppies (Fuller & Clark, 1966).

In another set of experiments, beagles and terriers who had been isolated from weaning to 15 weeks of age were compared with pet-reared siblings (Fuller & Clark, 1968). Over a period of 5 weeks each subject was observed 25 times in the arena and records made of behaviors such as responding to a handler's call, following a moving person, playing with toys, and interacting with another puppy of the same age.

A quantitative scoring system was devised for two aspects of behavior, *response intensity* to key stimuli and *activity index*. The components and ordinal values of each measure are shown on the figures depicting the results of the study. The response indices of pet-reared (control) beagles and terriers were practically identical (Fig. 8.2). In both breeds, isolates were inhibited in responding during early tests. Intensity scores rose with repeated exposure to the arena; the rate of increase was much higher in terriers. At the end of the study isolated beagles were still inhibited compared with controls. On the second measure pet-reared terriers were more active than similarly treated beagles throughout the observations (Fig. 8.3). During early tests, isolates of both breeds

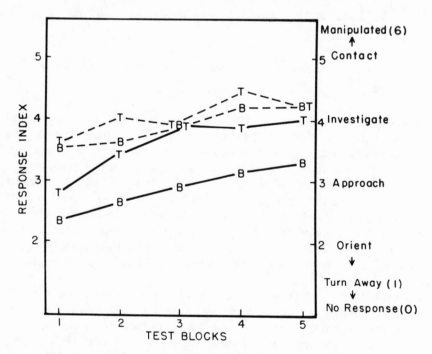

FIG. 8.2. Effect of isolation experience on mean intensity of response. B, beagles; T, terriers. Solid lines, isolated; dashed lines, pet-reared controls.

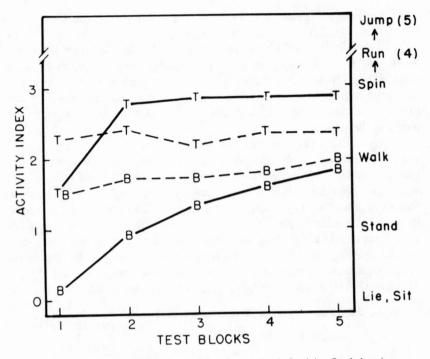

FIG. 8.3.   Effect of isolation experience on mean level of activity. Symbols as in
Fig. 8.2.

were inactive, as had been anticipated from previous experiments and, as expected, activity increased in both breeds with repeated trials. Beagles were
approximately equal with their controls by the fifth week. Terriers, however,
were significantly more active than their controls by the second week.

At first glance this might appear to be a disordinal interaction. Experiential
deprivation makes terriers more active; it makes beagles less so. In terms of overt
behavior this is certainly true. But there are other ways of looking at the data. In
both breeds isolation produced an initial aversion to the arena, reduced activity,
and decreased the intensity of response to the test stimuli. Both breeds recovered
with experience, but the terriers improved more rapidly and more completely. To
be sure, the isolated terriers overshot and became more active than their controls;
isolated beagles were less active than their controls. In a sense, however, isolation acted similarly on both breeds in exaggerating the characteristics that differentiated them when they were raised conventionally. Normally active and inves-
tigative terriers became more so. Deliberate and cautious beagles became
hesitant and timid. It appeared that isolation had made both breeds more emotionally reactive, but they expressed the change in divergent genotype-typical
patterns.

How Important Are Genotype-Environment Interactions?

Many examples of GE interactions have been reported in the behavior-genetics literature. A representative sample is contained in Table 8.1. It is not difficult to find these interactions if one looks for them. One need only expose a number of diverse selected or inbred lines to experiential deprivation, environmental enrichment, or stresses such as electric shock, malnutrition, or extreme temperature. It is unlikely that all the strains will respond identically to these treatments and the result is a GE interaction. The results of most of these studies, perhaps all, are compatible with the variable threshold model already presented.

How important are these interactions to the science of ethology? Natural species, except in some instances of sexual dimorphism or differentiation into castes, are not as phenotypically diverse as beagles and terriers, or C57BL and DBA mice. Behavior geneticists work with selected and inbred lines whose genotypes and phenotypes deviate in many ways from those of their free-living ancestors. Domestication may have had less effect on the basal repertoire of a species than has been commonly believed. Nevertheless, it has affected the intensity and frequency of forms of behavior that are important for survival. Quantitative as well as qualitative changes in behavior have implications for fitness. Behavior geneticists probably have overemphasized the importance of GE interactions and underrated the essentially uniform patterns that characterize a species. Thiessen (1972) has criticized our overconcentration on variability for its own sake and asserted that much of it is "genetic junk." Sex, age, seasonal factors, population density, and quality of habitat may be more potent than intraspecific genetic diversity in producing behavioral variation. But if GE interactions have been overrated by behavior geneticists they may have been underrated by ethologists.

All adequately studied species turn out to be genetically heterogeneous. Environmental effects observed on a small sample of these genotypes should not be applied uncritically to an entire species. When a sufficiently large number of individuals are observed, variability in response is usually found. If any part of this variability is a function of genetic differences we have the essential condition for natural selection and evolution. Theory of the evolution of behavior outruns empirical data on the relation between genes and adaptive behavior.

The capacity to adapt behaviorally to environmental change can evolve through two distinct mechanisms. In one case genetic differences are associated on a one-to-one basis with specific responses. The left-hand side of Fig. 8.4 depicts this situation. Genotypes a and b are associated with different phenostable courses of development. G_a is more fit in environment A; G_b in environment B. Genotypes may also exist that favor maximum adaptive phenolability. This situation is shown on the right side of the figure. Genotype G_o (o for options) is able to respond to environmental change and to produce viable progeny in either

TABLE 8.1

Examples of Genotype-Environment Interactions

Subjects	Treatments	Behavioral Measure	Results	Reference
Six strains of rat in diallel cross.	Pre- and post natal environment.	Open field defecation score. (Emotionality).	Variable maternal effects on offspring emotionality.	Broadhurst & Jinks, 1966.
McGill bright and dull rats.	Impoverished, standard and enriched rearing.	Performance in Hebb-Williams maze.	Both strains poor in impoverished environment. Both good in enriched. Brights excel in standard.	Cooper & Zubek, 1958.
C57BL and Swiss albino mice.	Reared by mother vs. fostered on rat.	Conflict among offspring.	Conflict reduced in fostered C57; no effect on Swiss.	Denenberg, 1977.
Four inbred strains of mice in diallel cross.	Undisturbed, mildly stressed, strongly stressed.	Activity and defecation in open field.	Extreme variation in treatment effects.	Henderson, 1967.
C57BL/6J and BALB/cJ mice.	Maintenance at 5° or $22^\circ C$.	Size of nest in home cage.	BALBs make larger nests at 5°. Little change in C57 nests.	Lynch & Hegmann, 1973.
C3H/HeJ and JK mice.	Enriched vs. standard rearing.	Hoarding of food pellets.	Enrichment increased hoarding in JK; no effect on C3H.	Manosevitz, Campenot & Swecionis, 1968.
Three inbred strains of mice.	Enriched vs. standard rearing.	Open field activity & defecation; activity in running wheel.	Rearing influences both activity measures; no effect on defecation.	Manosevitz & Montemayor, 1972.

G-E Interaction and Selection

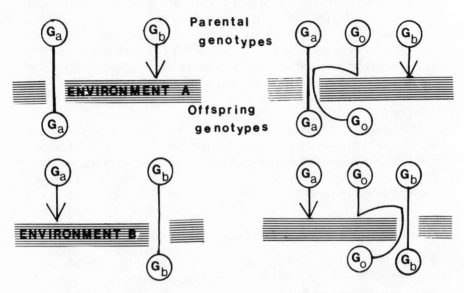

FIG. 8.4.   Genotype–environment interaction and selection. Left: Genotypes G_a and G_b compete in each of two environments, A and B. In A, G_a is more fit; a shift to B decreases fitness of G_a and increases the fitness of G_b. Right: The same two environments with a third competing genotype, G_o, an opportunistic genotype. G_o is able to propagate in either environment and is favored over G_a and G_b when the environment fluctuates between A and B.

environment. If a species occupies a constant habitat (A, for example) G_a will be favored over both G_b and G_o. If the environment fluctuates between A and B, G_o is favored over both competitors. Whether G_o is associated with the behavioral phenotype appropriate for A or for B depends on environmental input. But note that the phenolability associated with the capacity to respond adaptively to environmental challenges is as genotype dependent as the fixed action patterns that have interested ethologists for decades. In fact, phenolability is most characteristic of animals with large brains, and it must require a more complex genetic foundation than is needed for the organization of stereotyped responses. Behavior geneticists might well devote more attention to genotypic influences on the lability of behavior than on its fixed characteristics.

GENETIC ARCHITECTURE AND BEHAVIORAL
PHYLOGENY

Ethology is grounded in evolutionary concepts. Roe and Simpson's (1958) *Behavior and Evolution* brought together the views of many distinguished scientists and is still an excellent introduction to behavioral phylogeny. The value of the comparative method in deducing the phylogeny of innate behavior patterns, particularly communication by sound, posture, and gesture, was emphasized earlier by Lorenz (1950). However, there are problems with the application of evolutionary theory to behavior. The validity of the very idea of behavioral phylogeny has been criticized (Klopfer, 1976) on the ground that it is impossible to distinguish between similarity due to common ancestry (homology) and that attributable to common selective pressures (analogy). His warning is appropriate, but it need not lead to the abandonment of the consideration of the role of behavior in evolution.

In this presentation we are not concerned as much with homology and phylogeny as with the role of behavior as a component of fitness and, therefore, as a directive force in evolution. Consideration of this topic requires a brief outline of theories relating the genetic architecture underlying variation in a behavioral character to the character's role in natural selection.

Mendelian Approaches to Behavior Adaptation

We can distinguish between approaches based on one or two loci (Mendelian analysis) and those based on biometrical analysis of quantitative variability. The latter type of variation is considered to depend basically on Mendelian principles, but the numerous loci that are involved are not individually identifiable. Mather (1949) called these polygenes and we now apply the term polygenic inheritance to many behavioral traits. Both the Mendelian and the biometric approaches are important to behavior genetics. Benzer's studies of mutants with well-defined behavioral peculiarities have advanced our knowledge of neural development and function in *Drosophila*. The neurological mutants of mice have contributed to our understanding of brain development and function in mammals (Sidman, 1970). Coat color mutants, such as albinos, have been studied frequently by behavior geneticists. Albinos do differ from pigmented animals in many aspects of behavior (Fuller & Thompson, 1978), and even normally pigmented carriers of the albino allele are impaired in some phases of development (Henry & Haythorn, 1975). However, neurological mutants and albinos are so rare in natural populations that they cannot be classified as true polymorphisms. We know that genetic polymorphisms are extremely common and the possibility that some of them have behavioral significance is well worth investigating.

This research approach may seem to contradict the views of Mayr (1970), who writes:

> An individual, the target of selection, is not a mosaic of characters each of which is the product of a given gene. Rather, genes are merely the units of the genetic program that govern the complicated process of development, ultimately resulting in the phenotypic character. To consider genes as independent units is meaningless from the evolutionary point of view because the individual as a whole is the unit of

selection. To regard genes as independent units is meaningless from the physiological viewpoint because genes interact with each other in producing the various components of the genotype [p. 162].

For Mayr, the genotype has properties not deducible from the sum of its components. I agree. But he also acknowledges the possibility of switch genes where an allelic substitution at a single locus may shift the phenotype so that the direction and intensity of selection pressure is modified. A familiar example is industrial melanism, the replacement of light-colored variants of several species of moths by darker variants in areas where pollution has altered the color of the bark of the trees on which the moths rest.

How much evidence is there for switch genes that change behavior drastically enough to influence evolutionary processes? Sociobiologists seem to assume that such genes are abundant enough to be important. Trivers (1971) writes: "Assume that the altruistic behavior of an altruist is controlled by an allele (dominant or recessive) a_2, at a given locus, and that there is only one alternative allele, a_1, at that locus and that it does not lead to altruistic behavior [p. 36]." I have doubts that a simple allelic substitution could switch such complex behaviors as altruism and selfishness or change the mating system of a species from polygamy to monogamy. It is hard to imagine how such a switch gene would operate physiologically. Of course, any allelic substitution that affects sensory, neural, or motor functions will have some effect on behavior, but a single mutation seems unlikely, by itself, to shift behavior into promising new channels. However, there are reports suggesting that some genetic polymorphisms have behavioral correlates that affect selective processes indirectly.

Krebs, Gaines, Keller, Myers, and Tamarin (1973) found large fluctuations in the frequency of alleles at the transferrin (*Tf*) and leucine aminopeptidase (*LAP*) loci in the vole, *Microtus pennsylvanicus,* that were associated with changes in population density. They believe that the gene frequency changes over time were indicative of an association between the alleles at these loci and the ability to survive and propagate in stressful situations. Specifically, they hypothesized that females heterozygous at the *Tf* locus were apt to migrate during periods of increasing population density. Homozygotes, particularly for the Tf^c allele tended to remain in their original territory. The implication is that the *Tf* locus has something to do with social interactions, but, to my knowledge, this relationship has not been verified by direct observation of individual encounters. Krebs et al.'s data were obtained from field experiments supplemented by genotype determination of captured animals (Myers & Krebs, 1971).

A group of immunogeneticists have applied a very different approach to the genetics of social behavior (Andrews & Boyse, 1978; Yamaguchi, Yamzaki, & Boyse, 1978; Yamzaki, Yamaguchi, Andrews, Peake, & Boyse, 1978). Andrews and Boyse claim to have mapped two loci related to mating preference in house mice. They hypothesize that one of these, *Ris* (recognition of identity signal) provides a cue, probably olfactory, by which males can distinguish preferred females. *Ris* is assigned a location in the *Qa–Tla* region of the major histocompatibility region of the mouse. Another locus, *Rir* (recognition of identity response) is assumed to determine whether male response is positive or negative. The genetic procedures are complex and ingenious, but I have reservations regarding the validity of their conclusions. Males were given five to six repeated tests to determine their sexual preference. The significance of deviations from the null hypothesis (random choice) was apparently computed on the basis of the total number of tests rather than on the number of individuals. The receptivity of the rejected females was confirmed by pairing them with other males, but one cannot exclude the possibility that factors other than identity recognition by males bias the possibility of a female being bred. In spite of these experimental problems, the hypothesis that identity recognition based on cellular characteristics is important in mating choice has evolutionary implications. More stringent statistical analysis and more adequate behavioral controls are needed.

Isolation between closely related taxa need not imply a behavioral barrier. An example that involves a physiological difference with a relatively simple genetic basis has been reported in the lacewing flies, *Chrysopa carnea* and *C. downsei* (Tauber, Tauber, & Nechols, 1977). In the laboratory these species interbreed readily. In nature, crossing is not observed. *C. carnea* produces three generations per summer; *C. downsei* produces one, then enters diapause. Standard Mendelian crosses were made between the two species and the occurence of diapause in the hydrids was observed. The data are consistent with control of diapause by two loci. The *C. downsei* phenotype (diapause) requires recessive-allele homozygosity at both loci. This case should probably not be called behavioral isolation in the strict sense, as the two species will mate when they are artificially manipulated. It does demonstrate that relatively simple genetic systems may have substantial evolutionary consequences.

QUANTITATIVE INHERITANCE: GENETIC ARCHITECTURE AND EVOLUTION

Let us now consider deductions that can be made regarding behavioral evolution from knowledge of the genetic architecture that regulated quantitative characteristics. It is well known that variation in such traits is largely polygenic. The theory relating genetic architecture to the evolution of behavior has been explicated by several investigators (Broadhurst & Jinks, 1966; Bruell, 1967; Henderson, 1978). Essentially, stabilizing selection for an intermediate expression of a trait, and directional selection for high or low values involve different genetic strategies. Stabilizing selection requires either additivity of gene effects, so that a heterozygote is midway between the two homozygotes; or a balance between dominant genes at different loci, some with increasing, some with decreasing action on the phenotype. Traits with such a history respond well to selection in the laboratory and are said to be highly heritable. In contrast, selection for the extreme manifestations of a trait favors alleles with dominant effects in one direction only. Such a system ensures that the favored alleles will be effective in a single as well as in a double dose. Over time, recessive alleles will tend to disappear from the selected population, and the heritability of the trait will become low. Clearly, it is most efficient for a species to have those traits most directly related to fitness regulated by dominant genes. Falconer (1960) noted that in domestic animals heritability is lower for traits such as fertility and viability, than for traits desired by breeders for purely economic considerations (percent of butterfat in milk, size of eggs).

Accepting these principles, one can work backward from determination of the genetic architecture of a trait to a reconstruction of the role of that trait in natural selection. High unidirectional dominance is evidence that extreme values of that trait increase fitness significantly. Balanced bidirectional dominance implies a history of selection for intermediate values of the trait. High additive variance indicates that a trait has not been important for fitness, or that selective pressures are inconsistent and that the ability to vary phenotypes rapidly in response to a change in environment is valuable.

There is an increasing number of studies in which the genetic architecture of a behavioral characteristic in domesticated strains has been used as the basis for inferences regarding the relation of that trait to fitness. Bruell (1964, 1967) demonstrated that F_1 hybrids between inbred strains of mice almost always surpassed both parental strains on activity tests. The poorer performance of the parents was attributed to the inevitable, random loss of some dominant alleles during inbreeding. These losses generally involve different loci in any two strains; thus, in their hybrid offspring the deficiency of one is compensated by the contribution of the other. As might be expected, Bruell found that heterosis was more pronounced in crosses whose parents were distantly related than in crosses between lines with similar origins.

More sophisticated modes of genetic analysis were used by Broadhurst and Jinks (1966), who measured open-field activity and defecation in six lines of rats and all possible F_1 hybrids. This diallel-cross design provides estimates of many genetic parameters. Their genetic findings were: (1) genes enhancing developmental phenostability tend to be dominant; (2) dominance also occurs for rapid decrease of emotional defecation on successive tests in the open field; and (3) exploratory activity is also regulated by dominant genes, but these may have either a negative or a positive sign. From these observations they concluded that selection has operated to: (1) buffer the fetal and infant rat from external stress; (2) promote a rapid return to normal emotional status after arousal; and (3) strike a balance between bold curiosity that could lead to disaster, and timidity that could cause exclusion from essential resources. None of this is surprising, but it is satisfying to find that the genetic findings are consonant with our expectations of what organisms should do in order to prosper.

The genetic approach to reconstruction of evolutionary history is gaining ground, but there is difference of opinion regarding the best way to proceed. For example, Lynch (1977) concluded that: "because inbred mice are not weakling, analysis of crosses between them may represent a more efficient method for detecting genetic architecture than inbreeding of wild populations [p. 534]." Smith (1978) disagrees on the basis of his comparison of the heritability of shuttle avoidance in selected wild mice and in inbred strains. He doubts that it is safe to deduce the selection history of a hypothetical parent population from the genetic architecture of contemporary inbred strains of largely unknown origin. Obviously, more research is needed to settle this matter.

Henderson (1978) made an effort to escape the bind of first accepting the basic hypothesis of the relation between genetic dominance and fitness of a trait, then conducting a genetic experiment and, from its results deducing the fitness value of that trait. He deduced from observation that a particular trait was related to fitness and then conducted a breeding experiment to determine its genetic architecture. Specifically, he reasoned that mice should be selected for low levels of motor activity during infancy because they are safest when they remain in their nest. If they are accidentally dislodged their chances of being retrieved are greater if they remain quiet and await their mother. He predicted that, in direct contradiction to the situation in adults, low activity would be dominant in infant mice. A triple-test cross confirmed his prediction; the heritability of activity was low and inactivity was dominant. Hybrid infants were less active than inbred infants; the opposite was true in the Bruell studies of adults previously described. The offspring of newly trapped wild mice were even less active than domestic hybrids. This was attributed by Henderson to a relaxation of selection for infant immobility over many generations of laboratory rearing.

In a very different species, Cheng, Shoffner, Phillips, and Shapiro (1979) studied imprinting to visual and auditory stimuli in five groups of mallards (*Anas platyrhyncos*) ranging from a long-established game-farm line to 25%, 50%, 75%, and 100% wild ancestry. Wild ducklings imprinted better than domesticated ducklings, and heterosis was found in the hybrids, as would be predicted from the hypothesis that traits with adaptive value show little additive variation. Efficient imprinting seemed to be correlated with differences in arousal during early exposure to stimulation.

Confirmations of a prediction do not prove the universal value of a method. However the approach of Henderson and Chang et al. is promising and has considerable potential. As Henderson states, the method is most powerful when predictions can be made that a particular kind of behavior will differ in its contribution to fitness at different parts of its life-span, or in dissimilar environments. The method is obviously limited to species in which multiple distinct lines or strains are available. Principles derived from such studies are, however, likely to have general application.

It is only fair to state that the usefulness of this approach to defining adaptive behavior has been challenged on the basis that the experimental subjects available for analysis of genetic architecture are not representative of natural populations, and that a history of inbreeding distorts the genetic structure typical of a species under natural selection (Maxson, 1973). These warnings must be kept in mind but much of the data obtained in this way seems to substantiate accepted ideas of the characteristics of fitness-enhancing behavior. There is, of course, a possible circularity in such experiments. One can hypothesize that high levels of behavior B_1 are adaptive and thus should have increased fitness during recent evolutionary history. Suppose now that an experiment is performed and low levels of B_1 are found to be inherited in dominant fashion. One could conclude that the original hypothesis was wrong, and that actually low levels of B_1 enhance fitness. Alternatively one might consider the possibility that B_1 has little to do with fitness but, in the population studied, it is correlated genetically with another behavior, B_2, that has been selected on the basis of enhancing fitness. The best way out of such circularity is good ethological studies relating B_1 and B_2 independently to survival and propagation.

Another problem with the attempt to deduce the selective importance of a particular form of behavior is the choice of the method of measuring it. Behavior geneticists have been sure for years that in an open field C57BL mice were more active than BALB/c mice. Whitford and Sipf (1975) found no important differences between these strains when the height of their open field was reduced from 36 inches to 1 inch by insertion of a plexiglass sheet. It is clear that open-top, open-field activity is very different from contact-top, open-field activity. I suspect that the 1-inch ceiling comes closer than the traditional high-sided box to the life-style of real mice. No matter how sophisticated the genetic and technical procedures may be, they are useful only to the extent that they are relevant to the animal's adaptation in its natural environment.

SUMMARY AND CONCLUSIONS

Ethologists and behavior geneticists have many interests in common. Ethologists deal with a great range of species in natural or seminatural conditions. They go far beyond descriptive natural history in a search for general principles that will explain the origins and functions of an animal's ethogram. All this information on behavioral phenotypes is essential for a behavior geneticist working with a particular species. In reciprocity the techniques of behavior genetics may be useful in three areas of ethological research.

1. The explanation of individual and group differences in species-typical behavior.
2. The degree of flexibility in developmental programs that are encoded in the genotype and fixed at fertilization. Both stability and lability of development are joint functions of genotypes and environments.
3. The deduction of the adaptive value of specific forms of behavior by analysis of the genetic architecture underlying variation in these behaviors.

REFERENCES

Andrews, P. W., & Boyse, E. A. Mapping of an H-2 linked gene that influences mating preference in mice. *Immunogenetics*, 1978, *6*, 265–268.

Barnett, S. A. *A study in behaviour*. London: Methuen, 1963.

Benzer, S. Genetic dissection of behavior. *Scientific American*, 1973, *229*(3), 24–37.

Boice, R. Burrows of wild and albino rats: Effects of domestication, outdoor raising, age, experience, and maternal state. *Journal of Comparative and Physiological Psychology*, 1977, *91*, 649–661.

Broadhurst, P. L., & Jinks, J. L. Stability and change in the inheritance of behaviour in rats: A further analysis of statistics from a diallel cross. *Proceedings of the Royal Society, B.*, 1966, *165*, 450–472.

Bruell, J. H. Heterotic inheritance of wheel-running in mice. *Journal of Comparative and Physiological Psychology*, 1964, *58*, 159–163.

Bruell, J. H. Behavioral heterosis. In J. Hirsch (Ed.), *Behavior-genetic analysis*. New York: McGraw–Hill, 1967.

Cheng, K. M., Shoffner, R. N., Phillips, R. E., & Shapiro, L. J. Early imprinting in wild and game-farm mallards (*Anas platyrhyncos*): Genotype and arousal. *Journal of Comparative and Physiological Psychology*, 1979, *93*, 929–938.

Connor, J. L. Genetic mechanisms controlling the domestication of a wild house mouse population (*Mus musculus*, L.). *Journal of Comparative and Physiological Psychology*, 1975, *89*, 118–130.

Cooper, R. M., & Zubek, J. P. Effects of enriched and restrictive early environments on the learning ability of bright and dull rats. *Canadian Journal of Psychology*, 1958, *12*, 159–164.

Denenberg, V. H. Interactional effects in early experience research. In A. Oliverio (Ed.), *Genetics, environment and intelligence*. Amsterdam: North–Holland, 1977.

Dewsbury, D. A. Copulatory behavior of deer mice (*Peromyscus maniculatus*): I. Normative data, subspecific differences and effects of cross-fostering. *Journal of Comparative and Physiological Psychology*, 1979, *93*, 151–160. (a)

Dewsbury, D. A. Factor analysis of measures of copulatory behavior in three species of muroid rodents. *Journal of Comparative and Physiological Psychology*, 1979, *93*, 868–878. (b)

Dewsbury, D. A., & Lovecky, D. V. Copulatory behavior of old field mice (*Peromyscus polionotus*) from different natural populations. *Behavior Genetics*, 1974, *4*, 347–355.

Ebert, P. D., & Hyde, J. S. Selection for agonistic behavior in wild female *Mus musculus*. *Behavior Genetics*, 1976, *6*, 291–304.

Falconer, D. S. *Quantitative genetics*. New York: Ronald Press, 1960.

Feldman, M. W., & Lewontin, R. C. The heritability hangup. *Science*, 1975, *190*, 1163–1168.

Fuller, J. L. Hereditary differences in the trainability of purebred dogs. *Journal of Genetic Psychology*, 1955, *87*, 229–238.

Fuller, J. L. Effects of experiential deprivation upon behaviour in animals. *Proceedings of the Third World Congress of Psychiatry* (Vol. 3) Toronto: University of Toronto Press, 1964. (a)

Fuller, J. L. Measurement of alcohol preference in genetic experiments. *Journal of Comparative and Physiological Psychology,* 1964, *57,* 85–88. (b)

Fuller, J. L. Experiential deprivation and later behavior. *Science,* 1967, *158,* 1645–1652.

Fuller, J. L., & Clark, L. D. Genetic and treatment factors modifying the postisolation syndrome in dogs. *Journal of Comparative and Physiological Psychology,* 1966, *61,* 251–257.

Fuller, J. L., & Clark, L. D. Genotype and behavioral vulnerability to isolation in dogs. *Journal of Comparative and Physiological Psychology,* 1968, *66,* 151–156.

Fuller, J. L., & Thompson, W. R. *Foundations of behavior genetics.* St. Louis: Mosby, 1978.

Henderson, N. D. Prior treatment effects on open field behavior in mice: A genetic analysis. *Animal Behaviour,* 1967, *15,* 364–376.

Henderson, N. D. Genetic dominance for low activity in infant mice. *Journal of Comparative and Physiological Psychology,* 1978, *92,* 118–125.

Hendrichs, H. Die soziale Organisation von Säugentierepopulationen. *Saugetierkundliche Mitteilungen, München,* 1978, *26,* 81–116.

Henry, K. R., & Haythorn, M. M. Albinism and auditory function in the laboratory mouse. I. Effect of single gene substitutions on auditory physiology, audiogenic seizures and developmental processes. *Behavior Genetics,* 1975, *5,* 137–149.

Jerison, H. J. *Evolution of the brain and intelligence.* New York: Academic Press, 1973.

Katzev, R. D., & Mills, S. K. Strain differences in avoidance conditioning as a function of the classical CS–US contingency. *Journal of Comparative and Physiological Psychology,* 1974, *87,* 661–671.

Kawamura, S. The process of sub-culture propagation among Japanese macaques. *Journal of Primatology* (English ed.), 1962, *2,* 43–60.

King, J. A. Behavioral comparisons and evolution. In A. Oliverio (Ed.), *Genetics, environment and intelligence.* Amsterdam: North–Holland, 1977.

King, J. A., Price, E. O., & Weber, P. L. Behavioral comparisons within the genus Peromyscus. *Papers of the Michigan Academy of Science, Arts, and Letters,* 1968, *53,* 113–136.

Klopfer, P. H. Evolution, behavior and language. In M. E. Hahn & E. C. Simmel (Eds.), *Communicative behavior and evolution.* New York: Academic Press, 1976.

Krebs, C. J., Gaines, M. S., Keller, B. L., Myers, J. H., & Tamarin, R. H. Population changes in small rodents. *Science,* 1973, *179,* 35–41.

Lorenz, K. Z. The comparative method in studying innate behavior patterns. *Symposia of the Society for Experimental Biology, 4.* New York: Academic Press, 1950.

Lorenz, K. Z. *Evolution and modification of behavior.* Chicago: University of Chicago Press, 1965.

Lynch, C. B. Inbreeding effects upon animals derived from a wild population of *Mus musculus. Evolution,* 1977, *31,* 526–537.

Lynch, C. B., & Hegmann, J. P. Genetic differences influencing behavioral temperature regulation in small mammals. II. Genotype–environment interactions. *Behavior Genetics,* 1973, *3,* 145–154.

Manosevitz, M., Campenot, R. B., & Swencionis, C. F. Effects of enriched environment upon hoarding. *Journal of Comparative and Physiological Psychology,* 1968, *80,* 319–324.

Manosevitz, M., & Montemayor, R. J. Interaction of environmental enrichment and genotype. *Journal of Comparative and Physiological Psychology,* 1972, *79,* 67–70.

Mather, K. *Biometrical genetics.* London: Methuen, 1949.

Mayr, E. *Populations, species and evolution.* Cambridge, Mass.: Harvard University Press, 1970.

Maxson, S. C. Behavioral adaptations and biometrical genetics. *American Psychologist,* 1973, *28,* 268–269.

McGuire, T. R., & Hirsch, J. Behavior-genetic analysis of *Phormia regina:* Conditioning, reliable individual differences, and selection. *Proceedings of the National Academy of Science, USA,* 1977, *74,* 5193–5197.

Miller, D. B. Social displays of mallard ducks (*Anas platyrhynchos*): Effects of domestication. *Journal of Comparative and Physiological Psychology,* 1977, *91,* 221–232.

Moss, R., & Watson, A. Inherent changes in the aggressive behaviour of a fluctuating Red Grouse (*Lagopus lagopus scoticus,* Lath.) population. *Ardea,* 1980, *68,* 113–120.

Myers, J. H., & Krebs, C. J. Genetic, behavioral and reproductive attributes of dispersing field voles, *Microtus pennsylvanicus* and *M. ochroaster. Ecological Monographs,* 1971, *41,* 53–78.

Plomin, R. J., & Manosevitz, M. Behavioral polytypism in wild *Mus musculus. Behavior Genetics,* 1974, *4,* 145–157.

Ploog, D., Hupfer, K., Jurgens, U., & Newman, J. D. Neuroethological studies of vocalization in squirrel monkeys with special reference to genetic differences in calling in two subspecies. In M. A. B. Brazier (Ed.), *Growth and development of the brain.* New York: Raven Press, 1975.

Plotkin, H. C., & Odling-Smee, F. J. Learning, change and evolution: An enquiry into the telemony of learning. In J. S. Rosenblatt, R. A. Hinde, C. Beer, & M-C. Busnel, *Advances in the study of behavior* (Vol. 10). New York: Academic Press, 1979.

Price, E. O. Effect of early outdoor experience on the activity of wild and semi-domestic deermice. *Developmental Psychobiology,* 1969, *2,* 60–67.

Price, E. O., & Loomis, S. Maternal influence on the response of wild and domestic Norway rats to a novel environment. *Developmental Psychobiology,* 1973, *6,* 203–208.

Roe, A., & Simpson, G. G. *Behavior and evolution.* New Haven: Yale University Press, 1958.

Rowell, T. E. Variability in the social organization of primates. In D. Morris (Ed.), *Primate ethology.* New York: Doubleday, 1969.

Scott, J. P., & Fuller, J. L. *Genetics and the social behavior of the dog.* Chicago: University of Chicago Press, 1967.

Selander, R. K. Behavior and genetic variation in natural populations. *American Zoologist,* 1970, *10,* 53–66.

Seligman, M. E. B., & Hager, J. L. *Biological boundaries of learning.* New York: Appleton–Century–Crofts, 1972.

Sidman, R. E. Cell proliferation, migration and interaction in the developing mammalian central nervous system. In F. O. Schmitt (Ed.), *The neurosciences: Second study program.* New York: Rockefeller University Press, 1970.

Smith, R. H. Selection for shuttle avoidance in wild *Mus musculus. Behavior Genetics,* 1978, *8,* 269–274.

Stacey, P. B., & Bock, C. E. Social plasticity in the acorn woodpecker. *Science,* 1978, *202,* 1298–1300.

Tauber, C. A., Tauber, M. J., & Nechols, J. R. Two genes control seasonal isolation in sibling species. *Science,* 1977, *197,* 592–593.

Thiessen, D. D. A move towards species-specific analyses in behavior genetics. *Behavior Genetics,* 1972, *2,* 115–126.

Trivers, R. L. The evolution of reciprocal altruism. *Quarterly Review of Biology,* 1971, *46,* 35–57.

Whitford, F. W., & Sipf, S. G. Open-field activity in mice as a function of ceiling height: A genotype–environment interaction. *Behavior Genetics,* 1975, *5,* 275–280.

Williams, G. C. A defense of monolithic sociobiology and genetic mysticism. *Behavioral and Brain Sciences,* 1981, *4,* 257.

Yamaguchi, M., Yamazaki, K., & Boyse, E. A. Mating preference tests with the recombinant congenic strain BALB–HTG. *Immunogenetics,* 1978, *6,* 261–264.

Yamazaki, K., Yamaguchi, M., Andrews, P. W., Peake, B., & Boyse, E. A. Mating preference of F_2 segregants of crosses between MHC-congenic mouse strains. *Immunogenetics,* 1977, *6,* 253–259.

9 Genetics of Social Behavior in Nonhuman Animals

J. P. Scott

Bowling Green State University

Definitions. Social behavior is behavior that either is stimulated by or has a stimulating effect on one or more members of the same species. It is normally expressed within a social system (i.e., as an interaction between two or more individuals). Social behavior thus characteristically involves mutual stimulation and response, with reciprocal causation. The unit of expression of social behavior is therefore not that of an individual but the interaction between two or more individuals, the behavior of each being dependent on that of the other. This makes difficulties for genetic analysis based on assumptions of independent assortment, as the expression of social behavior is dependent not only on the interaction of genes within an individual but also on interaction with genes from another individual (Scott, 1977; Scott & Fuller, 1965).

Theoretical Problems of Analysis

This problem can be approached by considering the nature of social systems and how they may be affected by genetic variation. This is in part an empirical problem but also one that may be approached theoretically through examining the nature of the social relationships that are the basic units of measurement.

Characteristics of a Social Relationship. The dyadic relationship is the least complex of social systems and therefore should ordinarily be the unit of choice in experimental and theoretical work. Such a relationship has all the general characteristics of a social system:

363

1. A primary characteristic of social systems is interaction and feedback. This implies two-way causation. In the early stages of formation of a relationship, the behavior involved is highly variable and relatively unorganized with respect to that of the other individual involved.

2. Interaction leads to mutual adaptation (i.e., a functional relationship is established).

3. The individuals involved tend to repeat a successful mutual adaptation with less and less variation. Repetition leads to the stabilization of the relationship through habit formation, after which it becomes highly predictable.

4. A stable relationship has certain characteristics, one of which is unequal reciprocal effects, even if the behavior involved appears identical. For example, in the leader–follower relationship between a lamb and its mother, both respond to each other but the mother has more effect on the behavior of the lamb than the reverse.

5. In many cases the behavior of two individuals in a well-developed relationship becomes qualitatively different. In a dominance–subordination relationship in hens, the dominant hen pecks and the subordinate hen avoids. Using a somewhat different terminology, the two individuals adopt different social roles.

6. Implied in the foregoing description is the fact that relationships change. The formation of a relationship is a process, and the results of measuring it will depend on when the process is measured. Obviously, the ideal way to measure a relationship is to do so continuously throughout its development.

Genetic variation theoretically can lead to different performances in each of the aforementioned characteristics. Factors other than genetic variation will produce important effects also: relative age of the members of a relationship and their previous experience, to mention only two. Assuming that these latter factors are kept relatively constant, and assuming the reality of genetic action on a particular social behavior, we can predict that each relationship developed by an organism with another organism will be different from every other relationship, because each relationship will involve a different pair of genotypes. It follows that variation in the nature of social relationships will be proportional to the genetic variation in the population concerned. An interesting question is: whether the nature of interaction on this level is additive or multiplicative. From the nature of behavioral adaptation one would predict the latter, that social interaction would magnify the effects of genetic variation.

Techniques for Analyzing the Effects of Genetics on Social Systems. Variation in social systems is obviously not an ideal tool for the analysis of genetic systems. A simple anatomical or physiological trait is far more convenient to use if one is interested in the placement of genes on chromosomes and the like. Therefore, proper emphasis in this field should be placed on the manipulation of genetics as an independent variable, with social systems as the dependent variable. Genetics then becomes a tool for the analysis of social systems.

Standard genetic techniques depend on two general situations. In one the genetic organization of the population is known or partially known. Under this condition the techniques fall into two classes, both employed under standard conditions of rearing: (1) the measurement of strain differences; and (2) cross-breeding experiments between two or more populations with known characteristics.

Adapting the strain comparison technique to the study of social systems can be done at varying levels of complexity. The first is the situation in which there are only two genetically different kinds of individuals, which may come from two different strains or even more simply may represent the two sexes within a strain. An obvious design is to combine the two types into every possible dyadic combination (e.g., male–male, female–female, and male–female combinations) and observe the system that develops in each combination. The same design can be applied to any number of inbred strains, or other genetically distinct populations, and the expected outcome is that the systems resulting from each type of interaction should be different.

A simplified version of this design is to compare dyads composed only of similar individuals (Fuller & Hahn's, 1976, homogeneous sets design). The results are thus limited to those systems that result from the interaction of identical or highly similar genotypes and do not give the full range of potential systems produced by the genotypes.

The preferred technique would then be to combine genotypes in all possible ways. This again can be done in two techniques. One is to measure an individual in only one dyad, but to duplicate the number of dyads so as to obtain a reasonably large sample. A second method is to use the round-robin technique, in which every individual is paired with every other individual in the population. The latter technique has the advantage that it gives a picture of a more inclusive system of linked dyads and has considerable intrinsic interest with respect to the analysis of the organization of larger social groups such as litters of dogs or flocks of hens. The former technique is preferable in that assigning the causes of variance is a much simpler matter (Kraemer & Jacklin, 1979). It is an empirical question as to whether the dyads in which the same individual is involved are developed independently from each other. In dog groups the same animal may be dominant in one relationship and subordinant in another, whereas in fighting mice an individual tends to be either a winner or a loser in all relationships.

Stewart (1974, pp. 267–276) has used the model of the diallel cross to set up all possible dyadic combinations of genotypes within and between different strains. In this technique it is possible to separate the effects of combining similar and dissimilar genotypes.

A second general method of manipulating known genetic variables is the crossbreeding experiment between two or more strains with known characteristics. The classical method is the improved Mendelian experimental design involving reciprocal crosses and the collection of data on F_1's, F_2's, and backcrosses in both directions, preferably with repeated backcrosses. The result is three kinds of homogeneous (nonsegregating) populations (the two parent strains and the F_1's) and three heterogeneous (segregating) populations, (the F_2's and the two backcross populations). Applying the technique of measuring dyadic systems developed within each of these populations, it would be expected that: (1) the systems developed within each population would differ from those in other populations; and (2) the systems developed in heterogeneous populations would be more variable than those developed in homogeneous populations.

A more modern crossbreeding technique is the diallel cross, where F_1 hybrids are obtained from all possible paired combinations within a group of several strains. Applied to social behavior, the simplest technique is to use only those systems that are developed within populations. This method is particularly well adapted to the analysis of genetic differences among inbred strains, as it gives a more general picture than can be obtained from detailed crosses between two strains. Genetically produced variance should occur only between pairs of populations rather than within them. Using reciprocal crosses controls not only for sex-linked inheritance but also for the possibility of mother–offspring effects. In such a cross, one could also measure the systems developed between individuals from different genotypic populations, which would give more comprehensive results but which might be quite difficult to interpret on an overall basis. A still more recent method is that of recombinant inbred strains (described in the section on Agonistic Behavior).

Situations also occur in which genetic variation is uncontrolled and its relevance to a particular characteristic is suspected but is actually unknown. This is the usual situation that confronts an experimenter at the outset of genetic research. The appropriate technique is selection. If the nature of the population can be changed as the result of selection, it may be concluded that genetics has an effect. Elaborate designs for selection experiments are available (McClearn & DeFries, 1973).

Applied to the problem of social systems, the results of selection will vary according to the stage of development of the system in which the selection is practiced. Most selection experiments have been done in the initial stages of the formation of a relationship, a time when the behavior of the individuals concerned is most variable. Also, behavior at the outset of a relationship may be only poorly correlated with its final expression in a well-developed social system. It would obviously be more meaningful to select from well-developed social systems, but this technique has been heretofore avoided in order to save time.

The other problem is what to select. This is somewhat arbitrary, depending on the interests of the experimenter, who could select either for some characteristic of the system as a whole, or for the role of the individual, or for both. Actually, it is not possible to select for individual characteristics independently of the system in which they are expressed. An example would be the selection for dominance in agonistic behavior. The expression of dominance is dependent on the expression of subordination by another individual. This question is related to the more general problem of what to measure in a system.

Measuring the Characteristics of a Social System.   Because a social system is very complex, it follows that the most meaningful experiments will include several measures of what seem to be the outstanding characteristics of the system. Among others we can list: (1) differences in the form and frequency of expressed behavior patterns by individuals; (2) the amount of variation in the initial interaction; (3) the type of adaptation that is reached through mutual interaction (e.g., in the case of agonistic behavior, peaceful interaction, mutual threat, and varying degrees of dominance); (4) the time it takes to reach a stable organization; (5) the degree of inequality or differentiation of behavior, as in the measurement of dominance; and (6) changes in the nature of any of these measurements in the course of development of the social system. Such developmental changes frequently give hints regarding the way in which genetic differences produce effects on social systems.

Techniques for Avoiding the Measurement of Expression of Behavioral Interaction in Social Systems.   Most experimenters in the past, including myself, have used special techniques in order to facilitate genetic analysis. The most common technique is that of the standard tester, (Fuller & Hahn, 1976), who may be a trained person, a trained animal, or an animal from an inbred strain that shows no genetic variation. The effect is to set up dyadic systems in which there is genetic variation in only one of the members, thus facilitating genetic analysis.

Training a person or animal to react in a standardized fashion, however, has the effect of inhibiting feedback, one of the basic characteristics of a social relationship. Scott & Fuller (1965) found that such a design had an unpredicted and undesirable side effect. Dogs that interacted with a tester whose behavior was standardized simply did not develop the complex social relationships of which dogs are capable in normal situations involving mutual causation and feedback, and this in turn limited capacities for training and performance.

A second method is to measure behavior only in the initial stages of the formation of a system. This has the advantage that the initial behavior may sometimes have a major effect on the final organization of the system (a mouse that attacks first has a good chance to win and become dominant), and it measures behavior at a time when it is relatively unmodified by learning and experience. However, as Scott and Fuller (1965) found, breed differences frequently do not appear in the early stages of the development of a relationship, as the animals simply have not yet made a choice from their potential behavioral repertoire.

Systems of Social Behavior: Some Needed Research

From a survey of social behavior in the animal kingdom (Scott, 1950, 1956) has developed the following list of major behavioral systems:

Investigatory
Shelter Seeking–Shelter Building

Ingestive
Sexual
Epimeletic (care giving)
Et-epimeletic (care soliciting)
Agonistic (evolved from defensive behavior)
Allelomimetic
Eliminative

Any of these behavioral systems may be incorporated into a social system, but only five—sexual, epimeletic, et-epimeletic, agonistic, and allelometic—are always social in their functions.

Interaction between individuals showing these behaviors results in the following major social relationships: sexual, dominance–subordination, care–dependency, and leader–follower. The behavior and genetic research that has been done within the context of these relationships has depended largely on the choice of the species. Behavior geneticists have tended to favor those species in which the genetics has been well worked out and is easily manipulable, but these are not always highly social species. Sexual behavior and sexual relationships are universal in animals that reproduce sexually. Agonistic behavior and dominance–subordination relationships are widely found in the vertebrate animals and occasionally in some of the higher invertebrates, and the practical interest in aggression and violence has also helped to focus research in this area. Care–dependency relationships between mothers and offspring are universal among mammals and almost universal among birds. Again, a considerable amount of research has been done on this behavior.

On the other hand, allelomimetic behavior (defined as similar behavior with some degree of mutual imitation) is seldom found in the insects and is missing in many mammals such as the mouse. Even where allelomimetic behavior occurs, leader–follower relationships are not always clear-cut or prominent. What little research has been done in this important area has been confined to the dog, where allelomimetic behavior is a major part of social behavior but where leader–follower relationships are developed chiefly in connection with human activities.

The following compilations of behavior-genetics research on social behavior therefore reflect these limitations. Research on sexual and epimeletic behavior has been done in a wide variety of species, that on agonistic behavior on a smaller number, and that on allelomimetic behavior is almost nonexistent.

Issues in Research on Behavior Genetics and Social Organization

Importance of Genetic Variation.   As with heritability, there is no general answer to this question. However, it is important to know whether there are many cases where genetic variance produces major modifications of social organization, or whether such effects are generally trivial. As pointed out previously, there are theoretical reasons for supposing that social organization may result in magnification of what are basically relatively small differences produced by genetics.

An important practical question is the effect of sex on variations in social behavior and social organization. For example, in mammals there is considerable evidence to show that both sexes have the capacity to exhibit almost all patterns of social behavior but that these capacities may be either enhanced or suppressed by associated anatomical and experiential conditions. On the other hand, sex in insects is associated with clear-cut differentiation of behavioral capacities.

Genetic Variation in Signaling Behavior.   Ethologists have been concerned with what they have called "species-specific" behavior, which is defined not as behavior specific to a species, but as behavior that is universal among species members. They frequently assume that such behavior is invariable, but they rarely measure variation among individuals. They usually assume in addition that interspecific variation is genetically determined, but in most cases there is no way of experimentally testing this assumption.

Behavior geneticists, on the other hand, have been chiefly concerned with genetics as it affects intraspecific variation and have analyzed all sorts of behavior, social and otherwise. Their results justify the general conclusion that any kind of behavior that has been extensively studied shows important genetic variation, and social behavior is no exception.

Herein lies an issue. The types of behavior that the older ethologists have described as fixed action patterns are almost entirely patterns of social behavior, and particularly signaling behavior. Are such patterns really exempt from genetic variation, and if so, why?

One explanation is that in behaviors that require fine coordination between individuals to be effective, as in many sorts of social behavior, there should be consistently strong selection against deviants. This would apply especially to behavior that conveys information, as variable signals would be unclear. Another possibility is that genes that cause such variation are unusually stable and seldom mutate. If so, it would be necessary to explain why these should be different from all other classes of genes, and one explanation would be the selection of genotypes in which such mutations are rare.

The Physiological Genetics of Social Behavior.   Another issue involves the ways in which genes affect social behavior and social organization. The mere demonstration of genetically induced variance is always unsatisfactory, as genetic processes may or may not be reversible, may be restrictive or facilitating, and may produce their observable effects on different levels. This is especially true of social behavior, as genetic differences appear either on the physiological level, as in emotional reactions, or only on the level of social interaction, or both.

Evolutionary Issues.   Finally there are a group of issues that have evolutionary implications. One of these is the issue of major versus minor gene effects. In characteristics in which so much nongenetic variation is possible, it can be argued that selection would be effective only with respect to major gene effects. On the other hand, minor genes still might be effective in nonselective processes of evolutionary change such as random drift.

Then there is the issue of the reciprocal effects between genetics and social organization. Genetic variation should produce variation in social organization, but social organization also can, through the process of selection, have an effect on genetic variation. Assuming such reciprocal effects, one prediction would be that stability of the genetic composition of the population would be maintained.

The fact that the expression of social behavior always requires interaction with at least one other individual of a separate genotype questions the generally accepted hypothesis that the unit of natural selection is always the individual organism (a single genotype). Basically, evolutionary change depends not only on the differential survival of individuals but on the survival of their respective progeny. In a sexually reproducing species, a given parent does not transmit its own genic system to its progeny but randomly chosen halves of that system that must combine with similarly chosen halves from one or more others. Thus, more than one genetic system is always involved in selection. This leads to procedures well known in selective animal breeding, those of progeny testing and the selection of mating pairs ("nicking") rather than of individuals per se.

The foregoing considerations apply to any characteristic showing genetic variation. In addition, the behavioral interactions representing two or more genotypes raise other complex problems regarding the process of selection. Differential production and survival of offspring could result from either competitive or cooperative interactions but should be achieved in different ways, through individual selection in one case and selection of combinations of individuals in the other. Further ramifications of polysystemic theory applied to evolutionary processes are discussed elsewhere (Scott, 1981 pp. 129–157).

REFERENCES

Fuller, J. L., and Hahn, M. E. Issues in the genetics of social behavior. *Behavior Genetics*, 1976, *6*, 391–406.

Kraemer, H. C., & Jacklin, C. N. Statistical analysis of dyadic social behavior. *Psychological Bulletin*, 1979, *89*, 217–224.

McClearn, G. E., and DeFries, J. C. *Introduction to Behavior Genetics*. San Francisco: Freeman, 1973.

Scott, J. P. The social behavior of dogs and wolves: An illustration of sociobiological systematics. *Annals of the New York Academy of Science*, 1950, *51*, 1001–1122.

Scott, J. P. The analysis of social organization in animals. *Ecology*, 1956, *41*, 385–402.

Scott, J. P. Social genetics. *Behavior Genetics*, 1977, *7*, 327–346.

Scott, J. P. The evolution of function in agonistic behavior. In P. F. Brain & D. Benton (Eds.),

Multidisciplinary approaches to aggression research. Amsterdam: Elsevier/North Holland, 1981. pp. 129–157.

Scott, J. P., & Fuller, J. L. *Genetics and the social behavior of the dog.* Chicago: University of Chicago Press, 1965.

Stewart, J. M. Genetic and ontogenetic determinants of agonistic behavior in animal societies: Prototypic experiments with mice. In J. deWit & W. W. Hartup (Eds.), *Determinants and origins of aggressive behavior.* The Hague: Mouton, 1974.

10 Epimeletic and Et-Epimeletic Behavior in Animals

John C. Gurski

Fort Hays State University

Purpose

This chapter reviews studies on genetic effects, interspecific and intraspecific, in parent–offspring and pre- and postweaning behaviors. It is divided into four subsections, followed by summary and conclusion statements. The first subsection is devoted to the definition of terms, and a discussion of behaviors that are commonly used as dependent measures. The second subsection reports on invertebrate species. The third subsection reviews vertebrate investigations of interspecies and intraspecies differences. Emphasis in those sections is on factors found to be significantly affected in the natural or laboratory behavioral repertoires of several species. Most studies of breed or strain differences included here are from the laboratory. The fourth subsection considers the question of the existence of a parental "instinct."

Terms and Definitions

The study of maternal–filial interactions can be separated into two mutually influenced systems: epimeletic or care-giving, and et-epimeletic or care-soliciting behaviors (Scott, 1968). Epimeletic behavior is provided postnatally in most species by biologically related conspecifics. In numerous instances care is shared by other members of the principal caregiver's group.

Epimeletic Behavior. Epimeletic behavior consists of two main subsystems that promote survival and safety for the offspring: provisioning and/or shelter giving. Provisioning may be physiologically mediated before the embryo is released into the external environment. The considerable effects of the prenatal environment in mannals and attempts to understand and control them by behavior geneticists are considered in later sections. Postnatally, nutrients are also provided to the offspring from internally produced secretions. The behaviors promoting the effective provisioning of the young (e.g., nursing in mammals) have been the subject of behavioral genetic investigation. Provisioning may also be through exogenous means, usually of two forms: (1) the parent may provide freshly caught prey; or (2) in the case of some wasps and other insects, the prey upon which the newly hatched larvae feed is immobilized and stored in the nest. Another form of exogenous provisioning is the regurgitation of partially digested or undigested food for the young, a form practiced by insects, avians, and carnivores.

Shelter giving is the other major component of epimeletic behavior. It involves several different types of behaviors in species that are postnatally altricial. (Altricial offspring are those that are relatively immobile and that cannot provision themselves for a period of time following birth). Shelter-giving behaviors generally involve the construction and maintenance of a hive, nest, or rearing site. The major component of this subsystem of epimeletic behavior is nest building, which takes two forms depending on whether or not the parent remains with the young to provide nutrients. When the parent remains, the nest is often of less permanent construction. An exception is found in the social and eusocial (Wilson, 1972) insects, where the nest is maintained permanently and rebuilt if destroyed or moved. In those species where the parent oviposits and departs, the nest is more enclosed and is often provisioned before the parent departs.

Shelter is either temporary or permanent. Temporary structures are maintained until the young are fledged or weaned. Permanent shelters outlast at least one set of offspring; an example is the beehive. Associated with the nest are behaviors that maintain it. The nest is cleaned; dead offspring may be removed or refuse may be disposed. The nest may be partially rebuilt.

A second component of shelter giving is the retrieval of the offspring to the nest. Offspring move from the nest or are removed from it by conspecifics or by predators. Parents or others may carry the young back to the nest and deposit it there. In numerous studies, retrieving has been used as an experimental test of strength of maternal responsiveness to altricial young, which do not usually leave the nest at early ages. Retrieving tests can be conducted also when at later ages altricial young spontaneously leave the nest site.

The third component of shelter giving is the defense of the nest and its inhabitants from conspecifics or predators. It is possible to broaden this category to include defense of the offspring from temperature extremes; included in this broadened definition, then, is huddling or shielding of the offspring by the parent.

Et-Epimeletic Behaviors. The et-epimeletic, care-soliciting system is generated by the behaviors of the offspring when they are hungry, or when they are isolated from caregiver, conspecifics, and/or nest site. Regarding nutrition, the young may solicit feeding from the parent; also, the young of many species may approach the provisioning parent in search of food. The young of several vertebrate species exhibit distress calls or vocalizations when involuntarily separated from the nest or if the direction to the nest is lost. These calls have been experimentally separated into distress from physical discomfort and distress due to separation from parent(s), littermate(s), or nest. Gurski, Davis, and Scott (1980) demonstrated that in the dog physical distress can be signaled from shortly after birth when the pup is placed in an open box with aluminum floor at room temperature. Separation distress, which is elicited when the pup is placed in a warm and physically congenial environment that mimics the contact cues of the litter but not its social quality, does not occur until well into the third week after birth. Although breed and hybrid effects were observed, they were not as strong as the developmental effects across breeds. Distress vocalizations have been studied in rodents, which emit ultrasounds that are influenced by the physical environment (e.g., cold stress) as well as the social environment. There have been no studies of the rodent that specifically discriminate discomfort due to cold from that due to social isolation.

Postweaning Tests. In mammals, especially laboratory colonies, postweaning tests of spontaneous activity, "emotionality," and developmental levels are used: weight of young at selected postweaning intervals, the open-field test of spontaneous activity, and defecation in the open field. Some researchers have used water escape and hole-in-wall tests; measures of approach or avoidance learning have been at times employed (Reading, 1966). Additionally, postweaning measures of "attachment" of offspring to parents have employed distress vocalization as tests of attachment in those species that call to each other.

Not all investigators in the area are in complete agreement on the validity of the various measures. For example Archer (1973) examined emotionality in "novel environment" tests and offered criticisms of open-field ambulation as well as defecation. Some alternative tests were proposed in that review. But in spite of the controversy surrounding its use as a measure, and the high heritability associated with activity levels in different strains of mice and rats, use of the open-field test is still generally accepted and scores so derived provide a reliable comparison between studies. Each animal is tested for 3 to 5 minutes in the enclosure, which is divided evenly into sections, the moving of the animal from section to section being counted over the period of testing. Other measures such as latency to move, latency to move to the center of the field, rearing on hindpaws, etc., have been used at various times.

Invertebrate Investigations

Organization. The only group in which extended care of offspring is practiced, and in which genetical investigation has progressed, is the bees (Apoidea). Consequently this section concentrates on pertinent studies in bees. Some comparative information would be of value in understanding the advantages of breeding experiments in the bee; this information is presented first.

There are some 20,000 species of bees, ranging in social organization from solitary to highly social. The literature on the bee is adequate for genetical investigation (see for example Michener, 1969). Colonies are structures based on degrees of sociality and methods of provisioning. Only instances of progressive provisioning of the offspring will fit the given definition of epimeletic behavior, as mass-provisioned larvae are enclosed in cells until it is time for them to become part of the colony. Also, progressively provisioned offspring are part of the colony structure from their first days. Colonies can be monogynous (one queen) or polygynous. Communal nests occur; defense and centralization of location seem to be the causes, and not provisioning, as each female most likely builds its own nest and independently provisions its offspring. Some communal colonies share in provisioning (Michener, 1969). Progressive provisioning occurs only in *Bombus* (bumblebee) and *Apis* (honeybee) to any consistent degree.

Because provisioning is the most complex epimeletic behavior that is practiced in this group, it is the one on which this review concentrates. The young are fed different diets depending for the most part on the caste to which they are destined. The "jelly" secretion that all young are initially fed is a varied proportion of clear and white components. Queens are fed these components in a 1:1 ratio, workers in a 3:1 ratio. The queen-designate's ratio remains the same throughout the provisioning period; workers at some point begin to be fed a mixture of pollen, honey, and jelly. Worker larvae may have their diet changed to 1:1 jelly exclusively if the colony is in need of a queen and none is available; the result will be a female with enlarged ovaries. Queens are usually produced in large numbers; most are killed.

Kerr (1950) provides evidence that queens are heterozygous at two loci. Genetic effects appear to regulate caste; environmental factors effectuate it. Determination of caste can have both environmental and genetic components. Current theory (Kerr, 1974; Michener, 1969) assumes that queens must be heterozygous at both of two loci (*AaBb*). Homozygosity at either locus produces workers. This hypothesis has been tested and accounts for the 3:1 worker–queen production ratio in, for example, *Melipona*. Among the important environmental influences are diet (already discussed) and allometric differences in abdominal ganglia. Workers have five ganglia, queens and *some* workers (thought to be genotypic gynes) have four ganglia.

The worker caste is the one that rears the brood. They cooperate in provisioning of their sisters and brothers (the drones, with whom they share on the average 25% of their genes). Because experimental control of mating frequencies is possible through artificial insemination it is possible to study the effects on bee behavior of various within-and-between-strain matings. Hamilton's (1964) theory of the genetics of selection for altruistic behavior seems to be generally accepted in the area if one assumes that group selection is an important vehicle for genetic change. This is no less the case in the explanation of altruism in the *Hymenoptera*. Together with the haplodiploid genetic organization of the bee's hereditary mechanism, the theory is invoked by Rothenbuhler, Kulincevic, & Kerr (1968) to explain the frequent occurrence of cooperative rearing in this order. However, elsewhere Rothenbuhler (1975) cautions that bees will drift from colony to colony, so it cannot be automatically assumed that they are all equally related.

Genetical Investigation. Very little has been done on the genetics of brood care in the social insects, although the methods for observation and colony control are available (Gaul, 1941; Wilson, 1972). Species such as those of the *Halictinae,* including the bumblebee, stingless bee, and honeybee, are candidates for genetic investigation through either selection or environmental manipulation because their life cycles and chromosome structure have been thoroughly investigated (Rothenbuhler, 1975; Rothenbuhler et al., 1968; Wilson, 1972). Methods of instrumental insemination have been thoroughly developed for *Apis mellifera* (Mackensen & Tucker, 1970) and allow for studies of brood care at the individual or whole colony level. Gary (1962, 1963) has also developed a successful method to allow seminatural mating flights. Rothenbuhler (1967) cites several studies that lay the groundwork for an intensive study of honeybee mating behavior.

Investigations in the Bee. *Apis mellifera* presumably originated in southeastern Asia and has spread from that location and diversified geographically and racially. Some 25 geographic races are recognized (Rothenbuhler et al., 1968). Although provisioning is the most commonly found epimeletic behavior in the bee, certain aspects of nest hygiene should also be considered in this respect. The nest-cleaning behavior of the bee has been investigated genetically (Rothenbuhler 1964a, 1964b). Hygienic behavior helps resist disease, especially American foulbrood. Removal of larvae consists of ingesting them soon after their death. The Brown line is disease resistant; the Van Scoy line is not. They are designated hygienic and nonhygienic, respectively. Using a single-drone inbred-queen strategy it was found that strains of bees resistant to American foulbrood inherit the behavior of uncapping and removing dead brood from cells of the nest. This result is congruent with a two-locus hypothesis: The hygienic line was homozygous for a recessive uncapping allele and a recessive removing allele. There are other environmental factors also involved in hygienic behavior. These are related to nectar supply, with less plentiful supplies being positively related to poor hygienic behaviors.

It has also been found that there is an age difference in the latency to perform hygienically under conditions of variation in food supply. (Momot & Rothenbuhler, 1971). Elsewhere, other researchers have crossbred Caucasian and African bees and studied hygienic behavior in the F_2 hybrid. At all data points but the first, the offspring perform more like the hygienic African parent; this argues for the higher incidence of hygienic over nonhygienic behaviors in unselected hybrids (Kerr, 1974).

Summary

The components of invertebrate epimeletic behavior, provisioning and nest building, show wide degrees of between-species variability: Nests vary structurally and in size and complexity; provisioning is slightly less diverse, with most species practicing either mass provisioning or progressive provisioning.

However, it must be added that the presence of et-epimeletic behaviors in this phylum is not established. Because the definition given earlier strongly implies that the two components are mutually conditioning, the behaviors reviewed in this section must be taken in perspective. Until it is established that there is care soliciting in invertebrates, we cannot conclude that there is compatibility between the mechanisms of offspring care in this and in the vertebrate phylum, except at the survival level.

Vertebrate Investigations

In this section the major intraspecific and, where possible, interspecific differences in vertebrate maternal–filial behaviors are considered. Intraspecific differences focus on studies of prenatal and postnatal maternal effects. Methods of investigation and analysis, strain differences, and fostering effects are emphasized.

Epimeletic behavior is actually quite rare in vertebrates, in the sense that it applies to the behaviors of parents toward young following hatching or parturition, except in birds and mammals. Although many species of fish build nests, and the males usually guard them (e.g., *Tilapia macrocephala* and other cichlids), this care is extended to the young in only a very few species. From fish to mammals prenatal care increases in complexity and duration; mortality rates of the young decrease dramatically. Hormonal factors may decrease in importance where parental figures are more diverse.

Interspecific Patterns: Some Paradigms

Mouse and Rat. In the mouse, females housed socially exhibit anestrus, which develops into the estrous state if they are later exposed to male urine (Marsden & Bronson, 1970). No strong correlations between nest building and either temperature or estrous cycle are found however (Lisk, Pretlow, & Friedman, 1969). Hormonal effects associated with pregnancy strongly affect variables like nest size; males implanted with progesterone are only slightly less responsive than females. Defense of the nest and young have been related also to lactation hormones in the mouse (Svare & Gandelman, 1973).

Long ago Leblond (1940) reported that behaviors of the rat and mouse are similar. Members of the genus *Rattus* build nests, lick the anogenital area, huddle, nurse, retrieve, and defend the young. These same behaviors were listed by Wimer and Fuller (1968) as characteristic of the members of genus *Mus*. Subsequent ethological studies of rat maternal behavior confirms Leblond's original statement (Rosenblatt & Lehrman, 1963). In a recent study Rosenblatt and his colleagues explored variables affecting the onset of maternal behavior in the rat; Szechtman, Siegel, Rosenblatt, and Komisaruk (1977) demonstrated that a mild tail pinch was followed by anogenital licking; in virgins the onset of all components of maternal behavior followed intermittently presented tail pinches and bore a frequency-dependent relationship to the acceleration of the onset of these components. There are no between-strain comparisons of this phenomenon in the literature at present. Concerning other factors involved in maternal-offspring relations, it has been observed that maternal rats emit a pheromone that has an origin in liver bile (Marsh, 1977). Nulliparae secrete this pheromone but males produce it only when injected with female liver bile.

Possibly the most comprehensive review of the hormonal basis of maternal behavior in the rat is that of Moltz (1971), who has detailed his own experiments and those of his colleagues. He reports that in the primiparous rat ovariectomy will reduce maternal behaviors to zero in about 50% of the subjects. The procedure drastically reduces circulatory levels of estrogen and progesterone and, by a feedback process, prolactin. On the basis of this and other evidence Moltz hypothesizes that there exist varying thresholds of arousal to circulating levels of these hormones. Primiparae are more strongly influenced by hormones to initiate maternal caregiving; in multiparae it was found that the experiential factor of having reared one or more litters normally overrode the reduction of hormones, resulting in a less drastic reduction of epimeletic behaviors.

More recently the differences in responsiveness to the young by primiparous and multiparous rat dams have also been demonstrated to be influenced by temperature changes (Wright & Bell, 1977, 1978). Cold stress at three time intervals (0-, 2-, or 5-min) increased attention to the young in primiparae but not in multiparae, whereas temperature increases show the opposite effects. Differential effects on lactation as a function of parity were also noted. In postweaning open-field tests the stressed offspring of primiparae had shorter movement latencies than did their control group, and those of stressed multiparae had longer latencies than their controls.

Finally, the close correspondence in the epimeletic and et-epimeletic systems of rat and mouse is further supported by a series of experiments reviewed by Denenberg (1970), in which he and his colleagues reared mouse pups using rat "aunts." These mice showed lowered activity in the open field, lowered incidence of fighting, and lowered plasma corticosterone levels in a novel environment. In a subsequent study comparing hybrid mice reared by rats to those reared by foster mouse parents, survival rates were higher in the fostered pups but weaning weights were higher in the rat-reared group (Denenberg, Paschke, & Zarrow, 1973).

Felines.   Members of the cat family are mainly asocial with the exception of the lion *Panthera leo*. These genera follow a mammalian pattern of birth and rearing similar to that of the rodent family. Although very much remains to be learned about this family, studies by Bertram (1975) and Rudnai (1973) of the lion and by Schneirla, Rosenblatt, and Tobach (1963) and Haskins (1977) of the cat support the foregoing statement of similarity to other mammals. Et-epimeletic vocalizations increase maternal solicitude and appear to be an influence on the mother for at least 30 days postpartum (Haskins, 1977). Maternal behaviors most affected are retrieving, nursing, and vocalization of mother toward the young, all of which are increased.

A major difference among the members of this family is shown in the tendency to rear the young cooperatively. The pride is the unit of selection of the lion; the dams rear the young communally after the neonatal period, the first few weeks. They will suckle and guard the young in turn (Bertram, 1975). Domesticated cats, being primarily asocially organized solitary predators, are not often observed caring for the young communally.

Canines.   In the dog, maternal–offspring relationships are interactive and formative as well as mutually influential. Breed characters that differentiate social relationships are considered in a subsequent section.

The dog is a genetic gold mine. Being a descendant of a medium-sized strain of wolves, the dog is the most variable of all mammals (Stewart & Scott, 1975). Epimeletic behavior in the dog is similar in all breeds that have been studied (Rheingold, 1963), but there are indications of individual and strain variations. Scott and Fuller (1965) found that 75% of Basenji mothers would retrieve their day-old puppies as contrasted to 38% among cocker spaniels. The F_1 hybrids were intermediate, scoring 63%. The results are limited by small numbers.

Pups are altricial and poikilothermic during the first few weeks. The dam stays with the litter almost constantly for the first 2 weeks; when she begins to spend more time away they vocalize for a few minutes, then bunch together and settle down. As they become mobile they approach the mother until about the fourth week, when nursing is terminated. The mother is a passive nurse, but she is attentive to the removal of excreta by licking; anogenital licking stimulates urination and defecation for the first few weeks, until pups gain voluntary control over bowel and bladder functions. The dam rarely retrieves the pup to the nest; she walks backward, and the pup follows her.

A behavior not often seen in the domesticated dog but present in the wolf and wild dog (Kleiman, 1967) is regurgitation of solid food for the young by the dam and other pack members. In the domestic dog this behavior may show reduced penetrance through selection against it by humans (Scott & Fuller, 1965). Estrous cycle patterns in the dog are genetically determined at one or two per year (Stewart & Scott, 1975). An attempt was made to select for two annual estrous cycles in silver foxes (Belyaev & Trut, 1975). In the selected line a reorganization in the desired direction was paralleled by decreased fertility during breeding season, with 30% failing to produce litters. Four factors accounted for the reduced fertility pattern: (1) failure to mate; (2) failure to produce; (3) failure to lactate; and (4) cannibalism. The results were considered as specific effects of the selection. This finding indicates that: (1) estrous cycles may be influenced somewhat; but (2) the systemic aspects of caregiving are strongly organized and are detrimentally affected when only one character is selected.

Primates.   Harlow, Harlow, and Hansen (1963) stated very early that maternal-offspring interactions should be studied as a system. The many studies by Harlow and his students have contributed significantly to research in primates of this system. In fact, the laboratory prototype for epimeletic–et-epimeletic study in primates is virtually the study of one animal, the rhesus macaque (*Macaca mulatta*). Those aspects of maternal behavior that are important to survival of the young are generally the same as in the genera discussed earlier. A significant behavior not seen elsewhere is intense physical contact, with either mother or offspring clinging to the other. Bodily contact is of importance in increasing the likelihood and availability of the maternal food supply. A mother separated from her own young will adopt another infant; if the adoption fails, it is attributed, among other factors, to the infant's failure to cling to the mother.

As the rhesus infant reaches its 20th week, the mother initiates periods of separation from it (Hinde, 1969). Before then the mother is less often observed rebuffing the infant's approaches. Her role is critical in promoting a weakening of the system that earlier insured the survival of her infant until it could provision itself. Whereas parental caregiving prepared the way for bonding behavior at first, as well as providing the opportunity for social learning (Brueggeman, 1973), rejection redirects that bond toward other members of the troop.

An important interspecific difference in response to separation has been demonstrated in a series of experiments by Kaufman and Rosenblum (Kaufman,

1973; Kaufman & Rosenblum, 1969; Rosenblum, 1971, 1973). In those studies it was observed that the asocial pigtail (*M. nemestrina*) and social bonnet (*M. radiata*) caregiving and soliciting interactions were quite different when the mother and infant were separated. Whereas the bonnet infant was adopted by one of the other adult females in the group shortly after the separation, the pigtail infant showed profound distress and remained isolated for at least the first postseparation week. The response to separation was also highly correlated with the structure of their respective groups; the postparturient bonnet female reunites with her group immediately, allowing others access to the infant; the pigtail female separates from the group and will even attack those below her on the dominance hierarchy if she is approached by them. Bonnet infants are less restrained in exploration and play than are the pigtail infants; less intense physical punishment is inflicted by the bonnet mother.

Individual Versus Multiple Caretaker Systems

An important variable making for interspecific differences in epimeletic and et-epimeletic patterns is whether individual or cooperative caregiving is practiced. A recent paper (Gurski & Scott 1980) reviewed this area in human and infrahuman mammals. Summarizing that review, it was found that multiple mothering is a widespread phenomenon in mammals. There are exceptions to that conclusion, notably the hamster, herd animals, and a few primate species. Where multiple mothering is found it was hypothesized to be a function of the group's sociality and relative stability. It was also hypothesized that multiple caretakers served a function in promoting species survival by familiarizing the young with several troop members, ones that could serve as substitute dependence figures if the mother was injured or killed, or if she was separated from the young.

Mice pups don't seem to be affected by reduced contact with biological mothers (Gurski, 1977). Inbred strains will mother their own strain's young or young of other strains with equal frequency. Rats will care for mice, which are distantly related genetically (Denenberg, 1970). Most experiments in which multiple mothering has been observed have not given the extent of relationship or addressed questions of gene flow, and so cannot be accepted as evidence for *or* against Hamilton's (1964) theory of altruism. The time at which adoption proceeds successfully is important; it appears that the closer to postpartum day the adoption is attempted, the more successful it is. The changes in size and appearance of offspring, whether they are altricial at birth, and in the mother the composition of milk, all are factors that contribute to successful onset and continuation of multiple mothering (Gurski, 1977).

Bertram's (1975) work with the lion seems to point to a genetic influence, as females cooperatively rear their young at stages in the period postpartum to weaning. The females within the pride are normally closely related to each other, the males are genetically unrelated. Males rarely provide any care for the young.

This species is synchronous in its estrous cycle. A recent study of the effects of synchrony as a reproductive strategy indicates that it should reduce desertion by the male (Knowlton, 1979). The data from the lion appear to fit the model.

Canids such as wolves also provision and care for the young cooperatively. The extent of relationships among members of the pack in the wild has not been studied; the relationship of wolves in captivity should be quite variable. Even though it appears that altruism is shown in the care of the young (Rabb, Woolpy, & Ginsburg, 1967), opting for selection in favor of the group regardless of relationship, the original activity may have been influenced by the degree of relationship among members of the pack.

Animals that move in herds and are themselves preyed upon do not usually provide care for young other than their own regardless of genetic relationship to the offspring or mother, although experimental studies have resulted in successful adoption and/or fostering by nonrelated females in sheep (Hersher, Richmond, and Moore, 1963; Shillito & Hoyland, 1971; Smith, Van-Toller, & Boyes, 1966). The window through which this can be successfully accomplished is very small and needs to be precisely controlled. Selection has obviously gone against multiple mothering in the herd animals, the result being a generally exclusive parent–offspring relationship reinforced by strong social attachment between them.

In primates multiple caretaking systems are commonplace. There has been very little done to test the genetics of the systems. A possible exception is that of caretaking patterns in the bonnet (*Macaca radiata*) and pigtail (*M. nemestrina*) macaque monkey (Kaufman, 1973; Kaufman & Rosenblum, 1969; Rosenblum, 1971, 1973). Although the species are different, they are closely related and differ very little in their respective ecologies. In their caretaking behavior they are opposites; the bonnet shares the responsibility among other members of the group, whereas the pigtail is exclusive in its care of the young by the biological mother.

Human examples of caregiving are quite varied. The motives for caregiving are social and philosophical (Eiduson, Cohen, & Alexander, 1973; Hazan, 1973). The success of multiple caretaking systems (e.g., kibbutzim, communes, day-care centers) hinges on commonality of social goals and agreement among the caregivers upon the methods of socialization of the young. Informally, care is shared among family members even though the members of nuclear families may think of themselves as practicing an essentially individual mode of caretaking. We (Gurski & Scott, 1980) have theorized that multiple-care systems function to select for the group's survival, in that these systems provide alternative dependency and attachment figures in cases where the principal caregiver is unavailable temporarily or permanently. The coming together of the infant with other members of its group in a relationship of dependence increases the possibility of forming attachments to these alternative caregivers.

Effects of Separation or Change in Environment

A factor that produces interspecific differences is that of environmental manipulation of the system of caregiving; this can be accomplished through the separation of infant from mother with or without the opportunity for subsequent adoption, or through observation of species in different ecological niches.

When separated from the parent, the young of the species emits a distress call, a component of et-epimeletic behavior. Of course not all species vocalize; for those species that do not, it is assumed that et-epimeletic and attachment patterns do not need to be stimulated in that manner.

Several mammalian species distress-vocalize. For example, rat pups vocalize to cold stress in the ultrasonic range (Bell, 1974; Bell, Nitschke, Bell, & Zachman, 1974). Ultrasounds were found to be effective directional cues for localization of pups. Pup odors interact with vocalization to determine the speed at which females initiated searching; the presence of odors also increased the attention to ultrasounds when the two cues were presented together (Smotherman, Bell, Hershberger, & Coover, 1978). When the pups were removed from the nest, which eliminated their odors as a factor, the latency to initiate retrieval was affected, as well as the direction toward which the dam oriented and the maintenance of predisposition to continue retrieving.

Separation for an extended interval also affects the young rat. Hofer (1973) reported on separation of 2-week-old Wistar pups from the dam for 18 hours. Waking, activity levels, elimination, and self-grooming increased compared to those of the control group. In the guinea pig, Pettijohn (1979) examined separation-induced distress vocalization. Under maximum separation from familiar social and physical surroundings, the distress vocalization rate remained high for the first 4 weeks and then linearly declined to near zero by the 12th week; when the mother was present the vocalization rate was extremely low, and when she was absent the rate was moderate if the infant was isolated in the home box and high if the subject was in an unfamiliar box.

Recently we (Gurski, Davis, & Scott, 1980) examined the onset of separation distress to isolation in the dog. We were able to reduce vocalizations to zero in the neonatal, and to a lesser extent in the transitional stage by the combination of a soft, confining area and controlled temperature. A comparison was made to the vocalizations of the same pup in a standard testing box with aluminum floor. Although vocalization to the uncomfortable standard box was high from Day 3 (first test day), separation-distress vocalization did not begin until Day 11 and remained at a low level until Day 21, when it rose rapidly. This indicates to us that the process of attachment in the dog is initiated in the transition period but does not reach a high level until some 2 weeks later. Contrary to the research with the monkey by Harlow and his colleagues, in the dog contact comfort alone does not reduce vocalization once the period of primary socialization begins. The importance of the differences between the process of attachment in dog and monkey was demonstrated by Mason and Kenney (1974), who fostered rhesus juveniles to dogs; of eight monkeys so treated, seven were clinging within 4 hours. Later, given a choice between an unfamiliar monkey and a familiar dog the fostered monkey preferred the dog significantly more often. Mason and Kenney did not test the dog's preference.

The work of Kaufman and Rosenblum (already mentioned) is also important in the present context. In their studies it was demonstrated that species show different responses to the separation tests, with some species being organized socially to accommodate permanent or temporary separation of mother and young. As may be recalled, the possibility existed for spontaneous adoption in both species, but only the bonnet macaque infant was adopted, at least in the first week; the pigtail infant did not seek to be adopted, nor was it approached by other maternal females, during that time.

When more rigorous manipulations of the maternal–filial system are attempted, differences among species are noted. Gurski (1977) reported that inbred mice (DBA/2J, C3H/FeJ, C57Bl/6J) adapted to daily rotation of either litter or mother in either a within- or between-strain design that brought them back to the home cage only every third day, over a period of 21 days. Compared to nonrotated controls, the rotated litters showed a slightly lower preference for any female, including the biological mother. In a test of attachment, using dependent measures of proximity to and following of the three maternal figures (mother and two surrogates) and a nonfamiliar female of its strain, the weanling pups showed slightly lower overall scores on the attachment test. They also distributed those preferences over the three maternal figures that had cared for them. Pups who had mothers rotated to them scored at the control group level except that they too distributed their preferences over the dams that had cared for them. Other measures (e.g., weaning weights, open-field, etc.) showed no differences from their control group for any of the pups. I concluded that in the mouse the epimeletic–et-epimeletic systems remained robust under conditions for which the system could have never been selected.

Compared to the results with the mouse, some early research with the rat (Denenberg, Ottinger, & Stephens, 1962; Ottinger, Denenberg, & Stephens, 1963) showed that, under conditions of daily rotation, rat pups showed lowered weaning weights and higher mortality rates than nonrotated controls. Because there are no studies of communal rearing of young in the rat, although there are several of these studies in the mouse (described previously), it is possible that a difference here exists between rat and mouse; the rat caregiving system appears more vulnerable to repeated separation than that of the mouse.

Griffin (1967) reported a study in which he rotated rhesus macaques among four different mothers, every 2 weeks for 32 weeks. In all cases the mother and infant accepted each other within a few hours. Aside from a greater variability in behavior, in the rotated groups the mother–infant system withstood the radical changes during the study. No lasting effects or deficits in social behavior were observed.

The literature on species from the same genus that have adapted to different ecological niches (e.g., *Peromyscus* [King, 1963]) shows some differences in epimeletic behavior. Nest-building tests of northern region inhabitants and those of the southern region show that the former build larger nests and use more cotton than do the latter. Subspecies differences in nipple-clinging time by the pup, a ubiquitous behavior in *Peromyscus,* where *P. bairdii* cling considerably longer than does *P. gracilis,* may be related to the former's arboreal habitat; the latter is a surface inhabitant. Conversely, *P. gracilis* is higher in nest defense than is *P. bairdii,* and the former weans the pups earlier. In primates, Poirer (1972) reported that north Indian and Nilgiri langur females share in the care of the troop's young from the day of birth of the infant. These two examples of the effects of ecological adaptations are surely paralleled by different socialization processes in the young.

Intraspecific Differences: Prenatal

This section reviews the genetical investigations of intraspecific (strain and breed) differences in epimeletic and et-epimeletic behaviors. Prenatal effects are considered first; then a review of findings on postnatal maternal effects is made. Before the research into the effects of prenatal manipulations is considered. I discuss briefly the prenatal maternal environment.

Maternal Environment. From a genetic point of view, contributions of the maternal environment begin at the moment of conception. Two main modes of influence are cytoplasmic inheritance and intrauterine environment. By cytoplasmic inheritance is meant the contribution to the developing fetus of extranuclear material in the fertilized ovum. This is a hereditary mechanism in that the makeup of the cytoplasm is determined by both natural selection and prior genotypic characters. Broadhurst (1967b) suggested that a probable mechanism by which this type of maternal inheritance could come about in mammals lies in the difference in the amount of cytoplasmic material contributed by sperm and ovum, the latter's contribution being many times greater.

The second prenatal effect of major proportions, the influence of the intra-uterine environment, is still a genetically based one, but more subject to environmental variability than is cytoplasmic inheritance. The nutritional habits of the mother will affect the development of the fetus and the general health of both mother and offspring during pregnancy. Recently, reviews of mouse cytogenetics (Miller & Miller, 1975) and embryogenetics (McLaren, 1976) have appeared in the literature. If a maternal effect is established phenotypically, it becomes necessary to analyze it further, and in particular to investigate the possibility that it is due to a cytoplasmic involvement rather than to an intrauterine effect. In such experiments the fertilized ovum must be transplanted to the uterus of another organism (e.g., from the uterus of a mouse of one strain to the uterus of a female member of a second strain). This transplantation is a difficult control procedure requiring precise synchrony as to physiological readiness of the donor and host. The technique was outlined by McLaren and Michie (1956).

Methods of Investigation and Analysis. The genetical research on the mouse is greater in amount than for any other laboratory mammal. Consequently it is the mouse that has been and continues to be the most studied in the area of prenatal effects. The calendar of development and phenotypic characters has been worked out in detail (E. L. Green, 1968; M. C. Green, 1975). As the behavioral aspects of epimeletic and et-epimeletic systems are complex and multifaceted, assumptions of multiple-factor inheritance (Roderick & Schlager, 1968) should underlie their genetic aspects. In multiple-factor inheritance it is not possible to determine the mode of inheritance of each allele, but it is possible to study the relative contributions to the phenotype of genetic and environment factors. Factors in the environment shared by the principals can affect the offspring phenotype. The proper use of experimental design and analysis can improve our understanding of those aspects that we can control. Consequently considerable attention has been given to proper methods of investigation and analysis of results in this area.

The most powerful method of investigation of prenatal genetic and environmental factors is the diallel cross (Broadhurst, 1967b). This method has been employed because it allows for analysis of the between-strain, within-strain, and environmental effects of a set of independent variables over several separate strains, which can be as genetically similar or diverse as needs permit. Analysis of the diallel cross is made using analysis of variance (ANOVA) based on the biometrical model (Broadhurst, 1967a, 1967b; Mather & Jinks, 1971). ANOVA allows for a test of reciprocal differences, enabling the assessment of prenatal maternal effects.

Most experimental tests of the prenatal maternal environment on postpartum and postweaning behaviors involve manipulations on the pregnant dam, often coming under the rubric "stress." Joffe (1969) reviewed the major studies of his own and others of irradiation, teratogens, and various external stimulations (see the following). Others (Deitchman, Kapusinski, & Burkholder, 1977) have handled pregnant females, compared them to a nonhandled group from the same strain, and found differences in grooming of the young as a function of this relatively noninvasive method of "stress." Finally, earlier research on selection (Falconer, 1955) and analysis of quantitative traits like milk yield of mother and weaning weight of offspring (Falconer, 1960) provides suitable paradigms for further research. The investigation of intraspecies prenatal effects in breeds or strains has not always concentrated on the epimeletic or et-epimeletic systems. More discussion goes on in the literature pertaining to the control of the maternal environment through various means than to the observation and measurement of maternal mediators of offspring postweaning differences. Yet it is understood that these differences are either maternally mediated or are the result of an interaction between the maternal and offspring genotypes. As better control over the outcomes of various experimental procedures is gained, researchers may then give more attention to the molecular analysis of the experimental conditions and begin a genetical analysis of the behaviors involved.

Genetical Investigations. Early research on prenatal genetic effects was conducted by Thompson (1957) and Thompson and Olian (1961). In the later study, inbred mice were injected with adrenaline during pregnancy. The predominant effect was a regression of offspring activity scores toward the mean of all three strains: A/Jax (low active) increased activity, BALB/Ci (low/medium active) demonstrated no change, and C57BL/6 (high active) decreased activity.

Weir and DeFries (1964) continued this line of investigation, pointing out the risk in generalizing the effects of prenatal maternal stress on offspring from results on a single genotype. Using BALB/cJ and C57BL/6J, their findings generally confirmed Thompson's work. Offspring of stressed females of the high-activity line (C57) were less active than their controls; offspring of stressed females from the low-activity strain were more active than their controls. In that scores from open-field ambulation and percentage of distance traveled to the center were both in the direction of the mean of all scores from the offspring, predictions of the general effects of stressors of any particular type on a genotype, and of any clear genetic effect that could be generalized to strains other than those tested, are not possible from the study. The tendency for stressors to reduce "emotionality" in one strain and increase it in another does not make for safe speculation. The effects could have also been mediated postnatally, as the offspring were not cross-fostered.

Another study that tested the postnatal results of stress-treated postnatal experience in the dam on open-field behavior in the rat is that of Ottinger et al (1963). The study in question was also important because it demonstrated that cross-fostering on a daily basis between two litters and mothers was much less adverse in its effects than shocking plus cross-fostering (Denenberg et al., 1962). Gurski (1977) addressed Weir and DeFries (1964) more directly: C57BL/6J mothers were rotated on a daily basis to litters of their own strain, or to DBA/2J and C3H/FeJ litters, for either a within-strain or between-strain effect of multiple mothering on offspring behaviors. The open-field scores did not differ among the groups or from controls. Only the tendency to follow and maintain close proximity to the biological or other dam was affected; multiply mothered weanlings tended to distribute social contacts more equally toward all dams with which they had significant contacts of dependence. This result argues against cross-fostering being a significant stressor in the mouse and points up a possible species difference between mouse and rat with respect to multiple mothering (Gurski & Scott, 1980).

In an extension of Weir and DeFries' (1964) experiment with BALB/cCrgl and C57BL/Crgl strains, dams were mated in all possible diallel combinations and stressed during pregnancy; offspring activity was tested at 48 days (DeFries, 1964). The paternal strain interacted with prenatal treatment, such that offspring of BALB fathers were on the whole more active than those of C57BL fathers, females were more active than males, and a heterotic effect was found in the hybrid offspring. Fulker (1970) has provided an interesting reanalysis of the three studies just described. He concluded that the maternal genotype countered the expected additive effects by serving as a physiological buffer against stress effects, going in the opposite direction of those effects and canceling them. Neither the fetal nor maternal genotype is solely causative; rather, they are of necessity interactive.

By far the best comprehensive review available on prenatal determinants is Joffe's (1969) book, which details important early research and control methods for the genetical studies of prenatal maternal environment. The review is focused primarily on laboratory investigations of rodents. Joffe's (1965) work with Maudsley Reactive (MR) and Nonreactive (MNR) strains is also described, in which he performed a 2 × 2 diallel cross after administering pregestational stress to the females. Half of the premating-stressed females were then stressed during pregnancy. All litters were fostered to MNR untreated females, to equate postnatal variables. Results on 96 offspring from 24 litters tested in the open field at 100 days showed that the offspring of MNR fathers decreased, and those of MR fathers increased in activity. Only premating stress effects were judged causative, as the premating, postmating, and premating-stressed groups were not different in scores. The avoidance-conditioning scores of the combination groups were higher than the simple premating group, however, indicating that the effect was a complex one. Very little was actually reported of maternal–offspring postnatal interactions which may have changed and influenced the postweaning activity: Joffe reports only that stressed mothers who fostered nonstressed offspring affected those offspring, as they were significantly inferior to other groups in learning.

Because of the necessity for controls for each and every aspect of these studies, requiring large numbers of animals and many hours of data gathering on control animals, studies of this type have not been numerous in the literature. Also very little in the way of observation is reported on the ways in which maternal and offspring behaviors are changed during their significant interactions (postpartum to weaning). Remedying this gap in our knowledge would provide much needed information toward understanding what it is that a particular environmental effect does to a genotype, and to a genotype-by-genotype interaction. A recent study seems to point in this direction. Burkholder (1977) used prenatal maternal crowding in CD strain Charles River rats as a stress variable. He reported that intermediate levels of crowding (8 per group) increased offspring *independent* activity over that of the low (2) and high (16) crowding groups. However, offspring *dependent* behaviors decreased as a function of prenatal crowding group, with the low group exhibiting the highest scores. (Offspring independent behaviors are initiated without any solicitation; offspring dependent behaviors are solicited by offspring.)

It is clear that if offspring activity is affected, maternal variables must be implicated. More baseline data on significant strain-specific epimeletic and et-epimeletic behaviors need to be gathered and compared with data gathered on stressed dams and litters before postweaning effects can be fully understood. In order for this research to be conducted, epimeletic behavior must be hypothesized to be a dynamic system with strain-specific differences. These differences can be observed in the offspring of crossed strains. An example is Lee's (1973) study of intraspecific differences in nest building in the mouse. His calculations of narrow and broad sense heritabilities from data on C57BL/6J and BALB/cJ parents, and reciprocal F_1, F_2, B_1, and B_2 offspring led him to hypothesize that nest building has undergone natural selection and possesses adaptive maternal as well as thermoregulatory significance.

Eventually researchers in this area may need to turn to other genera. Rasmussen (1975) states that the genus *Peromyscus* includes 40 species native to Central and North America, many of which are interfertile. His earlier (Rasmussen, 1968) research on size inheritance in *P. maniculatus* and *P. polionotus*, including data on backcrosses, was consistent with a model of prenatal maternal influences. These species and subspecies differences should show enough variability in epimeletic and et-epimeletic behaviors to be of value in genetical investigations.

Birds

There are wide variations among birds in the developing components of epimeletic and et-epimeletic behavior; only brief mention of this group has been made here. A number of recent sources are available that consider interspecies differences (e.g., Silver, 1977; Skutch, 1976). Because there are fewer studies available of intraspecies differences in birds where genetic systems are specifically manipulated, less can be said about them here than about mammals.

The most frequently cited group on which genetic studies have been made is that of the domestic fowl. It is the most commercially exploited of all animal species; having experienced the greatest artificial selection pressure, the domestic fowl shows rapid and usually robust adaptation. Ferguson (1968) states that fowl have a high degree of social organization, a statement supported by Guhl (1953, Guhl & Fischer, 1969).

Wild fowl show a marked propensity to brood the young and yield 45–60 eggs annually (Ferguson, 1968). Comparing this behavior to that of domestic strains of poultry, we find very little that the two strains have in common. Because it reduces egg production (Fuller & Thompson, 1978), this propensity to "broodiness" has been subjected to negative selection in commercial circles.

Hutt (1949) reported that a flock of Rhode Island Reds reduced broodiness from 91% to about 2%. The hormonal basis in prolactin production is known, although the genetic basis of broodiness is still unknown.

Fuller and Thompson (1978) have reviewed some earlier studies that relate broodiness to sex of parent. The data they cite find a positive relationship between the sex of sire and broodiness. Because broodiness cannot be totally related to sex of parent, some autosomal factors were also implicated. Broodiness is probably a quantitative character, multiply and variably influenced by a number of different gene paths.

Inasmuch as broodiness responds to negative selection, is it also possible to select for higher percentages of broodiness in those animals in whom it has earlier been reduced? Hess, Petrovich, and Goodwin (1976) used the Japanese quail, a bird that has been propagated by artificial incubation in laboratory settings because of earlier reported difficulties in inducing parental behavior. They successfully induced broodiness in the parents in one generation and increased it in succeeding generations through a severe selection program. Orcutt and Orcutt (1976) elsewhere reported that domestic strains are usually reluctant to nest and rarely demonstrate any reproductive behavior past courtship, copulation, and occasional nest building. They report also that broodiness has been induced in lab stocks. The female chose a secluded site as in the wild. Eggs laid after the nest was built were always laid in it. The hen became very active and vocalized frequently on hatching day. She then spent much time brooding the young.

In a preceding section it was stated that more research must be done on strain-specific components of epimeletic behavior, considering it as a dynamic system that may possibly show differences if approached at an appropriate level of analysis. A classic example of this type of strategy is illustrated by Dilger's (1962a, 1962b) study of two closely related species of lovebird, *Agapornis roseicollis* and *A. fischeri*. In comparing the nest-building behavior of the hybrids with the parents, Dilger found that hybrids have difficulty in preparing nests, having received incompatible combinations of traits from the parents. However, after 3 years of experience the hybrids improved greatly: Incompatible behaviors were suppressed. To date there are no followup studies on F_2 or backcross offspring; a polygenic mode of inheritance was hypothesized but not confirmed.

Intraspecific Differences: Postnatal

Postpartum maternal effects are easily observable. Two possible factors that may result in different phenotypes coming from the same genotypes are: (1) the particular mother or mothers that rear the offspring; and (2) the number of concurrent offspring in the family unit. In species where multiple births occur, the amount (and possibly the quality) of maternal care must vary with the size of the litter. If the same offspring have surrogate caregivers, as in adoption or in fostering experiments in the laboratory, the influences of all caregivers affect the resultant phenotype. One other important component is that of parity, the effect of previous pregnancies on the maternal response. Gurski (1975) noted that the rate of cannibalization of the young decreased with successive pregnancies in C57BL/6J, C3H/HeJ, and DBA/2J dams. The influence of the maternal environment is central to the course of early development in the young. Because there are variations among mothers and even identifiable variations within the same mother from litter to litter, it is possible that for some species the effects of one mother might be offset by the effects contributed by maternal care given by foster mothers, or by more than one mother, to the same offspring.

The remainder of this section, devoted to postnatal effects, reviews behavioral investigations of intraspecific differences in the mouse, rat, rabbit, dog, and herd animals. Although the most useful information for the reader would be the ways in which various strains differ on the components of the two major systems to which this review is devoted, the available data do not always provide this information as completely as is desired. Instead the strategy here is to cite those references where differences have been investigated. From these results a direction may be indicated for future research.

Mouse. Gandelman (1973d) investigated the ontogeny of maternal responsiveness in female Rockland–Swiss albino mice. He found that, when females were exposed to neonatal pups, there was a linear progression from 11% responsiveness at 22 days to 80% at 72 days. Pups were ignored most often at 22 days; pup killing was highest at 32 days and decreased thereafter. Gandelman hypothesized that "spontaneous" maternal behavior may develop longitudinally, and that olfaction and sexual maturity (ca. Day 46) are strongly implicated in maternal responsiveness.

Priestnall (1973) tested the effects of handling the mother or litter on subsequent maternal attention to the young in CFLP mice (born and reared in England). He recorded nest-related behaviors, activity levels, grooming, ingestion of food, nursing, licking, and drinking of water. Licking and grooming patterns were higher in the litter-handled (LH) groups. Maternal handling was ineffective in changing any observed behaviors from control levels. Whereas levels of litter licking were initially higher in LH groups, they were not different from control levels at the fourth hour of observation. These results are confirmed by the work of Sherrod, Connor, and Meier (1974), who found handling effects do not persist nor do they strongly affect overall levels of maternal activity or maternal–offspring interactions. Cross-fostering resulted in almost immediate acceptance (at 8 days) of the pups, with no increase in pup killing. Sherrod et al. (1974) concluded that maternal responses may mediate handling effects, a finding similar to that of Fulker (1970), discussed previously.

Strain differences in nursing under communal conditions were investigated by Werboff, Steg, and Barnes (1970) in A/J and C57BL/6J mice. Communal nursing enhanced the growth of offspring in both strains. A/J pups showed higher activity levels, especially in pups from groups composed of mixed-strain mothers and littermates. The weaning weights of the C57 pups from these groups was highest. The effects of communal nursing were concluded to be strain specific, related to genotype of dams *and* pups. With further investigation it might be possible to determine exactly how these effects were mediated.

Methods of cross-fostering (between-strain) and infostering (within-strain) have been employed to determine postnatal maternal-offspring interactions. They have been useful in determining effects on subsequent offspring behaviors but the in vivo mediators are not as clear. Reading (1966) used BALB/c and C57BL/6 mice and found that rearing by BALB/c dams significantly increased weaning weights over those of pups reared by C57BL/6 dams. A significant cross-fostering effect was also found in postweaning behavioral tests. BALB/c pups reared by C57BL/6 dams (B/C) were more active in the open field than the B/B group. B/C groups had longer hole-in-wall latencies than B/B, but C/C were slower in water escape tests than C/B groups. The results were further modified by lower defecation scores in the open field of C/C than C/B, and of B/C than B/B males and females. One possible contribution of the dams to these results comes from earlier research by Ressler (1962): he reported that BALB/c mice handle their offspring more than do C57BL/10 mice. Retrieving scores of the two strains are also different and are influenced by their pup-handling scores. BALB/cJ and C57BL/6J dams also differ in nest building; compared to the latter strain the former are high users of available materials. Lynch and Hegmann (1972, 1973) tested males and females from A/J, BALB/cJ, C3H/HeJ, and DBA/1J matings, and reciprocal cross F_1 offspring from BALB/cJ × C57BL/6J matings. Substantial strain and offspring genetic effects were reported in nest-building scores. The F_1 scores indicate a heterotic effect and the F_2 scores show a regression to the mean of the F_1 and midparent scores. Although hybrid animals built larger nests, the mean scores of their offspring were smaller, regressing toward the foundation population values. Whether or not a single or larger order model of inheritance is appropriate cannot be determined conclusively, although it would seem that an additive (polygenic) model more parsimoniously fits the data. As more complex models of analysis based on factor analytic and causal analytic assumptions take their place in the methodology of behavior geneticists working in this area (such as in the work of Knowlton, 1979 and McNair & Parson, 1979) these questions may be answered. What is needed is a combination of the foregoing analyses applied to molecular analyses of maternal and offspring behaviors during the postnatal–preweaning schedule. Behavior is composed of elements, themselves composed of reflex, species-specific factors. These are theoretically based on genotype, which should be determinable. Most research in this area up to now has remained at a level of analysis too general to be of use in determining the *behaviors* of importance in maternal–offspring dyads. Until the contribution of development to the whole has been determined for offspring behaviors after weaning it remains as a confounding variable that masks the contribution of genotype.

Concerning et-epimeletic behaviors in the mouse, Nitschke, Bell, and Zachman (1971) found differences among BALB/cJ, C3H/HeJ, and C57BL/6J pups recorded at several periods during prenatal development. Strain differences in signaling rate, peak frequencies, and signal duration were recorded; C57BL/6J pups emit fewer signals at 2°C and showed an increase in signal duration.

Several areas require further investigation in the mouse before a degree of specificity can be claimed. None of the components of epimeletic and et-epimeletic patterns has been clearly defined genetically in the mouse, although several studies have provided direction in the formation of a systematic research effort.

Rat. Three very good basic reports describe maternal–filial interactions in the rat (Barnett, 1975; Rosenblatt & Lehrman, 1963; Wiesner & Sheard, 1933). The effect of handling in Sprague–Dawley rats has been investigated by Villescas, Bell, Wright, and Kuffner (1977). (Palm [1975] estimated that over 50% of the inbred rat strains are descendants of Wistar albinos.) The study demonstrated different epimeletic patterns, higher activity levels in the open field, and higher body weight at weaning.

Nutritional factors have recently been investigated in the rat (Massaro, Levitsky, & Barnes, 1977; Misanin, Zawacki, & Kreiger, 1977). Pups of malnourished dams fostered to normal-diet dams received more attention, behavioral development was affected by sex of pup, and the dams reduced overall exposure of the pups to other factors in the environment. Retrieval times prior to contact were not different, but once the dams had investigated the pups in a two-choice (own versus fostered) situation the foster pups were more quickly retrieved to the nest. This form of unequal maternal attention might compensate for the effects of undernutrition, although much more research is needed before a conclusion can be made.

Rat dams orient to pup cues (Smotherman et al., 1978). These cues are auditory (Bell, 1974; Bell et al., 1974) and olfactory (Conely & Bell, 1977). Whether these findings can be analyzed along strain-specific lines awaits future research.

Peromyscus. The process of parturition (Layne, 1968) and care (Eisenberg, 1968) in *Peromyscus* has been described. Classification of hybridization has been discussed (Hooper, 1968). Tests of postnatal patterns in *P. gossipinus* and *P. leucopus* reveal a difference in amount of time spent by the dam during early postnatal and later preweaning latencies to retrieve, number of pups retrieved, and total retrieving times. Different intraspecific litter survival rates were observed; the Gainesville (*P. gossipinus*) group showed longer retrieving times and higher defense scores.

Peromyscus offspring exhibit subspecies differences in distress vocalization to cold "stress" (Hart & King, 1966). Vocalization of *P. maniculatus bairdii* and *P. m. gracilis* were studied by tape recording and sound spectrographic analysis. Subspecific and age differences were found for several vocalization parameters. Vocalizations were assumed to help the mother locate her young. This hypothesis has some support from Bell's work with ultrasounds (discussed earlier). That the vocalizations subside at about 10 days, when pups begin to provision themselves, indicates an et-epimeletic function for these signals.

Rabbit. Six factors figure importantly in reproduction in the rabbit (Sawin & Crary, 1953). All are involved with nest construction or defense. Rabbits do not retrieve pups removed from the nest (Ross, Sawin, Zarrow, & Denenberg, 1963). They do exhibit considerable strain differences in several of the nest-related factors that have been studied. The strain differences are complex and need further refinement before any conclusion can be drawn regarding the contribution of genotype to these scores. Over 50 domestic breeds are now available, and because rabbits breed well under a variety of conditions, their sex is easily distinguishable at birth, and their ontogeny is well studied, they might be the species of choice for the investigation of the physiological aspects of epimeletic behavior in mammals (Fox, 1975).

Dog. A paradigmatic approach incorporating investigation of genetic and environmental differences in a number of different behavioral systems is reported by Scott and Fuller (1965). (In addition to epimeletic and et-epimeletic systems the project studied agonistic, allelomimetic, and reproductive systems.) Studies of caregiving and care soliciting have been conducted on several breeds (Gurski et al., 1980; Pettijohn, Wong, Ebert, & Scott, 1977; Rheingold, 1963; Scott & Fuller, 1965; Scott, Stewart & DeGhett, 1973). Rheingold (1963) detailed a study of epimeletic patterns in her now classic volume in this area: She tested cocker spaniels, Shetland sheep dogs (brown, black, and merle), and beagles. Rheingold reported that cockers were highest in nursing time, possibly because it was easiest to nurse from them; however beagles, which are morphologically similar to cockers, were the least likely to be nursing at any particular observation period. Cockers also "toilet-trained" the litter earliest (stopped anogenital licking (ca. Day 17); Beagles were next earliest (Day 25); Sheltie merle dams were still licking at Day 39. There are developmental and breed differences for cleaning of feces compared to licking of urine: Cockers stop at around Day 26, beagles around Day 29, Sheltie merles at Day 35, and Sheltie browns at Day 67. Although the dog retrieves pups in a different manner from some other species considered previously, there are breed differences: Shelties retrieved pups for a longer time than cockers or beagles; the Sheltie merle dam "herded" its pups by remaining between them and the observer. Punishment of pups, a parental behavior not previously considered because it has not usually been included as part of epimeletic behavior, is reported by Rheingold to show considerable variability among breeds as to its onset and intensity of administration. Scott and Fuller (1965) also studied African Basenjis and fox terriers. They reported that Basenjis were equal to cockers in percentage of time away from their young at 4 weeks. The F_1 generation spent less time with their pups than their parents had with them; a better milk supply was hypothesized to be the critical factor. When the purebred dams were observed nursing under undisturbed conditions, the percentage nursing, from highest to lowest, was ordered as follows: Shelties, cockers, fox terriers, beagles, and Basenjis. Basenji dams retrieved pups most often, followed by Shelties, Basenji × cocker F_1's, fox terriers, cockers, and beagles. A genetic analysis of generational data revealed that the maternal environment was a significant effect in several tests of the offspring. A recent study by Nesbitt (1975) of a feral dog pack on a wildlife refuge confirmed Scott and Fuller's (1965) description of domesticated dogs as similar to the wolf in social behavior given the proper conditions. The pack is the social unit of these feral dogs; they rear the young somewhat separately, as does the wolf (Rabb et al., 1967), but the surviving young are integrated into the pack at an early age.

Studies of separation-distress vocalization and its relationship to genetic and environmental variables have continued at Scott's laboratory in Bowling Green. Scott et al. (1973) studied the effects of separation on vocalization rates and separated the absence of familiar individuals and fear of strange ones as compatible additive effects. Gurski et al. (1980) have extended and refined the original work in this area to permit investigation of distress due to physical discomfort and that due to isolation from nest, littermates, and mother.

Forty-eight puppies from four genetic backgrounds were the subjects. Beagles, Telomians, and their reciprocal F_1 hybrids were divided into four groups and isolated for short (10-min) periods under conditions that were defined as uncomfortable (U) and comfortable (C). The first condition was produced by a standard 60-cm cubical box, used in previous studies for testing separation responses of older puppies. The floor was a sheet of aluminum and the walls were bare plywood; the subjects were tested at 21°C (approx.). The C condition was produced by placing pups in a short vertical section of stovepipe, lined with disposable diapers that were discarded after each test in order to keep the area free from odors. The stovepipe diameter was slightly larger than the body length of the puppies; larger diameter pipes were introduced to meet this requirement as the puppies grew. This environment was placed in a ventilated, partially soundproofed incubator kept at a constant 30°C temperature. Vocalizations were picked up by a microphone installed inside the incubator; an observation glass allowed partial visual contact with the subject from the darkened experimental room. The four experimental groups were composed of pups who were run in a counterbalanced order in the C and U conditions every other day, beginning at 3, 11, 17, or 25 days. The C condition had been demonstrated in pilot work to reduce vocalizations to zero in the neonatal period for the first time. Although the U condition resulted in an immediate high rate of vocalization even in the youngest group, vocalizations in the C condition remained at zero until 11 days; from that day until Day 31 they increased curvilinearly to approximately the same level as the U condition (Fig. 10.1). Although genetic differences in the separation response tended to be minor compared to those in other characters (Scott & Fuller, 1965), hybrids with beagle mothers were the highest of all.

The study demonstrated that discomfort due to physical and social factors are separable; this confirmed an earlier study by us that demonstrated the motivational distinctiveness of the separation reaction (Davis, Gurski, & Scott, 1977). The research confirmed the findings of a genetic difference between breeds and, most importantly, gave a more precise definition of the onset of the critical period for primary socialization in the dog than was previously possible. Significant genetic differences were reported in hybrids, which vocalized at higher rates than their purebred parents, and in reciprocal hybrids, which vocalized somewhat more when the dam was from the beagle population. Pettijohn et al. (1977) demonstrated that beyond the fourth week distress vocalization is

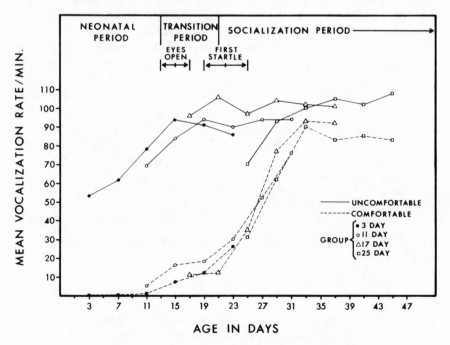

FIG. 10.1. Mean distress vocalization scores of puppies in two different conditions during first 7 weeks of life (see text for description of study).

reduced at different rates in different breeds: Beagles who were given food ate it and reduced their vocalization; Shelties and Telomians (a Malaysian breed not previously available in America, now being bred and studied at Scott's and at other labs) did not eat or reduce vocalizations. However, Shelties and Beagles were induced to reduce vocalizations by sitting next to a mirror; Telomians still did not reduce vocalizations.

Although time-consuming and more expensive than research with the mouse, studies on epimeletic and et-epimeletic behaviors in the dog provide more clear-cut genetic differences than perhaps any of the studies so far considered in this area, with the possible exception of the rabbit; the latter species is not as variable or complex as is the dog, and so possibly somewhat less generalizable to humans and therefore less interesting. The advantage that the dog offers is as a function of a longer preweaning period, during which many types of independent studies can be done, and the dog's closer similarity to humans in the socialization process. These advantages also can be exploited without the extensive time period that would be required for genetical investigation in primates.

Sheep. This section ends with a review of the herd animal, specifically sheep. Building upon the earlier work of Hersher et al. (1963), Smith et al. (1966) tested the critical period hypothesis, in latency to adopt a strange lamb, in ewes of Clum-Forest, Half Breed, and Mule. The researchers reported that ewes of all three breeds will accept the first two or three lambs presented to them, including an unrelated lamb if, after the separation from her own lamb, the ewe is exposed to both her own and the strange lamb. For comparison with these breeds, Shillito and colleagues (Baldwin & Shillito, 1974; Shillito & Hoyland, 1971) studied maternal care and its abolishment in Soay sheep, a type they term the most primitive (i.e., outbred) of domestic sheep in Europe. In that breed, some but not all ewes made responses to all new lambs even though they had lambs of their own. Attachment, however, was determined to be dependent on mutual recognition. Olfactory cues seemed more critical than visual ones to identification of the lamb at close quarters. In eight pregnant anosmic (olfactory bulbectomized) ewes, four of five ewes tested postpartum did not properly lick the lamb after birth, and six of eight fed strange lambs.

There are more problems to be investigated in the herd animals concerning the critical period for adoption in precocial animals. The studies described represent only a beginning.

Is There a Parental Instinct?

If any set of behaviors could be considered instinctual, in the sense of being genetically encoded to appear, given the appropriate animal in the appropriate circumstances, it would be the behaviors composing the systems reviewed in this section. Proper provisioning and sheltering must be provided in order for the species to survive, for without the survival of a certain proportion of each new generation the species is doomed to extinction. Consequently, parental and/or maternal behaviors have been termed instinctual by more than one worker in the field (Leblond & Nelson, 1936; Seitz, 1958; Smith, 1965). Perhaps it is possible to examine the issue here.

Although analysis of strain and breed contributions to the separate components of epimeletic and et-epimeletic behaviors is still incomplete, and much remains to be done to determine the contributions of prenatal and postnatal additive, dominance, and epistatic factors, there can be little argument that species members behave in a generally similar fashion. Members of the same genera are morphologically similar; they are generally similar in behavior as well. There are, however, several critical factors that work together to produce adaptive parent–offspring interactions. Physiological factors are primary among these; several studies have implicated olfaction as critical to acceptance of the offspring by the mouse dam (Gandelman, 1973a; Gandelman, Zarrow, & Denenberg, 1971; Zarrow, Gandelman, & Denenberg, 1971). Mouse dams who were bulbectomized and then presented with young either rejected or otherwise neglected them and would have deserted them had there been an opportunity to leave the testing environment; or they cannibalized the young, usually within 24 hours after presentation. Most dams did not build proper nests. In sheep, Baldwin and Shillito (1974) found that ewes who were similarly treated did not properly lick the young after birth; they fed strange lambs and generally did not properly identify their young or form a typical ewe–lamb relationship.

The role of hormones, a second set of physiological variables in timing the onset of estrus, parturition, and lactation, has been investigated in detail in the rat (Moltz, 1971) and in the mouse (Gandelman, 1972). There has been little work done, however, in relating them to different strains and breeds. A potentially important contribution to this area can be made by separating the genetic from developmental sources of variation. Because lactation and nest building show strain and breed differences the possibility exists that these differences are partially mediated by differences in levels of circulating hormones. It is only potentially possible at present to separate the genetic from developmental factors, the result of early experiences, in any clear way.

Hormonal factors also may be involved in nest building in the mouse; progesterone was shown to mediate nest building of the brood type in nonpregnant females (Lisk et al., 1969); however, presentation of live 1-day-old pups was also sufficient to evoke nest building in ovariectomized females, even though the odor and sight of pups of the same age who were not alive was not sufficient to elicit the brood-nest behavior (Gandelman, 1973b). Finally, testosterone propionate (TP), given in the latter half of pregnancy in Rockland–Swiss mice, reduced maternal nest building, and TP with progesterone reduced nest building in nulliparae that had earlier built nests in the presence of young (Gandelman, 1973e). It was hypothesized that TP may affect neural tissue involved in nest building rather than inhibiting progesterone's effects, because ovariectomized females (no circulating progesterone) were observed to build large nests in the presence of the young. The foregoing studies indicate that hormones that circulate during the postparturient interval do affect provisioning and sheltering behaviors; those that compete with the effects of these hormones work contrary to what usually occurs when pups are presented.

Gandelman and others have investigated pup killing and cannibalism due to hormonal imbalances in the mouse, finding that postparturient dams treated with TP killed as many pups as did males (Gandelman, 1972). Virgin mice that had previously exhibited maternal behavior began killing pups after TP treatment. TP was not effective in inducing pup killing or cannibalism in lactating dams (Gandelman & Davis, 1973). Maternal behavior was also lost in postparturient Rockland–Swiss albino mice after being treated with estradiol benzoate, even though they had previously exhibited maternal behaviors (Gandelman, 1973c).

The adaptive significance of cannibalism has not been adequately defined. Because it reduces the contribution of the parent's genes to the next generation, it would appear to be an aberration rather than a trait favored by natural selection. In fact, selection would be expected to act against cannibalism. Why, then, does it persist?

A recent study of the hamster, a rodent that shows high cannibalism rates in otherwise untreated dams when the nest box is disturbed, related pup cannibalism to the normal profile of maternal behavior. Day and Galef (1977) stated that more than 75% of hamster dams cannibalized under several conditions during their first postpartum day. Cannibalism rates were observed to change concomitant with reduction in litter size and internal changes. Day and Galef hypothesized that cannibalism is an organized part of hamster maternal behavior that allows her to adjust her litter size to environmental conditions.

Though this may be so in the hamster, Fox (1968) relates that puerperal sows will mutilate and destroy young piglets, showing cannibalism in some cases even days after parturition. Horses sometimes reject and savage foals. Harlow's work with surrogate-reared rhesus macaques revealed that in many cases the mother will maltreat and nearly kill the infant. In zoo animals cannibalism has been observed in the bear and in wild swine.

Can higher rates of cannibalism be selected? Very little work has been done to answer this important question. Hauschka's (1952) study of the mouse still remains as the only experimental effort in the area. That study revealed strain and environmental (isolation of A/Ha female during pregnancy) differences. There was a slight response to selection over 13 generations: From 22.6 ± 7.8 % in the base population, this rate increased to $29.2 \pm 2.0\%$ in the $F_{12,13}$ dams. The degree of inbreeding or rate of litter succession was not correlated with cannibalism or mutilation rates. Progressive deterioration of care with age of dam was observed, however. Hauschka postulated a partially dominant pattern, indicated by its frequency and penetrance, in the A/Ha × C3H/Strong F_1 hybrid offspring.

Also, females that mutilate were selective, predictable, and precise about the anatomical targets of mutilation: Three different dams that reared four litters (a total of 25 pups) showed only left ear mutilations ($p = 3.0 \times 10^{-8}$). Digit removals were also frequently but less often observed. Hauschka discounted diet because of negative results from other strains eating the same diet. The ritual nature of the mutilation patterns also weakens a deficiency hypothesis.

From the foregoing it is apparent that maternal responsiveness is not a mechanical reaction, even in rodents. To the previously cited genetic and physiological factors can be added a third: Maternal responsiveness develops longitudinally. Gandelman's (1973d) demonstration that females show a linear progression of responsiveness to the young implicated factors other than just ovarian development.

The evidence demonstrates that epimeletic and et-epimeletic systems are complex and variable. Several levels of organization are involved: genetic, physiological, developmental, and social. There are complex interactions between levels such that functions of one cannot be considered predominant over all the others. Even though the epimeletic and et-epimeletic systems involve the interaction between at least two genotypes in a social setting and thus the social organization of the systems would seem to be predominant, it need not always be so; maternal rats have been observed retrieving pups and then their own tails after all the pups had been retrieved (Barnett, 1975).

The two systems of behavior reviewed in this section appear to be multiprocess ones, with several important internal, external, and interactive inputs, most of which must be operating in the normal range, in order for peak efficiency of parental investment to be realized. Also no one level of organization can malfunction and cause all the others to malfunction, although impaired functioning at one level would affect the workings of other systems. However the levels would be expected in most cases to provide compensatory changes, because of the considerable importance to species survival. Only when several important factors are not operating within the normal range does the system fail and the young do not survive. To some extent the evolution of multiple caregiver systems can reduce that result; it cannot totally prevent its occurrence.

Summary and Conclusions

The literature reviewed in this chapter is useful as a beginning in understanding the range of caretaking and care-soliciting behaviors that characterize species.

The invertebrate investigations provide information on the ways that various amounts of parental investment and sociality affect primarily the epimeletic behaviors. Many genetical analyses are needed, done in a systematic manner, with attention to strain origins and designs utilizing backcrosses and F_2's.

The vertebrate investigations reveal that the epimeletic and et-epimeletic systems are mutually conditioned and interactive. Under several conditions at several levels of organization and function, changing the parameters of one system changes the responses of complementary factors in the other system. Genetic effects are far from being completely understood. There is a need for careful genetical analyses of the ways in which interfertile species, and strains as well, differ in their expression of epimeletic and et-epimeletic patterns.

At present, research in this field is moving in several desirable directions. Early research aimed at controlling the effects of the maternal part of the interactions, as in much of the research on prenatal stress. Another early goal was to infer a maternal contribution by assessing offspring activities at a late period. Neither of these approaches increases our direct understanding of maternal mediators that could underlie significant genetic differences between interfertile groups. Research efforts aimed at a more molecular analysis of differences in the profile of epimeletic patterns in different strains and breeds will be the foundation upon which a systematic genetical analysis of epimeletic behavior can be built.

Considerably more specificity is needed in the molecular analysis of behavior, especially in situations where there has been little opportunity for learning to occur. These can then be compared to the later expression of behavior in similar situations (e.g., retrieving in primiparous versus multiparous dams). An analysis of the various components of each system should take into account more than single measures. The dependent measures need refining and more attention needs to be paid to how different strains and breeds differ on the adjunctive behaviors they emit while being tested. Methods of analysis at the multivariate level may be much more valuable for understanding the profile of these behaviors than are simple univariate analyses of one score (e.g., retrieving time or type of nest built), that omit most of the data on individual differences.

In past studies the offspring has been the dependent variable of preference in most cases. This trend has only provided inferential data on maternal mediators of development, and very little on the ways in which dams of different strains express their epimeletic pattern, although there were exceptions, as noted throughout. That strategy needs to be continued, but at another level of refinement. As discussed here and elsewhere (Scott, 1977) most behaviors are expressed in social systems. The study of offspring behaviors and how they change is one aspect of this, the study of maternal behaviors is the other aspect, and the interaction between the two is the third aspect. As more reliable information about the molecular differences between strains is gathered, a systematic analysis of genetic contributions to the behaviors can then be made.

REFERENCES

Archer, J. Tests for emotionality in rats and mice: A review. *Animal Behavior*, 1973, *21*(2), 205–235.

Baldwin, B. A., & Shillito, E. E. The effects of ablation of the olfactory bulbs on parturition and maternal behavior in Soay sheep. *Animal Behavior,* 1974, *22,* 220–223.

Barnett, S. A. *The rat: A study in behavior.* Chicago: University of Chicago Press, 1975.

Bell, R. W. Ultrasounds in small rodents: Arousal-produced and arousal-producing. *Developmental Psychobiology,* 1974, *7*(1), 39–42.

Bell, R. W., Nitschke, W., Bell, N. J., & Zachman, T. A. Early experience, ultrasonic vocalizations, and maternal responsiveness in rats. *Developmental Psychobiology,* 1974, *7*(3), 235–242.

Belyaev, D. K., & Trut, L. M. Some genetic and endocrine effects of selection for domestication in silver foxes. In M. W. Fox (Ed.), *The wild canids.* New York: Van Nostrand Reinhold, 1975.

Bertram, B. C. R. The social system of lions. *Scientific American,* 1975, *232*(5), 54–65.

Broadhurst, P. L. Biometrical analysis of behavioral inheritance. *Science Progress,* 1967, *55,* 123–139. (a)

Broadhurst, P. L. An introduction to the diallel cross. In J. Hirsch (Ed.), *Behavior-genetic analysis.* New York: McGraw-Hill, 1967. (b)

Brueggeman, J. A. Parental care in a group of free-ranging rhesus monkeys (*Macaca mulatta*). *Folia primatologia,* 1973, *20,* 178–210.

Burkholder, J. H. The effects of prenatal maternal crowding on subsequent maternal and social behavior to CD strain Charles River rats. Dissertation *Abstracts International,* 1977, *38*(2-B), 952.

Conely, L., & Bell, R. W. Neonatal ultrasounds elicited by odor cues. *Developmental Psychobiology,* 1977, *11*(2), 193–197.

Davis, K. L., Gurski, J. C., & Scott, J. P. Interaction of separation distress with fear in infant dogs. *Developmental Psychobiology,* 1977, *10*(3), 203–212.

Day, C. S., & Galef, B. G. Pup cannibalism: One aspect of maternal behavior in golden hamster. *Journal of Comparative and Physiological Psychology,* 1977, *91*(5), 1179–1189.

DeFries, J. C. Prenatal maternal stress in mice: Differential effects on behavior. *Journal of Heredity,* 1964, *55,* 289–295.

Deitchman, R., Kapusinski, D., & Burkholder, J. Maternal behavior in handled and nonhandled mice and its relation to later pup's behavior. *Psychological Reports,* 1977, *40*(2), 411–420.

Denenberg, V. H. The mother as a motivator. In W. J. Arnold & M. M. Page (Eds.), *Nebraska Symposium on Motivation,* 1970. Lincoln: University of Nebraska Press, 1970.

Denenberg, V. H., Ottinger, D. R., & Stephens, M. W. Effects of maternal factors upon growth and behavior of the rat. *Child Development,* 1962, *33,* 65–71.

Denenberg, V. H., Paschke, R. E., & Zarrow, M. X. Mice reared with rats: Effects of prenatal and postnatal maternal environments upon hybrid offspring of C57B1/10J and Swiss Albino mice. *Developmental Psychobiology,* 1973, *6*(1), 21–31.

Dilger, W. Behavior and genetics. In E. Bliss (Ed.), *Roots of behavior.* New York: Harper & Row, 1962. (a)

Dilger, W. The behavior of lovebirds. *Scientific American,* 1962, *208,* 88–98. (b)

Eiduson, B. N., Cohen, J., & Alexander, J. Alternatives in child-rearing in the 70's. *American Journal of Orthopsychiatry,* 1973, *43*(5), 720–731.

Eisenberg, J. F. Behavior patterns. In J. A. King (Ed.), *Biology of Peromyscus (rodentia).* Lawrence, Kans.: American Society of Mammalogists (Special Publ. No. 2), 1968.

Falconer, D. S. Patterns of response in selection experiments with mice. *Cold Spring Harbor Symposium on Quantitative Biology,* 1955, *20,* 178–196.

Falconer, D. S. *Introduction to quantitative genetics.* New York: Ronald Press, 1960.

Ferguson, W. Abnormal behavior in domestic birds. In M. W. Fox (Ed.), *Abnormal behavior in animals.* Philadelphia: W. B. Saunders, 1968.

Fox, M. W. *Abnormal behavior in animals.* Philadelphia: W. B. Saunders, 1968.

Fox, R. R. The rabbit, *Oryctolagus cuniculus.* In R. C. King (Ed.), *Handbook of genetics* (Vol. 4). New York: Plenum Press, 1975.

Fulker, D. W. Maternal buffering of rodent genotypic responses to stress: A complex genotype–environment interaction. *Behavior Genetics*, 1970, *1*, 119–124.

Fuller, John L., & Thompson, W. R. *Foundations of behavior genetics*. New York: Mosby, 1978.

Gandelman, R. Induction of pup killing in female mice by androgenization. *Physiology and Behavior*, 1972, *9*, 101–102.

Gandelman, R. Development of cannibalism in male Rockland–Swiss mice and the influence of olfactory bulb removal. *Developmental Psychobiology*, 1973, *6*, 159–164. (a)

Gandelman, R. Induction of maternal nest building in virgin female mice by the presentation of young. *Hormones and Behavior*, 1973, *4*, 191–197. (b)

Gandelman, R. Maternal behavior in the mouse: Effects of estrogen and progesterone. *Physiology and Behavior*, 1973, *10*(1), 153–155. (c)

Gandelman, R. The ontogeny of maternal responsiveness in female Rockland–Swiss albino mice. *Hormones and Behavior*, 1973, *4*(3), 257–268. (d)

Gandelman, R. Reduction of maternal nest building in female mice by testosterone propionate treatment. *Developmental Psychobiology*, 1973, *6*(6), 539–546. (e)

Gandelman, R., & Davis, P. G. Spontaneous and testosterone-induced pup killing in Rockland–Swiss mice: The effects of lactation and the presence of the young. *Developmental Psychobiology*, 1973, *6*(3), 251–257.

Gandelman, R., Zarrow, M. X., & Denenberg, V. H. Olfactory bulb removal eliminates maternal behavior in the mouse. *Science*, 1971, *171*, 210–211.

Gary, N. E. Queen honeybee attractiveness as related to mandibular gland secretion. *Science*, 1962, *136*, 773–774.

Gary, N. E. Observations of mating behavior in the honeybee. *Journal of Apicultural Research*, 1963, *2*, 3–13.

Gaul, A. T. Experiments in housing vespine colonies, with notes on the homing and toleration instincts of certain species. *Psyche, Cambridge*, 1941, *48*(1), 16–19.

Green, E. L. Breeding systems. In E. L. Green (Ed.), *Biology of the laboratory mouse* (2nd ed.). New York: Dover Publications, 1968.

Green, M. C. The laboratory mouse, *Mus musculus*. *Handbook of genetics* (Vol. 4). New York: Plenum Press, 1975.

Griffin, G. A. The effects of multiple mothering on the infant–mother and mother–infant affectional systems. *Dissertation Abstracts International*, 1967, *28*(3-B), 1226–1227.

Guhl, A. M. Social behavior of the domestic fowl. *Technical Bulletin of the Kansas Agricultural Station*, 1953, No. 73.

Guhl, A. M., & Fischer, F. J. The behavior of chickens. In E. S. Hafez (Ed.), *The behavior of domestic animals* (2nd ed.) London: Bailliere, 1969.

Gurski, J. C. *Multiple mothering in mice: A laboratory study*. Unpublished doctoral dissertation, Bowling Green State University. 1975.

Gurski, J. C. *Multiple mothering in mice: Effects on maternal and offspring behaviors*. Paper presented at the 75th meeting of the American Psychological Association, San Francisco, August 1977.

Gurski, J. C., Davis, K., & Scott, J. P. Interaction of separation distress with contact comfort and discomfort in the dog. *Developmental Psychobiology*, 1980, *13*(5), 463–467.

Gurski, J. C., & Scott, J. P. Individual vs. multiple mothering in mammals. In W. P. Smotherman & R. W. Bell (Eds.), *Maternal influences and early behavior*. Jamaica, N.Y.: Spectrum Publications, 1980.

Hamilton, W. D. The genetical evolution of social behavior. *Journal of Theoretical Biology*, 1964, *1*, 1–52.

Harlow, H. F., Harlow, M. K., & Hansen, E. W. The maternal affectional system of rhesus monkeys. In H. L. Rheingold (Ed.), *Maternal behavior in mammals*. New York: Wiley & Sons, 1963.

Hart, F. M., & King, J. A. Distress vocalizations of young in two subspecies of *Peromyscus maniculatus*. *Journal of Mammalogy*, 1966, *47*, 287–293.

Haskins, R. Effect of kitten vocalizations on maternal behavior. *Journal of Comparative and Physiological Psychology*, 1977, *91*(4), 830–838.

Hauschka, T. A. Mutilation patterns and hereditary cannibalism. *Journal of Heredity*, 1952, *43*(3), 115–123.

Hazan, B. Introduction. *In* A. I. Rabin & B. Hazan (Eds.), *Collective education in the kibbutz*. New York: Springer Publishing Co., 1973.

Hersher, L., Richmond, J. B., & Moore, A. U. Maternal behavior in sheep and goats. In H. L. Rheingold (Ed.), *Maternal behavior in mammals*. New York: Wiley & Sons, 1963.

Hess, E. H., Petrovich, S. B., & Goodwin, E. B. Induction of parental behavior in Japanese quail (*Coturnix coturnix japonica*). *Journal of Comparative and Physiological Psychology*, 1976, *90*(3), 244–251.

Hinde, R. A. Analyzing the role of the partners in a behavioral interaction—mother–infant relations in rhesus macaques. *Annals of the New York Academy of Science*, 1969, *159*, Art. 3, 681–695.

Hofer, M. A. Maternal separation affects infant rats' behavior. *Behavioral Biology*, 1973, *9*(5), 629–633.

Hooper, E. T. Classification. In J. A. King (Ed.), *Biology of Peromyscus (rodentia)*. Lawrence, Kans.: American Society of Mammalogists (Special Publ. No. 2), 1968.

Hutt, F. B. *Genetics of the fowl*. New York: McGraw-Hill, 1949.

Joffe, J. M. Genotype and prenatal and premating stress interact to affect adult behavior in rats. *Science*, 1965, *150*, 1844–1845.

Joffe, J. M. *Prenatal determinants of behavior*. New York: Pergamon, 1969.

Kaufman, I. C. Mother–infant separation in monkeys: An experimental model. In J. P. Scott & E. Senay (Eds.), *Separation and depression*. Washington, D.C.: AAAS, 1973.

Kaufman, I. C., & Rosenblum, L. A. Effects of separation from mother on the emotional behavior of infant monkeys. *Annals of the New York Academy of Science*, 1969, *159*, Art. 3, 681–695.

Kerr, W. E. Genetic determination of castes in the genus *Melipona*. *Genetics*, 1950, *35*, 143–152.

Kerr, W. E. Advances in bee cytology and genetics. *Annual Review of Entomology*, 1974, *19*, 253–268.

King, J. A. Maternal behavior in *Peromyscus*. In H. L. Rheingold (Ed.), *Maternal behavior in mammals*. New York: Wiley & Sons, 1963.

Kleiman, D. G. Some aspects of social behavior in the *Canidae*. *American Zoologist*, 1967, *7*, 365–372.

Knowlton, N. Reproductive synchrony, parental investment, and the evolutionary dynamics of sexual selection. *Animal Behaviour*, 1979, *27*(4), 1022–1033.

Layne, J. M. Ontogeny. In J. A. King (Ed.), *Biology of Peromyscus (rodentia)*. Lawrence, Kans.: The American Society of Mammalogists (Special Publ. No. 2), 1968.

Leblond, C. P. Nervous and hormonal maternal factors in the maternal behavior of the mouse. *Journal of Genetic Psychology*, 1940, *57*, 327–344.

Leblond, C. P., & Nelson, W. D. Parental instinct in the mouse, especially after hypophysectomy. *Anatomical Record* (Suppl.), 1936, *66*, 29–30.

Lee, C. T. Genetic analyses of nest-building behavior in laboratory mice (*Mus musculus*). *Behavior Genetics*, 1973, *3*, 247–256.

Lisk, R. D., Pretlow, R. A., & Friedman, S. M. Hormonal stimulation necessary for elicitation of maternal nest-building in the mouse (*Mus musculus*). *Animal Behavior*, 1969, *17*, 730–737.

Lynch, C. B., & Hegmann, J. P. Genetic differences influencing behavioral temperature regulation in small mammals. I. Nesting by *Mus musculus*. *Behavior Genetics*, 1972, *2*(1), 43–53.

Lynch, C. B., & Hegmann, J. P. Genetic differences influencing behavioral temperature regulation in small mammals. II. Genotype–environment interactions. *Behavior Genetics*, 1973, *3*(2), 143–154.

Mackensen, O., & Tucker, K. W. *Instrumental insemination of queen bees.* (Agricultural Handbook No. 390). Washington, D.C.: U.S. Government Printing Office, 1970.

Marsden, H. M., & Bronson, F. H. Estrous synchrony in mice: Alteration by exposure to male urine. *Science,* 1070, *144,* 1469.

Marsh, L. L. Emission of the maternal pheromone in the nulliparous female and the adult male rat. *Dissertation Abstracts International,* 1977, *38*(2-B), 950.

Mason, W. A., & Kenney, M. D. Redirection of filial attachments in Rhesus monkeys: Dogs as mother surrogates. *Science,* 1974, *183,* 1209–1211.

Massaro, T. F., Levitsky, D. A., & Barnes, R. H. Protein malnutrition induced during gestation: Its effect on pup development and maternal behavior. *Developmental Psychobiology,* 1977, *10*(4), 339–345.

Mather, K., & Jinks, J. L. *Biometrical genetics.* Ithaca, N.Y.: Cornell University Press, 1971.

McLaren, A. Genetics of the early mouse embryo. *Annual Review of Genetics,* 1976, *10,* 361–388.

McLaren, A., & Michie, D. Studies on the transfer of fertilized mouse eggs to uterine foster-mothers. I. Factors affecting the implantation and survival of native and transferred eggs. *Journal of Experimental Biology,* 1956, *33,* 394–416.

McNair, M. R., & Parson, G. A. Models of parent–offspring conflict. III. Intra-brood conflict. *Animal Behaviour,* 1979, *27*(4), 1200–1209.

Michener, C. D. Comparative social behavior of bees. *Annual Review of Entomology,* 1969, *14,* 299–342.

Miller, O. F., & Miller, D. A. Cytogenetics of the mouse. *Annual Review of Genetics,* 1975, *9,* 285–303.

Misanin, J. R., Zawacki, D. M., & Krieger, W. G. Differential maternal behavior of the rat dam toward natural and foster pups: Implications for nutrition research. *Bulletin of the Psychonomic Society,* 1977, *10*(4), 313–316.

Moltz, H. The ontogeny of maternal behavior in some selected mammalian species. In H. Moltz (Ed.), *The ontogeny of vertebrate species.* New York: Academic Press, 1971.

Momot, J. P., & Rothenbuhler, W. C. Behavior genetics of nest cleaning in honeybees. VI. Interactions of bee age, genotype and nectar flow in mixed colonies. *Journal of Apicultural Research,* 1971, *10*(1), 11–21.

Nesbitt, W. H. Ecology of a feral dog pack on a wildlife refuge. In M. W. Rox (Ed.), *The wild canids.* New York: Van Nostrand Rheinhold, 1975.

Nitschke, W., Bell, R. W., & Zachman, T. Distress vocalizations of young in three inbred strains of mice. *Developmental Psychobiology,* 1971, *5*(4), 363–370.

Orcutt, F. S., Jr., & Orcutt, A. B. Nesting and parental behavior in domestic common quail. *Auk,* 1976, *93*(1), 135–141.

Ottinger, D. R., Denenberg, V. H., & Stephens, M. W. Maternal emotionality, multiple mothering, and emotionality at maturity. *Journal of Comparative and Physiological Psychology,* 1963, *56,* 313–317.

Palm, J. E. The laboratory rat, *Rattus norvegicus.* In R. C. King (Ed.), *Handbook of genetics* (Vol. 4). New York: Plenum Press, 1975.

Pettijohn, T. F. Attachment and separation distress in the infant guinea pig. *Developmental Psychobiology,* 1979, *12*(1), 73–81.

Pettijohn, T. F., Wong, T. W., Ebert, P. D., & Scott, J. P. Alleviation of separation distress in three breeds of young dogs. *Developmental Psychobiology,* 1977, *10,* 373–381.

Poirer, F. E. Introduction. In F. E. Poirer (Ed.), *Primate socialization.* New York: Random House, 1972.

Priestnall, R. Effects of handling on maternal behavior in the mouse (*Mus musculus*): An observational study. *Animal Behavior,* 1973, *21,* 383–386.

Rabb, G. B., Woolpy, J. H., & Ginsburg, B. E. Social relationships in a group of captive wolves. *American Zoologist,* 1967, *7,* 305–311.

Rasmussen, D. I. Genetics. In J. A. King (Ed.), *Biology of Peromyscus (rodentia).* Lawrence, Kans.: American Society of Mammalogists (Special Publ. No. 2), 1968.

Rasmussen, D. I. The genus *Peromyscus.* In R. C. King (Ed.), *Handbook of genetics* (Vol. 4). New York: Plenum Press, 1975.

Reading, A. J. Effect of maternal environment on the behavior of inbred mice. *Journal of Comparative and Physiological Psychology,* 1966, *62*(3), 437–440.

Ressler, R. H. Parental handling in two strains of mice reared by foster parents. *Science,* 1962, *137,* 129–130.

Rheingold, H. L. Maternal behavior in the dog. In H. L. Rheingold (Ed.), *Maternal behavior in mammals.* New York: Wiley & Sons, 1963.

Roderick, T. H., & Schlager, G. Multiple factor inheritance. In E. L. Green (Ed.), *Biology of the laboratory mouse* (2nd ed.). New York: Dover Publications, 1968.

Rosenblatt, J. S., & Lehrman, D. S. Maternal behavior in the laboratory rat. In H. L. Rheingold (Ed.), *Maternal behavior in mammals.* New York: Wiley & Sons, 1963.

Rosenblum, L. A. Infant attachment in monkeys. In H. R. Schaffer (Ed.), *The origins of human social relations.* New York: Academic Press, 1971.

Rosenblum, L. A. Maternal regulation of infant behavior. In C. P. Carpenter (Ed.), *Behavior regulators of behavior in primates.* Lewisburg, Pa.: Bucknell University Press, 1973.

Ross, S., Sawin, P. B., Zarrow, M. X., & Denenberg, V. H. Maternal behavior in the rabbit. In H. L. Rheingold (Ed.), *Maternal behavior in mammals.* New York: Wiley & Sons, 1963.

Rothenbuhler, W. C. Behavior genetics of nest cleaning in honey bees. I. Responses of four inbred lines to disease-killed brood. *Animal Behavior,* 1964, *12,* 578–583. (a)

Rothenbuhler, W. C. Behavior genetics of nest cleaning in honey bees. IV. Responses of F_1 and backcross generations to disease-killed brood. *American Zoologist,* 1964, *4,* 111–123. (b)

Rothenbuhler, W. C. Genetic and evolutionary considerations of social behavior of honeybees and some related insects. In J. Hirsch (Ed.), *Behavior-genetic analysis.* New York: McGraw–Hill, 1967.

Rothenbuhler, W. C. The honey bee, *Apis mellifera.* In R. C. King (Ed.), *Handbook of genetics* (Vol. 3). New York: Plenum Press, 1975.

Rothenbuhler, W. C., Kulincevic, J. M., & Kerr, W. E. Bee genetics. *Annual Review of Genetics,* 1968, *2,* 413–438.

Rudnai, J. *The social life of the lion.* Wallingford, Pa.: Washington Square East, 1973.

Sawin, P. B., & Crary, D. D. Genetic and physiological background of reproduction in the rabbit. II. Some racial differences in the pattern of maternal behavior. *Behavior,* 1953, *6*(2), 128–146.

Schneirla, T. C., Rosenblatt, J. S., & Tobach, E. Maternal behavior in the cat. In H. L. Rheingold (Ed.), *Maternal behavior in mammals.* New York: Wiley & Sons, 1963.

Scott, J. P. *Early experience and the organization of behavior.* Belmont, Calif.: Wadsworth, 1968.

Scott, J. P. Social genetics. *Behavior Genetics,* 1977, *7*(4), 327–346.

Scott, J. P., & Fuller, J. L. *Genetics and the social behavior of the dog.* Chicago: University of Chicago Press, 1965.

Scott, J. P., Stewart, J. M., & DeGhett, V. J. Separation in infant dogs: Emotional responses and motivational consequences. In J. P. Scott & E. C. Senay (Eds.), *Separation and depression: Clinical and research aspects.* Washington, D.C.: AAAS, 1973.

Seitz, P. F. D. The maternal instinct in animal subjects. I. *Psychosomatic Medicine,* 1958, *20*(3), 215–226.

Sherrod, K. B., Connor, W. H., & Meier, G. W. Transient and enduring effects of handling on infant and maternal behavior in mice. *Developmental Psychobiology,* 1974, *1*(1), 31–37.

Shillito, E. E., & Hoyland, V. J. Observations on parturition and maternal care in Soay sheep. *Journal of Zoology* (London). 1971, *165,* 509–512.

Silver, R. (Ed.). *Parental behavior in birds.* New York: Academic Press, 1977.

Skutch, A. F. *Parent birds and their young.* Austin: University of Texas Press, 1976.

Smith, F. V. Instinct and learning in the attachment of lamb and ewe. *Animal Behaviour,* 1965, *13,* 84–86.

Smith, F. V., Van Toller, C., & Boyes, T. The 'critical period' in the attachment of lambs and ewes. *Animal Behavior,* 1966, *14,* 120–125.

Smotherman, W. P., Bell, R. W., Hershberger, W. A., & Coover, G. D. Orientation to rat pup cues: Effects of maternal experiential history. *Animal Behaviour,* 1978, *26,* 264–273.

Stewart, J. M., & Scott, J. P. The dog, *Canis familiaris.* In R. C. King (Ed.), *Handbook of genetics* (Vol. 4). New York: Plenum Press, 1975.

Svare, B., & Gandelman, R. Postpartum aggression in mice: Experiential and experimental factors. *Hormones and Behavior,* 1973, *4*(4), 323–334.

Szechtman, H., Siegel, H. I., Rosenblatt, J. S., & Komisaruk, B. R. Tail-pinch facilitates onset of maternal behavior in rats. *Physiology and Behavior,* 1977, *19*(6), 807–809.

Thompson, W. R. Influence of prenatal maternal anxiety on emotionality in young rats. *Science,* 1957, *125,* 698–699.

Thompson, W. R., & Olian, S. Some effects on offspring behavior of maternal adrenalin injections during pregnancy in three inbred mouse strains. *Psychological Reports,* 1961, *8,* 87–90.

Villescas, R., Bell, R. W., Wright, L., & Kuffner, M. Effect of handling on maternal behavior following return of pups to the nest. *Developmental Psychobiology,* 1977, *10*(4), 323–329.

Weir, M. W., & DeFries, J. C. Prenatal maternal influence on behavior in mice: Evidence of a genetic basis. *Journal of Comparative and Physiological Psychology,* 1964, *58,* 412–417.

Werboff, J., Steg, M., & Barnes, L. Communal nursing in mice: Strain-specific effects of multiple mothers on growth and behavior. *Psychonomic Science,* 1970, *19*(5), 269–271.

Wiesner, B. P., & Sheard, N. B. *Maternal Behavior in the Rat.* London: Oliver and Boyd, 1933.

Wilson, E. O. *The insect societies.* Cambridge, Mass: The Belknap Press of Harvard University, 1972.

Wimer, R. E., & Fuller, J. L. Patterns of behavior. In E. L. Green (Ed.), *Biology of the laboratory mouse* (2nd ed.). New York: Dover Publications, 1968.

Wright, L. L., & Bell, R. W. Interactive effects of parity and pup stress on the maternal behavior of *Rattus norvegicus. Developmental Psychobiology,* 1977, *10*(4), 331–337.

Wright, L., & Bell, R. W. Interactive effects of parity and early pup stress on the open-field behavior of laboratory rats. *Developmental Psychobiology.* 1978, *11*(5), 413–418.

Zarrow, M. X., Gandelman, R., & Denenberg, V. H. Lack of nest building and maternal behavior in the mouse following olfactory bulb removal. *Hormones and Behavior,* 1971, *2,* 227–238.

11 The Genetics of Agonistic and Sexual Behavior

Janet S. Hyde

Denison University

Genetic factors are sources of both intraspecific uniformity and intraspecific variability in behavior; nowhere are these twin principles more evident than in sexual behavior and agonistic behavior. For example, genetic factors may control the courtship pattern that is uniformly displayed by all or nearly all members of a particular species of bird, but genetic factors may also control variability in the length of copulation time.

The genetic mechanism underlying this duality is straightforward. In sexually reproducing species, some loci are polymorphic (two or more alleles are possible), whereas other loci are monomorphic (only one allele exists). Behaviors that show intraspecific variability are probably controlled by polymorphic loci, whereas those showing intraspecific uniformity are probably controlled by the monomorphic loci.

Ethologists have been most concerned with the study of intraspecific uniformity in behavior ("species-specific behavior"). Most behavior geneticists have been concerned with intraspecific variability in behavior. Accordingly, the latter phenomenon is the focus of the present review.

Behavior geneticists have had to deal with serious measurement problems in their study of agonistic and sexual behaviors. Because both behaviors are social, they of necessity involve at least two genotypes; hence the observed behavior reflects two genotypes, not one. Thus the attempt to obtain a score for a single genotype is difficult. For example, if one wants a measure of the aggressiveness of the C57BL strain of mice, should one do that by pairing them with other C57BL's; or should the score be the average of scores when paired with a variety of other inbred strains; or should the score result from pairing with a standard inbred opponent? Reflecting this complexity, a variety of behaviors and testing

409

situations have been used to measure agonistic behavior. Within-strain pairings have been used, although they have been criticized, as discussed in a later section. A round-robin technique, in which members of one strain are successively paired with members of all other strains, has also been used. Finally, a standard opponent from an inbred strain not itself being measured has also been used. Because isolation reliably increases the aggressiveness of mice, animals are often reared in isolation prior to testing, although in some studies they have not been. The "dangler" technique has also been used with mice—a standard opponent is dangled passively in front of the test mouse, eliciting attack but not itself attacking. Although many of these methods deal with "spontaneous" aggression that results when two animals are simply placed with each other, other types of agonistic behavior, such as competition for food, have also been investigated. Finally, although most of the research has concerned intermale aggression, there has been some investigation of agonistic behavior in females, particularly of maternal or postpartum aggression and, in one research program, of isolation-induced interfemale aggression.

The genetic analysis of sexual behavior poses similar problems, inasmuch as the behavior results from the interaction of two animals and thus two genotypes. Although the basic copulatory pattern is fairly uniform for a species, variables such as mating speed and duration of copulation show the kind of variability necessary for behavior-genetic analysis. Sexual behavior has received some highly elegant genetic analyses because it is a behavior, unlike agonistic behavior, in which the geneticist's dream species, *Drosophila melanogaster*, engages.

AGONISTIC BEHAVIOR

Mice

The most extensive genetic analyses of agonistic behavior have been done with mice (see an earlier review by McClearn & DeFries, 1973).

Strain Differences.   Differences among inbred strains in agonistic behavior were demonstrated early. Ginsburg and Allee (1942) found differences among inbred strains when the males were paired in a successive "round-robin" situation, with C57BL/10s being the most aggressive, C3Hs intermediate, and BALB/cs the least aggressive as measured by fights won. Scott (1942) ranked these strains in a different order with respect to the tendency to initiate attacks against an intruder, with C3H being first and C57BL last. It was later discovered that the C3H strain has defective vision (Fuller & Wimer, 1966). Lee (1971) found that a C3H and two others that carried the RD (rodless retina) gene showed fewer attacks in response to a simulated attack by a dangled male than did three sighted strains. The C57BL, and BALB strains always showed significant differences in agonistic behavior, though not necessarily in the same direction (Bauer, 1956; Scott, 1966).

Southwick and Clark (1968), using a different measurement technique, found differences among 14 inbred strains in the aggressiveness of the males. After a period of housing in isolation, males were placed in groups of four of the same strain, and observations of several aspects of agonistic behavior (tail rattling, chasing, attacking, fighting, and social grooming) were made. Large strain differences were found on every measure, as well as on a composite score. However, this technique for measuring strain differences in aggression has been criticized (Fuller & Hahn, 1976). Essentially it confounds strain of attacker with strain of victim and gives no information on the outcome of pairing two mice of different strains. That is, because we may find that there is more fighting in a group of C57BLs than in a group of As, that does not necessarily mean that if a C57BL were paired with an A, the C57BL would be the attacker and the A the victim. Stewart (1974) studied dyadic pairs in all combinations of inbred strains A/J, BALB/cJ, C57BR/cdJ, DBA/2J, and SWR/J. A factorial analysis of variance of attack behavior demonstrated that the genotypes of both attacker and victim had important effects on the elicited behavior. The amount of attacking between unlike strains significantly exceeded that among individuals of the same strain.

The violent fighting observed between males under the conditions described is qualitatively different from that elicited in other situations (Scott & Fredericson, 1951) and has been labeled "spontaneous" or "isolation-induced" aggression. Contrastingly, competition for a single food pellet involves struggling for possession but no directed attacks on the other individual. Fredericson and Birnbaum (1954) found that whereas BALB/cs would peacefully share a food pellet, C57BL/10s were strongly competitive. Within strains both sexes reacted similarly. These strain differences were not related to spontaneous fighting. In intermale spontaneous combats, BALB/cs fought sooner and more intensely than the other strain (Fredericson, Story, Gurney, & Butterworth, 1955).

Inbred strain differences in social dominance have also been demonstrated. Lindzey, Manosevitz, and Winston (1966) used three measures of dominance: tube dominance, food competition, and spontaneous fighting (aggression). They found large and reliable strain differences in all three measures; tube dominance appeared to be inversely related to food competition and aggression.

Lindzey, Winston, and Manosevitz (1961) measured tube dominance in three inbred strains and later repeated the study using different sublines of the three original strains (Lindzey, Thiessen, Blom, & Tucker, 1969). The later study did not replicate some of the findings with the original strains.

Selective Breeding. Lagerspetz (1964; Lagerspetz & Lagerspetz, 1971; Lagerspetz, Tirri, & Lagerspetz, 1968; Lagerspetz & Wuorinen, 1965) has conducted a program of artificial selection for male aggressiveness, beginning with a heterogeneous stock of albino mice as a foundation population. Results are available through the 19th generation of selection. The criterion for selection was a rating of aggressiveness on a 7-point scale. The genetic design was bidirectional selection with no control line. That is, one line was selected for high aggressiveness and a second line was selected for low aggressiveness. Inbreeding was practiced systematically in some of the early generations.

The results showed a response to selection in both directions; that is, it was possible to select both for higher levels of aggressiveness and for lower levels. The limits of selection appear to have been reached by the seventh generation of selection, perhaps because inbreeding had been practiced earlier, eliminating much of the genetic variability.

The genetic design—bidirectional selection—is relatively simple compared with the more elaborate six-line selection designs that are currently favored for selection programs (see, for example, the study of female aggressiveness described later). The design lacks an unselected control line and replicates of the lines. Nonetheless, some interesting results can be obtained. For example, in a study of possible physiological pathways mediating the genetic influences, Lagerspetz et al. (1968) reported that males of the aggressive line had lower serotonin levels in the forebrain, higher noradrenaline levels in the brainstem, and heavier testes than males from the unaggressive line. Unfortunately, because this is based on the study of only two lines, there is a possibility that the demonstrated correlations are accidental. Another study in this research program, demonstrating genotype–environment interactions, is discussed in a later section.

In another selection program, van Oortmerssen and Bakker (1981) selected for short and long attack latencies in male mice descended from four males and three females trapped in the wild. Inbreeding as well as selection were practiced in the first three generations. Selection for short attack latencies was successful, with a realized heritability of .30. Four attempts to select for long attack latencies all failed because, for unknown reasons, the lines died out within two generations.

Maternal Effects and Cross-Fostering.   Findings of differences among inbred strains or between lines selected in opposite directions, are generally interpreted as indicating that aggression is subject to genetic influence (i.e., that genetic variation creates variation in aggressive behavior). However, such an inference is not completely justified. In particular, inbred strain comparisons confound genotype of subject with genotype of mother. Thus, differences in aggressiveness between strain *A* and strain *B* could be due to: (1) genotypic differences in aggressiveness; (2) postnatal maternal effects (i.e., differences between strain *A* and *B* in mothering); or (3) prenatal maternal effects, due to differences in intrauterine environment between strain-*A* females and strain-*B* females.

The design used to test for the last possibility is the intrauterine transplant. I was not able to find any studies using this technique to clarify genetic influences on aggressive behavior.

The technique used to document postnatal maternal effects is the cross-fostering study. Ginsburg and Allee (1942) were the first to use such a technique in the study of aggressive behavior. Using strains C57BL/10 and BALB/c and a split-litter technique, half the pups were reared by their own mothers and half were reared by a foster mother of the other strain. Fostering appeared to have no effect, thus suggesting that strain differences in aggressiveness could be attributed to genotype of the subject.

Fredericson (1952b) cross-fostered entire litters of C57BL/10 and BALB/c strains. When the young were 38 days of age he induced food competition between them and their foster parents. Neither strain of young showed altered behavior, but those of the highly competitive C57 strain induced similar behavior in their peaceful foster parents.

Both designs confound strain differences in maternal behavior with rearing own versus strange young. The necessary control is the "in-fostered" group, that is, pups reared by a foster mother of their own strain. This necessary additional group was included in a study by Southwick (1968). Thus his study included three rearing conditions: pups reared by their own mother (control), pups fostered by another mother of their own strain (in-fostered), and pups fostered by another mother of the other strain (cross-fostered). The findings indicated that genotype and maternal effects interact. CFWs were highly aggressive no matter what their rearing conditions, and As were generally unaggressive. However, As reared by CFW mothers were more aggressive than control As or in-fostered As. Thus maternal effects depended on the strain of the pups.

Lagerspetz and Wuorinen (1965) performed a cross-fostering experiment in conjunction with their selective breeding for aggressiveness. Mice from the high line were significantly more aggressive than those of the low line regardless of the nature of their preweaning maternal care.

Recombinant Inbred Strains. The new technique of recombinant inbred strains permits sophisticated genetic analyses of behavior. This technique has been applied to the analysis of aggressive behavior by Eleftheriou, Bailey, and Denenberg (1974). They tested fighting after isolation rearing in males from inbred strains C57BL and BALB, the reciprocal F_1 hybrids, and seven recombinant inbred strains formed by inbreeding of the various genotypes formed in the F_2 (Bailey, 1971). Two of the recombinant inbred strains scored outside the range of the original inbreds; on the basis of this result, the authors concluded that the fighting behavior was controlled by alleles from at least two loci. Evidence from reciprocal crosses also suggested the possibility of a cytoplasmic factor from the ovum influencing the behavior.

Messeri, Eleftheriou, and Oliverio (1975) studied the genetics of tube-dominance behavior in inbreds C57BL and BALB, using the techniques of recombinant inbred strains and phylogenetic tree analysis. They concluded that the behavior was influenced by genes at at least three loci.

Genotype–Environment Interactions.   Greenberg (1972) tested for the effects of temperature and population density on aggression in two inbred strains of mice. He found that the environmental variables did not interact with strain differences.

On the other hand, Lagerspetz (1964) did find a genotype–environment interaction effect on aggressiveness. Experience with defeats reduced aggressiveness in both high and low selected line animals; victories increased the aggressiveness of high-line animals, but had no effect on low-line animals.

Genotype and Pheromones.   The importance of pheromones in aggressive encounters has been well documented (Bronson, 1971). Kessler, Harmatz, and Gerling (1975) showed that genetic factors are involved in the pheromone–aggression relationship. They found that urine from intact DBA males, when swabbed on castrates, elicited more aggression than did urine from intact C57BR or CBA males. Therefore it appears that genes may modify the quality of aggression-eliciting pheromones in male mouse urine, thus affecting aggressive behavior.

Aggression and Fitness.   Agonistic behavior is believed to be important in an adaptive, evolutionary sense because of its relationship to Darwinian fitness (DeFries & McClearn, 1970); that is, if the most aggressive, dominant males sire most of the offspring, this means that they have high fitness. Horn (1974) investigated genetic factors in relation to aggression and fitness. He grouped RF, BALB, DBA, and C57BL males in a seminaturalistic environment with C57BL females. He found that the RF males were the most aggressive, and that they sired 95.6% of the offspring. Results indicated that three factors, unrelated to aggressiveness, accounted for the reproductive disadvantage of BALB and C57BL males: differences in general fertilizing ability, effects of grouping, and pregnancy blockage. However, these factors could not explain the reproductive disadvantage of the DBA males; hence it appeared that differences in aggression were important determinants of differences in fitness between RF and DBA males. Therefore, one way that genetic factors may influence an animal's fitness is by their influence on the animal's agonistic behavior. These findings are relevant to the topic of sexual selection, which is discussed in a later section.

Female Aggression.   Genetic influences on four types of aggression in female mice have been investigated: (1) food competition; (2) "naturally occurring," isolation-induced interfemale aggression; (3) maternal aggression (aggression by a female during the postpartum period, also called "postpartum aggression''); and (4) female aggression following neonatal administration of androgen.

Most experts on the genetics of aggressive behavior have claimed that female mice from inbred strains seldom if ever fight violently (McClearn & DeFries, 1973). However, Fredericson (1952a) found that female pairs of the C57BL/10 strain competed over food with the same intensity as male pairs. And Ebert (1976) found that approximately 25% of a population of wild-trapped female mice fought when paired with another female, after being reared in isolation (compared with none of the inbred females tested). This rather marked difference between inbreds and wilds must be attributed to genetic differences, owing either to selective effects in the domestication process, or to inbreeding.

Ebert and Hyde (1976) then selectively bred for female aggression in mice. The foundation population consisted of wild-trapped mice, within-family selection was practiced, and six selection lines were formed: a line selected for high aggression, a line selected for low aggression, an unselected control line, and a replicate of each of these lines. Selection was done on the basis of a 5-point rating scale, based on observation of a 7-minute paired encounter with a C57BL female, which never attacks. Initial results were reported through the fourth generation of selection (Ebert & Hyde, 1976), and selection progressed through the 11th generation (Hyde & Sawyer, 1980). By the 10th generation, separation of the lines was good, with nearly 50% of females in the high lines fighting, and zero to 10% in the low lines. Realized heritabilities over eight generations of selection were .12 and .14 for the two high lines and .34 and .46 for the two low lines. Male aggression did not show a correlated response to selection for female aggression (Hyde & Ebert, 1976). These results are in agreement with those of the opposite experiment—in selection for male aggression, female aggression showed no correlated response (van Oortmerssen & Bakker, 1981). Maternal aggression, however, showed a definite correlated response (Hyde & Sawyer, 1979). In a comparison of rearing in isolation versus group housing, isolation increased the aggressiveness of the females in all lines, and there was no line-by-rearing condition (genotype–environment) interaction (Hyde & Sawyer, 1980). There are estrous-cycle variations in female aggressiveness, and these may provide a clue as to the adaptive nature of the behavior (Hyde & Sawyer, 1978).

St. John and Corning (1973) studied maternal aggressiveness and male aggressiveness in four inbred strains and a heterogeneous line. The results indicated that maternal aggressiveness was high in lines in which the males were aggressive and maternal aggressiveness was low in lines in which the males were relatively unaggressive. These results indicate that maternal aggression is influenced by genotype and may be under the same genetic control as male aggression.

Vale, Ray, and Vale (1972) found that female mice from different strains, given neonatal administrations of androgen, differed in their adult aggressive behavior. In particular, females from strains in which the males were aggressive themselves showed high levels of adult androgen-induced aggression, whereas females from strains in which the males are unaggressive showed little aggression. Vale et al. (1973a) obtained similar results with neonatal administration of estrogen. Therefore, the effect of neonatal hormone exposure on female mice depends on their genotype.

Single-Locus Effects.   Several investigators have found evidence of single-locus effects on aggressive behavior. Ciaranello, Lipsky, and Axelrod (1974) studied two BALB sublines and concluded there was evidence of a single-locus effect on aggressiveness. Kessler, Elliot, Orenberg, and Barchas (1977) studied isolation-induced aggression in BALBs, As and the F_1 and F_2 crosses and backcrosses; their data were consistent with high aggressivity being a single autosomal recessive trait.

In an elegant program of research aimed at actual mapping of effects, Ginsburg, Maxson, Selmanoff, and others have documented the effects of the Y chromosome on aggression in mice (Maxson, Ginsburg, & Trattner, 1979; Selmanoff, Jumonville, Maxson, & Ginsburg, 1975; Selmanoff, Maxson, & Ginsburg, 1976). The original evidence came from differences between reciprocal crosses of DBAs and C57BLs. However, the observed results were also consistent with other models. To test for such alternative interpretations, Maxson et al. (1979) developed a congenic stock of C57BL mice with the DBA Y chromosome by a backcross system of breeding. The results for that stock indicated that the Y chromosome from the DBA strain, by itself, does not have an incremental effect on aggression; rather, there appears to be an interaction between Y-chromosome effects and autosomal effects on aggression.

A novel approach is that of Schroder (1980), who compared the aggressiveness of male mice resulting from 600 R of gamma-ray irradiation of their paternal spermatozoa. These males were more aggressive than unradiated controls. Because such irradiation produces heterozygosity for translocations, the increased aggressiveness was attributed to chromosomal rearrangements. Werner and Schroder (1980) have similarly studied changes in courtship of male guppies after irradiation.

Dogs.   Scott and Fuller (1965) conducted an extensive research program on agonistic behavior in dogs in the context of two social relationships, dog + human and dog + dog. The general design of their experiments included repeated measurements as behavior developed from birth through 1 year of age. The genetic variable was manipulated by making comparisons among five pure breeds and subsequently conducting a Mendelian experiment involving two of the breeds chosen for maximum differences.

Dog breeds differ widely in agonistic behavior, and their sample included two highly aggressive breeds: the wirehaired fox terrier and the Basenji. An intermediate strain was the Shetland sheep dog, whose behavior is easily inhibited by training. At the opposite extreme were the cocker spaniel and the beagle, both of which have the reputation of being extremely peaceful. The crossbreeding experiment was done between the Basenji (the so-called African barkless dog) and the cocker spaniel.

Agonistic behavior was measured in three situations. Two of these involved the dog + human relationship with the experimenter acting as a standard tester. The third involved dog + dog relationships, in a modified homogeneous sets design. Each dog was tested in combination with every other animal in the litter in which it grew up, a reasonably natural situation that gave a picture of the organization of the total litter. Thus animals were tested only within their own breed or hybrid population.

In the handling test the human handler approached and handled dogs in a standard fashion and all immediate reactions were noted. The test was administered at 5, 7, 9, 11, 13, 15, and 52 weeks, the testing being done alternately by male and female handlers familiar to the dogs. The test thus measured changes in the developing dog + human relationship. Under the conditions of rearing, the puppies invariably became subordinate to humans and never showed any serious fighting toward them. Two aspects of agonistic behavior were therefore scored: playful fighting, and fear and avoidance behavior.

With respect to playful fighting, all breeds were similarly low at 5 weeks of age. The amount of such behavior increased in all breeds through 15 weeks. At this age the cockers showed decidedly lower amounts of this behavior than the other four breeds. All breeds showed lower scores as adults than as puppies, with wirehaired terriers at the top. Results at 15 weeks were analyzed in the crossbreeding experiment. There was considerable overlap among the parent breeds. F_1's and F_2's followed the Tryon distribution in which F_1's and F_2's were similar. Differences could be explained on the basis of a two-factor hypothesis.

Fuller's emotional reactivity test was also administered in the context of the dog + human relationship. A standard tester interacted with a dog in a strange room, chiefly using various sorts of threatening behavior. Overt behavioral reactions were measured, many of which could be considered agonistic. In addition, heart rates and sinus arrhythmia were also measured. The five pure breeds were compared at 17, 34, and 51 weeks of age. There was considerable overlap between breeds in every measure, but the magnitude of variance accounted for by breed differences generally increased with age. Eighty-three percent of Basenjis were biters and 93% of cockers were nonbiters at one year. Beagles, cockers, and Shetland sheep dogs underwent marked reductions in biting compared with the earliest scores at 17 weeks, whereas there was little change in Basenjis and terriers.

The arrhythmia index, associated with a slow heart rate, was equally high in Basenjis and cockers and lower in the other breeds. Breed differences increased as the animals grew older, accounting for 66% of the variance at 1 year. Heritability indices were generally higher in physiological measures than in the purely behavioral ones.

In the third testing situation litters of puppies were allowed to compete for a single bone each week, beginning at 2 weeks of age, and were tested in paired-dominance situations at 5, 11, 15, and 52 weeks of age. The principal datum was a measure of the relationship: complete dominance, defined as a dog keeping possession of the bone for 8 of 10 minutes and being able to take it from the other pup at will. Other measures of individual behavior were included, the most important being the number of barks by both animals. Unavoidably, this also is a measure of the relationship.

Developmental scores indicated that dominance was being actively organized at 11 weeks and was largely stable by 15 weeks. At all ages wirehaired terriers showed more complete dominance than any other breed, but the differentiation was otherwise not clear-cut. The effect of breed was seen primarily in male–female relationships. By 52 weeks of age all breeds showed a majority of males dominant over females, but this difference was sizable only in the Basenji and fox terrier breeds, where distinct male dominance appeared at 11 weeks, and by 52 weeks no females were dominant over males. That is, in the homogeneous sets design, breed differences appeared only in pairs of unlike sex.

Sex also interacted with weight. Heavier dogs became dominant only in male–male relationships and male–female relationships. This was correlated with the fact that in the development of dominance relationships, males almost invariably got into serious fights, whereas females tended to settle the issue of dominance on the basis of threats and vocalization. When complete dominance was compared in segregating and nonsegregating populations the relationships in the segregating populations showed significantly greater proportions of complete dominance in all combinations of like and unlike sex pairs. This confirmed the hypothesis that genetic variation should affect variation in social relationships, in this case in the degree of differentiation between members of a dyad.

Barking (Scott, 1976; Scott & Fuller, 1965) is an example of signaling behavior and in a wild species would be described as a fixed action pattern. In the bone competition situation, the maximum amount of barking and breed divergence occurred at 11 weeks and so was analyzed at that age. Two measures of barking were analyzed, the tendency to bark or not bark given the opportunity, and the tendency to bark to excess. The average number of barks was highest in cocker spaniels and next highest in beagles, terriers and Shelties being low and Basenjis close to zero. The percentage of animals barking amounted to approximately 20% in Basenjis and 68% in cockers. Data in the hybrids were consistent with the hypothesis of one dominant gene for barking.

On the other hand, the data concerning the number of animals barking to excess can be explained by the hypothesis of a single gene with no dominance. Different genetic mechanisms appear to underlie these two aspects of variation in barking. With respect to issues raised at the outset of this section, this example demonstrates that a high degree of genetic variance can exist in the quantitative aspects of a so-called species-specific trait. The fact that so many of these and other differences can be explained with one- or two-factor genetic hypotheses supports the notion that in behavioral selection major gene effects may be important, and that a polygenic hypothesis is not a universal one.

Other Species

Most research on competition among insects has centered on sexual selection in *Drosophila*. Sexual selection is reviewed in a separate section later.

One study of bees is worth noting. Stort (1975) defined aggressiveness as the number of stings in the gloves of the experimenter. He studied aggressiveness in Africanized bees, Italian bees, their F_1 cross, and backcrosses. The results indicated a fairly simple, two-locus pattern of inheritance of the behavior. Strictly speaking, the behavior is social only in the sense that it involves group defense against another species.

The genetics of social dominance and aggression in chickens has been investigated by means of selective breeding (Craig & Baruth, 1965; Craig, Ortman, & Guhl, 1965; Guhl, Craig, & Muellen, 1960; Ortman & Craig, 1968). Guhl et al. (1960) and Craig et al. (1965) practiced bidirectional selection for social dominance for five generations. The resulting lines differed substantially in the following ways, for both sexes: (1) high-line birds had more frequent social interactions; (2) interactions were more physically severe between highs than between lows; and (3) interline contests were usually won by high-line animals.

Investigating the physiological mechanisms involved, Ortman and Craig (1968) administered testosterone to castrate males and females from the selection lines. Line differences were maintained. The authors concluded that selection had not changed endogenous androgen levels in the lines; instead, selection had altered the animals' responsiveness to social stimuli.

By comparison with the extensive research on genetics and agonistic behavior in mice, relatively little has been done with rats. Boreman and Price (1972) compared the social dominance of wild and domestic Norway rats and their F_1 hybrids. Domestic rats were dominant both in spontaneous and competitive interactions, although this may have been an artifact of their greater uninhibitedness or their greater body size. The authors concluded that wild and domestic rats may have evolved different social behaviors to maximize fitness in their respective "natural" environments.

SEXUAL BEHAVIOR

Drosophila

Genetic variations in mating speed and in mating duration have been the primary variables subject to behavior-genetic analysis, and such studies are discussed in the following section. It is also worth noting that quantitative analysis, using oscilloscope recordings, of courtship sounds in *Drosophila* are possible (Ikeda,

Takabatake, & Sawada, 1980). Courtship sequences in hybrids between *D. melanogaster* and *D. simulans* have also been analyzed (Wood, Ringo, & Johnson, 1980).

Selective Breeding.   Manning (1961) selected pairs of *D. melanogaster* for fast and slow mating speeds. He formed replicated high lines (selected for fast mating speed) and replicated low lines (selected for slow mating speed), and an unselected control line. After 25 generations, the mean mating speed was 3 minutes in the fast lines and 80 minutes in the slow lines. Over the early generations of selection, the realized heritability was .30.

In a further analysis, Manning performed reciprocal crossing of the lines, in essentially a diallel design. Crossing of high with low lines yielded intermediate mating speeds for both reciprocal crosses, indicating that both sexes had been affected by selection. The high line also had a lower level of arena activity than the low line.

Manning (1963) next attempted to determine whether the same genes were controlling mating speed in males and females. To do this, he practiced single-sex selection for mating speed. There was no response to selection for fast-mating males, nor for slow-mating females. The lack of response to selection for fast mating in males suggests that the alleles controlling fast mating speed may have been fixed by natural selection, because fast mating speed has obvious adaptive value. This is consistent with the results of a diallel analysis by Fulker (1966), which is discussed later.

Manning (1968) also selected for slow mating speed in *D. simulans,* a sibling species of *D. melanogaster.* Females, but not males, responded to selection. Females selected for slow mating speed failed to become receptive, and indeed exhibited repellent behaviors toward courting males.

Kessler (1968, 1969) selected for fast and slow mating speeds in *D. pseudo-obscura;* after 12 generations of selection, he performed a diallel cross of the resulting selection lines. The results indicated that slow-mating females reduced mating speed whenever they were involved; fast-mating and control females did not differ significantly from each other. Fast-mating males speeded up all matings in which they were involved; control and slow-mating males did not differ. There is some discrepancy between these results and those of Kaul and Parsons (1966), who found strong male determination of mating speed. The discrepancy is probably accounted for by variations in methods used (Ehrman & Parsons, 1976).

Having reviewed this extensive literature on genetics of mating speed, Ehrman and Parsons (1976, p. 172) reached the following conclusions: (1) male mating speed is selected for rapidity of copulations; (2) for a given species, fast matings tend to be controlled by the genotype of the male, whereas slow matings are controlled by the genotype of the female (Parsons, 1974); (3) mating speed is associated with fertility and number of progeny; and (4) in relation to other components of fitness encompassing the entire life cycle, mating speed is a most important component. As there is no more recent evidence contradicting these conclusions, they appear still to be valid.

Diallel Cross.   Fulker (1966) performed a diallel analysis of mating speed as measured by copulation frequency of male *D. melanogaster*. There was a strong directional dominance for high mating frequency; a minimum of five loci were estimated to be involved. The trait was highly correlated with fitness. The results indicated a history of strong natural selection for fast (rather than intermediate or slow) mating speed, a point made by Ehrman and Parsons (1976).

As already noted, Manning (1961) and Kessler (1968, 1969) also performed diallel analyses in conjunction with selection studies.

Single-Locus Effects.   Hall (1978) has studied the effects of a recessive, mutation, "fruitless," that renders males behaviorally sterile. Mutant males court wild-type males more frequently (7 times more) and in a more sustained fashion (100 times more) than wilds court wilds. The mutant males do court females, but attempt copulation in less than 1% of the trials and never do copulate; thus they are behaviorally sterile. Mutant males also stimulate wild-type males to court them with high frequency, and pheromonal effects seem to be involved. This is an interesting example of a major-gene effect on sexual behavior, modifying it in the direction of male homosexuality. On the other hand, Dow (1976) has found that a mutation at the yellow locus is responsible for enhanced female receptivity. Sciandra and Bennett (1976) also found evidence of single-locus effects, both on male and female sexual behaviors. Von Schilcher and Manning (1975) found evidence of X-chromosome determination of courtship songs in *D. melanogaster*.

Other Insects

Cowan and Rogoff (1968) selectively bred for a high number and low number, respectively, of mating strikes made by male houseflies, *Musca domestica*. Selection continued for four generations, and two high and two low lines were formed. One low line was lost due to failure to breed (a result that is not too surprising, given what they were being selected for). The remaining low line averaged about 6 strikes per hour, and the high lines averaged about 21 strikes per hour by the F_4 generation.

Bentley and Hoy (1972) have done a detailed analysis of genetic control of courtship sounds in crickets. Male crickets produce a courtship song that varies from one species of cricket to the next and serves the purpose of attracting conspecific females. Bentley and Hoy studied the sonic patterns of the Australian field crickets *Teleogryllus commodus* and *T. oceanicus* and their F_1 reciprocal hybrids, produced in the laboratory. The hybrid song patterns were intermediate between those of the parental species; therefore, a simple single-locus, dominant–recessive genetic mechanism can be ruled out. Backcrosses also resulted in intermediate inheritance, eliminating the possibility of single-locus control with incomplete penetrance. Therefore, the behavior must be polygenically controlled. There was also evidence for X-linked determination of one aspect of the sound pattern, as evidenced by differences between the reciprocal F_1 crosses.

In contrast to the male, the female cricket does not produce a courtship song. She does, however, move toward the sound of a male conspecific. Hoy and Paul (1973) found that *T. oceanicus* females preferred the song of *T. oceanicus* males, and *T. commodus* females preferred the song of *T. commodus* males. Interestingly, F_1 females preferred the song of F_1 males over that of either parent species. As noted previously, the calling song of the hybrid is distinct from that of the parent species. The preference of F_1 females for the song of F_1 males suggests that the female's response is somehow genetically coupled to the male's production of the song. This should lend itself to some elegant genetic and physiological investigations.

Mice

Levine, Barsel, and Diakow (1966) documented strain differences in the sexual behavior of mice. ST males had a significantly higher rate of mating success than CBA males; mating patterns differed to some extent, also.

McGill (1970; see also McGill, 1962, 1965, 1969; McGill & Blight, 1963) has analyzed male sexual behavior in mice in inbred strains C57BL, DBA, and their F_1 hybrid. The approach was multivariate, with 14 variables being measured, including latency to first mount, number of thrusts with intromission preceding ejaculation, and number of mounts without intromission. There were significant differences between the two inbred strains on 10 of these 14 measures. Data from the F_1 indicated that different modes of inheritance might be responsible for different aspects of sexual behavior. For example, there was evidence of dominance on seven of the measures, intermediate inheritance on three of them, and overdominance on three. When a cross between two other inbred strains was carried out, the results were different, so that the evidence on modes of inheritance was specific to the strains sampled. Further, the results were specific to the particular kinds of environmental conditions under which the study was carried out. A complete biometrical analysis, including results from F_1, F_2, and backcross matings, was also performed.

Investigating hormone–genotype interactions, McGill and Tucker (1964; see also Tucker & McGill, 1966) studied the sexual behavior of male DBAs, C57BLs, and their F_1 hybrid, before and after castration. Before castration, DBA males showed higher sex drive (more ejaculations) than C57BL males, and the F_1's resembled the DBAs. Following castration, both inbreds lost their ejaculation capacity almost immediately (none ejaculated after 8 days postcastration). There was considerable hybrid vigor, though, as evidenced by the F_1's having their last ejaculation a median of 28 days after castration. Later research demonstrated that the postcastration retention of the ejaculatory reflex was dependent on genotype, but the prolonged retention of the reflex by F_1 hybrids was due not to heterosis but rather to heterozygosity at some specific loci (McGill & Manning, 1976). The role of experience was also demonstrated; in F_1 hybrids, precastration sexual experience was not essential to ejaculation after castration, but such experience did increase the frequency of ejaculation after castration (Manning & Thompson, 1976).

Vale and Ray (1972) did a three-strain diallel analysis of male mouse sexual behavior. Seventeen variables were measured. The mode of inheritance and extent of genetic control varied from one variable to the next. The proportion of variance accounted for by genetic factors ranged from 9% (mean time between mounts with intromission; number of times male licked female's genitalia) to 62% (number of "roots").

Dewsbury and Lovecky (1974) studied the copulatory behavior of *Peromyscus* from two natural populations, one in South Carolina and one in Florida. The basic copulatory pattern was identical for the two populations. However, there were some quantitative differences, particularly in intromission frequency.

There is some evidence of Y-chromosome effects on male mouse sexual behavior (Weir and Hogle, 1973), paralleling the research on Y-chromosome effects on aggression.

One approach to understanding the adaptive value of certain behavior patterns is the comparison of inbred strains with random-bred wild stock. The assumption is that the inbred strains should have suffered inbreeding depression, and thus the behavior of wilds, compared with inbreds, should reflect greater adaptation and fitness. This approach was used by Dewsbury, Oglesby, Shea, and Connor (1979). They found that inbred males, compared with random-bred wilds, showed shorter latencies to the first mount and intromission, but longer latencies to ejaculate and more preejaculatory mounts and thrusts. They concluded that rapid initiation of copulation in a novel environment may not be adaptive, but that rapid ejaculation once copulation is initiated is adaptive (a point apparently not recognized by recent sex advice manuals).

Humans

The hypothesis that homosexuality in humans is genetic was advanced by Kallman (1952a, 1952b). He found perfect concordance for homosexuality among all the identical twin pairs he studied. Among the nonidentical twins the degree of concordance was not statistically significant. Several criticisms of this study could be made. First, the identical twin pairs were reared together, not apart, so that the concordance for homosexuality could be explained by their common environment rather than their common genes. Second, nonidentical twins have on the average half of their genes in common so that one would expect, on a genetic basis, a moderate degree of concordance in them, rather than the zero concordance found by Kallman. Further, Heston and Shields (1968) failed to replicate Kallman's findings. And numerous case studies of individual identical twin pairs discordant for homosexuality have been reported (McConaghy & Blaszczynski, 1980; see review in Zuger, 1976). Thus the hypothesis of the genetic determination of homosexuality in humans is not generally accepted.

Two more general issues are raised by this area of research. The research generally has been based on an oversimplified measurement technique and conceptualization of homosexuality. It rests on the notion that two types—heterosexual and homosexual—suffice to categorize all people. It fails to make use of more complex rating scales such as the 7-point scale introduced by Kinsey more than 30 years ago (Kinsey, Pomeroy, & Martin, 1948). Second, there has been insufficient attention given to the evolutionary implications of a genetic theory. If there were a gene or genes for exclusive homosexuality, one can scarcely imagine genes against which natural selection would act more strongly. It would be unlikely that they would persist in the population. Of course, this does not rule out the possibility of genes for bisexuality, for the person could reproduce and pass the genes on through heterosexual behavior, while also engaging in homosexual behavior.

Other Species

Clark, Aronson, and Gordon (1954) have studied the sexual behavior of the platyfish, the swordtail, and their F_1, F_2, and backcross hybrids. Duration of copulation was higher for swordtails (2.39 sec) than for platyfish (1.36 sec), but frequency of copulation was higher for platyfish. The percentage of copulations resulting in insemination was also higher in platyfish. In general, the copulatory behavior of F_1's was either intermediate or closer to that of the swordtails. No simple mode of inheritance could account for the data, so that polygenic inheritance seems most likely. Clark et al. also did some preliminary consideration of the mechanisms of reproductive isolation of these species in natural settings.

Siegel (1965, 1972) studied male mating behavior in chickens by performing bidirectional selection for cumulative number of completed matings; a randombred control line was also maintained. Mass selection was practiced for 11 generations with no apparent plateau in response. Realized heritability was .16 in the high line and .32 in the low line, although response was actually fairly symmetrical because selection intensity was greater in the high line. Other measures of mating behavior showed a correlated response to selection, as did semen volume.

Cook, Siegel, and Hinkelmann (1972) attempted a more complete understanding of the genetic mechanisms involved by crossing the lines in all possible combinations (essentially a diallel analysis) at the 10th generation. The results indicated that dominant and/or epistatic factors and sex-linked loci influenced low-frequency mating behavior, in contrast to a more additive genetic basis for high-frequency mating behavior. The authors hypothesized that two genetic systems are involved. There was no evidence that aggressiveness had shown a correlated response to selection for mating ability.

McCollum, Siegel, and vanKrey (1971 studied the effects of androgen replacement on these selection lines; they found that the effect on sexual behavior depended on genotype.

Cunningham and Siegel (1978; see also Sefton & Siegel, 1975) selectively bred for male mating behavior in Japanese quail. The criterion for selection was cumulative number of completed matings by a male in eight 8-min observation periods. Five selection lines were formed: two high lines, two low lines, and a random-bred control line. There was little response to selection in the high lines until the sixth generation, with progress from then until the twelfth generation, the last to be reported. In the low lines, the response continued for all 12 generations. Reverse selection was done on a sample of subjects, for two generations beginning at the eighth generation. The high lines responded, showing that additive genetic variance remained in them. However, the low lines did not respond to reverse selection. Cloacal gland size and aggressiveness appeared to have shown correlated responses to selection.

Goy and Jakway (1959) and Jakway (1959) studied sexual behavior in both males and females from two inbred strains of guinea pigs, the F_1, F_2, and backcross generations. There was evidence of dominance and heterosis for some measures (but see Broadhurst & Jinks, 1963, for a discussion of measurement problems).

McLean, Dupeire, and Elden (1972) demonstrated strain differences in the mating behavior of male Sprague–Dawley, Long–Evans, and Wistar rats. Wistar males exhibited significantly more copulatory responses than the other two strains, which in turn did not differ from each other.

Studies of sexual behavior in dogs indicate that the general patterns of copulation are similar in all breeds, although there is considerable variation in performance. Scott estimates that 25–50% of initial attempts in both sexes are failures.

Extensive individual variations occur in the length of the female dog's estrous cycle. Other major variations occur with respect to the length between cycles, the regularity of recurrence, and correlations with seasonal changes. In European breeds, female usually run two cycles per year at intervals of 6 months or more, and estrus may occur at any time of year. In the African Basenji and Australian dingo, as well as in the Indian pariah dog, annual seasonal cycles occur with a peak frequency at the time of the autumnal equinox. As Oppenheimer and Oppenheimer (private communication) observe, this is adaptive in that it results in concentrating birth at the coolest seasons of the year in these tropical climates.

The onset of estrus is associated with declining day lengths (Fuller, 1956). When dingoes are transported to the Northern Hemisphere they switch cycles to the new autumn season.

Scott, Fuller, and King (1959) analyzed cycles in a cocker–Basenji cross and concluded that the annual cycle of the Basenji was inherited as a one-factor recessive. F_1's were more variable than either parent strain, having apparently inherited both the tendency toward 6-month cycle of the cockers and the tendency of the Basenji to respond to declining day lengths.

Sexual Selection

Darwin defined sexual selection as: (1) competition among members of one sex for members of the other sex; and (2) differential preference by members of one sex for members of the other sex (Trivers, 1972). For the majority of species, this amounts to males competing among themselves for females, and females preferring some males over others. This results in selection for different physical and behavioral traits in males and females. Both agonistic behavior and sexual behavior are involved. Most of the behavior-genetic research in this area has used either *Drosophila* or mice.

Drosophila.   Female *Drosophila* prefer some males—and are more likely to mate with them—than others. In particular, females prefer males whose genotype is in the minority, as opposed to males of the majority genotype (see reviews by Ehrman & Probber, 1978; Petit & Ehrman, 1969). This phenomenon is called the rare-male mating advantage, and is an example of frequency-dependent mating. Frequency-dependent mating has been found in seven *Drosophila* species (Borisov, 1970; Petit & Ehrman, 1969; Spiess, 1968; Spiess & Spiess, 1969), and in the flour beetle (Sinnock, 1969, 1970). However, it does not appear to occur in houseflies (Childress & McDonald, 1973).

From an evolutionary, population–genetic point of view, the rare-male mating advantage ensures the maintenance of genetic variability (polymorphisms) in the population, by ensuring that rare genotypes increase in frequency. Indeed, in more naturalistic situations, involving mass matings, Ehrman (1970) demonstrated that, beginning with an initial 80:20 ratio of genotypes, the ratio over generations converged toward 50:50. Such mating preferences are advantageous when there is heterosis (Lacy, 1979).

Effects on the female of prior experience and age have been studied (Pruzan, 1975; Pruzan & Ehrman, 1974). Averhoff and Richardson (1974) have provided evidence of pheromonal mediation of the rare-male mating advantage, although Bryant (1979) has offered an alternative interpretation of the results, and the observations of Powell and Morton (1979) yielded data opposite to those of Averhoff and Richardson. Other research has been directed toward mechanisms involved in the rare-male mating advantage. Ehrman (1970) has demonstrated that olfactory cues are involved.

Mice. The studies on agonistic behavior and fitness are relevant to sexual selection. Levine (1958) housed ST and CBA males together with ST females and found that ST males won 25 of 27 fights and sired 83% of the offspring. This documents the kind of competition between males, and resulting reproductive advantage of the successful competitors, postulated in the theory of sexual selection. However, later studies showed that changes in the experimental situations could change the results (Levine, Barsel, & Diakow, 1965).

DeFries and McClearn (1970) found social dominance and Darwinian fitness to be closely related. They grouped three males from different inbred strains (chosen from A, BALB, C57BL, DBA, I, and C3H) with three BALB females in a three-cage dominance-testing situation. A and BALB males were most dominant and DBAs were most subordinate. In 18 of the 22 groupings, the dominant male sired all the litters.

Yanai and McClearn (1972) found that female mice have mating preferences; in particular, females of strains C57BL and DBA both preferred males of the other strain, not their own. The authors concluded that such preferences would have evolutionary significance in reducing inbreeding. Yanai and McClearn (1973a) replicated this, but found that BALB and C3H females had no such preferences. Yanai and McClearn (1973b) extended these studies by looking at females from reciprocal crosses of BALBs and DBAs, which were fostered by BALB or DBA males. They found that the most important influence in the mating preference of a female was the genotype of her foster father. Females reared by DBA males preferred males of other strains; BALB fathers had no effect on female mating preference.

In conclusion, there is evidence supporting the theory of sexual selection. Males do compete with each other, and the winners are more reproductively successful. Females prefer some males over others, and the pattern of their preference seems to lead to maintenance of genetic variability in the population. However, these results are not found in all species, nor in all genotypes of a species.

CONCLUSION

The research reviewed here leaves little doubt that there are genetic influences on both agonistic and sexual behaviors, in the sense that genetic variations produce behavioral variations. This has been amply documented by findings of differences among inbred strains and by the success of artificial selection experiments.

It seems important to emphasize this point (a trite one to behavior geneticists), in view of the recent publicity given to sociobiology (Wilson, 1975). Sociobiologists typically emphasize, as did the ethologists before them, genetic determination of intraspecific uniformity in behavior, particularly of aggressive and sexual behaviors. This may lead to the conclusion that there is a kind of evolutionarily produced "species determinism" of aggressive and sexual behaviors. Behavior geneticists have ample evidence of a much different phenomenon—the enormous genetically produced *variation* in aggressive and sexual behaviors within a species. It is also worth noting that there are evolutionary mechanisms—for example, the rare-male mating advantage—that seem to serve the precise purpose of maintaining such genetic and behavioral variability.

With regard to genetic mechanisms, most research finds evidence of polygenic control. Examples are the work of Eleftheriou et al. (1974) and Messeri et al. (1975), using recombinant inbred mice to study aggressive behavior, and that of Fulker (1966), using a diallel analysis of copulation frequency in *Drosophila*. However, though polygenic mechanisms seem to be the norm, there are also demonstrations of single-locus effects, and even some sophisticated attempts at mapping. Examples are Kessler et al.'s (1977) Mendelian analysis of aggression in mice, Scott and Fuller's (1965) analysis of barking in dogs, and Hall's (1978) investigation of the fruitless mutant causing behavioral sterility in *Drosophila;* the work on Y-chromosome effects on aggression in mice is a beginning at mapping. There is a great need for replication and extension of work on single-locus effects and mapping.

Other sophisticated approaches include the analysis of hormone–genotype interactions (McGill & Manning, 1976) and pheromone–genotype interactions (Kessler et al., 1975). Such analyses are important first steps in elucidating the pathways between genotype and behavior.

A multivariate approach (Vale & Ray, 1972) needs to be maintained in future research. It is clear that the ordering of inbred strains can differ substantially for different measures of agonistic behavior. Similarly, Vale and Ray (1972) demonstrated that the mode of inheritance and the extent of genetic control can vary from one measure of mouse sexual behavior to the next.

The issues of adaptation and evolution are becoming increasingly important in research. An elegant approach for future research is the formation of an a priori prediction—for example, for the outcome of an inbred strain versus wild-trapped comparison—on the basis of evolutionary considerations and then the testing of this prediction.

The application to humans of the research with nonhumans reviewed here is a dangerous pastime, one this author is not eager to undertake. However, a number of general principles and phenomena might be mentioned that are relevant to such extrapolations to humans. One is the evolution of the menstrual cycle in humans, compared with the near-universal estrous cycle in nonhuman mammals. This implies that human reproductive and sexual behaviors differ considerably from that of other species. The evolution of the big brain in humans means that learning and cultural evolution will be of relatively greater importance in our agonistic and sexual behavior, compared with these behaviors in other species, which will be more under endocrine and genetic control, a point noted by Beach (1947) many years ago. Finally, the evolution of long-term pair-bonding between males and females, not to mention monogamy, means that aggression is unlikely to have the immediate consequences for reproductive fitness that it does in other species (e.g., the most aggressive male mice siring 95.6% of the offspring, as found by Horn, 1974). Nonetheless, it seems likely that there are genetically produced variations in human sexual and aggressive behaviors. It may be that it is worthwhile to attempt to track them down in research—for example, a discovery of genetic variation in human sexual desire and an understanding of the biochemical intermediaries between such genotypes and behaviors might be very helpful in treating problems of sexual desire, which are proving difficult to treat using standard behavior therapy techniques (Kaplan, 1979).

REFERENCES

Averhoff, W. W., & Richardson, R. H. Pheromonal control of mating patterns in *Drosophila melanogaster*. *Behavior Genetics*, 1974, *4*, 207–225.

Bailey, D. W. Recombinant inbred strains. *Transplantation*, 1971, *11*, 325–327.

Bauer, F. J. Genetic and experimental factors affecting social relations in male mice. *Journal of Comparative and Physiological Psychology*, 1956, *49*, 359–364.

Beach, F. A. Evolutionary changes in the physiological control of mating behavior in mammals. *Psychological Review*, 1947, *54*, 297–315.

Bentley, D. R., & Hoy, R. R. Genetic control of the neuronal network generating cricket (*Teleogryllus gryllus*) song patterns. *Animal Behavior*, 1972, *20*, 478–492.

Boreman, J., & Price, E. Social dominance in wild and domestic Norway rats (*Rattus Norvegicus*). *Animal Behavior*, 1972, *20*, 534–542.

Borisov, A. I. Disturbances of panmixia in natural populations of *Drosophila funebri* polymorphores by the inversion of II-1. *Genetika*, 1970, *6*, 61–67.

Broadhurst, P. L., & Jinks, J. L. The inheritance of mammalian behavior re-examined. *Journal of Heredity*, 1963, *54*, 170–176.

Bronson, F. H. Rodent pheromones. *Biology of Reproduction*, 1971, *4*, 344–357.

Bryant, E. H. Inbreeding and heterogamic mating: An alternative to Averhoff and Richardson. *Behavior Genetics*, 1979, *9*, 249–256.

Childress, D., & McDonald, I. C. Tests for frequency-dependent mating success in the housefly. *Behavior Genetics*, 1973, *3*, 217–223.

Ciaranello, R. D., Lipsky, A., & Axelrod, J. Association between fighting behavior and catecholamine biosynthetic enzyme activity in two inbred mouse sublines. *Proceedings of the National Academy of Science, USA*, 1974, *71*, 3006–3008.

Clark, E., Aronson, L. R., & Gordon, M. Mating behavior in two sympatric species of xiphophorin fishes: Their inheritance and significance in sexual isolation. *Bulletin of the American Museum of Natural History*, 1954, *103*, 135–226.

Cook, W. T., Siegel, P. B., & Hinkelmann, K. Genetic analyses of male mating behavior in chickens. II. Crosses among selected and control lines. *Behavior Genetics*, 1972, *2*, 289–300.

Cowan, B. D., & Rogoff, W. M. Variation and heritability of responsiveness of individual male houseflies, *Musca domestica*, to the female sex pheromone. *Annals of the Entomological Society of America*, 1968, *61*, 1215–1218.

Craig, J. V., & Baruth, R. A. Inbreeding and social dominance in chickens. *Animal Behavior*, 1965, *13*, 109–113.

Craig, J. V., Ortman, L. L., & Guhl, A. M. Genetic selection for social dominance ability in chickens. *Animal Behavior*, 1965, *13*, 114–131.

Cunningham, D. L., & Siegel, P. B. Response to bidirectional and reverse selection for mating behavior in Japanese quail *Coturnix coturnix japonica*. *Behavior Genetics*, 1978, *8*, 387–398.

DeFries, J. C., & McClearn, G. E. Social dominance and Darwinian fitness in the laboratory mouse. *American Naturalist*, 1970, *104*, 408–411.

Dewsbury, D. A., & Lovecky, D. V. Copulatory behavior of old-field mice (*Peromyscus polionotus*) from different natural populations. *Behavior Genetics*, 1974, *4*, 347–355.

Dewsbury, D. A., Oglesby, J. M., Shea, S. L., & Connor, J. L. Inbreeding and copulatory behavior in house mice: A further consideration. *Behavior Genetics*, 1979, *9*, 151–164.

Dow, M. A. The genetic basis of receptivity of yellow mutant *Drosophila melanogaster* females. *Behavior Genetics*, 1976, *6*, 141–143.

Ebert, P. D. Agonistic behavior in wild and inbred *Mus musculus*. *Behavioral Biology*, 1976, *18*, 291–294.

Ebert, P. D., & Hyde, J. S. Selection for agonistic behavior in wild female *Mus musculus*. *Behavior Genetics*, 1976, *6*, 291–304.

Ehrman, L. Stimulation of the mating advantage in mating of rare *Drosophila* males. *Science*, 1970, *167*, 905–906.

Ehrman, L., & Parsons, P. A. *The genetics of behavior*. Sunderland, Mass.: Sinauer, 1976.

Ehrman, L., & Probber, J. Rare *Drosophila* males: The mysterious matter of choice. *American Scientist*, 1978, *66*, 216.

Eleftheriou, B. E., Bailey, D. W., & Denenberg, V. H. Genetic analysis of fighting behavior in mice. *Physiology and Behavior*, 1974, *13*, 773–777.

Fredericson, E. Aggressiveness in female mice. *Journal of Comparative and Physiological Psychology*, 1952, *45*, 254–257. (a)

Fredericson, E. Reciprocal fostering of two inbred mouse strains and its effect on the modification of inherited aggressive behavior. *American Psychologist*, 1952, 7, 241–242. (b)

Fredericson, E., & Birnbaum, E. A. Competitive fighting between mice with different hereditary backgrounds. *Journal of Genetic Psychology*, 1954, *85*, 271–280.

Fredericson, E., Story, A. W., Gurney, N. L., & Butterworth, K. The relationship between heredity, sex, and aggression in two inbred mouse strain. *Journal of Genetic Psychology*, 1955, *87*, 121–130.

Fulker, D. W. Mating speed in male *Drosophila melanogaster:* A psychogenetic analysis. *Science*, 1966, *153*, 203–205.

Fuller, J. L. Photoperiodic control of estrus in the Basenji. *Journal of Heredity*, 1956, 47, 179–180.

Fuller, J. L., & Hahn, M. E. Issues in the genetics of social behavior. *Behavior Genetics*, 1976, *6*, 391, 406.

Fuller, J. L., & Wimer, R. E. Neural sensory and motor functions. In E. L. Green (Ed.), *Biology of the laboratory mouse* (2nd ed.). New York: McGraw–Hill, 1966.

Ginsburg, B. & Allee, W. C. Some effects of conditioning on social dominance and subordination in inbred strains of mice. *Physiological Zoology*, 1942, *15*, 485–506.

Goy, R. W., & Jakway, J. S. The inheritance of patterns of sexual behavior in female guinea pigs. *Animal Behavior*, 1959, *7*, 142–149.

Greenberg, G. The effects of ambient temperature and population density on aggression in two inbred strains of mice, *Mus musculus*. *Behaviour*, 1972, *42*, 119–130.

Guhl, A. M., Craig, J. V., & Muellen, C. D. Selective breeding for aggressiveness in chickens. *Poultry Science*, 1960, *39*, 970–980.

Hall, J. C. Courtship among males due to a male-sterile mutation in *Drosophila melanogaster. Behavior Genetics*, 1978, *8*, 125–141.

Heston, L., & Shields, J. Homosexuality in twins: A family study and a registry study. *Archives of General Psychiatry*, 1968, *18*, 149–160.

Horn, J. M. Aggression as a component of relative fitness in four inbred strains of mice. *Behavior Genetics*, 1974, *4*, 373–381.

Hoy, R. R., & Paul, R. L. Genetic control of song specificity in crickets. *Science*, 1973, *180*, 82–83.

Hyde, J. S., & Ebert, P. D. Correlated response in selection for aggressiveness in female mice. I. Male aggressiveness. *Behavior Genetics*, 1976, *6*, 421–428.

Hyde, J. S., & Sawyer, T. F. Estrous cycle fluctuation in aggressiveness of house mice. *Hormones and Behavior*, 1978, *9*, 290–295.

Hyde, J. S., & Sawyer, T. F. Correlated characters in selection for aggressiveness in female mice. II. Maternal aggressiveness. *Behavior Genetics*, 1979, *9*, 571–578.

Hyde, J. S., & Sawyer, T. F. Selection for agonistic behavior in wild female mice. *Behavior Genetics*, 1980, *10*, 349–360.

Ikeda, H., Takabatake, I., & Sawada, N. Variation of courtship sounds among three geographical strains of *Drosophila mercatorum. Behavior Genetics*, 1980, *10*, 361–376.

Jakway, J. S. Inheritance of patterns of mating behavior in the male guinea pig. *Animal Behavior*, 1959, *7*, 150–162.

Kallman, F. J. Comparative twin study on the genetic aspects of male homosexuality. *Journal of Nervous and Mental Disease*, 1952, *115*, 283–298. (a)

Kallman, F. J. Twin and sibship study of overt male homosexuality. *American Journal of Human Genetics*, 1952, *4*, 136–146. (b)

Kaplan, H. S. *Disorders of sexual desire*. New York: Simon and Schuster, 1979.

Kaul, D., & Parsons, P. A. Competition between males in the determination of mating speed in *Drosophila pseudoobscura. Australian Journal of Biological Science*, 1966, *19*, 945–947.

Kessler, S. Speed of mating and sexual isolation in *Drosophila. Nature*, 1968, *220*, 1044–1045.

Kessler, S. The genetics of *Drosophila* mating behavior. II. The genetic architecture of mating speed in *Drosophila pseudoobscura. Genetics*, 1969, *62*, 421–433.

Kessler, S., Elliot, G. R., Orenberg, E. K., & Barchas, J. D. A genetic analysis of aggressive behavior in two strains of mice. *Behavior Genetics*, 1977, *7*, 313–321.

Kessler, S., Harmatz, P., & Gerling, S. A. The genetics of pheromonally mediated aggression in mice. I. Strain differences in the capacity of male urinary odors to elicit aggression. *Behavior Genetics*, 1975, *5*, 233–238.

Kinsey, A. C., Pomeroy, W. B., & Martin, C. E. *Sexual behavior in the human male*. Philadelphia: Saunders, 1948.

Lacy, R. C. Adaptiveness of a rare male mating advantage under heterosis. *Behavior Genetics*, 1979, *9*, 51–54.

Lagerspetz, K. Studies on the aggressive behavior of mice. *Ann. Acad. Scie. Fenn.* (B), 1964, *131*, 1–131.

Lagerspetz, K. M. J., & Lagerspetz, K. Y. H. Changes in the aggressiveness of mice resulting from selective breeding, learning, and social isolation. *Scandinavian Journal of Psychology*, 1971, *12*, 241–248.

Lagerspetz, K. Y. H., Tirri, R., & Lagerspetz, K. M. J. Neurochemical and endocrinological studies of mice selectively bred for aggressiveness. *Scandinavian Journal of Psychology*, 1968, *9*, 157–160.

Lagerspetz, K., & Wuorinen, K. A. A cross-fostering experiment with mice selectively bred for aggressiveness and non-aggressiveness. *Inst. Psychol., Univ. Turku*, No. 17, 1965.

Lee, C. The effect of the RD locus on the agonistic behavior of mice. *Acta Psychologica Tai-wanica*, 1971, *13*, 1–6.

Levine, L. Studies on sexual selection in mice. I. Reproductive competition between albino and black-agouti males. *American Naturalist*, 1958, *92*, 21–26.

Levine, L., Barsel, G. E., & Diakow, C. A. Interaction of aggressive and sexual behavior in male mice. *Behaviour*, 1965, *25*, 272–280.

Levine, L., Barsel, G. E., & Diakow, C. A. Mating behavior of two inbred strains of mice. *Animal Behavior*, 1966, *14*, 1–6.

Lindzey, G., Manosevitz, M., & Winston, H. D. Social dominance in the mouse. *Psychonomic Science*, 1966, *5*, 451–454.

Lindzey, G., Thiessen, D. D., Blum, S., & Tucker, A. Further observations on social dominance in mice. *Psychonomic Science*, 1969, *14*, 245–246.

Lindzey, G., Winston, H., & Manosevitz, M. Social dominance in inbred mouse strains. *Nature*, 1961, *191*, 474–476.

Manning, A. The sexual behavior of two sibling *Drosophila* species. *Behaviour*, 1959, *15*, 123–145.

Manning, A. The effects of artificial selection for mating speed in *Drosophila melanogaster*. *Animal Behavior*, 1961, *9*, 82–92.

Manning, A. Selection for mating speed in *Drosophila melanogaster* based on the behavior of one sex. *Animal Behavior*, 1963, *11*, 116–120.

Manning, A. The effects of artificial selection for slow mating in D. simulans. *Animal Behavior*, 1968, 16, 108–113.

Manning, A., & Thompson, M. L. Postcastration retention of sexual behavior in the male BDF1 mouse: The role of experience. *Animal Behavior*, 1976, *24*, 523–533.

Maxson, S. C., Ginsburg, B. E., & Trattner, A. Interaction of Y-chromosomal and autosomal gene(s) in the development of intermale aggression in mice. *Behavior Genetics*, 1979, *9*, 219–226.

McClearn, G. E., & DeFries, J. C. Genetics and mouse aggression. In J. K. Knutson (Ed.), *The control of aggression*. Chicago: Aldine, 1973.

McCollum, R. E., Siegel, P. B., & vanKrey, H. P. Responses to androgen in lines of chickens selected for mating behavior. *Hormones and Behavior*, 1971, *2*, 31–42.

McConaghy, N. & Blaszczynski, A. A pair of monozygotic twins discordant for homosexuality: Sex-dimorphic behavior and penile volume responses. *Archives of Sexual Behavior*, 1980, *9*, 123–132.

McGill, T. E. Sexual behavior in three inbred strains of mice. *Behaviour*, 1962, *19*, 341–350.

McGill, T. E. Studies of the sexual behavior of male laboratory mice: Effects of genotype, recovery of sex drive, and theory. In F. A. Beach (Ed.), *Sex and behavior*. New York: Wiley, 1965.

McGill, T. E. An enlarged study of genotype and recovery of sex drive in male mice. *Psychonomic Science*, 1969, *15*, 250–251.

McGill, T. E. Genetic analysis of male sexual behavior. In G. Lindzey & D. D. Thiessen (Eds.), *Contributions to behavior-genetic analysis: The mouse as a prototype*. New York: Appleton, 1970.

McGill, T. E., & Blight, W. C. The sexual behaviour of hybrid male mice compared with the sexual behavior of males of the inbred parent strains. *Animal Behavior*, 1963, *11*, 480–483.

McGill, T. E. & Manning, A. Genotype and retention of the ejaculatory reflex in castrated male mice. *Animal Behavior*, 1976, *24*, 507–518.

McGill, T. E., & Tucker, G. R. Genotype and sex drive in intact and in castrated male mice. *Science*, 1964, *145*, 514–515.

McLean, J. H., Dupeire, W. A., III, & Elden, S. T. Strain differences in the mating behavior of Sprague–Dawley, Long–Evans, and Wistar male rats. *Psychonomic Science*, 1972, *29*, 175–176.

Messeri, P., Eleftheriou, B. E., & Oliverio, A. Dominance behavior: A phylogenetic analysis in the mouse. *Physiology and Behavior*, 1975, *14*, 53–58.

Oppenheimer, E. C., & Oppenheimer, J. R. *The Indian pariah dog (*Canis familiaris*) in West Bengal: Activity patterns, reproduction, and mortality*. Private communication, 1974.

Ortman, L. L., & Craig, J. V. Social dominance in chickens modified by genetic selection–physiological mechanisms. *Animal Behavior*, 1968, *16*, 33–37.

Parsons, P. A. Male mating speed as a component of fitness in Drosophila. *Behavior Genetics*, 1974, *4*, 395–404.

Petit, C., & Ehrman, L. Sexual selection in Drosophila. *Evolutionary Biology*, 1969, *3*, 177–223.

Powell, J. R., & Morton, L. Inbreeding and mating patterns in Drosophila pseudoobscura. *Behavior Genetics*, 1979, *9*, 425–429.

Pruzan, A. Effect of age, rearing and mating experiences on frequency dependent sexual selection in *Drosophila pseudoobscura*. *Evolution*, 1975, 29.

Pruzan, A., & Ehrman, L. Age, experience, and rare-male mating advantages in Drosophila pseudoobscura. *Behavior Genetics*, 1974, *4*, 159–164.

St. John, R. D., & Corning, P. A. Maternal aggression in mice. *Behavioral Biology*, 1973, *9*, 635–639.

Schroder, J. H. Increase in aggressiveness of male mice after irradiation of paternal spermatozoa with 600 R of gamma rays as dependent on fertility. *Behavior Genetics*, 1980, *10*, 387–400.

Sciandra, R. J., & Bennett, J. Behavior and single-gene substitution in Drosophila melanogaster. I. Mating and courtship differences with *w*, *cn* and *bw* loci. *Behavior Genetics*, 1976, *6*, 205–218.

Scott, J. P. Genetic differences in the social behavior of inbred strains of mice. *Journal of Heredity*, 1942, *33*, 11–15.

Scott, J. P. Agonistic behavior in mice and rats: A review. *American Zoologist*, 1966, *6*, 683–701.

Scott, J. P. Genetic variation and the evolution of communication. In M. Hahn & E. Simmel (Eds.), *Communicative behavior and evolution*. New York: Academic Press, 1976.

Scott, J. P. & Fredericson, E. The causes of fighting in mice and rats. *Physiol. Zool.*, 1951, *24*, 273–309.

Scott, J. P., & Fuller, J. L. *Genetics and the social behavior of the dog*. Chicago: University of Chicago Press, 1965.

Scott, J. P., Fuller, J. L., & King, J. The inheritance of annual breeding cycles in hybrid Basenji–cocker spaniel dogs. *Journal of Heredity*, 1959, *50*, 255–261.

Sefton, A. E., & Siegel, P. B. Selection for mating ability in Japanese quail. *Poultry Science*, 1975, *54*, 788–794.

Selmanoff, M. K., Jumonville, J. E., Maxson, S. C., & Ginsburg, B. E. Evidence for a Y-chromosomal contribution to an aggressive phenotype in inbred mice. *Nature*, 1975, *253*, 529–530.

Selmanoff, M. K., Maxson, S. C., & Ginsburg, B. E. Chromosomal determinants of intermale aggressive behavior in inbred mice. *Behavior Genetics*, 1976, *6*, 53–69.

Siegel, P. B. Genetics of behavior: Selection for mating ability in chickens. *Genetics*, 1965, *51*, 1269–1277.

Siegel, P. B. Genetic analyses of male mating behavior in chickens. I. Artificial selection. *Animal Behavior*, 1972, *20*, 564–570.

Sinnock, P. Non-random mating in Tribolium castaneum. *Genetics* (suppl.), 1969, *61*, 55.

Sinnock, P. Frequency dependency and mating behavior in Tribolium castaneum. *American Naturalist*, 1970, *104*, 469–476.

Southwick, C. H. Effect of maternal environment on aggressive behavior of inbred mice. *Communications in Behavioral Biology*, 1968, *1*, 129–132.

Southwick, C. H., & Clark, L. H. Interstrain differences in aggressive behavior and exploratory activity of inbred mice. *Communications in Behavioral Biology*, 1968, *1*, 49–59.

Spiess, E. B. Low frequency advantage in mating of *Drosophila pseudoobscura* karyotype. *American Naturalist*, 1968, *102*, 363–379.

Spiess, L. D., & Spiess, E. B. Minority advantage in interpopulational matings of *Drosophila persimiles*. *American Naturalist*, 1969, *103*, 155–172.

Stewart, J. M. Genetic and ontogenetic determinants of agonistic behavior in animal societies: Prototypic experiments with mice. In J. deWit & W. W. Hartup (Eds.), *Determinants and origins of aggressive behavior*. The Hague: Mouton, 1974.

Stort, A. C. Genetic study of the aggressiveness of two sub species of *Apis mellifera* in Brazil. IV. Number of stings in the gloves of the observer. *Behavior Genetics*, 1975, *5*, 269–274.

Trivers, R. L. Parental investment and sexual selection. In B. Campbell (Ed.), *Sexual selection and the descent of man*. Chicago: Aldine, 1972.

Tucker, G. R. & McGill, T. E. Hormonal reactivation of castrated male mice of different strains. *Psychological Reports*, 1966, *19*, 810.

Vale, J. R., & Ray, D. A diallel analysis of male mouse sex behavior. *Behavior Genetics*, 1972, *2*, 199–209.

Vale, J. R., Ray, D., & Vale, C. A. The interaction of genotype and exogenous neonatal androgen: Agonistic behavior in female mice. *Behavioral Biology*, 1972, *7*, 321–334.

Vale, J. R., Ray, D., & Vale, C. A. Interaction of genotype and exogenous neonatal estrogen: Aggression in female mice. *Physiology and Behavior*, 1973, *10*, 181–184. (a)

Vale, J. R., Ray, D., & Vale, C. A. The interaction of genotype and exogenous neonatal androgen and estrogen: Sex behavior in female mice. *Developmental Psychobiology*, 1973, *6*, 319–328. (b)

van Oortmerssen, G. A., & Bakker, T. C. M. Artificial selection for short and long attack latencies in wild *Mus musculus domesticus*. *Behavior Genetics*, 1981, *11*, 115–126.

Von Schilcher, F., & Manning, A. Courtship song and mating speed in hybrids between *Drosophila melanogaster* and *Drosophila simulans*. *Behavior Genetics*, 1975, *5*, 395–404.

Weir, J. A., & Hogle, G. A. Influence of the Y chromosome on sex ratio and mating behavior in PHH and PHL mice. *Genetics*, 1973, *74*, S294.

Werner, M., & Schroder, J. H. Mutational changes in the courtship activity of male guppies (*Poecilia retuculata*) after X- irradiation. *Behavior Genetics*, 1980, *10*, 427–430.

Wilson, E. O. *Sociobiology: The new synthesis*. Cambridge, Mass: Harvard University Press, 1975.

Wood, D., Ringo, J. M., & Johnson, L. L. Analysis of courtship sequences of the hybrids between *Drosophila melanogaster* and *Drosophila simulans*. *Behavior Genetics*, 1980, *10*, 459–466.

Yanai, J., & McClearn, G. E. Assortative mating in mice. I. Female mating preference. *Behavior Genetics*, 1972, *2*, 173–183.

Yanai, J., & McClearn, G. E. Assortative mating in mice. II. Strain differences in female mating preference, male preference and the question of possible sexual selection. *Behavior Genetics*, 1973, *3*, 65–74. (a)

Yanai, J., & McClearn, G. E. Assortive mating in mice. III. Genetic determination of female mating preference. *Behavior Genetics*, 1973, *3*, 75–84. (b)

Zuger, B. Monozygotic twins discordant for homosexuality: Report of a pair and significance of the phenomenon. *Comp. Psychiatry*, 1976, *17*, 661–669.

12

Sociobiology and Behavior Genetics

John L. Fuller

State University of New York at Binghamton

The objective of this chapter is to examine the relations between two active areas of research, sociobiology and behavior genetics. It is somewhat surprising that these two disciplines have evolved in almost total isolation from each other, as both have their roots in behavioral biology with emphasis on genetics. To be sure, behavior geneticists have devoted their attention primarily to characteristics of individuals such as emotionality and learning, but they have also investigated social behavior such as fighting, mating, and caretaking. In general, behavior geneticists have been content to adopt the techniques developed for the study of physical characteristics to behavioral phenotypes. It is sociobiology[1] that has generated and emphasized comprehensive principles: kin selection, inclusive fitness, cost-benefit analysis based on parental investment, and evolutionary stable strategies. The major concerns of behavior geneticists have been: (1) quantification of the genetic contribution to behavioral differences between and within populations; and (2) utilization of genetic variation as a tool to investigate the development of individual behavior. Evolutionary considerations have not been completely neglected, but they have not had a major role in most behavior-genetic research.

Wilson (1975) defined the content of sociobiology as ⅓ invertebrate zoology, ⅓ vertebrate zoology, and ⅓ population biology. As major inputs of the field he

---

[1]In the strict sense sociobiology is a discipline and not a theory. Conceivably a sociobiologist could conduct behavioral research without reference to evolutionary theory. I shall, however, follow common practice and use the term to refer to a body of theory that seeks to explain social behavior and organization by principles of population genetics and natural selection.

named: *modifiability of individual behavior,* demographic analysis, equilibrial population densities, *gene flow between populations,* and *coefficients of relationship.* The italicized items are also prominent in behavior-genetic research. Wilson stated explicitly that sociobiology in the past had been too closely associated with ethology and behavioral physiology. In contrast, behavior genetics seems to be involved more and more with these disciplines. Does this mean that sociobiology and behavior genetics should go their separate ways? I do not believe so and herein argue that the two disciplines can be mutually supportive.

Sociobiology focuses on the role of evolution through natural selection as the most important determinant of social behavior and organization. In Wilson's terminology its interest is in the ultimate (remote) rather than the proximate (immediate) causation of social phenomena. The idea of natural selection acting upon heritable behavior patterns is not new. Darwin (1859/1927) gave much attention to the role of selection in the production of "instincts." But the sociobiology that we know today had to await the development of theoretical population genetics, the accumulation of detailed information on the social behavior of diverse species, and the analysis of the selection process in terms of the costs and benefits of alternative ways of behaving. These alternatives are the "strategies" of sociobiology.

If sociobiology had remained restricted to arguments about the role of group selection in animals (Maynard Smith, 1976; Wynne-Edwards, 1972), and the role of haplodiploidy in the evolution of sociality in hymenoptera (Hamilton, 1964), it would never have been featured in leading newsmagazines or been debated by a variety of scientists, and adherents of diverse political and social beliefs (Wade, 1976). But the extension of sociobiological theory to humans by Wilson (1975) produced strong reactions and disputes that have not yet been settled. Some of the flavor of these debates may be found in Gregory, Silvers, and Sutch (1978) and in Barlow and Silverberg (1980). Among the participants in these symposia were ethologists, psychologists, anthropologists, philosophers, and a few geneticists. Ruse (1979) provides an informative and critical appraisal of these and other contributions to the great sociobiology polemic. This chapter does not review the literature of attack and counterattack in detail. Instead it looks at the relationships between sociobiology and behavior genetics in terms of their objectives, methods, and data bases.

Somewhat surprisingly, full-time behavior geneticists have not been well represented at sociobiology symposia despite the weight given to genetics in sociobiological research. Nevertheless, three examples can be cited. Fuller (1978) pointed out that the evolutionary models of sociobiology require: (1) considerable genetic heterogeneity within a population; (2) correlations between genotypes and behaviors; (3) correlations between gene-associated behavioral variation and Mendelian fitness. He emphasized the noncongruence of the units of heredity and those of behavior. Genes are structures that code for molecules. Behavior is a function, a flow of reactions that is somewhat arbitrarily divided into units. Thiessen (1979) asserted that: "behavior genetics, through its systematic methodologies can add credibility to the theoretical structure of sociobiology . . . or at least set up predictions that can be tested in the field or laboratory [p. 193]." DeFries (1980) concentrated on behavior–genotype correlations and presented evidence for the heritability of emotional and social behavior in mice, and cognitive abilities in humans.

Throughout the sociobiology debates there are signs of confusion concerning the nature of the relationship between genes and behavior. Difficulties arise from the failure to recognize two distinct and potentially independent issues. The first deals with the degree to which a genotype limits the developmental potential of its possessor. Regarding this point there are two extreme views: (1) organisms (humans at least) come into the world as a *tabula rasa* upon which any program of social behavior can be imposed by proper manipulation; and (2) organisms have an inherited biological nature, shaped by natural selection, that predisposes them to develop behaviorally along lines that have been successful in the past. Most biologists and many psychologists would probably accept the second alternative for nonhuman animals despite the existence of individual differences among members of a species. For humans, the concept of species-typical behavior is still controversial. Behavioral differences among cultures and among social classes within a culture are great. One anthropologist (Washburn, 1978) asserts that: "evolutionary, comparative and sociobiological approaches to human social behavior are far more likely to cause confusion than to be helpful [p. 71]." I would merely point out that any mammalian group that does not match its social organization to its reproductive physiology and the duration of the dependency of its offspring is bound to become extinct.

A very different issue is the degree to which genetic variation within a species is a factor in prescribing social status and biological fitness of individuals. As good a scientist as Wilson (1980) is sometimes vague about the difference between genes as agents for directing individual ontogeny, and gene variants as causes of behavioral diversity in a population. He writes that a possibility for the present state of human social behavior is: "genetic variability still exists, and . . . at least some behavioral traits have a genetic foundation [p. 296]." Can one imagine a behavioral trait that does not have a genetic foundation? Without a genotype there is no organism and no behavior. I believe that Wilson meant to say that variability in the phenotypic expression of social behavior of humans is often attributable to *differences* in their genotypes. He could not have meant that behavior can be divided into genetic and nongenetic categories.

In the following sections I review some sociobiological concepts from the perspective of an experimenter in behavior genetics and suggest possible contributions of experiments with genetically defined subjects to the evolutionary theories of sociobiology. The choice of topics is somewhat arbitrary, but I hope that it will serve the purpose of strengthening the relationship between these two disciplines.

SOCIAL BEHAVIOR AND ORGANIZATION

Social behavior and social organization are the main interests of sociobiology. Wilson (1975) defines a society as a group of individuals belonging to the same species and organized in a cooperative manner. "Reciprocal communication of a cooperative nature, transcending mere sexual activity is the essential, intuitive criterion of a society (p. 7)." The definition is reasonably precise but, as Wilson realizes, it leaves open the question of how large and how well organized a group must be to warrant designation as a society. Social behavior is performed by individuals and can be investigated genetically like other individual characteristics. Social organization is a characteristic of groups including such factors as size, coherence, openness, composition by sex and age, and dominance structure. Although the genotypes of its members undoubtedly affect such group characteristics, and genotypic differences may underlay status and role distinctions, social organization is not amenable to genetic analysis to the same degree as social behavior. For the present, social organization is best left to ethologists and sociobiologists, while geneticists consider social behavior as an individual phenotype. Because by definition social behavior involves more than one individual, the difficulties of measurement are challenging (Fuller & Hahn, 1976).

Social behavior is directed at other individuals of the same species, often, but not always, involving interactions among members of the same group. The rejection of a strange bachelor male by a troop of baboons is social behavior even though the intruder is not a group member. Mating is social behavior even though in some species male and female meet briefly for copulation and never interact again. Wilson's designation of a society as a cooperative organization does not imply that there are no conflicts among its members or that the members are selected to behave in a way that maximizes the welfare of their group rather than themselves.

SOCIOBIOLOGY: THE CENTRAL PRINCIPLES

It is clearly impossible in a brief space to review sociobiology in detail, but in the following section the main concepts are identified and discussed. Additional information representative of the current state of the field may be found in Dawkins (1976), Barash (1977), and Clutton-Brock and Harvey (1978).

Individual Selection

Sociobiology is classically Darwinian in assigning the motive power for evolution to natural selection. Species are heterogeneous with respect to genotypes and phenotypes that are, to a significant degree, correlated with each other. The genotypes of relatives, particularly those of parents and offspring and of siblings, are also positively correlated. Individuals vary with respect to their success in leaving progeny who carry copies of their genes. To the degree that an individual's relative reproductive success (fitness) is dependent on the genes it carries, that individual will contribute a disproportionate (high or low) number of its genes to the following generation. Because the contribution of a gene to fitness varies with the physical and social environment in which its possessor competes for resources, the stability of the gene pool is dependent on environmental stability.

Behaviorists tend to look at natural selection as competition between individuals whose behavioral phenotypes are functions of their genotypes. The situation can be regarded also as a competition between alleles, organisms being only the means through which "selfish genes" can be replicated, matched against each other, and transmitted to another organism (Dawkins, 1976). Though dramatic, Dawkin's terminology may be misleading rather than helpful. Purpose should not be implied for genes any more than to the impersonal environmental forces of natural selection.

Putting these ideas together, the fundamental theorem of sociobiology can be expressed as: Genes influence important characteristics of social behavior that, as a result of natural selection, maximize the probability that copies of these genes will be represented in later generations. This theorem is probably acceptable to most behavior geneticists and ethologists. There is abundant evidence that the basic evolutionary postulates are valid and one might reasonably wonder why controversy persists. Discussion of this matter here is deferred until the relations between sociobiology and behavior genetics have been considered in some detail.

Kin Selection

The explanation of altruistic behavior has long posed problems for natural selection theory. If an altruist, A, performs an act that benefits a recipient, R, but reduces A's chances of survival and future reproduction, selection should operate to reduce the probability of such acts. The problem evaporates, however, if one looks at natural selection as operating upon gene frequencies rather than the welfare of individuals. Provided that R is related to A, any act of A that helps R will increase the probability that copies of A's genes will be passed on. Obviously, the closer the relationship between A and R, the higher the probability that A's act will benefit A's genes. In general, selection will favor helping behavior when its cost in fitness, k, is less than $1/r$, where r is the coefficient of relationship between A and R (Hamilton, 1963). The consideration of kin selection leads to a definition of the *inclusive fitness* of an individual as the sum of *personal fitness* (through offspring) and that portion of the fitness of relatives that is based on shared genes.

The validity of this explanation of the evolutionary basis of altruism is supported by a number of observations (Barash, 1977). Caretaking and other helping behavior is more common among closely related than distantly related individuals. Altruism is also more common in nondispersing species that form social groups whose members are likely to be related. Highly social species showing altruism are apt to discriminate against intruders and to recognize group members individually.

Reciprocal Altruism

An additional explanation for altruism is reciprocity (Trivers, 1971). A helps R at some risk with the result that A's behavior increases the probability that R or some other individual will help A in the future. When A and R belong to the same group it is very difficult to distinguish between kin selection and reciprocity. However, a case has been made for the reciprocal model for the evolution of cleaning symbioses in coral-reef fishes. The small cleaners are spared from predation, and even sheltered, by large predatory fish who benefit from the removal of ectoparasites that nourish the cleaners. Similar arguments have been made for the evolution of alarm calls in birds that may benefit other birds at some risk to the alarmist.

Trivers argues that reciprocity may be more important than kin selection in the development of human altruism. In his view, selection has favored a complex regulating system that leads to individual differences in degree of altruism and tendency to cheat. All individuals are considered to be potentially altruistic or cheating (nonreciprocating beneficiaries). Humans should be sensitive to the costs and benefits of an altruistic act, as well as the characteristics of cheaters. Thus: "selection would favor developmental plasticity of those traits regulating altruistic and cheating tendencies, and responses to these tendencies in others (Trivers, 1971, p. 53)." This is far from the idea of genetic determinism. In fact, the model does not require that there be alternative alleles for altruism and nonaltruism. Genes affecting the information-processing capacity of the brain would have general effects on the ability to appraise the costs of any behavior, whether or not altruism is involved. The importance of Triver's model is that it shows how social organization could influence behavioral evolution independently of kin selection, and without invoking mechanisms of group selection that contravene well-established principles of population genetics.

BEHAVIORAL STRATEGIES

Behavioral strategies are a central concern of sociobiology. The word *strategy*, taken over from military usage, is defined as a plan or action for achieving a particular objective. Its root is the Greek *strategia*, the title of a general. Sociobiologists do not claim that an animal (or a gene) deliberately and logically chooses the best way to achieve a desired goal. Rather they assume that the processes of natural selection, operating randomly rather than directly, favor bearers of alleles facilitating behavior that improves the bearer's reproduction relative to that of nonbearers. Presumably genes do this by their influence on the structure of the nervous system and its plasticity in response to external stimulation. I consider later some of the problems of using strategies as phenotypes in behavior-genetic analysis and suggest an alternative approach. For the present, I use "strategy" in its sociobiological sense.

In discussing the genetics and evolution of social strategies it is convenient to recognize two classes of social interactions, cooperative and competitive (Fuller & Hahn, 1976). In cooperative interactions all participants have the possibility of increasing their inclusive fitness. Examples are group affiliation, mating, care-giving and care soliciting. In competitive interactions some participants, occasionally all, are likely to lose fitness. Competition occurs when critical resources are in short supply. Even in cooperative interactions all participants need not have identical interests. Each should be selected to attempt to obtain the benefits of a relationship at minimal cost. This theme, based largely on the concept of parental investment (Maynard Smith, 1977; Trivers, 1972), recurs frequently in the sociobiological explanations of cooperative interactions.

Strategies in Cooperative Interactions

Cooperative interactions generally involve relations between mates, parents, and offspring, or other closely related individuals. The majority of animals reproduce sexually, and an individual of one sex must to some degree cooperate with an individual of the opposite sex. If sex had not evolved naturally, sociobiology and behavior genetics would not exist, but it is only recently that evolutionists have given serious consideration to the selective pressures that lead to sexuality. Cloning and self-fertilization seem to be very efficient ways to reproduce, and evolutionary biologists have sought for explanations for the prevalence of sexual reproduction that occurs even in species that are potentially asexual. The generally accepted explanation for the triumph of sexuality is that the benefits of increased genetic variability among sexually produced offspring outweigh the disadvantage of requiring two organisms in order to propagate (Williams, 1975).

Given the existence of sex there is tremendous variation in the mating behavior of animals. Sexual activity is commonly seasonal, but it differs from species to species. Age of sexual maturity, duration of fertility, number of offspring per clutch or per litter—each varies. These variations appear to be adaptive in relation to the habitat of a species and its ability to modify its own environment. There can be no doubt that natural selection has molded the genetic substrate underlying rates of growth, endocrine interactions, and sensitivity to key stimuli that induce mating at the optimal time. Clearly, an individual unresponsive to these physical and social stimuli would be unlikely to pass on its genes. Once a satisfactory coordination of the complex reproductive process has been achieved through natural selection one would expect that the genic balance underlying the process would be evolutionarily stable.

Who Mates With Whom.   For the great majority of animal species mating involves male solicitation and display, with female discrimination and a degree of choice. Natural selection is postulated to favor male behavior that overcomes female reluctance to mate and coordinates her physiological cycles so that a mating will be fertile. In a complementary manner, females should be selected to choose males whose appearance and vigor promise that their offspring will survive to maturity and thus insure the continued survival of the female's own genes. A female's opportunity to choose is constrained when there is a shortage of males, or when some males are excluded from the mating competition because of their subordinate status. Exceptions to these rules are important, for they support the rule that the discriminating sex is the one for whom reproduction involves the greater future costs (Wilson, 1975, p. 326).

Parental Investment.   An important concept for explaining the evolution of different mating patterns in the two sexes is *parental investment,* defined by Trivers (1972) as an investment by a parent in one offspring that increases that offspring's chance of survival at the cost of the parent's ability to invest in other offspring. Often a male's contribution to parenthood is limited to the time and energy expended in courtship and copulation and the production of a copious supply of small sperm cells. After a rest he can repeat the process with another female. Females are burdened with the production of a smaller number of large, energy-rich ova, and often with the costs of incubation, pregnancy, and caring for their immature young. Once committed by a fertile copulation a female's prospects for further reproduction in the immediate future are greatly reduced.

There is great variation among species in the relative parental investments of the sexes. In many birds and in some carnivorous mammals, males participate in sheltering, brooding, and feeding their young. In such cases the parental investments of the sexes are more nearly equal. Comparative studies of mating systems demonstrate an association between the relative costs of parenthood in males and females and the species-typical family organization. In species where male investment is low, polygyny and promiscuity are the common mating systems. Monogamy, whether permanent or restricted to a single breeding period, is rare in mammals, but common in birds where it is found in 90% of the species. It is always associated with a male's substantial contribution to the rearing of young. The sociobiological explanation of this general principle is that the need of the young for postnatal care is the selective force that shapes the mating system and the degree of mate fidelity. When young have a long period of immaturity during which they cannot forage for food, it pays males better to collaborate in the rearing of a small number of offspring from one female than to impregnate additional females and desert them promptly.

Evolution of Strategies.   The literature on reproductive strategies is large, growing, and sometimes includes such titillating words as: cuckoldry, desertion, enticement, divorce, and prostitution. Although these terms have ethical implications for humans, sociobiology regards their animal analogues as the outcome of gene selection, and the best possible solutions, given the biological needs of a species and its available resources, for both sexes. All this is orthodox evolutionary theory and is supported by numerous observations. Examples may be found in Orians (1969), Wilson (1975), Barash (1977), and Crook (1977). One may ask, however, whether the evolutionary hypothesis based on correlations of habitat demands and behavior is falsifiable in Popper's sense (Ruse, 1979; Turner, 1967, pp. 147–152). Given the assumption that natural selection has shaped behavior because it is adaptive (increases fitness), then by definition the behavior observed must be adaptive. If at first glance behavior such as nest desertion or rearing a nest parasite (a cowbird, for example) appears to be maladaptive, an ingenious believer can always suggest plausible explanations for indirect benefits.

The inability to test a hypothesis experimentally is not a fatal defect in science. Much of astronomy and geology rests upon a patient gathering of data regarding the universe and the earth as they appear today, and the testing of historical hypotheses against present-day traces of remote events. Biological evolution is more complex than physical evolution but the same analytical procedure is adaptable to both. This brings up the question of whether the techniques of behavior genetics that deal with the relations between genes and behavior in present-day animals can be used to validate sociobiological explanations for individual behavioral strategies. Can we demonstrate by experiment that the difference between monogamy and polygamy has a genetic basis? that animals can sense the degree of their relationship to conspecifics and confine their altruism appropriately to close relatives? I, personally, have little doubt that the basic principles of sociobiology are correct, although its applications to specific cases, and especially to humans, sometimes seem forced. As Barash (1977) points out there is no unitary adaptive significance to social behavior. A species-typical behavioral repertoire is a compromise based on environmental forces that pull in different directions. The possibilities of adaptation are also limited by phylogenetic inertia (Wilson, 1975), which restricts the tolerable extent of genetic change. Once a successful genotype has been assembled, a novel allelic substitution is more likely to impair than to improve fitness. Yet, given time, changes do occur, more complex structures and behaviors evolve until almost every habitat is occupied by some living creatures.

Behavior-Genetic Analysis of Mating Strategies.   Mate choice is a phenotype amenable to behavior-genetic analysis. What is the best strategy for assuring maximum representation of one's genes in the next generation? If the goal is simply to transmit the maximum number of copies of one's genes to one's progeny, the answer would be to mate with near relatives who already carry copies of many of these genes. The evidence is overwhelming, however, that such inbreeding results in decreased rather than improved fitness (Lerner, 1958; Ralls, Brugger, & Ballou, 1979; Roughgarden, 1979). Two explanations, not mutually exclusive, have been offered: (1) inbreeding increases the probability of homozygosity for deleterious recessive genes; (2) the advantages of heterosis at some loci are lost. Either explanation leads to the prediction that natural selection should act against inbreeding. Matings outside the bounds of a species are even more deleterious than inbreeding. Most are infertile. If young are produced, their viability and their fertility are impaired. The logical conclusion is that natural selection should favor matings between unrelated individuals of the same species.

There is a wide choice of mechanisms by which outbreeding within a species might be assured. The random dispersal of family groups over a wide territory greatly reduces the probability of encountering and mating with a relative. Another type of strategy involves a refinement of simple recognition of conspecific individuals of the opposite sex. If specific individuals or classes of individuals can be recognized there are several possibilities. Males and females might be genetically programmed to be more aroused sexually by individuals with characteristics, pheromones for example, that are different from their own. In species where the young remain for some time in a family group, male and female siblings may develop dominance relationships that are incompatible with reciprocal sexual arousal. For students of behavioral development these proximate influences upon mating preference are of great interest.

The behavioral bases of mating preferences and sexual isolation have been investigated intensively in *Drosophila* and to some extent in *Mus*. Although there are important differences in details of mating and courtship of *Drosophila* species (Spieth, 1952), the general pattern involves promiscuous advances by males and discriminative passivity on the part of females. Mating speed has been widely studied from the genetic point of view. Parsons (1974) has summarized the results as follows: (1) natural selection favors vigorous courtship and rapid mating by males; (2) mating speed is largely controlled by males, but in slow matings the discriminative role of females becomes more important; (3) mating speed is correlated positively with fertility and fecundity and is probably the most important component of individual fitness in most *Drosophila* species.

Is mating speed an isolated characteristic that has been selected for itself alone, or is it part of a more general set of attributes? The latter interpretation seems to be correct. Pyle (1978) showed that mating speed is correlated significantly with courtship duration, general activity, and the size and number of the sensory aristae. He distinguishes between behavioral correlations based on task similarity and those resulting from genetic associations such as pleiotrophy, linkage, and the incidental selection of genes that compensate for the costs of selecting for a generally beneficient trait. This last category emphasizes the point that the total genotype is involved in the determination of fitness. Pyle's data do not enable him to choose among the possibilities.

Even if there were no discrimination involved in mating, variation in reproductive success would have evolutionary consequences. In fact, however, *Drosophila* matings are not random with respect to genotype. The restriction of mating, whether partial or complete, between potentially interfertile populations is called sexual or ethological isolation (Petit & Ehrman, 1969). According to Mayr (1970): "ethological barriers to random mating constitute the largest and most important class of isolating mechanisms in animals [p. 58]." The behavior-genetic analysis of such mechanisms has obvious implications for sociobiological explanations of mating strategies. An excellent example is Ehrman's (1965) study of homogametic and heterogametic matings in five races of *Drosophila paulistorum*. She predicted that natural selection would favor the evolution of isolating mechanisms between races occupying contiguous areas, but not between races that rarely come into contact. Her experiments confirmed the prediction; isolation coefficients were consistently higher in pairings between sympatric than between allopatric races.

Frequency-dependent mating is another behavioral characteristic that has implications for sociobiological and evolutionary theory. In general this involves a shift in the relative contributions to fitness of two alleles (A and a) depending on their frequencies. For example, male a/a homozygotes might be favored when their frequency is less than .1, but avoided when they are more abundant. Under such circumstances the frequency of a would tend to stabilize at .32 ($.1^{\frac{1}{2}}$). In *Drosophila* frequency-dependent mating has been observed in mutant-bearing strains, in chromosomal variants (Ehrman & Petit, 1968), and in wild-type strains (Tardif & Murnik, 1975).

How might natural selection program a female fruit fly to mate with a white-eyed male if it is rare, and with a red-eyed male when white-eyed males are common? Humans do assign values to objects based on their rarity, but this type of appraisal seems to be beyond the capacity of a *Drosophila*. So far as we know they cannot count. An explanation has been offered by Spiess and Schwer (1978). They found a strong minority-mating advantage for two eye-color mutants. After careful observation of many courtships they proposed that a female's first contact with a male suitor usually terminates before copulation. She becomes "habituated" to the phenotype of this male and is less likely to be aroused by another male of the same type. A different type of male, provided it is sexually competent, has a mating advantage. Inasmuch as most females have their first courtship experience with majority males, the result is a minority advantage in actual copulation. This example shows how behavior-genetic analysis can sharpen our ideas of the manner in which natural selection might affect gene frequencies.

The extent of minority advantage in mating among animals in general is undetermined. An attempt to demonstrate it in houseflies was unsuccessful (Childress & McDonald, 1973). Houseflies do not engage in elaborate, prolonged courtship before copulation, and thus the opportunity for phenotypic evaluation, as required by the Spiess and Schwer model, does not exist. The ultimate explanation for ethological isolation, minority-advantage, and other mating strategies is probably similar for all *Drosophila* species in which these have been observed. Natural selection must favor mating systems that preserve enough genetic homogeneity to allow adaptation to a variety of environmental challenges, without losing the advantages of a coadapted genotype. A combination of sociobiological, ecological, and behavior-genetic approaches seems the best hope for further progress in understanding the evolutionary process.

Theoretical models for the genetic consequences of different mating strategies have been tested in flour beetles (promiscuous mating) and houseflies (permanent pairing) (Taylor 1975). Equations were developed on a theoretical basis for the probability of female mating over time for the two strategies. Observations showed that the actual course of mating in homogeneous populations of both species followed the predicted equations. In heterogeneous populations of flies the equations also predicted fairly well the proportions of heterogametic and homogametic pairings. Taylor concluded that pair formation, assuming equal numbers of males and females with all females mating, could lead only to assortative mating with changes in genotypic proportions rather than in gene frequency. With promiscuous mating there is a greater opportunity for sexual selection. One might then ask why pair-bonding strategies have evolved in some species. It is probable that the benefits of biparental care to survival of progeny sometimes outweigh the disadvantages of reduced intensity of sexual selection.

It is clear that selection for reducing inbreeding need not involve genes that modify mating preferences directly. Inbreeding depression is better regarded as a selective force that may affect dispersal patterns, ease of habituation to a prospective partner, and any physical or behavioral characteristic that allows individuals to differentiate between prospective partners. Theories involved only with "ultimate" explanations for avoidance of inbreeding are insufficient to specify how it occurs in the world about us.

Strategies in Nature. Laboratory experiments are helpful, but we would also like to know how inbreeding is avoided in animals living under natural conditions. A study by Koenig and Pitelka (1979) on the acorn woodpecker demonstrates how complex the matter can be. This species has the unusual trait of communal nesting and, when food is abundant, of occupying the same nesting site from year to year. This might seem to be a life-style conducive to inbreeding. In fact inbreeding is rare because acorn woodpeckers follow three "rules": (1) A female will not mate with a male from her group who was reproductively active at the time of her hatching; (2) when a female leaves her nest, she breeds; (3) individuals leave their natal colony alone or with others of their own sex. Given these rules the probabilities of close inbreeding are rather small even though the birds may be unable to recognize their siblings once they leave the natal nest.

Acorn woodpeckers also provide a good example of the flexibility of social organization in a taxonomic group not noted for its adaptability. In some areas they are known to adopt a different life-style, mating in pairs instead of communally, and migrating seasonally instead of defending a year-round territory (Stacey & Bock, 1978). There is no evidence that the two cultures represent two genotypes, although the possibility has not been formally excluded. Migration seems to be an adaptation to a deficient food supply; permanent holding of territory depends on a stable supply of acorns. This kind of plasticity of social organization poses a problem for simple models of natural selection. One position might be that "communal" genes and "pair-bond" genes are a balanced polymorphism maintained by a patchy environment. Enough intermixture of genes from the two life-styles occurs to prevent genetic separation. Another possibility is that the neural basis for the rules of communal and pair mating is present in all individuals. Selection has operated to sensitize acorn woodpeckers to cues that guide them to adopt the mating strategy that is best under the circumstances. Only a genetic analysis can decide which hypothesis is correct. Their strategy is not unique. A surprisingly similar mode of reducing inbreeding is found in chimpanzees (Pusey, 1980).

Strategies in Competitive Interactions: Agonistic Behavior

Agonistic behavior can be defined as a social interaction in which the fitness of one participant increases at the expense of the other. The concept might be expanded to conflicts between groups—human wars are an outstanding example—but the present discussion is restricted to interactions between two competitors. If there is a genetic contribution to competitive skills, "loser" genes decrease in frequency due to natural selection. It is assumed here that genes do play such a role, and that natural selection has a major role in shaping the strategies of agonistic behavior. Empirical evidence for the inheritance of aggression is presented later.

Game-Theory Approach.   A major contribution to the analysis of agonistic behavior was made by Maynard Smith (1974, 1976) who applied the theory of games to specify the best strategies for the resolution of two-participant conflicts. There are various ways of winning contests. The least costly is to bluff an opponent until it retreats. If the opponent fails to retreat, the conflict may escalate to physical combat. The winner is usually the stronger and more experienced of the contestants. Even the winner in such a struggle pays a cost in energy consumption and possible injury. Given these possibilities we may ask whether natural selection favors any particular choice of options.

Maynard Smith described an imaginary species that is a mixture of hawkish and dovish individuals in the proportions of p and $1 - p$, respectively. A hawk has a 50:50 chance of victory in a fight with another hawk; it wins against a dove by a brief display. In contests between doves there is no physical combat, but one participant withdraws after a variable period of display. A dove has a 50:50 chance of winning such a contest by outlasting its opponent in a show of power. The expected payoffs for each type of encounter are as follows:

	In an encounter with a	
The average payoff to:	Hawk	Dove
a hawk is:	$\frac{1}{2}(V - C_a)$	V
a dove is:	0	$\frac{1}{2}V - C_d$

The entries in this table represent units of fitness. $V =$ the value of the contested resource; $C_a =$ the cost of energy expended in attack plus the risk of injury; $C_d =$ the cost of time and energy devoted to display. A mixture of hawks and doves will remain in numerical balance when the average payoffs for each strategy are equal. Assuming, as previously stated, that in a hawk–hawk or dove–dove encounter the contestants have equal chances of winning, the proportion of hawks, p, is $(2C_d + V)/(2C_d + C_a)$. The proportion of doves is $1 - [(2C_d + V)/(2C_d + C_a)]$. The same relations hold if any individual can behave as a hawk or a dove and does so with the probabilities of p and $1 - p$, respectively. Maynard Smith showed that, given these two strategies, the proportion of hawks and doves remains constant from generation to generation. In his terms there is a frequency-dependent evolutionary stable strategy (EES). As long as V, C_d, and C_a remain constant there will be no shifts in the proportions of the two phenotypes. Dawkins (1980) has pointed out that an EES is not necessarily the best strategy. It is one that is immune to cheating when there are a named set of alternatives. A new strategy that changes the rules of the game is always conceivable, and it may disrupt the stable state.

The validity of any model depends on the accuracy with which its assumptions correspond with actual behavior and its consequences. Seeking for greater verisimilitude Maynard Smith and Price (1973) found that a third option, retaliation, introduced into a population of hawks and doves, would replace both of the original strategies. A retaliator plays dove against dove, and hawk against hawk.

Intuitively this makes sense. If a retaliator can win its objective without attacking, fine. If an opponent attacks, fight back. The payoff matrix now becomes:

In an encounter with a

The average payoff to:	Hawk	Dove	Retaliator
a hawk is:	$\frac{1}{2}(V - C_a)$	V	$\frac{1}{2}(V - C_a)$
a dove is:	0	$\frac{1}{2}V - C_d$	$\frac{1}{2}V - C_d$
a retaliator is:	$\frac{1}{2}(V - C_a)$	$\frac{1}{2}V - C_d$	$\frac{1}{2}V - C_d$

It turns out that retaliator is stable against invasion by a hawk mutant; it is thus an ESS. In a population containing only doves and retaliators there would be no phenotypic distinction between them, no selection in either direction, and their initial proportions would remain constant except for random drift.

The game-theory approach has simple elegance and is influential in contemporary sociobiology. Let us now consider some problems in its application to agonistic behavior. Maynard Smith assumes that the behavioral differences among competitors represent genetic variation. He writes of "invasion by a mutant, *H*," which implies that an allelic change at a single locus can switch behavior from the dovish to the hawkish mode. His assumption is undoubtedly a simplification for the sake of clarity of presentation, and his model could be adapted to a polygenic system. A more important point is that the payoff matrices and the deductions from them are also applicable to a genetically uniform population that hands down traditional modes of behaving from parent to offspring. A balanced ratio of hawks and doves could be maintained if offspring tended to behave like father or mother through imitation, tutelage, or indirect environmental pressures. Actually I doubt that genotypic differences are irrelevant to success in conflict, or that in most species (*Homo sapiens* is an obvious exception), parents indoctrinate their offspring with their own strategies. I merely point out that game theory without empirical data does not exclude this interpretation.

An alternative game-theory model of agonistic behavior follows. Rather than separating individuals into phenotypic classes this matrix deals with the choice of alternative types of behavior by an individual. It is assumed that every mature member of a species has the neurological and physical equipment to execute all species-typical forms of agonistic behavior. Although details vary greatly among species the general pattern is to attempt to win a contested resource by inducing a competitor to conclude that the cost of competing is too high and thus to withdraw. In the two-participant case each attempts to gain the resource by display. Intensity of display increases over time but it need not escalate to physical attack. Mutual display has two possible consequences: (1) one contestant withdraws leaving the other as victor; (2) a contestant may call its opponent's possible bluff by an attack. The attacked individual then has two alternatives: (1) withdrawal without combat; (2) retaliation with an outcome that is settled by physical prowess. The payoff matrix is:

	In a contest where one opponent chooses to:	
The payoff for the other of:	Call a bluff and escalate.	Wait for opponent withdrawal.
Calling a bluff is:	$\frac{1}{2}(V - C_a) - C_d$	$V - C_d$
Waiting for withdrawal is:	$-C_d$	$\frac{1}{2}V - C_d$

Assuming, following Maynard Smith, that the frequencies of bluffing– withdrawal and bluffing–attack should be selected so that each mode of reaction produces equal benefits, one may solve for the frequency, p, with which a bluff should be called. The result is $p = V/C_a$. The outcome is not surprising. The higher the cost of attack, the less is the probability of benefiting from calling what may be a bluff. The higher the value of the potential benefit, the greater is the incentive to attack if necessary. Costs and benefits are not fixed values but depend on the nature of the contested resource, and the contestant's physical state and current needs. These may change markedly during the course of a conflict. The implication of the bluff-call model is that the contestants have some way of appraising relative costs and benefits of a particular response. Sociobiology assumes that natural selection has provided animals with the means to do this without the aid of a formal logic or lessons in economic theory.

Assessment and Choice of Strategy.   Parker (1974) proposed a model of assessment based on the idea that animals in a conflict should act to maximize their gain in fitness from existing resources. In particular he postulated that each potential combatant matches its own prowess (he calls it resource-holding potential, RHP) against that of its opponent and determines the probable cost of a challenge in units of fitness. Assuming that prospective combatants differ in RHP due to size, age, experience, and tactical advantage, he predicted the following outcomes: (1) Closely ranked opponents often resort to physical combat; (2) when opponents are mismatched the one with the lower RHP should withdraw. These rules are shown to define an ESS. Many, perhaps most, contests are in fact asymmetric with respect to RHPs and are settled by displays, threats, and retreat by the weaker party (Maynard Smith & Parker, 1976).

The game-theory approach and the idea of an ESS provide a rational basis for understanding how agonistic behavior, including aggressive displays and signs of submission, might have evolved. The models differ radically from classical ethological explanations of display as a means of reducing the probability of injury and thus helping to perpetuate the species. Maynard Smith, Parker, and others of their persuasion explain the settling of conflicts by threat rather than by fighting as the outcome of individual rather than group selection. I agree. Nevertheless, the most ingenious theoretical models need experimental and observational verification.

Empirical Studies in Natural Populations.   It is easy to write an equation in which the probability of a given strategy is a function of its costs and its benefits in terms of fitness, but in general we have no reliable way of translating the effects of a particular strategy into a numerical value. Parker (1978) has gone a long way with quantification of mating strategies in dung flies where males compete for the available females. Insects may well be the group of choice for testing hypotheses of the genetic regulation of strategies. Here, however, I consider two studies with vertebrates.

The first example is the work of Clutton-Brock, Albon, Gibson, and Guiness (1979) on fighting among red deer (*Cervus elaphus* L.) males in an island preserve. The object of the project was to test the concept that cost–benefit relationships regulate the frequency and severity of fighting. Most of the predictions based on this hypothesis were confirmed. There was one important exception. The expectation that young stags should fight less because they have much to lose and little to gain was not verified. Perhaps in stags the wisdom to assess risks and benefits accurately and apply the results rationally comes only with maturity. Or perhaps the experience that young stags gain while fighting in a losing cause pays off by a gain in their future ability to compete. The important point is that most of the predictions of the cost–benefit model were verified. On the whole, red deer stags act "logically," but not to the degree that contests were often settled without fighting. Assessments of opponents were not very accurate; many contests escalated to vigorous attack and counterattack, eventually becoming battles of attrition. Almost 100% of stags sustained some injury during fights; 20–30% of these were serious although their effect on lifetime reproductive success appeared to be small.

How important is the choice of a strategy to the Darwinian fitness of these deer? Clutton-Brock and his associates are uncertain. Very old and very young stags have fewer females in their harems than do stags in their prime. Extensive records of individuals are needed to determine if there is a correlation between lifelong reproductive success and choice of strategy at various stages of the life cycle. Even then, significant correlations would not prove that the differences in behavior were determined genetically. That would require family studies that are difficult, if not impossible, to carry out in a free-living species with a polygynous mating system and frequent shifts of females from one harem to another.

Evidence for the association of genetic variation in natural animal populations with differences in aggressive behavior and reproductive success is very scanty. One example has been reported recently. Moss and Watson (1980) removed eggs from a fluctuating population of red grouse at different stages of the population cycle. The eggs were hatched in a laboratory; the chicks were then reared under standard conditions and tested later for aggression. The intensity of agonistic behavior in the laboratory-reared birds correlated negatively with the population density at the time the eggs were collected. Breeding experiments conducted in captivity showed that differences in the strength of aggression were heritable. The authors concluded that in a natural habitat selection seems to favor the nonaggressive genotype until high population densities are reached; then the direction of selection shifts and the population declines. These data support the idea that aggression has an important role in determining fitness and that populations may be polymorphic with respect to "aggression genes." In the grouse example, the relation between high aggression and fitness reverses as the social environment (population density) changes. Until more evidence is available, however, we should be cautious and refrain from assuming that the periodic changes in agonistic behavior within a population necessarily imply changes in gene frequency. Behavioral development is flexible enough to permit considerable variation in choice of strategies without invoking genetic polymorphism.

Empirical Studies in Laboratories.   The immediately preceding remarks may be overly conservative. The assumption of genetic neutrality of aggressive behavior runs contrary to a substantial body of evidence from laboratory studies. Mice have been selected for high and low aggressiveness in a domestic stock (Lagerspetz, 1961) and from a stock of recent feral origin (Ebert & Hyde, 1976). Inbred lines of mice differ greatly in frequency and intensity of fighting (Lagerspetz & Lagerspetz, 1974; Scott, 1966), indicating that the populations from which these strains were derived were polymorphic for alleles that influence aggression. Selection for aggression was also successful in domestic fowl (Craig, Biswas, & Guhl, 1969; Craig, Ortman, & Guhl, 1965).

Unfortunately for the purpose of testing sociobiological concepts of selection, much of the research with inbred lines of mice has employed contests between individuals of the same strain and age (homogeneous sets as defined by Fuller & Hahn, 1976). Behavioral differences between participants in such contests have no genetic significance although they may yield important information of the effects of experience. Even here, however, few observations have been made on contests deliberately designed to be asymmetrical. A tacit assumption seems to be that results from intrastrain contests can be used to rank strains on an aggression scale without actually matching animals of different strains. The assumption appears invalid as shown by the wide differences in rank order for the same strains when observed in different laboratories (Lagerspetz & Lagerspetz, 1974). The absence of observations on contests planned to be asymmetrical means that we have little information on possible RHP assessment in mice despite a considerable amount of research on the agonistic behavior of this species.

Observations on interstrain competition are not entirely lacking. In several experiments with mice, Darwinian fitness has been assayed by determining the proportion of offspring sired by males of different strains who are competing for the same female(s) (DeFries & McClearn, 1970; Horn, 1974; Kuse & DeFries, 1976; Levine, Barsel, & Diakow, 1966; Yanai & McClearn, 1972, 1973a, 1973b). In all but Horn's study two or more types of males were confined with a female and paternity was determined by the coat color of the offspring. Horn's subjects lived in a simulated natural environment; paternity was determined by starch-gel electrophoresis of hemoglobin and serum enzymes. Despite the differences of procedure all studies found that strains did vary in reproductive success in a competitive situation, although differences were not entirely attributable to social dominance based on fighting ability. Fertilizing ability and female preferences also played a role.

Although it is clear that genes play an important part in producing individual differences in the agonistic behavior of mice, rearing and social experience are also potent influences. (Cairns, 1972; Denenburg, 1977; Denenberg, Hudgens, & Zarrow, 1964). Currently there is a developing interest in demonstrating genotype–environment interactions in experiments in which both genotypes and experiences are manipulated (Simon, 1979).

Maintaining Genetic Heterogeneity.   There are two basic hypotheses for the maintenance of genetic heterogeneity in a population (Powell & Taylor, 1979). One holds that polymorphisms are maintained by heterosis. Heterozygotes (*Aa*) are more fit than either *AA* or *aa;* hence the frequencies of *A* and *a* are balanced. Such a system involves costs, possibly substantial, because of decreased fitness in the two types of homozygotes. A second hypothesis postulated that genetic heterogeneity is maintained by environmental heterogeneity. Selection pressures differ from niche to niche within the species range. Some genotypes do better in niche A; others in niche B. Within this general theory there are two variants. Levene (1953) postulated random distribution of genotypes to niches, differential survival based on the suitability of the genotype–niche match, and random mating among the survivors. This model is appropriate to a plant that disperses its seeds widely over a patchy environment. With mobile animals the Powell–Taylor habitat-choice model is an attractive alternative. It postulates that individuals are sensitive to the external and internal milieu, and that they select from available niches that one for which they are genetically preadapted. Laboratory and field tests with *Drosophila persimilis* support the hypothesis. Powell and Taylor deal specifically with habitat choice, but their model can be adapted to social behavior. Such a model would require that animals assess conflict situations as postulated by Parker (1974) and that they match their own genetically influenced RHP with the demands of various conflict strategies. If situations are varied enough to require different strategies for success, genetic heterogeneity would be maintained.

ENVIRONMENTALLY BALANCED GENOTYPES:
AN ALTERNATIVE TO ESS

Up to this point I have followed contemporary usage in writing of behavioral strategies and genes that determine, or at least influence an individual's choice of strategy. Here I confess to uneasiness about using a strategy as a phenotype amenable to genetic analysis. I have no quarrel with the use of the word *strategy* as a simile. It facilitates speaking and writing about complex sequences of behavior that are directed at particular ends. There are recognizable analogies between a red deer stag protecting his harem, and a military commander defending a village against insurgent guerilla forces.

In employing simile one must be careful not to imply too much. What sociobiologists call a strategy is not a structure, but an association of related processes presumably brought together through natural selection. A strategy is much more complex than the phenotypes generally studied by geneticists, even by those who have a strong interest in behavior. My concern is with the tacit implication that a *strategy* is a relatively fixed behavioral phenotype with a characteristic genetic underpinning. Thus Maynard Smith (1976a) writes of "invasion of a population by a hawk mutant." I have no difficulty conceiving of a gene substitution that would affect an animal's propensity to attack, but this need not imply alternative alleles for hawk and dove strategies. Indeed, an animal's choice of strategy is often dictated more by circumstances than by genotype. Allelic substitutions simply influence the probability of various options in a specific situation.

I propose that natural selection operates to produce environmentally balanced genotypes (EBGs), an idea akin to Dobzhansky's (1962) concept of the co-adapted genotype. I have also borrowed from Schneirla's (1965) distinction between approach and withdrawal processes, and Wright's (1934) hypothesis of polygenically determined developmental thresholds.

Schneirla's basic idea was that weak stimulation induces approach and strong stimulation leads to withdrawal. He considered the distinction important with respect to ontogenesis and also, as is often unrecognized, to phylogenesis. According to Schneirla (1965):

> The emergence of such processes (stimulus filtering and functional plasticity) must have rested, above all, on conditions of natural selection favoring the survival of animals equipped for consistent approaches to low intensity sources and withdrawal from high intensity sources of stimulation . . . mutations determining thresholds of response in the directions specified must have had a heavy weighting to survival [p. 4].

The Threshold Model

Wright's developmental threshold model originated from his work on polydactyly in guinea pigs. Fuller, Easler, and Smith (1950) applied his idea to the transmission of audiogenic seizure susceptibility in mice. The results of crosses between a seizure-susceptible and a seizure-resistant strain did not fit Mendelian expectations, but the data could be explained by postulating that animals of given strains and their F_1 hybrids were characterized by genotype-specific mean values of an underlying physiological trait that affected seizure susceptibility. Individuals below the threshold level of this underlying trait were resistant; those above threshold were susceptible. Although each genotype has a characteristic mean value of the trait, it shows considerable variation which may, as a first approximation, be represented by a normal curve. (Fig. 12.1). Given a standard test most animals in strain R are resistant; most in strain S are susceptible; their F_1 contains both susceptible and resistant animals. The threshold model does not contradict Mendelian inheritance of subunits of the physiological system underlying seizure susceptibility. In fact it predicts that with the proper genetic material it should be possible to locate some of the participating genes on particular chromosomes. Seyfried, Yu, and Glaser (1980), using recombinant inbred strains from a cross between a susceptible and resistant strain of mice, have validated the basic threshold model and identified an association between one of

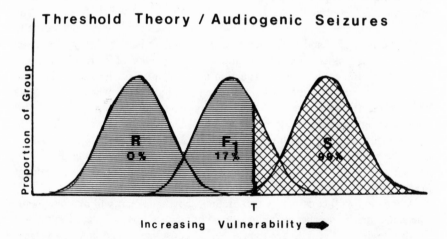

FIG. 12.1. Threshold theory of audiogenic seizure susceptibility in mice. Mice may be divided into two phenotypic classes depending on the presence or absence of a convulsive response to an intense high-frequency sound. Whether a particular animal is susceptible or not is assumed to depend on its position with respect to some (as yet unspecified) continuously distributed physiological variable. Genetically identical mice tend to have similar vulnerability at the same age, but this varies normally about a characteristic genotypic mean. However, all animals of genotype R are resistant because none of them exceed the seizure threshold T. Most mice of genotype S are susceptible because they are above T. F_1 hybrids are exactly intermediate to R and S on the underlying vulnerability scale, but are more like R phenotypically. The F_1 would be more sensitive to agents altering the threshold, as its genotypic mean falls close to T.

the genes affecting susceptibility and the *Ah* locus, whose position unfortunately is not yet certain.

Thresholds and Agonistic Behavior.   The threshold hypothesis as applied to audiogenic seizures can be applied to agonistic behavior. Assume three modes of responding to a conflict with another individual—withdrawal, display, and attack, each with its own physiological underpinning and genetically influenced threshold. Within each of these modes of reacting there are graded responses, but these can be neglected in a preliminary presentation. The basic assumption is that the probability of an animal's adopting one of these modes is a function of the animal's assessed value (*V*) of the resource at stake and the assessed cost (*C*) of winning the competition by force. When *C* is much greater than *V*, withdrawal occurs; if *C* is much less than *V*, attack is favored when display is not sufficient to make the opponent yield. Display has a dual function: to induce the opponent to withdraw peacefully and to improve the accuracy of assessment of the cost of a physical challenge. Figure 12.2 depicts these relationships graphically.

In Figure 12.3 the threshold concept is applied to agonistic behavior in a hypothetical animal. The probabilities of withdrawal (*W*), display (*D*), and attack (*A*) are shown to depend on simultaneous evaluations of *V* and *C*. When *V* < *C*, the choice is between *W* and *D;* when *V* > *C*, it is between *D* and *A*. The probability of each choice between these pairs of alternatives varies according to the assessment of *V* and *C* by each participant. There is a necessary interaction between the two assessments. For example, the evaluation of the relative costs of attack or continued display is a function of the display of the antagonist. The figure shows the probabilities of each strategy for individuals whose assessments accurately reflect the true *C/V* relationship. At Point 1 where costs are still relatively high in comparison with resource value, there is equal advantage in continuing display or withdrawing. At Point 2 a similar balance exists where costs are relatively low and there is equal advantage in continuing display or attacking. At Point 1, the hanger-on has an even chance of winning a bargain; at Point 2 the attacker has an even chance of winning at some cost.

A daredevil will consistently underrate costs; a faintheart will overrate them. The same threshold concept applies for both but curves are displaced to the left for daredevils; to the right for fainthearts. Because costs, and possibly values, may change over the course of conflict, an individual's strategy may change at any time. On the whole the system is adaptive, because when resource value is less than cost of attack, or when it is approximately equal, display usually suffices to settle the conflict at minimal cost to both parties; when value exceeds cost, attack is most likely to be initiated by the stronger party. Value is based largely on the relative abundance of a resource in the environment. As a resource becomes more scarce the model predicts an escalation of conflict. If the scarcity is prolonged over a number of generations the genotypic balance that regulates

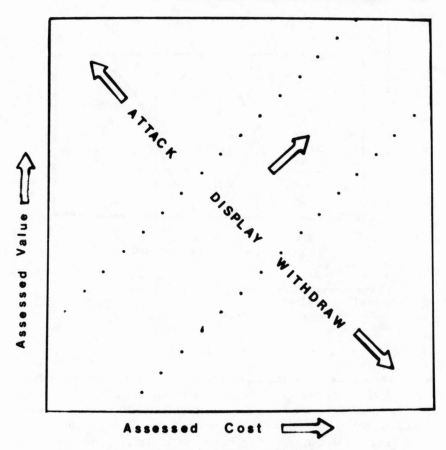

FIG. 12.2. A simple model of the probability of withdrawing, displaying, or attacking in a potential conflict situation based on an assessment of the costs of each option relative to the value of a contested resource.

thresholds would shift to match the choice of options to new conditions. Moss and Watson's red grouse may be an example of such a shift.

Advantages of EBG over ESS

A multiple-threshold model is as versatile as game theory in generating models to explain the choice of strategies in social interactions. It leads to the concept of an *Environmentally Balanced Genotype* (EBG), which regulates the points at which a stimulus complex elicits alternative modes of response in the most adaptive manner. Genes that affect androgen receptors in the hypothalamus, or the rate of synthesis of endorphins, would be good candidates for such a role. They would

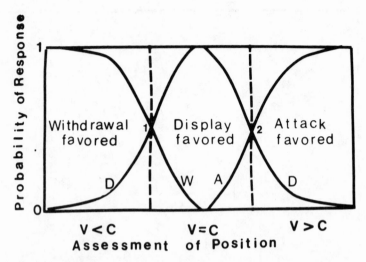

FIG. 12.3.   The dual-threshold model applied to the probabilities of withdrawal (W), display (D), and attack (A), based on a contestant's assessment of the costs and benefits of each. At 1, where costs are still relatively high compared with resource value, there is an equal probability of D and W. At 2 where value clearly exceeds assessed costs, there is an equal probability of D or A.

not be associated with specific action patterns, but for the probability of the elicitation of the alternatives. The EBG concept gives more attention than the ESS concept to the manner in which genes influence behavior and is thus more amenable to behavior-genetic analysis at the developmental–physiological level. The EBG hypothesis also emphasizes the fact that the stability of the gene pool is a function of the physical and social environment in which a choice of strategies must be made. A multiple-threshold system is flexible enough to allow adaptation to new situations without waiting for the slow processes of natural selection. It is not the probability of a particular strategy that is stabilized by genes, but a system of thresholds that match strategy to particular situations.

Behavior geneticists should be encouraged to turn from experiments on inbred strain differences and to test the concepts originated by sociobiologists. The multiple-threshold model and the EBG concept should be helpful in this effort. Perforce much of this work will be done with laboratory rodents, particularly mice, as large numbers of subjects and genetic controls are needed. Simon (1979) has called for more experiments directed toward defining features of the social and physical environment that bring out genotypic differences in fighting. In addition more attention should be given to age differences and to asymmetrical pairings. It is time to test empirically the genetic basis of the strategies and assessment capabilities so beloved of sociobiologists. If environments are as heterogeneous as I believe they are, optimal strategies will vary between niches.

This follows directly from the multiple-threshold model. Do the gene pools of these niches also vary in an apparently adaptive manner? Or, is it not possible that an EBG allows enough flexibility to permit individual adaptation to changing conditions and diverse habitats without the need for genetic selection?

HUMAN SOCIOBIOLOGY

As stated previously Wilsons's (1975) chapter on the biological and evolutionary roots of human social behavior aroused attention that would never have attended a purely zoological treatise. Some welcomed the attempt to unify human and animal social behavior by the evolutionary approach. However, negative reactions attracted more attention in the popular media. Four major disagreements are discernable in the writings of these critics:

1. Human behavioral development is so plastic that the concept of genetic control of social behavior and organization is erroneous. Genes may impose behavioral constraints but, at least for social characteristics, there is no universal, biologically determined human nature.

2. Because of the importance of cultural influences on human social organization the role of kin selection is not as important as in other animals. In many ways human societies are organized to favor group over individual selection.

3. Analogies between human and animal behavior are often far-fetched. Despite protestations to the contrary sociobiologists are guilty of anthropomorphism in the interpretation and description of animal behavior, and of zoomorphism in their explanations of human behavior.

4. Sociobiologists err in assuming that modal forms of social behavior and social organization must be adaptive, because otherwise they could not have evolved. The adaptational fallacy leads to the legitimizing of inequitable sharing of resources and power based on sexual, racial, or class membership. Belief in a biological basis for social stratification is often called "Social Darwinism" although it can be traced back to millenia before Darwin's birth.

Can human behavior-genetic research help to settle these issues? It appears that its greatest contribution might be in the first category: behavioral plasticity, including the relationship between genotype and phenotype, the nature of behavioral phenotypes, and the importance of genotype–environment interactions. Even here, good data are hard to find. Humans are not good subjects for controlled studies to separate genetic from environmental influences on social behavior. Nevertheless, research on the proximate influences of genes on social behavior and organization should have relevance for sociobiology, even though the contributions are apt to be methodological and conceptual rather than critical tests of evolutionary theory.

Constraints, Universality, and Polypheny (Polyethism)

Even critics of sociobiology accept the idea that biology imposes constraints on human behavior (Gould, 1976). Just what these constraints are is seldom specified but most psychologists and biologists would agree that human behavior is not infinitely malleable. True, humans live and survive in wandering tribal groups, stable villages, and ever-changing metropolitan areas. To a large extent their ability to adapt to these diverse social environments, and to invade all continents and climatic zones, is based more on their capacity to modify and control their personal environment, than on changing their needs and motivations. Everywhere they must manage to find mates, rear children, find shelter and food, cooperate and compete with one another.

Constraints on human physical performance are obvious although the limits may never be defined precisely. Four-minute miles are now common though they were once regarded as unattainable. This may be due more to improved running tracks and an increased number of competitors than to improvement in capacity resulting from better nutrition and coaching. The nature of constraints on intellectual performance is also not clear. We cannot solve complex algebraic equations as rapidly as a properly programmed computer, but we believe that we have an advantage over computers in artistic and literary creativity. There is no reason to suppose that the biological basis of intelligent behavior has changed much during recorded history, but there is also no proof that our brains have stopped evolving. The extent of biological constraints on social behavior, if indeed there are any, is controversial. One can conceive of them. The substitution of homosexual for heterosexual erotic behavior reduces individual Darwinian fitness. If homosexuality were universally adopted by an entire population, that group would become extinct. Excluding such fantasies, limits on social behavior in the real world must be very flexible. The diversity of cultures in terms of characteristics such as social stratification, marriage rules, and openness of membership indicates that many kinds of organizational behavior are compatible with survival. Rather than defining constraints we might better look for uniformities within the diversities.

Is There a Universal Human Ethogram?   The concept of a universal human nature that shows itself in the social behavior and organization of diverse cultures is relevant to the sociobiology debate. Are there social characteristics that can be found in all healthy members of *Homo sapiens* at appropriate developmental stages? Does the historical and ethnological record demonstrate that these are truly universal? Can apparent exceptions be explained in a way that preserves the basic theorem of behavioral selection through natural selection? If there is such a biogram, of what does it consist?

It can be argued that any such species regularities are evidence of natural selection for a trait of adaptive value (Tiger & Fox, 1971). There are flaws in this deduction. A mildly maladaptive trait might be carried along as an associated cost of selection for an adaptive trait of greater significance for inclusive fitness. This hypothesis avoids the error of positing adaptivity to any existing social behavior on the grounds that it would have been selected against were it not adaptive.

A number of attempts have been made to characterize the human ethogram (or biogram). Wilson (1975) proposed that the human traits most likely to be conservative (in the sense that they were difficult to modify) would be those found in the majority of primates. Here he seems to be applying the idea of homology to behavior. His list included: aggressive dominance systems; males dominant over females; scaled intensity of social signals; prolonged maternal care; an extended juvenile period of socialization; and a matrilineal social organization. In conclusion Wilson remarks that ''conservative traits are not necessarily more genetic (i.e., more heritable) than labile ones.'' He seems to be confusing heritability (a measure of phenotypic variability of a trait in a specified population under given conditions) with phenolability (a measure of the potential modifiability of a trait in an individual of specified genotype).

Tiger and Fox (1971) affirm the existence of a human biogram with rules analogous to the grammar of a language. They hypothesize that children are able to learn readily only the behavioral rules consistent with this human biogrammar. These rules include: humans are political animals in the Aristotelian sense (i.e., they organize into groups for joint action); they compete for status within the group; sexual and mother–infant bonding occur; males tend to dominate females. At a more individual level Eibl-Eibesfeldt (1970) has sought to demonstrate universal human expression of emotion and intentionality at the level of facial expressions and body movements and postures. To these he adds an innate predisposition for ethical conduct. This seems to be the sociobiologist's reciprocal altruism.

The dispute between advocates of the existence of a universal human ethogram and those who assume that human nature is molded by the experience of growing up is sterile insofar as it increases our understanding of social behavior. One group emphasizes the similarities of cultures; the other points to differences between them. Although cultural groups often differ in allelic frequencies (Cavalli-Sforza & Bodmer, 1971) there is no evidence that any of these variants have anything to do with regulation of the kinds of behavior studied by sociobiologists, ethnologists, and behavior geneticists.

My own philosophical bent inclines me to favor the idea of a universal ethogram based on a large brain and a prolonged period of infancy. These brains have selected to learn most efficiently those behaviors that have proved to be useful in promoting fitness in the past. Natural selection has also seen to it that adaptive modes of responding feel good and are readily reinforced. Behavior genetics, as currently practiced, can do little to identify the total genetic basis of such a biogram. In a sense every locus may contribute to it. A more manageable question is whether individual variations in expressing this biogram are completely determined by the cultural milieu or have some genotypic basis (Fuller, 1978).

Behavioral Polypheny.   The concept of behavioral polypheny is the opposite of the idea of a universal biogram. Of course it is possible to take a middle position, but such a pose does not settle the nature–nurture issue. There is plenty of room for argument as to whether nature or nurture was most important in producing a specific deviation from the species norm. Physical and behavioral variability in *Homo sapiens* is great. Only a few domestic animals such as the dog come close to such phenotypic diversity, and this is primarily the result of artificial selection. In dogs such selection has led directly or indirectly to differences in social behavior and organization (Fuller, 1953; Pawlowski & Scott, 1956). Social polypheny in a population could result from either genetic or experiential heterogeneity. Unfortunately few human genetic investigations have been directed specifically toward the traits that we collectively call social behavior. Nevertheless, findings on the genetics of intelligence, personality, assortative mating, and homosexuality have some relevance to the explanation of variations.

Comparisons of the degree of similarity in monozygotic and dizygotic cotwins, and of parent–offspring and sibling similarities in adopting families provide evidence for a genetic contribution to polypheny. Comparisons between monozygotic co-twins reared apart or together yield clues to the plasticity of a given genotype. Such studies are reported in detail elsewhere in this book; here a few selected examples that seem particularly relevant to sociobiological theory are reviewed.

Genetic Analysis of Human Polypheny

Intelligence.   Almost every study of the heritability of general intelligence has obtained positive results. The main controversies remaining are: (1) the nature of the phenotype; do genes work by facilitating the development of a general information-processing ability, or do they act on a number of specific perceptual and learning skills; (2) the relative importance of genetic and environmental factors in producing racial and social class differences in IQ scores; (3) the extent to which general intelligence is correlated with individual and inclusive fitness. The latter two issues are particularly relevant to sociobiology, but reliable evidence is limited in scope. Most studies show a positive correlation between IQ test scores and both income and social class (Jencks, 1972). However, the range of intelligence scores is almost identical in every socioeconomic group.

Nevertheless, the luck of the genetic draw at meiosis may have something to do with the social mobility of a matured zygote. Waller (1971a) showed that sons with IQ scores higher than their fathers' tended to move upward on a socioeconomic scale; those with IQs lower than their fathers' tended to move down. Such movements, however, do not prove that IQ affects biological fitness in terms of gene survival. All readers must be aware of the early eugenical predictions that the average level of intelligence in some Western countries must fall because dull people were having more children than bright ones. The evidence is mixed. Waller (1971b) found, in a white, Minnesota population, some evidence that ''natural selection is forcing an increase in mean IQ scores.'' In contradiction Udry (1978) reported an inverse relation between IQ and fertility in a broad sampling of urban, white, American women. The basis of the difference was not the desired size of family but a less consistent adherence to contraceptive procedures among the lower IQ women. Thus in a present-day, urbanized society genes affecting general intelligence may act indirectly upon biological fitness by modifying the spontaneity of copulation.

Based on the available data it is a reasonable deduction that treating humans as domestic animals and selecting for high tested intelligence would be successful. Unless, of course, the techniques necessary for managing such a breeding project turned out to be incompatible with the conditions necessary for the development of intelligent behavior. At this time it is impossible to determine if natural selection is based at all on intelligence. In most human societies there are niches suitable to individuals with very different rankings on an IQ scale. Individuals at any IQ level who fit themselves into a suitable niche may be equally fit. In such a situation the mix of high and low performers on IQ tests would remain stable as long as the proportions of distinct niches remained constant. Major technological and social innovations could alter the balance and have genetic consequences. This hypothesis is speculative, but does not seem unreasonable.

Aggressive Behavior.   Research on individual differences in personality and interests shows that here, too, genes play an important role. In general, heritabilities run lower for personality than for cognitive measures and are less stable over the life span. (Dworkin, Burke, Maher, & Gottesman, 1976). The aspects of personality most relevant for sociobiology are those manifested in aggression, sexual behavior including choice of mates, and altruism. That familial factors have a major influence on the incidence of violent aggression is well documented, but emphasis is usually given to cultural rather than genetic transmission from parents to offspring.

Unfortunately, psychologists studying human aggression often formulate genetic issues in terms of instincts versus learning theory (E.g., Lefkowitz, Eron, Walder, & Huesman, 1977). But although these investigators rejected the concept of innate aggressiveness, they did report correlations between overt aggression and a number of scales on the Minnesota Multiphasic Personality Inventory. Several of these scales have been found to have significant heritabilities (Fuller & Thompson, 1978, p. 347). Lefkowitz et al. also noted that aggression is positively correlated with failure in scholastic achievement and lower tested intelligence. No one would suggest that there are genes for violent and delinquent behavior, but arguments against innateness of aggression do not contradict hypotheses that some individuals are more readily induced than others to behave violently.

Twin studies of criminal and delinquent behavior have shown higher concordance in MZ than in DZ twins (Fuller & Thompson, 1978, p. 412.) The adoption method has also provided data supporting some genetic influence on criminality (Crowe, 1975; Hutchings & Mednick, 1975). There are problems in interpreting such data beyond those found in most twin and adoption studies. The sample includes only individuals whose misdeeds are a matter of record. Bright criminals are less likely to be caught. Reasons for inclusion in a sample may be as diverse as passing bad checks, prostitution, and assault (Crowe, 1975). Many of the probands could be psychiatrically diagnosed as psychopathic—a somewhat vaguely defined category. Psychopaths are often violent, but their violence is not usually elicited in situations and in forms that appear to be adaptive.

The most publicized application of genetics to aggressive social behavior is the postulated increased violence of XYY males based on their overrepresentation in penal and mental institutions. Although the phenomenon appears to be real, the majority of XYY males vanish into the general population and are unrecognized phenotypically (Kessler, 1975). On the whole it appears that a person's genotype does play some role in setting the probability that he or she will be convicted of a felony and behave violently toward others. The developmental paths intervening between the genotype and the violent phenotype are unknown. Social factors are obviously very important determinants. On the whole the research on genetics and sociopathy has more relevance for the prospects of early detection of individuals at risk than it does for sociobiological theory.

Aggression: Function, and Evolution.   It may be possible to design more informative studies bringing together the longitudinal and sociopsychological techniques of Lefkowitz and his associates with a genetic design. The difficulties of such a study are great; it would be expensive and would raise sensitive ethical issues. Without waiting for the results of this study I present here some personal views on the function, evolution, and possible future of aggressive behavior in humans.

Recorded history chronicles a continuous story of individual and group conflicts that often flare into bloody confrontation. Although anthropologists have recorded considerable differences among cultures in their level of violence, the idea that naturally peaceful humans have been corrupted by their society, and that we need only change society to eliminate violence, seems to be wishful thinking. The socially sanctioned group violence we call *war* is an exception (Benedict, 1934), but its persistence indicates that up to the present there has been no strong biological or cultural selection against it. Aggressive behavior is often reinforced by improved access to resources, which in turn increases biological fitness and the frequency of any genes that promote success in combat. I see no reason that our species should be immune to the selective forces that are considered to account for the prevalence of aggressive behavior in animals. Human conflicts are not as ritualized as those of Clutton-Brock's stags, but their function is the same.

This view definitely does not assert that humans have "innate" aggressive drives that must be expressed either in conflict or sublimated by substitute activities such as games. But as long as aggression pays off in benefits it will tend to persist whether its occurrence is determined by the rules of the society or by gene frequency. Cultures do vary in their addiction to violence, and this demonstrates that cultural changes might lead to a more peaceful society. Many of us are justifiably worried about today's level of violence, but it may be that we are actually better off than our prehistoric ancestors. Livingstone (1976, cited in Symons, 1979, 1980) estimated that about 25% of prehistoric human males died violent deaths in intraspecific combat. If he is correct, most of the world's people are less likely than their ancestors to be murdered despite great advances in the technology of homicide. Are we less violent because in many parts of the world resources per person have increased greatly and the threshold for killing is more rarely reached? Or is the reduction due to cultural factors such as religious exhortation, the establishment of peace-keeping organizations (police, courts), and possibly by education on the biological unity of our species? Or finally, have we changed genetically to a new EBG? I suspect that biological evolution and cultural change are involved in a symbiotic, interactive relationship that is too close to permit a quantitative evaluation of each factor separately. Cultural changes can certainly be more rapid, but are less permanent. Genetic changes are slower, but more stable.

Assortative Mating and Sex Differences. Assortative mating for heritable social characteristics will result in increased phenotypic variance, regardless of whether transmission is by genes or by indoctrination. In Western societies there are high spouse correlations for age, education, tested intelligence, height, skin color, and many other characteristics (Eckland, 1968; Vandenberg, 1971–72). Although by itself assortative mating does not alter gene frequencies, it may change phenotypic frequencies and thus affect natural selection. The importance of assortative mating for producing and maintaining social stratification is controversial, as are also its genetic implications. It is unlikely to disappear even in a period of enhanced personal mobility. A full treatment of the subject would be speculative and would include more population than behavior genetics. Thus I turn to the more entertaining issues of male–female behavioral differences, particularly in mating strategies.

Symons (1979) has argued persuasively that differences in male and female attitudes and emotional responses associated with sexual behavior are products of natural selection and thus based on genetic differences. The multiple review of his book in *The Behavioral and Brain Sciences* (1980) provides a cross-section of opinions on the validity of the sociobiological approach to human behavior. One comment (Eysenck, 1980) is particularly relevant to the relation between sociobiology and behavior genetics. Eysenck writes: "Symons follows in the footsteps of Wilson who also seemed unaware of the work of biologically minded psychologists, and whose remarks concerning behavioral genetics evidence lack of knowledge . . . of the methods and designs of modern work in that area [p. 186]." I agree, but must add that there is little existing behavior-genetic research that is specifically applicable to testing sociobiological theory. A similar point has been made by Thiessen (1972, 1979). Nevertheless some general issues merit discussion.

Symons proposes the interesting hypothesis that the emotional and dispositional aspects of male and female sexuality are less plastic (more innate?) than actual behavior. On this he is close to Searle (1978), who asserts that *intention* is a necessary part of human behavior. For example, Symons hypothesizes that, even in a happy monogamous marriage, the husband usually desires sexual variety more than the wife. In my terms (Fuller, 1979), he believes that inferred psychophenes (e.g., the desire for sexual novelty) are more evolutionarily stable than ostensible psychophenes (e.g., the number of females with whom a male copulates). An ostensible psychophene is a behavior that can be observed directly by any competent observer; an inferred psychophene is deduced from correlations among a number of behaviors that seem to have a common internal cause. In a general way, inferred psychophenes correspond to the traits of personality psychologists. Symon's idea runs counter to the tradition in genetics of studying phenotypes that are, potentially at least, directly observable by an investigator. Fantasies and desires depend on introspection and verbal report of unknown veracity, scarcely the kind of data desired by tough-minded scientists. If Symons is right, there are in sexual, and possibly in other social relationships, genetically based male–female differences in the emotional components of behavior.

Given the capacity of humans to learn new forms of behavior, and to be influenced by multiple inputs and prospective benefits, these innate differences can be overridden at some internalized cost. I am not sure that behavior genetics as now practiced can help much to test this provocative hypothesis of Symons, which has implications beyond the area of sexuality. The practical and ethical problems of obtaining reliable data on internalized states such as emotions and desires are great. The relationships between genes and such complex phenotypes as jealousy, lust, and romantic love are unknown. Presumably individuals of both sexes differ along these dimensions, and there may be considerable overlap as Symons suggests. Because it is fertile sexual unions that result in reproduction, it is external behavior rather than internal dispositions that determine gene transmission and evolution. But as every reader must admit to himself, overt behavior is only a partial guide to human nature. I find myself in agreement with Dewsbury (1980), who asserts that compared with animal studies human sociobiology has much less rigor and capacity for self-correction: "If the task of nonhuman sociobiology is difficult, that of human sociobiology may at present approach the impossible [p. 183]." However, I do not advocate discontinuance of human studies. "Impossible" problems have been solved. And even if sociobiological ideas cannot be rigorously proved or falsified, they help to meet the particularly human need to explain one's self as well as the external world.

Natural Selection in Humans: Individual or Group.   Sociobiological theory places great emphasis on individual selection. It rejects the idea that cooperative and altruistic behavior, which may lower the fitness of the altruist, evolved because they favored the survival of the altruist's group. The flaw in this hypothesis is that a cheater who accepts the benefits of other's altruism without its risks does better than the altruist. The dilemma was solved by the explication of kin selection and reciprocal altruism. Group selection is not theoretically impossible, but the conditions under which it occurs are relatively rare (Maynard Smith, 1976b). Groups must be small, must compete among each other at the level of demes, and must accept few immigrants. Groups as a whole must become extinct or decrease sharply in size.

Not all biologists agree on the overweening importance of individual selection for evolution. Gould (1980) decries emphasis on the natural selection of adaptive structures and behaviors as the only agent for evolutionary change. Genetic drift and fixation of neutral mutations are also involved. Though possible, these processes are characterized by a randomness that has no power to explain the diversity of behavioral adaptation in every phyla. They may interfere with reaching the ultimate goal of perfect phenotypic adaptation, but no knowledgeable person believes that evolution must lead to a perfect fit between organism and environment. A successful species, population, or individual is one that is more fit than any existing competitor.

A number of demographic investigations in small nonwesternized populations support the idea that achieved fertility is related to individual status. Irons (1980) concluded that the socioeconomic structure of the Yomut (an Iranian tribe) operated according to the theory of individual rather than group selection theory. He believes that quantitative methods, rare in social anthropology, permit valid tests of human sociobiological theory. Chagnon (1980) came to similar conclusions on the importance of kin selection in Amazonian Yanomamo Amerindians. Durham (1976) argued that primitive war, widespread in many parts of the world, allowed individuals to maximize their survival and reproduction by living in cooperative social groups and participating in collective aggression when access to scarce resources was at stake. In his view, warfare is not in the genes but in the culture, and it arose in the culture because it enhanced the ability of individuals who joined in group aggression to survive and reproduce in their environment. Cultural and inclusive fitness are seen as truly complementary and interactive. Livingstone (1980) makes the same point.

There are still supporters of group selection. Eibl-Eibesfeldt (1979) argues for it because: "nobody has yet been able to tell me that proportion of my genes I actually share with my brother in comparison with the proportion I share with any other group member [p. 50]." Williams (1980), in a paper supporting group selection in humans, places great emphasis on zygote survival rates, and argues that the small size of human families favors group over individual selection. But the zygotic fecundity of a mated pair is irrelevant. In a stable population of any species a pair averages two viable offspring, whether the number of fertilized ova is two or two million. Differential rather than absolute survival rates are the engines of evolution. Many group selectionist arguments are simply faulty population genetics.

An emphasis on individual selection does not imply that group structure and, in humans at least, group culture are not potent forces affecting the individual fitness of its members. It certainly does not deny that cultural patterns, whether or not they have specific genetic determinants, play a major role in competition between groups that may lead to coalescence, subjugation of one, or even to extinction of one. Group extinction, or near extinction, could alter gene frequencies in a species and qualify as group selection. Whether the genetic changes induced by such extinction would be related functionally to associated changes in social behavior and organization is uncertain. A likely answer is "No," but proving a null hypothesis is logically impossible and once extinction occurs there is no way of testing any hypothesis.

It seems likely that the human biogram, whatever it may be, has been shaped primarily by individual and kin selection, with possibly a small contribution of group selection. The mode of selection argument is, however, irrelevant to the two mainstreams of human behavior genetics: investigation of the sources of individual variation, and utilization of genetic differences to analyse behavioral and psychological development. Behavioral traits are too plastic to serve as markers for the generation-spanning demographic and genetic studies that could provide information on the detailed mechanisms of human evolution.

Zoomorphism and Anthropomorphism. The appeal to animal sociobiologists of colorful similes is well known. Thus one reads of adultery in mountain bluebirds (Barash, 1976), and of fratricide, infanticide, and suicide in birds (O'Connor, 1978). The use of such terms calls attention to similarities in animal and human social relationships, but it may lead to the false conclusion that they are similarly caused and serve the same functions. Taking an extreme negative view, Caplan (1980) wonders if it is possible to construct generalizations about behavior that apply to many species. He also worries about failures to distinguish between analogy and homology. I personally believe strongly enough in the unity of the biosphere to accept the idea that there are universal principles bridging taxonomic gaps although it may be difficult to distinguish between superficial similarities and fundamental processes. As for distinguishing between analogy and homology, it is difficult enough when applied to structures and does not appear to be very useful when applied to behavior. Actually, analogies in social behavior of distantly related taxa are more powerful evidence for selectionist theory than homologies would be.

It would be simple enough for sociobiologists to avoid charges of anthropomorphism by using less vivid language, although this might diminish the readership of their articles. The charge of zoomorphism in the interpretation of human behavior has deeper implications. The idea that human nature and behavior have been shaped by the impersonal forces of natural selection is anathema to many. Particularly disturbing is Wilson's (1975) suggestion that ethical principles are shaped in the same fashion. I return to this point later. A great deal of information gained from animals has, however, been transferred to human physiology at the cellular and organismal levels. Much of human learning follows rules that apply also to rats and monkeys, even though these species do not have languages and give no evidence of concern for trying to explain the causes of the phenomena with which they must cope in order to survive. It is obvious that the human nervous system evolved along a basic vertebrate plan. The question is: Are the human nervous system's functions so much more complex than those of other animals that a great gulf has arisen in the nature of further natural selection? Until *Homo sapiens* appeared on the scene selection acted upon the "hardware" of the nervous system. Finally the point was reached at which further selection was based on the "software" available to an individual.

Ethical and Social Implications.   Humans are aware of their motives and of the possible consequences of their actions; animals do not foresee the future—at least they cannot express their thoughts about it. If true, this statement implies that humans can to a degree control their own inclusive fitness. Even though animals may possess some self-awareness (Gallup, 1979; Griffin, 1976, 1978) the gap is so large that new principles must be educed to explain human evolution and cultural change. This apparently is the position of those who accept sociobiological theorizing for animals but reject it for humans.

As an example Gould's (1976) attack on contemporary sociobiology is based primarily on the allegation that it tends to justify social practices that are unethical. Violence, sexism and gross differences in economic staus are examples. Gould claims that the central theorems of sociobiology lead us to be tolerant of these ills because they are presented as products of our evolutionary history that were imbedded in our genotypes because they were adaptive. In contrast with such genetic determinism he asserts that flexibility of response is the major distinguishing characteristic of human behavior. Selection has favored the ability of humans to learn from their environment and to transmit their acquired knowledge nongenetically. The result is the near absence of fixed action patterns, particularly at the social level. Violence and caring for others are both biological only in the sense that humans inherit a brain that has the capacity to learn them. Gould uses this point to rebut the sociobiological emphasis on the maximization of self interest (by perpetuation of one's genes) as the force behind the evolution of sociality. However, it is only fair to note that much of sociobiology is concerned with the origins and maintenance of cooperative behavior. Whether one views helping others as true altruism or as enlightened selfishness is largely a matter of philosophical outlook. I see no way to disprove either hypothesis empirically. Gould's statement that there is no good evidence for genetic or environmental contributions to behavioral individuality implies either a lack of acquaintance with recent work in behavior genetics and developmental psychology, or an indictment of their practitioners. He would have been on stronger ground if he had asserted that their findings are mainly irrelevant to the social issues with which he is concerned.

This discussion of human sociobiology concludes with personal comments on its ethical and sociopolitical connotations. Some of the issues that have come to public attention are similar to those that have been raised about behavior genetics. One of these is race and class differences. Claims by behavior geneticists that racial differences in IQ are genetically determined have aroused much controversy (Loehlin, Lindzey, & Spuhler, 1975; Richardson & Spears, 1972). Sociobiology has tended to downgrade racial differences and to stress the universals of human nature that have been shaped by similar selective forces everywhere. In this sense it should not be called racist.

Sociobiology does not predict a classless society. In fact it places the motive power for human evolution in a competition for status that enhances reproductive potential. To the extent that success in competition has a genetic basis, genetic differences among social strata should arise. I must point out that the competition is not necessarily violent, and that maximizing inclusive fitness involves altruistic and cooperative as well as agonistic actions. Genetic studies have regularly shown that social classes differ in measured IQ and that the IQ is heritable. Whether such class differences have any evolutionary significance depends on whether there are class differences in reproductive rate. If not, and if there is interclass mobility, gene frequencies should not change. In fact, a pluralistic society with many available roles should promote genetic along with cultural diversity. Whether such diversity can be maintained without producing a hierarchical structure is a sociological rather than a genetic question.

The criticism that behavior genetics and sociobiology are rigidly deterministic is, in my opinion, based on unfamiliarity with these disciplines, or on a fear that, calling a behavioral trait "genetic" will lead people to conclude that it is un-modifiable. Whatever the basis of the deterministic charge, behavior geneticists and sociobiologists must continue to emphasize the coaction and interaction of genes and environments in producing a phenotype.

Behavioral sex differences are another matter. They have not been greatly emphasized in behavior genetics, except for spatial perception, where recent work has not supported the theory that an X-linked gene affects this ability (Guttmann & Shoham, 1979; Rose, Miller, Dumant-Driscoll, & Evans, 1979). Sex (or gender) differences play a greater role in sociobiology where male and female parental investment are considered to have been a major influence in producing male political dominance, and female assignment to domesticity and service roles. It is obvious in 20th-century Western cultures that women can perform comparably to men in every sphere except those requiring great physical strength. Yet there are still striking differences in male and female roles. Is this due to cultural bias or to biological differences? Or to some of each? Hormones do have effects on the brain, and genes could influence production of and sensitivity to them. Purifoy and Koopman (1979) found that women occupying managerial and technical positions (traditionally masculine) had higher con-centrations of serum androgens than did housewives and clerical workers (tradi-tionally feminine). It would be interesting to investigate the heritability of the hormonal differences. They could be effects rather than causes of a woman's choice of occupation.

A frequent criticism of sociobiology is that by explaining gender differences in behavior by natural selection it supports inequitable distribution of power and wealth based on sex. This claim may be valid in a political sense. In a society where sociobiological concepts are disseminated in the popular press, adherents of traditional sex-role assignment in the home and workplace may invoke science to support their position. However, the subjugation of women is greatest in those countries and subcultures where sociobiology and evolutionary theories are ei-ther unknown or considered sacrilegious. It seems unfair for critics to blame a field of science for social inequities that predate its birth by millennia. It does devolve on sociobiologists to make clear that a biological approach to social behavior does not imply genetic determinism and intractable human nature, or the total dominance of self-interest over the welfare of one's species. The con-cepts of inclusive fitness and reciprocal altruism involve responsibilities at least to one's kin and neighbors. The definition of "neighbor" changes as we become more mobile and encounter foreign cultures.

To a degree human sociobiology is an intellectual game played by academics hoping to enhance their personal fitness. Unfortunately the pace of evolution is slow, and behavior leaves no fossils. Thus it is difficult to judge the winner in a conflict between rival hypotheses. For example, consider the selective process that resulted in the shorter life-span of men as compared with women. Our species is unique in containing large numbers of postreproductive females. However, men retain the capability of producing viable gametes longer. Thus, the benefits of longevity in terms of inclusive fitness should be greater for men than for women, and selection should favor male longevity over female. Clearly this has not occurred. Men have approximately 11% fewer years to live than do their consorts. A possible sociobiological explanation is that women were selected to live longer because their participation in the rearing of grandchildren, nephews, and nieces increased their inclusive fitness. Old men were a liability because they were poor warriors, would not do housework, and their demise provided more resources for the children of the children whom they sired in their prime. There are other possibilities. The shortened life-span of males may be an unavoidable physiological cost of selection for the hormonal stimulation that converts a basically female phenotype to a masculine one. It could be attributed to having only one X chromosome whose deleterious recessive genes will be regularly expressed. Finally, gender-based differences in life-span can be attributed to variation in life stress imposed by a culture. These alternative hypotheses are not mutually exclusive, and neither support nor disprove a sociobiological explanation.

Because experimentation with humans is difficult, and to a great extent proscribed by ethical principles, sociobiological hypotheses are difficult to falsify. However, I am not convinced that in respect to falsifiability, sociobiology is any worse off than sociohistorical reconstructions of human prehistory that neglect biology (see Ruse, 1979, for a discussion). One may criticize and even ridicule some of its propositions, but sociobiology does provide a unified theory for explaining many facets of "human nature." According to its tenets we are the product of a long history of natural selection based on environmental pressures that induced a biopsychological fit to the characteristics of those environments. I prefer this view of my species to the statement in Genesis (1:27–28): "So God created man in his own image; in the image of God he created him." If we are godlike we are closer to the sensuous Greek and Roman deities than to the Judeo–Christian creator.

Religious leaders and moral philosophers have been the major sources of ethical ideas, but they do not speak with a single voice nor for all time. Technological changes (progress?) regularly produce new ethical problems. Few beneficient innovations are free of costs and many of these fall unequally upon various elements of society. With respect to such problems sociobiology asserts that we have, like other species, been selected to behave in a manner to improve our inclusive fitness. Ethical and social demands that interfere with this principle will be difficult to enforce. Such a general statement is not very helpful in dealing with such issues as: overpopulation, deterrence of crime, economic and social equity, resolution of group conflict, genetic engineering, the "right to die."

An evolutionary approach to the ultimate causation of complex social problems may be helpful, but proximate causes must be considered if these problems are to be dealt with effectively. Developmental and social psychology can benefit by inputs from behavior genetics and sociobiology. Ethical and social issues cannot become simply applied behavioral science, for this implies the technocratic horrors of Aldous Huxley's *Brave New World*. But without these inputs there is little hope for progress. As a first step it would be helpful to bring sociobiology and behavior genetics closer together.

REFERENCES

Barash, D. P. The male response to apparent female adultery in the mountain bluebird, *Sialia currucoides:* An evolutionary interpretation. *American Naturalist,* 1976, *110,* 1097–1101.

Barash, D. P. *Sociobiology and behavior.* New York: Elsevier, 1977.

Barlow, G. W., & Silverberg, J. (Eds.). *Sociobiology: Beyond nature/nurture?* Boulder, Colo.: Westview, 1980.

Benedict, R. F. *Patterns of culture.* New York: Houghton–Mifflin, 1934.

Cairns, R. B. Fighting and punishment from a developmental perspective. In J. K. Cole & D. D. Jensen (Eds.) *Nebraska symposium on motivation.* Lincoln, Neb.: University of Nebraska Press, 1972.

Caplan, A. L. A critical examination of sociobiological theory: Adequacy and implications. In G. W. Barlow & J. Silverberg (Eds.), *Sociobiology: Beyond nature/nurture?* Boulder, Colo.: Westview, 1980.

Cavalli-Sforza, L. L., & Bodmer, W. F. *The genetics of human populations.* San Francisco: Freeman, 1971.

Chagnon, N. A. Kin-selection theory, kinship, marriage and fitness among the Yạnomanö Indians. In G. W. Barlow & J. Silverberg (Eds.), *Sociobiology: Beyond nature/nurture?.* Boulder, Colo.: Westview, 1980.

Childress, D., & McDonald, I. C. Tests for frequency-dependent mating success in the housefly. *Behavior Genetics,* 1973, *3,* 217–223.

Clutton-Brock, T. H., Albon, S. D., Gibson, R. M., & Guiness, T. E. The logical stag: Adaptive aspects of fighting in red deer (*Cervus elaphus, L.*). *Animal Behaviour,* 1979, *27,* 211–225.

Clutton-Brock, T. H., & Harvey, P. H. *Readings in sociobiology.* San Francisco: Freeman, 1978.

Craig, J. V., Biswas, D. K., & Guhl, A. M. Agonistic behavior influenced by strangeness, crowding and heredity in female domestic fowl (*Gallus gallus*). *Animal Behaviour,* 1969, *17,* 498–506.

Craig, J. V., Ortman, L. L. & Guhl, A. M. Genetic selection for social dominance ability in chickens. *Animal Behaviour.* 1965, *13,* 114–131.

Crook, J. H. On the integration of gender strategies in mammalian social systems. In J. S. Rosenblatt & B. R. Komisaruk (Eds.), *Reproductive behavior and evolution.* New York: Plenum, 1977.

Crowe, R. R. An adoptive study of psychopathy: Preliminary results from arrest records and psychiatric hospital records. In R. R. Fieve, D. Rosenthal, & H. Brill (Eds.), *Genetic research in psychiatry.* Baltimore: Johns Hopkins, 1975.

Darwin, C. *The origin of species.* (Modern Readers Series) New York: MacMillan, 1927. (Originally published, 1859.)

Dawkins, R. *The selfish gene.* Oxford: Oxford University, 1976.

Dawkins, R. Good strategy or evolutionary stable strategy? In G. W. Barlow & J. Silverberg (Eds.), *Sociobiology: Beyond nature/nurture?* Boulder, Colo.: Westview, 1980.

DeFries, J. C. Genetics of animal and human behavior. In G. W. Barlow & J. Silverberg (Eds.), *Sociobiology: Beyond nature/nurture?* Boulder, Colo.: Westview, 1980.

DeFries, J. C., & McClearn, G. E. Social dominance and Darwinian fitness in the laboratory mouse. *American Naturalist*, 1970, *104*, 408–411.

Denenberg, V. H. Interactional effects in early experience research. In A. Oliverio (Ed.), *Genetics, environment and intelligence*. Amsterdam: North Holland. 1977.

Denenberg, V. H., Hudgens, G. A., & Zarrow, M. X. Mice reared with rats: Modification of behavior by early experience with another species. *Science*, 1964, *143*, 380–381.

Dewsbury, D. A. Methods in the two sociobiologies. *Behavioral and Brain Sciences*, 1980, *3*, 183–184.

Dobzhansky, T. *Mankind evolving*. New Haven, Conn.: Yale University, 1962.

Durham, W. H. Resource competition and human aggression: Part I. A review of primitive war. *Quarterly Review of Biology*, 1976, *51*, 385–415.

Dworkin, R. H., Burke, B. W., Maher, B. A., & Gottesman, I. I. A longitudinal study of the genetics of personality. *Journal of Personality and Social Psychology*, 1976, *34*, 510–518.

Ebert, P. D., & Hyde, J. S. Selection for agonistic behavior in wild female *Mus musculus*. *Behavior Genetics*, 1976, *6*, 291–304.

Eckland, B. K. Theories of mate selection. *Eugenics Quarterly*, 1968, *15*, 71–84.

Ehrman, L. Direct observation of sexual isolation between allopatric and sympatric strains of the different *Drosophila paulistorum* races. *Evolution*, 1965, *19*, 459–464.

Ehrman, L., & Petit, C. Genotype frequency and mating success in *willistoni* species group of *Drosophila*. *Evolution*, 1968, *22*, 649–658.

Eibl-Eibesfeldt, I. *Ethology: The biology of behavior*. New York: Holt, Rinehart, & Winston, 1970.

Eibl-Eibesfeldt, I. Human ethology: Concepts and implications for the sciences of man. *Behavioral and Brain Sciences*, 1979, *2*, 1–26; 50–57.

Eysenck, H. J. Sociobiology, standing on one leg. *Behavioral and Brain Sciences*, 1980, *3*, 186.

Fuller, J. L. Cross-sectional and longitudinal studies of adjustive behavior in dogs. *Annals of the New York Academy of Sciences*, 1953, *56*, 214–224.

Fuller, J. L. Genes, brains and behavior. In M. S. Gregory, A. Silver, & D. Sutch (Eds.), *Sociobiology and human nature*. San Francisco: Jossey–Bass. 1978.

Fuller, J. L. The taxonomy of psychophenes. In J. R. Royce & L. P. Mos (Eds.), *Theoretical advances in behavior genetics*. Ålphen aan den Rijn: Sijthoff and Noordhoff, 1979.

Fuller, J. L., Easler, C., & Smith, M. E. Inheritance of audiogenic seizure susceptibility in the mouse. *Genetics*, 1950, *35*, 622–632.

Fuller, J. L., & Hahn, M. E. Issues in the genetics of social behavior. *Behavior Genetics*, 1976, *6*, 391–406.

Fuller, J. L., & Thompson, W. R. *Foundations of behavior genetics*. St. Louis, Mo.: Mosby, 1978.

Gallup, G. G., Jr. Self-awareness in primates. *American Scientist*, 1979, *67*, 417–421.

Gould, S. J. Biological potential vs. biological determinism. *Natural History*, 1976, *85*(5), 1–22.

Gould, S. J. Sociobiology and the theory of natural selection. In G. W. Barlow & J. Silverberg (Eds.), *Sociobiology: Beyond nature/nurture?* Boulder, Colo.: Westview, 1980.

Gregory, M. S., Silvers, A., & Sutch, D. *Sociobiology and human nature*. San Francisco: Jossey–Bass, 1978.

Griffin, D. R. *The question of animal awareness*. New York: Rockefeller University Press, 1976.

Griffin, D. R. Humanistic aspects of ethology. In M. Gregory, A. Silvers, & D. Sutch (Eds.), *Sociobiology and human nature*. San Francisco: Jossey–Bass, 1978.

Guttman, R., & Shoham, I. Intrafamily invariance and parent–offspring resemblance in spatial abilities. *Behavior Genetics*, 1979, *9*, 367–378.

Hamilton, W. D. The evolution of altruistic behavior. *American Naturalist*, 1963, *97*, 354–356.

Hamilton, W. D. The genetical theory of social behaviour. I and II. *Journal of Theoretical Biology*, 1964, *7*, 1–52.

Horn, J. M. Aggression as a component of relative fitness in four inbred strains of mice. *Behavior Genetics*, 1974, *4*, 373–382.

Hutchings, B., & Mednick, S. A. Registered criminality in the adoptive and biological parents of registered male criminal adoptees. In R. R. Fieve, D. Rosenthal, & H. Brill (Eds.), *Genetic research in psychiatry.* Baltimore: Johns Hopkins University Press, 1975.

Irons, W. Is Yomut social behavior adaptive? In G. W. Barlow & J. Silverberg (Eds.), *Sociobiology: Beyond nature/nurture?* Boulder, Colo.: Westview, 1980.

Jencks, C. *Inequality: A reassessment of the effect of family and schooling in America.* New York: Basic Books, 1972.

Kessler, S. Extra chromosomes and criminality. In R. R. Fieve, D. Rosenthal, & H. Brill (Eds.), *Genetic research in psychiatry.* Baltimore: Johns Hopkins University Press, 1975.

Koenig, W. D., & Pitelka, F. A. Relatedness and inbreeding avoidance: Counterploys in the communally nesting acorn woodpecker. *Science,* 1979, *206,* 1103–1105.

Kuse, A. R., & DeFries, J. C. Social dominance and Darwinian fitness: An alternative test. *Behavioral Biology,* 1976, *16,* 113–116.

Lagerspetz, K. Genetic and social causes of aggressive behaviour in mice. *Scandinavian Journal of Psychology,* 1961, *2,* 167–173.

Lagerspetz, K. M. J., & Lagerspetz, K. Y. H. Genetic determination of aggressive behaviour. In J. H. F. vanAbeelen (Ed.), *The genetics of behaviour.* Amsterdam: North Holland, 1974.

Lefkowitz, M. M., Eron, L. D., Walder, L. O., & Huesman, L. R. *Growing up to be violent.* New York: Pergamon Press, 1977.

Lerner, I. M. *The genetic basis of selection.* New York: Wiley, 1958.

Levene, H. Genetic equilibrium when more than one ecological niche is available. *American Naturalist,* 1953, *87,* 311–313.

Levine, L., Barsel, G. E., & Diakow, A. C. Mating behavior in two inbred strains of mice. *Animal Behaviour,* 1966, *14,* 1–6.

Livingstone, F. B. Cultural causes of genetic change. In G. W. Barlow & J. Silverberg (Eds.), *Sociobiology: Beyond nature/nurture?* Boulder, Colo.: Westview, 1980.

Loehlin, J. C., Lindzey, G., & Spuhler, J. N. *Race differences in intelligence.* San Francisco: Freeman, 1975.

Maynard Smith, J. The theory of games and the evolution of animal conflict. *Journal of Theoretical Biology,* 1974, *47,* 209–221.

Maynard Smith, J. Evolution and the theory of games. *American Scientist,* 1976, *64,* 41–45. (a)

Maynard Smith, J. Group selection. *Quarterly Review of Biology,* 1976, *51,* 277–283. (b)

Maynard Smith, J. Parental investment: A prospective analysis. *Animal Behaviour,* 1977, *25,* 1–9.

Maynard Smith, J., & Parker, G. A. The logic of asymmetric contests. *Animal Behaviour,* 1976, *24,* 159–175.

Maynard Smith, J. & Price, G. R. The logic of animal conflict. *Nature, London,* 1973, *246,* 15–18.

Mayr, E. *Populations, species and evolution.* Cambridge, Mass: Harvard University Press, 1970.

Moss, R., & Watson, A. Inherent changes in the aggressive behaviour of a fluctuating red grouse (*Lagopus lagopus scoticus,* Lath.) population. *Ardea,* 1980, *68,* 113–120.

O'Connor, R. J. Brood reduction in birds: Selection for fratricide, infanticide, and suicide? *Animal Behaviour,* 1978, *26,* 79–96.

Orians, G. H. On the evolution of mating systems in birds and mammals. *American Naturalist,* 1969, *103,* 589–602.

Parker, G. A. Assessment strategy and the evolution of fighting behaviour. *Journal of Theoretical Biology,* 1974, *47,* 223–243.

Parker, G. A. Searching for mates. In J. R. Krebs & N. B. Davies (Eds.), *Behavioural ecology: An evolutionary approach.* Sunderland, Mass.: Sinauer, 1978.

Parsons, P. A. Male mating speed as a component of fitness in *Drosophila. Behavior Genetics,* 1974, *4,* 395–404.

Pawlowski, A. A., & Scott, J. P. Hereditary differences in the development of dominance in litters of puppies. *Journal of Comparative and Physiological Psychology,* 1956, *49,* 353–358.

Petit, C., & Ehrman, L. Sexual selection in *Drosophila*. *Evolutionary Biology*, 1969, *3*, 177–223.

Powell, J. R., & Taylor, C. E. Genetic variation in ecologically diverse environments. *American Scientist*, 1979, *67*, 590–596.

Purifoy, F. E., & Koopman, L. H. Androstenedione, testosterone and free testosterone concentration in women of varying occupations. *Social Biology*, 1979, *26*, 178–188.

Pusey, A. E. Inbreeding avoidance in chimpanzees. *Animal Behaviour*, 1980, *28*, 543–552.

Pyle, D. W. Correlated responses to selection for a behavioral trait in *Drosophila melanogaster*. *Behavior Genetics*, 1978, *8*, 333–340.

Ralls, K., Brugger, K., & Ballou, J. Inbreeding and juvenile mortality in small populations of ungulates. *Science*, 1979, *206*, *1101–1103*.

Richardson, K., & Spears, D. *Race and intelligence*. Baltimore: Penguin Books, 1972.

Rose, R. J., Miller, J. Z., Dumant-Driscoll, M., & Evans, M. M. Twin-family studies of perceptual speed ability. *Behavior Genetics*, 1979, *9*, 71–86.

Roughgarden, J. *Theory of population genetics and evolutionary ecology: An introduction*. New York: Macmillan, 1979.

Ruse, M. *Sociobiology: Sense or nonsense?* Boston: D. Reidel, 1979.

Schneirla, T. C. Aspects of stimulation and organization in approach/withdrawal processes underlying vertebrate behavioral development. *Advances in the Study of Behavior (Vol. 1)*. New York: Academic Press, 1965.

Scott, J. P. Agonistic behavior of mice and rats: A review. *American Zoologist*, 1966, *6*, 683–701.

Searle, J. R. Sociobiology and the explanation of behavior. In M. S. Gregory, A. Silvers, & D. Sutch (Eds.), *Sociobiology and human nature*. San Francisco: Jossey–Bass, 1978.

Seyfried, T. N., Yu, R. K., & Glaser, G. H. Genetic analysis of audiogenic seizure susceptibility in C57BL/6J × DBA/2J recombinant inbred strains of mice. *Genetics*, 1980, *94*, 701–718.

Simon, N. G. The genetics of intermale aggressive behavior in mice: Recent research and alternative strategies. *Neuroscience and Biobehavioral Reviews*, 1979, *3*, 97–106.

Spiess, E. B., & Schwer, W. A. Minority mating advantage of certain eye-color mutants of *Drosophila melanogaster*. I. Multiple-choice and single female tests. *Behavior Genetics*, 1978, *8*, 155–168.

Spieth, H. T. Mating behavior within the genus *Drosophila* (*Diptera*). *Bulletin of the American Museum of Natural History*, 1952, *99*, 395–474.

Stacey, P. B., & Bock, C. E. Social plasticity in the acorn woodpecker. *Science*, 1978, *202*, 1298–1300.

Symons, D. *The evolution of human sexuality*. New York: Oxford University, 1979.

Symons, D. The evolution of human sexuality: Precis. *The Behavioral and Brain Sciences*, 1980, *3*, 171–181.

Tardif, G. N., & Murnik, M. R. Frequency-dependent sexual selection among wild-type strains of *Drosophila melanogaster*. *Behavior Genetics*, 1975, *5*, 373–379.

Taylor, C. E. Differences in mating propensities: Some models for examining the genetic consequences. *Behavior Genetics*, 1975, *5*, 381–393.

Thiessen, D. D. A move towards species-specific analysis in behavior genetics. *Behavior Genetics*, 1972, *2*, 115–126.

Thiessen, D. D. Biological trends in behavior genetics. In J. R. Royce & L. P. Mos (Eds.), *Theoretical advances in behavior genetics*. Ålphen aan den Rijn: Sijthoff and Noordhoff, 1979.

Tiger, L., & Fox, R. *The imperial animal*. New York: Holt, Rinehart, & Winston. 1971.

Trivers, R. L. The evolution of reciprocal altruism. *Quarterly Review of Biology*, 1971, *46*, 35–57.

Trivers, R. L. Parental investment and sexual selection. In B. Campbell (Ed.), *Sexual selection and the descent of man*. Chicago: Aldine, 1972.

Turner, M. B. *Philosophy and the science of behavior*. New York: Appleton–Century–Crofts, 1967.

Udry, J. R. Differential fertility by intelligence: The role of birth planning. *Social Biology*, 1978, *25*, 10–14.

Vandenberg, S. G. Assortative mating or who marries whom? *Behavior Genetics*, 1971–72, *2*, 127–157.

Wade, N. Sociobiology: Troubled birth for a new discipline. *Science*, 1976, *191*, 1151–1155.

Waller, J. H. Achievement and social mobility: Relations among IQ score, education and occupation in two generations. *Social Biology*, 1971, *18*, 252–259. (a)

Waller, J. H. Differential reproduction: Its relation to IQ test score, education, and occupation. *Social Biology*, 1971, *18*, 122–136. (b)

Washburn, S. L. Animal behavior and social anthropology. In M. S. Gregory, A. Silvers, & D. Sutch (Eds.), *Sociobiology and human nature*. San Francisco: Jossey–Bass, 1978.

Williams, B. J. Kin selection, fitness and cultural evolution. In G. W. Barlow & J. Silverberg (Eds.), *Sociobiology: Beyond nature/nurture?* Boulder, Colo.: Westview, 1980.

Williams, G. C. *Sex and evolution*. Princeton, N.J.: Princeton University Press. 1975.

Wilson, E. O. *Sociobiology: The new synthesis*. Cambridge, Mass.: Harvard University Press, 1975.

Wilson, E. O. A consideration of the genetic foundations of human behavior. In G. W. Barlow & J. Silverberg (Eds.), *Sociobiology: Beyond nature/nurture?* Boulder, Colo.: Westview, 1980.

Wright, S. The results of crosses between inbred strains of guinea pigs differing in number of digits. *Genetics*, 1934, *19*, 537–551.

Wynne-Edwards, V. C. Animal dispersion in relation to social behavior. New York: Hafner, 1972.

Yanai, J., & McClearn, G. E. Assortative mating in mice. I. Female mating preference. *Behavior Genetics*, 1972, *2*, 173–183. II. Strain differences in female mating preference, male preference, and the question of possible sexual selection. *Behavior Genetics*, 1973, *3*, 65–74. (a) III. Genetic determination of female mating preference. *Behavior Genetics*, 1973, *3*, 75–84. (b)

Author Index

A

Abe, K., 53, 56, *81*
van Abeelen, J. H. F., 101, *113*
Ad Hoc Committee on Genetic Counseling, 182, *186*
Ahern, F. M., 253, *330*
Akhimova, M. N., 67, 69, *81*
Albersheim, P., 124, *153*
Albon, S. D., 451, *473*
Alepa, F. P., 52, *82*
Alexander, J., 383, *403*
Allan, A. M., 129, 135, 136, *151*
Allee, W. C., 410, 413, *430*
Allen, M. G., 200, *214*
Amacher, P. L., 274, *332*
Ambrus, C. M., 19, 30
Ambrus, J., 19, *30*
American Psychiatric Association, 191, *213*
Ames, B. N., 231, *331*
Amit, Z., 121, 124, *152*
Anandam, N., 127, *151*
Anastasi, A., 301, *326*
Anderson, S. M., 128, 130, *150, 152*
Anderson, V. E., 193, 196, *214*, 218, 224, 230, 234, 251, 252, 255, *326, 333*
Andrews, P. W., 355, *360, 362*

Aranow, L., 117, 118, *150*
Archer, J., 375, *402*
Aronson, L. R., 424, *429*
Asberg, M., 120, *148*
Ashton, G. C., 306, *327*
Averhoff, W. W., 426, *429*
Axelrod, J., 416, *429*
Axelrod, R., 323, 324, *326*
Ayala, F. J., 28, *30*, 225, *331*

B

Bagwell, M., 112, *115*
Bailey, D. W., 101, 108, 109, *113, 114, 115*, 126, 140, *149*, 413, 428, *429, 430*
Bajema, C. J., 245, *326*
Baker, H., 142, *148*
Baker, R., 139, *149*
Baker, S. W., 312, *326*
Bakker, T. C. M., 412, 415, *434*
Baldwin, B. A., 398, 399, *403*
Ballou, J., 444, *476*
Baran, A., 126, 140, *149*
Barash, D. P., 438, 440, 443, 468, *473*
Barcal, R., 52, *81*
Barchas, J. D., 142, *149*, 416, 428, *431*

Freeman, F. N., 267, *328, 332*
Frederickson, D. S., 168, *187*
Fredericson, E., 411, 413, *430, 433*
Freed, E. X., 127, *151*
Freeman, F. N., 262, *332*
Freeman, J. M., 173, *186*
Friedl, W., 79, *87*
Friedman, H. J., 133, 144, *153*
Friedman, R., 112, *114*
Friedman, S. M., 379, 399, *405*
Fujiya, Y., 37, *87*
Fukui, M., 137, *153*
Fulker, D. W., 246, 247, 249, *330,* 389, 392, *404,* 420, 421, 428, *430*
Fuller, J. L., 90, 96, 106, *114, 115,* 118, 122, 131, 133, 138, 147, 148, *149, 150,* 195, 206, 209, *213,* 248, 314, *328,* 338, 341, 348, 349, 354, *360, 361, 362,* 363, 365, 367, *370, 371,* 379, 390, 391, *404, 407, 408,* 410, 411, 416, 418, 426, 428, *430, 433,* 436, 441, 452, 454, 462, 464, 466, *474*

G

Gaines, M. S., 355, *361*
Galef, B. G., 400, *403*
Gallup, G. G., Jr., 469, *474*
Galton, F., 20, *30*
Gandelman, R., 312, *333,* 379, 392, 399, 400, 401, *404, 408*
Gandini, E., 307, *329*
Gant, N., 176, *186*
Garron, D. C., 310, *328*
Gartler, S. M., 176, *186*
Gary, N. E., 377, *404*
Gastaut, H. A., 38, *82*
Gastaut, Y., 38, *85*
Gaul, A. T., 377, *404*
Gerling, S. A., 414, 428, *431*
Gershon, E. S., 180, *186*
Gibbs, E. C., 36, *84*
Gibbs, F. A., 36, *84*
Giblett, E. R., 141, *149*
Gibson, G. E., 174, *186*
Gibson, R. M., 451, *473*
Ginsburg, B. E., 383, *406,* 410, 413, 416, *430, 432, 433*
Girgus, J. S., 316, *328, 333*
Glaser, G. H., 455, *476*

Glass, A. B., 56, *83*
Glukhova, R. C., 67, *83*
Goldberg, M. B., 305, *328*
Goldberger, A. S., 249, *329*
Goldstein, A., 117, 118, 121, *150, 151*
Goldstein, D. B., 128, 129, 135, 136, 137, *150*
Goodenough, D. R., 307, *329*
Goodwin, D. W., 120, *150, 153,* 179, *186*
Goodwin, E. B., 391, *405*
Gordon, M., 424, *429*
Gottesman, I. I., 75, 78, *83,* 193, 194, 195, 196, 198, 199, 200, 201, 202, 203, 205, 206, 207, *213,* 463, *474*
Gould, S. J., 460, 467, 469, *474*
Goy, R. W., 425, *430*
Gray, J. A., 72, 73, *83*
Green, E. L., 387, *404*
Green, M. C., 387, *404*
Greenberg, G., 414, *430*
Greenhill, L., 204, *215*
Gregor, A. J., 306, *332*
Gregory, M. S., 436, *474*
Greig, W. R., 52, *82*
Griffin, D. R., 469, *474*
Griffin, G. A., 385, *404*
Grubb, T. W., 211, *214*
Gruzelier, J., 80, *83*
Guhl, A. M., 390, *404,* 419, *430, 431,* 452, *473*
Guiness, T. E., 451, *473*
Gunderson, J. G., 206, 207, 209, *213*
Gurney, N. L., 411, *430*
Gurski, J. C., 375, 382, 383, 384, 385, 388, 392, 395, 396, 397, *403, 404*
Guthrie, R., 19, *30*
Guttman, R., 306, *329,* 471, *474*
Guze, S. B., 179, *186*

H

Haber, S. B., 99, 110, 112, *114, 115*
Hager, J. L., 339, *362*
Hahn, M. E., 365, 367, *370,* 411, *430,* 454, *474*
Haier, R. J., 78, *83,* 194, 195, *213*
Haldane, J. B. S., 12, 13, *30*
Hale, G. A., 316, *334*
Hall, C. S., 91, 95, *114*
Hall, J. C., 421, 428, *431*
Hamerton, J. L., 158, *186*

Subject Index

A

Activity
  genetic models, 98–100
  genetic variation, 94–98
  measurement of, 92–94
  types of, 90–92
Adaptation, 338
Adoption studies
  intelligence, 262–276, 281–298
  schizophrenia, 202–205
Aggression, *see* Agonistic behavior
Agonistic behavior, *see also* Sociobiology
  bees, 419
  chickens, 419
  deer, 451
  dogs, 416–419
  genetic models, 413, 416
  humans, 463–465
  Mus, 410–416
    in females, 414
    maternal effects, 412
    selective breeding, 411
    strain differences, 410–414
Alcohol, *see* Drugs

B

Bees
  agonistic behavior, 419
  care-giving, 376–378

Behavior genetics, themes of, 337
Behavioral polytypy, 339–345
  breeding studies, 343–345
  comparative studies, 342
  observational studies, 341
Broodiness, 390

C

Care-giving
  bees, 376–378
  components of, 373
  dogs, 381
  felines, 380
  genetic variation, 391–398
  group care, 382
  pathology in, 398–401
  primates, 381
  rabbits, 395
  rats, 379
Care-soliciting, 375, 384–386
  dogs, 384
  Mus, 384
  rats, 384
  separation stress, 384–386
Chickens
  agonistic behavior, 419
  mating behavior, 424
Crickets, mating behavior, 421
Cognitive abilities, *see also* Intelligence
  multivariate studies, 302–304

495